B.D. Ripley

ELEMENTS OF FUNCTIONAL ANALYSIS

The New University Mathematics Series

Editor:
Professor E. T. DAVIES
Department of Mathematics, University of Calgary, Alberta

This series is intended for readers whose main interest is in mathematics, or who need the methods of mathematics in the study of science and technology. Some of the books will provide a sound treatment of topics essential in any mathematical training, while other, more advanced, volumes will be suitable as preliminary reading for research in the field covered. New titles will be added from time to time.

BRICKELL and CLARK: *Differentiable Manifolds*
BROWN and PAGE: *Elements of Functional Analysis*
BURGESS: *Analytical Topology*
COOPER: *Functions of Real Variables*
CURLE and DAVIES: *Modern Fluid Dynamics* (Vols 1 and 2)
EASTHAM: *Theory of Ordinary Differential Equations*
MAUNDER: *Algebraic Topology*
PORTEOUS: *Topological Geometry*
ROACH: *Green's Functions: Introductory Theory with Applications*
RUND: *The Hamilton–Jacobi Theory in the Calculus of Variations*
SMITH: *Laplace Transform Theory*
SMITH: *Introduction to the Theory of Partial Differential Equations*
SPAIN: *Ordinary Differential Equations*
SPAIN: *Vector Analysis*
SPAIN and SMITH: *Functions of Mathematical Physics*
ZAMANSKY: *Linear Algebra and Analysis*

ELEMENTS OF FUNCTIONAL ANALYSIS

A. L. BROWN
Lecturer in Mathematics
University of Newcastle upon Tyne

AND

A. PAGE
Lecturer in Mathematics
University of Dundee

VAN NOSTRAND REINHOLD COMPANY
LONDON

NEW YORK CINCINNATI TORONTO MELBOURNE

VAN NOSTRAND REINHOLD COMPANY
Windsor House, 46 Victoria Street, London S.W.1

INTERNATIONAL OFFICES
New York Cincinnati Toronto Melbourne

Library of Congress Catalog Card No. 69–19851

First published 1970

Printed in Great Britain by
Butler & Tanner Ltd, Frome and London

Preface

This book provides an elementary introduction to the ideas and methods of functional analysis. The prerequisites for reading it are a willingness to accept the language of modern mathematics and a knowledge of elementary real analysis and linear algebra such as is usually obtained from courses given in the first year of English Universities and the second year of Scottish Universities. It will prove useful to undergraduate and postgraduate students whose courses include some functional analysis and also to some mathematicians in other fields of both pure and applied mathematics who need to acquire a knowledge of functional analysis. Almost all the material has been used by the authors in undergraduate and postgraduate (M.Sc.) courses given in the Universities of Newcastle upon Tyne and Dundee.

The mathematical use of the word *functional* as a noun dates from the early years of the century. The phrase *functional analysis* originally meant the analysis of functionals but, as modern analysis has developed, the use of the phrase has changed considerably. It is now used to refer, on the one hand, to a body of ideas and methods that have applications in many branches of mathematics, and on the other hand to a group of specialized subjects of recent origin each of which is a field of study in its own right. A unifying concept that lies at the heart of functional analysis is that of a *normed linear space*. In this book we have provided a systematic introduction to the theory of normed linear spaces and its applications. We have developed the theory far enough to include discussions of what N. Dunford and J. T. Schwartz, in their treatise *Linear Operators*, have called the three basic principles of linear analysis: viz., the Hahn–Banach theorem, the Principle of Uniform Boundedness and the Open Mapping Theorem.

The language in which the ideas and results of the theory of normed linear spaces are expressed is the language of *metric spaces*, so we have thought it natural to include in this book a detailed introduction to the relevant parts of the theory of metric spaces.

In more detail the outline of the contents of the book is as follows. In Chapter 1 we present for reference some definitions and results from set theory and elementary analysis without attempting to give a comprehensive review of either subject. In Chapter 2 we introduce the basic ideas and develop the language of metric spaces. Here the reader will find, in an abstract setting, many results already familiar from elementary analysis. In Chapter 3 we introduce the basic concepts in the theory of normed linear spaces and derive some of the elementary results of the theory. As an application of the general theory we give a discussion of an elementary (vector-valued) integral which makes the book self-contained with respect to integration. In Chapter 4 we return to metric spaces and give a detailed account of the various notions of compactness in metric spaces and some applications of compactness to analysis.

The remaining five chapters of the book are devoted to a deeper study of various aspects of the theory of normed linear spaces and bounded linear operators. In Chapter 5 we introduce the concept of a *linear functional*, prove the Hahn–Banach theorem on the existence of linear functionals, and derive some of its many consequences. In Chapter 6 we lay the foundations of the study of the spectral theory of bounded linear operators. This chapter contains a complete account of the Riesz–Schauder theory of compact linear operators on a normed linear space. In Chapter 7 we give an introduction to the Fréchet differential calculus and prove the Implicit and Inverse Function Theorems. This chapter may serve to introduce the reader to the rapidly growing field of nonlinear functional analysis. In Chapter 8 we return briefly to metric spaces in order to introduce the concept of a meagre set in a metric space and prove Baire's Theorem. We then use Baire's Theorem to obtain the Principle of Uniform Boundedness and the Open Mapping Theorem. We have included in this chapter a discussion of some perturbation theorems for linear operators which serves to illustrate the application of the general principles of the theory of normed linear spaces to a specific problem in functional analysis. In Chapter 9 we take up the study of Hilbert space. The theory of Hilbert space occupies a position somewhat apart from the rest of functional analysis and uses many techniques that are peculiar to itself. We introduce the reader to many of these special techniques and develop the theory far enough to obtain the Spectral Theorem for bounded self-adjoint operators. Chapters 7, 8 and 9 are essentially independent of each other and can be read in any order. Chapter 7 can be read even earlier for it does not depend upon any of the results of Chapter 5 and, with a very few references to Chapter 6, can be read after Chapter 4.

A functional analyst reading the above outline will notice that two important topics are not mentioned: general topology and the Lebesgue integral. In a book of this length it would be impossible to do justice to either of these topics. Also it would not be in keeping with the spirit in which this book was written to assume a knowledge of either general topology or the Lebesgue integral as a prerequisite for reading the book. By denying ourselves the use of the Lebesgue integral we have, of course, denied ourselves the use of the Banach spaces L^p which are, both historically and in applications, among the most important examples of Banach spaces. However, we hope that we have shown that it is possible to convey the spirit and power of the methods of functional analysis without introducing the Lebesgue integral. Our discussion of metric spaces was written with general topology always in mind, and anyone who has read and understood Chapters 2 and 4 should not find the step from metric spaces to topological spaces difficult.

A large number of exercises have been included. These fall broadly into three categories: those that are simply exercises, to help the reader to test his understanding of new ideas; those that develop the general theory further than we have taken it in the text; and those that illustrate the applications of functional analysis to classical analysis.

Each chapter is divided into sections which are numbered in a sequence within the chapter. Within each section definitions, theorems, lemmas, etc. are numbered in one sequence; certain displayed formulae are numbered in another sequence; and the exercises are numbered in a third sequence. The notation used in references between sections and chapters should be self-explanatory. The end of a proof of a theorem, lemma, etc., is indicated by the symbol □.

The list of references at the end of the book is not a bibliography: it contains only books to which we have made reference in the text. We have made no systematic attempt to attribute theorems to their author, nor have we given any references to the original research papers in which they appeared. We have, however, sometimes indicated the author of a major theorem together with the date of its first publication in the hope that this may give the reader some indication of the historical development of the subject.

The possibility of writing an elementary introduction to functional analysis which would be accessible to students at an early stage of their studies was originally suggested to us by Professor F. F. Bonsall and we began to write this book in collaboration with him when all three of us were at the University of Newcastle upon Tyne. Later Professor Bonsall, by then at the University of Edinburgh, found himself unable to continue with the project. We hope that his influence is reflected in the book, but while acknowledging our considerable debt to him we alone are responsible for the book in its final form.

We wish to thank the secretaries of the School of Mathematics in the

University of Newcastle upon Tyne for their generous help in preparing the typescript of this book, and especially Mrs Margaret Smith, who typed most of it.

1970

A. L. Brown
A. Page

Contents

CHAPTER 1

Preliminaries

This chapter is intended primarily for reference, but the discussion of sets, mappings and relations in Sections 1.1, 1.2 and 1.3 is self-contained —if concise and, at times, even terse—and should provide an adequate introduction to the notions of sets, mappings and relations for the reader who is not already familiar with current usage. The reader who is familiar with these notions should still read these three sections, in order to find out whether our notation and terminology differs from that to which he is accustomed. In Sections 1.4 and 1.5 we are concerned with rather technical results from set theory that will be needed occasionally in later chapters. The reader may omit these sections at a first reading and return to them when necessary.

In Section 1.6 we start from a convenient set of assumptions, and derive some fundamental properties of the real number system. The discussion of the real number system is not self-contained; it has been included simply to ensure that certain fundamental properties of real numbers are available for reference in the precise form in which we shall use them.

1.1 Sets

One of the prerequisites that we expect of the reader is a familiarity with the concept of a 'set' and with the elementary operations that can be performed with and on sets. Our discussion of sets is informal— there are no definitions or theorems in this section.

We are concerned with sets and members of sets which we shall usually call elements or points. We shall assume that the reader has a clear intuitive idea of the notion of a 'set' and of the notion of 'belonging to a set'. The words 'set', 'collection', 'aggregate' and 'family' are all synonymous.

With a few exceptions we shall adhere to the following notational conventions. Elements of sets will be denoted by lower case letters: $a, b, c, \ldots, x, y, z, \alpha, \beta, \gamma, \ldots$; sets will be denoted by capital letters:

$$A, B, C, \ldots, X, Y, Z.$$

We shall sometimes have to consider sets whose elements are themselves sets. As an aid to clarity of expression in such situations we shall always use the phrase 'family of sets' in preference to the phrase 'set of sets'. Families of sets will be denoted by capital script letters: $\mathscr{A}, \mathscr{B}, \mathscr{C}, \ldots$

When the symbol '$=$' appears between two symbols or sets of symbols it means that the symbol or set of symbols on the left denotes the same object as does that on the right. We shall use the word 'equal' in the same sense. In particular we remark that two sets are equal if and only if they have the same elements.

The symbols '$x \in A$' mean that x is an element of the set A (x is a member of the set A, or x belongs to A). The symbols '$x \notin A$' mean that x is not an element of the set A. We shall often write '$x, y \in A$' as an abbreviation for '$x \in A$ and $y \in A$'.

To obtain a satisfactory theory of sets it is necessary to admit that there is a set with no elements; such a set is said to be *empty* (or *void*). It is clear that there is exactly one empty set. In fact if A_1 and A_2 are empty and $A_1 \neq A_2$ then either A_1 or A_2 must have an element that does not belong to the other; since A_1 and A_2 have no elements this is impossible. (All arguments concerning the empty set follow this pattern.) We shall denote the empty set by $\emptyset$.

Given two sets A and B we shall say that A is a *subset* of B, and write $A \subseteq B$, if and only if each element of A is also an element of B. When $A \subseteq B$ and $A \neq B$ we shall say that A is a *proper subset* of B. When we use the word 'family' instead of 'set' we shall use the word 'subfamily' instead of 'subset'. It is clear that $\emptyset$ is a proper subset of each non-empty set. Also two sets A and B are equal if and only if both $A \subseteq B$ and $B \subseteq A$.

The basic operations in set theory are the processes by which new sets can be defined in terms of given sets. Essentially a set is defined by specifying a property of its elements; for instance we can, and shall, speak of the set of all integers and of the set of all even integers. If $S(x)$ is a proposition concerning x we shall denote by $\{x: S(x)\}$ the set of all x for which the proposition $S(x)$ is true. (Thus, for example, $\{x: x = 2n \text{ for some integer } n\}$ denotes the set of all even integers.) In the interests of brevity we shall sometimes modify the notation $\{x: S(x)\}$; the meaning of the modified notation should always be clear to the reader. (For instance we may denote the set of even integers by $\{2n: n = 0, \pm 1, \pm 2, \ldots\}$.) If $S(x)$ is a proposition which is meaningful for each element x of a set A we shall usually write $\{x \in A: S(x)\}$ instead of $\{x: x \in A \text{ and } S(x)\}$.

The following example, which is called the Russell paradox, after the mathematician and philosopher Bertrand Russell who discovered it in 1901, shows that the process of defining sets by properties of their elements needs further examination. Let

$$\mathscr{A} = \{A: A \text{ is a set and } A \notin A\}.$$

(The proposition, $S(A)$, used to define $\mathscr{A}$ is: 'the set A is not an element of itself'.) We have either $\mathscr{A} \in \mathscr{A}$ or $\mathscr{A} \notin \mathscr{A}$. If $\mathscr{A} \in \mathscr{A}$ then, by definition of $\mathscr{A}$, also $\mathscr{A} \notin \mathscr{A}$; and if $\mathscr{A} \notin \mathscr{A}$ then, by definition of $\mathscr{A}$, also $\mathscr{A} \in \mathscr{A}$. Consequently, the assumption that the proposition 'A is a set, and $A \notin A$' can be used to define a set, leads directly to a contradiction.

To eliminate from set theory contradictions such as Russell's paradox, it is necessary to make precise how a proposition can be used to define a set; this is the aim of axiomatic set theory. Whether or not an axiomatic set theory can be devised which is free from contradictions, and in which all the constructions with sets that occur naturally in mathematics are possible, is the concern of those interested in mathematical logic and the foundations of mathematics. The reader will find accounts of two axiomatic set theories which avoid all known paradoxes in the book of HALMOS [13] and the Appendix to that of KELLEY [18]. In common with most mathematicians we shall not follow the axiomatic development of set theory; we shall adopt the naïve point of view, and assume that, used in contexts that occur naturally in mathematics, the set theoretical constructions described informally in this chapter will not give rise to contradictions.

Let us now return to the description of the elementary operations that can be performed with sets. Let $\mathscr{A}$ be a family of subsets of a set X. We define

$$\bigcup_{A \in \mathscr{A}} A = \{x \in X: x \in A \text{ for at least one set } A \in \mathscr{A}\}$$

and

$$\bigcap_{A \in \mathscr{A}} A = \{x \in X: x \in A \text{ for all sets } A \in \mathscr{A}\}.$$

The set $\bigcup_{A \in \mathscr{A}} A$ is called the *union of the family* $\mathscr{A}$ and the set $\bigcap_{A \in \mathscr{A}} A$ is called the *intersection of the family* $\mathscr{A}$. The union and intersection of a family with only two members, A and B, will be denoted by $A \cup B$ and $A \cap B$, respectively. We shall say that two subsets A and B of X are *disjoint* if and only if $A \cap B = \emptyset$ and we shall say that a family $\mathscr{A}$ of subsets of X is *pairwise disjoint* if and only if $A \cap B = \emptyset$ whenever $A,B \in \mathscr{A}$ and $A \neq B$.

Given two subsets A and B of a set X we define

$$A \sim B = \{x \in X: x \in A \text{ and } x \notin B\}.$$

The set $A \sim B$ is called the *complement of B in A*. The following relations between unions, intersections and complements are readily verified from the definitions. Let $\mathscr{A}$ be a family of subsets of a set X and let B be a subset of X. Then

$$B \cap \bigcup_{A \in \mathscr{A}} A = \bigcup_{A \in \mathscr{A}} (B \cap A),$$

$$B \cup \bigcap_{A \in \mathscr{A}} A = \bigcap_{A \in \mathscr{A}} (B \cup A),$$

$$B \sim \bigcup_{A \in \mathscr{A}} A = \bigcap_{A \in \mathscr{A}} (B \sim A),$$

$$B \sim \bigcap_{A \in \mathscr{A}} A = \bigcup_{A \in \mathscr{A}} (B \sim A).$$

The last two equations are known as the De Morgan formulae. Strictly speaking, none of the symbols on the right hand sides of the above equations has been defined (see, however, p. 6). What they stand for is obvious: $\bigcup_{A \in \mathscr{A}} (A \cap B)$ is $\bigcup_{C \in \mathscr{C}} C$ where $\mathscr{C} = \{A \cap B \colon A \in \mathscr{A}\}$, for example. (Observe that the symbol $\{A \cap B \colon A \in \mathscr{A}\}$ is itself an abbreviation for $\{C \colon C = A \cap B$ for some $A \in \mathscr{A}\}$.)

1.2 Mappings

Perhaps even more important to us than the notion of 'set' is the notion of 'function' or 'mapping'. The terms 'function' and 'mapping' are traditionally conceived in the context of 'correspondences' and 'variables' and much of the associated terminology remains in common usage. However, it is now customary to define the term 'mapping' so that a mapping is a set whose elements are objects called 'ordered pairs'.

We shall not give a formal set theoretic definition of an ordered pair but be content with an explanation of what the term means. (The reader will find a formal definition in HALMOS [13, §6], and also in KELLEY [18, p. 259].) With two symbols x and y we associate a third symbol (x,y) called an *ordered pair*. The characteristic property of ordered pairs is the following: two ordered pairs (x,y) and (u,v) are equal if and only if $x = u$ and $y = v$. Given two sets X and Y we shall denote by $X \times Y$ the set of all ordered pairs of the form (x,y) where $x \in X$ and $y \in Y$. The set $X \times Y$ is called the *product* (or *Cartesian product*) of the sets X and Y.

Let us now use the idea of an ordered pair to give a formal definition of a mapping. A *mapping* is a non-empty set, f say, of ordered pairs with the property that, if $(x,y) \in f$ and $(x,z) \in f$, then $y = z$. The reader should note that not every set of ordered pairs is a mapping (see Section 1.3 below). Now let f be a mapping. We define two sets—

dom(f), called the *domain* of f, and range(f), called the *range* of f—by

$$\text{dom}(f) = \{x: (x,y) \in f \text{ for some } y\}$$

and

$$\text{range}(f) = \{y: (x,y) \in f \text{ for some } x\}.$$

Thus f is a subset of the product dom(f) $\times$ range(f). A mapping whose range has exactly one element is called a *constant mapping.*

The words 'function', 'functional', 'transformation' and 'operator' are all used as synonyms for the term 'mapping'. There is a convention in functional analysis, which we shall follow, that each of the above synonyms is reserved to denote mappings whose domains or ranges are special sets or types of sets.

Let us now try to reconcile the formal definition of a mapping given above with the traditional conception of a mapping as something that assigns to each element of a given set a definite element of another given set. Let f be a mapping and let X and Y be sets. We shall say that f is a *mapping of X into Y* if and only if dom(f) $= X$ and range(f) $\subseteq Y$, and we shall say that f is a *mapping of X onto Y* if and only if dom(f) $= X$ and range(f) $= Y$. If f is a mapping of X into Y then f is a subset of $X \times Y$, and to each element $x \in X$ there corresponds a unique element $y \in X$ such that $(x,y) \in f$. Traditionally y is called the *value of the mapping f at the element x* and the relation between x and y is denoted by $y = f(x)$ instead of $(x,y) \in f$. Using this notation we can write

$$\text{range}(f) = \{y : y = f(x) \text{ for some } x \in \text{dom}(f)\}$$

which we shall always abbreviate to range(f) $= \{f(x) : x \in \text{dom}(f)\}$. From now on we shall adopt this traditional notation. We remark that then $f = \{(x, f(x)) : x \in X\}$. In particular, if we take X to be a subset of R, the set of all real numbers, and $Y = \mathsf{R}$ we see that the object that we have formally defined to be the mapping f is what is called in elementary mathematics the 'graph of f'.

Since a mapping is a set, we can obtain a criterion for equality of mappings from the criterion for equality of sets. Let f and g be mappings. Suppose that $f = g$ and let $x \in \text{dom}(f)$. Then $(x,f(x)) \in f = g$ and hence $x \in \text{dom}(g)$ and $g(x) = f(x)$. Similarly if $x \in \text{dom}(g)$ then $x \in \text{dom}(f)$ and $f(x) = g(x)$. Thus dom(f) $=$ dom(g) and $f(x) = g(x)$ for all $x \in \text{dom}(f)$. Conversely it is obvious that if dom(f) $=$ dom(g) and $f(x) = g(x)$ for all $x \in \text{dom}(f)$ then $f = g$. Therefore two mappings are equal if and only if they have the same domain and take the same value at each point of their common domain. It follows that, to define a mapping of a set X into a set Y, it is sufficient to assign to each element x of X a unique element of Y. We shall frequently define mappings in this way. When the relation that assigns to an element x of X a unique element y of Y is determined by a complicated formula, $y = [x]$ say (where $[x]$ is a collection of symbols including x which defines a unique element of Y whenever the element x of X is specified), we shall denote the resulting mapping by $x \to [x]$. (In elementary mathematics, where the distinction between a mapping and its values is often not clearly drawn, this mapping would be denoted simply by $[x]$. We shall never use this convention.)

It often happens that a set is defined as the range of a mapping; such a set is said to be *indexed* by the domain of the mapping (which is then called an *index set*). Let the set X be indexed by a mapping with domain Λ. Since we are more interested in the range, X, of the mapping than in the mapping itself we shall usually denote the given mapping of Λ onto X by $\lambda \to x_\lambda$. Thus $X = \{x_\lambda \colon \lambda \in \Lambda\}$. Associated with certain special index sets there are special notational conventions. When Λ is the set N of all positive integers we shall write $\{x_n \colon n = 1, 2, \ldots\}$ instead of $\{x_n \colon n \in \mathsf{N}\}$; and when Λ is the set consisting of the first n positive integers (which we shall always denote by $\{1, 2, \ldots, n\}$ rather than $\{m \colon m$ is an integer and $1 \leq m \leq n\}$) we shall write $\{x_1, x_2, \ldots, x_n\}$ instead of $\{x_\lambda \colon \lambda \in \Lambda\}$; when n is a 'small' positive integer we shall even omit the indices: for example, $\{x\}$ denotes the set whose only element is x, and $\{x,y\}$ denotes the set whose only elements are x and y, etc.

There are also special notations for unions and intersections of indexed families of sets. Let $\{A_\lambda \colon \lambda \in \Lambda\}$ be an indexed family of subsets of a given set X. We shall denote the union and intersection of this family by $\bigcup_{\lambda \in \Lambda} A_\lambda$ and $\bigcap_{\lambda \in \Lambda} A_\lambda$ respectively. (With this notation the sets on the right-hand sides of the equations on p. 4 can now be interpreted as unions and intersections of families indexed by the given family $\mathscr{A}$.) We shall denote the union and intersection of an indexed family $\{A_n \colon n = 1, 2, \ldots\}$ by $\bigcup_{n=1}^{\infty} A_n$ and $\bigcap_{n=1}^{\infty} A_n$, respectively; and we shall denote the union and intersection of an indexed family $\{A_1, A_2, \ldots, A_n\}$ by $\bigcup_{j=1}^{n} A_j$ or $A_1 \cup A_2 \cup \ldots \cup A_n$ and $\bigcap_{j=1}^{n} A_j$ or $A_1 \cap A_2 \cap \ldots \cap A_n$, respectively.

Mappings whose domains are the set N of all positive integers are sufficiently important to merit a special name and notation: a mapping $n \to x_n$ of N into a set X is called a *sequence in* X and is denoted by (x_n) or $(x_n)_{n \geq 1}$. It is vital to maintain the distinction between a sequence (x_n), which is a mapping with domain N, and its range $\{x_n \colon n = 1, 2, \ldots\}$, which is a subset of X indexed by N.

There is an important way of obtaining new sequences from a given sequence. Let (x_n) be a sequence in a set X, and let (n_k) be a sequence in N such that $n_{k+1} > n_k$ for $k = 1, 2, \ldots$. The sequence $k \to x_{n_k}$ is called a *subsequence* of the sequence (x_n) and is denoted by (x_{n_k}). It is clear that every sequence is a subsequence of itself and that, if (z_n) is a subsequence of (y_n) and (y_n) is a subsequence of (x_n), then (z_n) is also a subsequence of (x_n).

Finally in this section we shall describe some of the operations that can be performed with and on mappings. Let f be a mapping of a set X into a set Y. For each subset A of X we define

$$f(A) = \{y \in Y : y = f(x) \text{ for some } x \in A\}$$

and for each subset B of Y we define

$$f^{-1}(B) = \{x \in X : f(x) \in B\}.$$

For brevity we shall always write $f(A) = \{f(x) : x \in A\}$. The reader will notice that $f(X) = \text{range}(f)$.

It follows at once from the definitions that

$$f(f^{-1}(B)) = B \cap \text{range}(f) \subseteq B$$

for each subset B of Y and that

$$A \subseteq f^{-1}(f(A))$$

for each subset A of X. It is easy to see that, in general, $A \neq f^{-1}(f(A))$. It follows also that if $\mathscr{A}$ is a family of subsets of X then

$$f\Big(\bigcup_{A \in \mathscr{A}} A\Big) = \bigcup_{A \in \mathscr{A}} f(A)$$

and

$$f\Big(\bigcap_{A \in \mathscr{A}} A\Big) \subseteq \bigcap_{A \in \mathscr{A}} f(A).$$

Again, in general, $f\big(\bigcap_{A \in \mathscr{A}} A\big) \neq \bigcap_{A \in \mathscr{A}} f(A)$. It is also easy to verify that if $\mathscr{B}$ is a family of subsets of Y then

$$f^{-1}\Big(\bigcup_{B \in \mathscr{B}} B\Big) = \bigcup_{B \in \mathscr{B}} f^{-1}(B)$$

and

$$f^{-1}\Big(\bigcap_{B \in \mathscr{B}} B\Big) = \bigcap_{B \in \mathscr{B}} f^{-1}(B).$$

Finally if B_1 and B_2 are subsets of Y then

$$f^{-1}(B_1 \sim B_2) = f^{-1}(B_1) \sim f^{-1}(B_2).$$

We shall now describe some ways of obtaining new mappings from old ones. Let f be a mapping of a set X into a set Y and let g be a mapping of Y into a set Z. We can define a mapping $g \circ f$ of X into Z by setting

$$(g \circ f)(x) = g(f(x))$$

for all $x \in X$. The mapping $g \circ f$ is called the *composition* of the mappings g and f. We have already used the idea of composition of mappings in defining a subsequence. In fact if (x_n) is a sequence in a set X and (n_k) is a sequence in N with $n_{k+1} > n_k$ for $k = 1, 2, \ldots$ then the subsequence

(x_{n_k}) is simply the composition of the mappings $k \to n_k$ and $n \to x_n$. It is clear that

$$(g \circ f)(A) = g(f(A))$$

for each subset A of X and that

$$(g \circ f)^{-1}(B) = f^{-1}(g^{-1}(B))$$

for each subset B of Z.

Now let f be a mapping of a set X into a set Y and let A be a subset of X. We can define a mapping $f|A$ of A into Y by setting

$$(f|A)(x) = f(x)$$

for all $x \in A$. The mapping $f|A$ is called the *restriction of f to A*. If f and g are mappings and $g = f|A$ for some subset A of dom(f) we say that f is an *extension of g*. The reader should note that, when mappings are regarded as sets of ordered pairs, a mapping f is an extension of a mapping g if and only if $g \subseteq f$. It is clear that if A is a subset of X then $f(A)$ is simply the range of $f|A$. Also

$$(f|A)^{-1}(B) = A \cap f^{-1}(B)$$

for all subsets B of Y.

A mapping f of a set X into a set Y is said to be *one-to-one* if and only if $f(x_1) \neq f(x_2)$ whenever $x_1, x_2 \in X$ and $x_1 \neq x_2$. Let f be a one-to-one mapping of X into Y. Then to each element $y \in$ range(f) there corresponds a unique element $x \in X$ such that $y = f(x)$. Thus we can define a mapping f^{-1} of range(f) onto X by setting $f^{-1}(y) = x$ whenever $y \in$ range(f) and $f(x) = y$. The mapping f^{-1} is called the *inverse* of the mapping f. We remark that for one-to-one mappings f the symbol $f^{-1}(B)$, where B is a subset of range(f), is ambiguous: it denotes both

$$f^{-1}(B) = \{x \in X : f(x) \in B\}$$

and

$$g(B) = \{g(y) : y \in B\},$$

where $g = f^{-1}$. However, the sets on the right-hand sides of the above equations are obviously equal so no confusion can arise from denoting them by the same symbol.

The *identity mapping* on a set X is the mapping i_X of X onto itself defined by $i_X(x) = x$ for all $x \in X$. If f is a one-to-one mapping of X onto Y then it is easy to see that $f^{-1} \circ f = i_X$ and $f \circ f^{-1} = i_Y$. Finally we observe that if f is a one-to-one mapping of X onto Y and g is a one-to-one mapping of Y onto Z then $g \circ f$ is a one-to-one mapping of X onto Z and $(g \circ f)^{-1} = f^{-1} \circ g^{-1}$.

1.3 Relations

A *relation* is a set of ordered pairs; thus a mapping is a special kind of relation—and, so far as we are concerned, by far the most important kind. We shall occasionally have to consider two other kinds of relation —called equivalence relations and partial orderings. Just as for mappings, there are traditional notations and terminologies for relations which were well established before it became usual to think of relations as sets of ordered pairs. We shall use these traditional notations.

If R is a relation and X is any set with $R \subseteq X \times X$ we shall say that R is a *relation on* X and we shall write xRy instead of $(x,y) \in R$. A relation R on a set X is said to be

(a) *reflexive* if and only if xRx for all $x \in X$,
(b) *symmetric* if and only if xRy implies yRx,
(c) *antisymmetric* if and only if xRy and yRx imply $x = y$,
(d) *transitive* if and only if xRy and yRz imply xRz.

A relation R on a set X is called an *equivalence relation* if and only if it is reflexive, symmetric and transitive. Let R be an equivalence relation on X. For each $x \in X$ the set $[x] = \{y \in X: xRy\}$ is called the *equivalence class of* x (determined by R). Since R is reflexive we have $x \in [x]$ for all $x \in X$. Therefore the equivalence classes determined by R are all non-empty and $\bigcup_{x \in X} [x] = X$. Also the family $\{[x]: x \in X\}$ of equivalence classes determined by R is pairwise disjoint. To see this, suppose that $x,y \in X$ and $[x] \cap [y] \neq \emptyset$. Choose $z \in [x] \cap [y]$. Then xRz and yRz and hence, by (b) and (d), we have yRx. If $u \in [x]$ then xRu and so, by (d), we have yRu and thus, by definition, $u \in [y]$. This shows that $[x] \subseteq [y]$. By symmetry $[y] \subseteq [x]$ and consequently $[x] = [y]$.

A pairwise disjoint family of sets whose union is X is called a *partition* of X. We have just seen that the family of equivalence classes determined by an equivalence relation on X is a partition of X. It is not difficult to see that every partition of X arises in this way from some equivalence relation on X. To see this let $\mathscr{A}$ be a partition of X. We can define a relation R on X by setting xRy if and only if the elements x and y of X belong to the same set in the family $\mathscr{A}$. We leave the reader to verify that R is an equivalence relation on X and that the sets in the family $\mathscr{A}$ are precisely the equivalence classes determined by R.

A relation $\leq$ on a set X is called a *partial order on* X if and only if it is reflexive, antisymmetric and transitive. If $\leq$ is a partial order relation on X we shall say that X is *partially ordered by* $\leq$. Each subset of a partially ordered set can be partially ordered in an obvious way.

The simplest example of a partial order relation is the identity relation defined by $x \leq y$ if and only if $x = y$. The identity relation is also an equivalence relation; it is the only equivalence relation which is also a partial order. A less trivial example of a partial order relation is the inclusion relation $\subseteq$ on the family of all subsets of a given set. Another familiar partial order relation is the usual relation $\leq$ of inequality on the set of all integers.

Let $\leq$ be a relation of partial order on a set X and let A be a subset of X. We say that A is *bounded above* if and only if there exists $x \in X$ such that $a \leq x$ for all $a \in A$; we call any such element $x \in X$ an *upper bound for* A. An upper bound x for A is called a *supremum for* A (or a *least upper bound for* A) if and only if $x \leq y$ for all upper bounds y for A. If x_1 and x_2 are suprema for A then $x_1 \leq x_2$ and $x_2 \leq x_1$ so $x_1 = x_2$. Thus A can have at most one supremum; when it exists we denote the supremum of A by $\sup A$. The terms *bounded below, lower bound for* A and *infimum for* A (or *greatest lower bound for* A) are defined analogously. The set A can have at most one infimum; when it exists we denote the infimum of A by $\inf A$. A set which is bounded above and below is said to be *bounded.*

Let us illustrate these definitions by examining the three partial orders introduced above. First consider the identity relation on a set X. Clearly no subset of X which has more than one element can be bounded above or below. Next consider the inclusion relation on the family of all subsets of a set X. Each family $\mathscr{A}$ of subsets of X is bounded above and below; a subset B of X is an upper bound for $\mathscr{A}$ if and only if $\bigcup_{A \in \mathscr{A}} A \subseteq B$ and a lower bound for $\mathscr{A}$ if and only if $B \subseteq \bigcap_{A \in \mathscr{A}} A$; it follows that $\bigcup_{A \in \mathscr{A}} A = \sup \mathscr{A}$ and $\bigcap_{A \in \mathscr{A}} A = \inf \mathscr{A}$. Finally consider the usual relation of inequality between integers. A subset of the set of all integers is bounded if and only if it is finite; the supremum and infimum of a bounded subset are the greatest and least elements, respectively, of the set.

A partially ordered set may contain elements x and y such that neither $x \leq y$ nor $y \leq x$. (Consider, for example, the inclusion relation on the family of all subsets of a set with two elements.) Partial orders for which this cannot happen are called linear orders. More precisely we shall say that a partial order relation $\leq$ on a set X is a *linear order* (or a *total order*) if and only if for each pair $x,y \in X$ we have either $x \leq y$ or $y \leq x$. When $\leq$ is a linear order relation on X we shall say that X is *linearly ordered by* $\leq$.

The identity relation on a set and the inclusion relation on the family of all subsets of a set are not linear order relations unless the sets concerned consist of a single point. The relation of inequality on the set of all integers is a linear order relation.

1.4 Countability

We assume that the reader knows what is meant by the terms 'a finite set', 'an infinite set' and 'the number of elements in a finite set', and we shall use without explicit mention the elementary properties of finite sets. We shall not give the set theoretic definitions of these terms nor shall we give formal statements and proofs of the properties of finite sets. The reader who is curious to see a formal treatment of 'finiteness' should consult the books of HALMOS [13, §12] or KELLEY [18, pp. 277–8]. We shall adopt the convention that the empty set is finite and has 0 elements.

An important class of infinite sets is the class of countably infinite sets: these are sets that can be put into one-to-one correspondence with the set of all positive integers, which we shall denote throughout by N. More precisely we say that a set X is *countably infinite* if and only if there is a one-to-one mapping of N onto X. If X is a countably infinite set then any one-to-one mapping $n \to x_n$ of N onto X is called an *enumeration of* X. If $n \to x_n$ is an enumeration of X then we have $X = \{x_n: n = 1, 2, \ldots\}$; thus a countably infinite set can be indexed by the set N. Often this is the only property of countably infinite sets that we shall use. When this is so we can avoid having to consider separately the cases of finite sets and countably infinite sets by indexing finite sets by N as follows: given a finite set $X = \{x_1, x_2, \ldots, x_n\}$ let $x_m = x_n$ for $m = n+1, n+2, \ldots$; then $X = \{x_m: m = 1, 2, \ldots\}$. A set which is either finite or countably infinite is said to be *countable*; and a set which is not countable is said to be *uncountable*. We shall exhibit an uncountable set in Theorem 1.6.6 below.

We now derive some properties of countable sets that we shall use in later chapters.

1.4.1 THEOREM *Each subset of a countable set is countable.*

PROOF Let A be a subset of a countable set X. If X is finite then so is A so we may suppose that X is countably infinite. Let f be a one-to-one mapping of N onto X. Then $f|f^{-1}(A)$ is a one-to-one mapping of the subset $f^{-1}(A)$ of N onto A. Consequently it is enough to prove that each subset of N is countable.

Let B be an infinite subset of N. We have to prove that B is countable so we have to define a one-to-one mapping g of N onto B. Let $g(1)$ be the least element of B and suppose that for each integer $n \geq 2$ we have chosen elements $g(2), g(3), \ldots, g(n)$ of B so that $g(j)$ is the least element of $B \sim \{g(1), g(2), \ldots, g(j-1)\}$ for $j = 2, 3, \ldots, n$. Since B is infinite the set $B \sim \{g(1), g(2),..., g(n)\}$ is non-empty; let $g(n+1)$ be its least element. By induction this defines a mapping g of N into B. Obviously $g(n+1) > g(n)$ for all $n \in \mathsf{N}$ so g is one-to-one and $g(n) \geq n$ for all $n \in \mathsf{N}$. It remains to prove that g maps N onto B. Suppose that $B \neq \{g(n): n = 1, 2,...\}$ and let k be the least element of $B \sim \{g(n): n = 1, 2,...\}$. In particular $k \in B \sim \{g(1), g(2),..., g(k)\}$ so, by definition of $g(k+1)$, we have $k \geq g(k+1)$. But we have seen that $g(k+1) \geq k+1$ so we have a contradiction. This proves that $B = \{g(n): n = 1, 2,...\}$ and hence that B is countably infinite. □

1.4.2 THEOREM *A mapping with countable domain has countable range.*

PROOF Let f be a mapping of a countable set X onto a set Y. If X is finite then Y is finite, so we may suppose that X is countably infinite. If g is any one-to-one mapping of N onto X then $f \circ g$ is a mapping of N onto Y. Consequently we may suppose that $X = \mathsf{N}$.

For each $y \in Y$ let m_y be the least element of $f^{-1}(\{y\})$ and let $Z = \{m_y: y \in Y\}$. By Theorem 1.4.1, Z is countable and therefore Y is also countable because $f|Z$ is a one-to-one mapping of Z onto Y. □

1.4.3 THEOREM *The set* $\mathsf{N} \times \mathsf{N}$ *is countably infinite.*

PROOF We leave the reader to verify that $(m,n) \to 2^{m-1}(2n-1)$ is a one-to-one mapping of $\mathsf{N} \times \mathsf{N}$ onto N. □

The following three theorems are the basic results about countability that we shall use in later chapters.

1.4.4 THEOREM *Let X and Y be countable sets. Then the set $X \times Y$ is also countable.*

PROOF We can index X and Y by N, say $X = \{x_n: n = 1, 2,...\}$ and $Y = \{y_n: n = 1, 2,...\}$. Then $(m,n) \to (x_m, y_n)$ is a mapping of $\mathsf{N} \times \mathsf{N}$ onto $X \times Y$ and, by Theorems 1.4.2 and 1.4.3, the set $X \times Y$ is countable. □

1.4.5 THEOREM *The union of a countable family of countable sets is itself countable.*

PROOF Let $\mathscr{A}$ be a countable family of countable sets and let $B = \bigcup_{A \in \mathscr{A}} A$. We can index $\mathscr{A}$ by N, say $\mathscr{A} = \{A_n: n = 1, 2,...\}$, and for each $n \in \mathsf{N}$ we can index A_n by N, say $A_n = \{a_{mn}: m = 1, 2,...\}$. Then $(m,n) \to a_{mn}$ is a mapping of $\mathsf{N} \times \mathsf{N}$ onto B and by Theorems 1.4.2 and 1.4.3 the set B is countable. □

1.4.6 THEOREM *The family of all finite subsets of a countable set is countable.*

PROOF Let A be a countable set and let $\mathscr{A}$ be the family of all finite subsets of A. If A is finite then so is $\mathscr{A}$, so we may suppose that A is countably infinite. Then obviously we may suppose further that $A = \mathsf{N}$. For each $n \in \mathsf{N}$ let $\mathscr{A}_n$ be the family of all subsets B of N with $B \subseteq \{1, 2, ..., n\}$. The family $\mathscr{A}_n$ is finite. Since $\mathscr{A} = \bigcup_{n=1}^{\infty} \mathscr{A}_n$, Theorem 1.4.5 shows that $\mathscr{A}$ is countable. □

1.5 Zorn's Lemma

Our aim in this section is to prove a result about partial order relations, now always called Zorn's lemma, which has far-reaching consequences in several branches of mathematics. Zorn's lemma will be used at two points in the text (in the proofs of Theorems 5.2.5 and 9.2.13).

In the proof of Zorn's lemma given below we make an assertion of the following type: given a non-empty family $\mathscr{A}$ of non-empty sets there is a mapping f with domain $\mathscr{A}$ such that $f(A) \in A$ for all $A \in \mathscr{A}$. Intuitively f chooses simultaneously an element from each set in the family $\mathscr{A}$. When the elementary part of set theory which is described in Sections 1.1 and 1.2 is put on an axiomatic basis (and in particular when the rules for defining sets, and hence mappings, are made precise) it is found that the existence of such a 'choice mapping' for $\mathscr{A}$ cannot be deduced from the axioms that suffice for the elementary part of set theory. Consequently the existence of a 'choice mapping' for $\mathscr{A}$ has to be postulated as an extra axiom of set theory—called the Axiom of Choice. We have explicitly mentioned the Axiom of Choice because it has excited more controversy than any of the other axioms of set theory since it was first formulated by E. Zermelo in 1908. A few mathematicians go so far as to refuse to use the Axiom of Choice. However, analysis, as we understand it, cannot be developed far without this axiom and we shall use it, explicitly and implicitly, without comment.

Let us now turn to the proof of Zorn's lemma. The proof given below is a slight modification of the proofs given in the books of HALMOS [13, §16] and DUNFORD and SCHWARTZ [8, p. 6]; the idea of the proof goes back to Zermelo. We begin with a preliminary lemma (whose proof does not require the Axiom of Choice).

1.5.1 LEMMA *Let $\leq$ be a partial order relation on a non-empty set X and f a mapping of X into itself which satisfy the following conditions:*

(a) *there is an element $0 \in X$ with $0 \leq x$ for all $x \in X$,*

(b) *each non-empty linearly ordered subset of* X *has a supremum,*
(c) $x \leq f(x)$ *for all* $x \in X$,
(d) *if* $x, y \in X$ *and* $x \leq y \leq f(x)$ *then either* $x = y$ *or* $y = f(x)$.

Then there is an element $x_0 \in X$ *with* $f(x_0) = x_0$.

PROOF For brevity we shall call a subset Y of X *admissible* if and only if it satisfies the following conditions:

(1) $0 \in Y$,
(2) $f(Y) \subseteq Y$,
(3) if Z is any non-empty linearly ordered subset of Y then $\sup Z \in Y$.

Obviously X is admissible. Let A be the intersection of the family of all admissible subsets of X. It is easy to see that A is admissible and that if B is any admissible subset of X with $B \subseteq A$ then $B = A$.

Suppose for the moment that we have proved that A is linearly ordered. Then, by (b), $x_0 = \sup A$ exists. By (2) $f(x_0) \in A$ and hence $f(x_0) \leq x_0$. Also by (c) $x_0 \leq f(x_0)$. Consequently $x_0 = f(x_0)$. Thus to prove the lemma we have only to prove that A is linearly ordered. Let

$$B = \{b \in A\colon \text{for each } a \in A \text{ either } b \leq a \text{ or } a \leq b\}.$$

Obviously A is linearly ordered if and only if $A = B$ so we have to prove that $A = B$. Since $B \subseteq A$, to prove that $B = A$ it is sufficient to prove that B is admissible.

We observe first that $0 \in B$ because, by (a), $0 \leq a$ for each $a \in A$. Thus B satisfies (1). Next let C be a non-empty linearly ordered subset of B and let $c = \sup C$. Given $a \in A$ we have either $b \leq a$ for all $b \in B$, in which case $c \leq a$; or $a \leq b_0$ for some $b_0 \in B$, in which case $a \leq c$. This shows that $c \in B$ and hence that B satisfies (3).

It is more difficult to prove that B satisfies (2). Let $b \in B$. We have to prove that $f(b) \in B$ so we have to prove that, for each $a \in A$, either $f(b) \leq a$ or $a \leq f(b)$. Let

$$C = \{a \in A\colon \text{either } a \leq b \text{ or } f(b) \leq a\}.$$

If we can prove that $C = A$ then it will follow that $f(b) \in B$ (because $a \leq b$ implies $a \leq f(b)$ by condition (c)). Since $C \subseteq A$, to prove that $C = A$ it is sufficient to prove that C is admissible. Since $0 \leq b$ we have $0 \in C$ so C satisfies (1). The verification that C satisfies (3) is similar to that given above for B—we leave the details to the reader. It is again more difficult to prove that C satisfies (2).

Let $a \in C$. Then one of the following relations holds: (α) $a \leq b$ and $a \neq b$, (β) $a = b$, (γ) $f(b) \leq a$. In case (β) we have $f(a) = f(b)$ and so $f(a) \in C$; and in case (γ) we have $f(b) \leq a \leq f(a)$ by (c) and so again

$f(a) \in C$. It remains to consider case (α). Since $f(a) \in A$ we have, by definition of B, either $f(a) \leq b$ or $b \leq f(a)$. By hypothesis $a \leq b$ and $a \neq b$ so, if also $b \leq f(a)$, (d) shows that $b = f(a)$. Thus in this case we must always have $f(a) \leq b$ and therefore $f(a) \in C$. This completes the proof that C satisfies (2) and hence that C is admissible. The proof of the lemma is now complete. □

Before we can state the main result of this section, which is always called Zorn's lemma, we need one more definition. We shall say that an element x of a set X partially ordered by $\leq$ is a *maximal element* of X if and only if $y \in X$ and $x \leq y$ imply that $y = x$.

1.5.2 THEOREM *Let X be a non-empty partially ordered set with the property that each non-empty linearly ordered subset of X has an upper bound. Then X has a maximal element.*

PROOF Suppose that the result is false. Then for each $x \in X$ there exists $y \in X$ with $x \leq y$ and $x \neq y$. Let $\mathscr{L}_0$ be the family of all non-empty linearly ordered subsets of X and let $\mathscr{L} = \mathscr{L}_0 \cup \{\emptyset\}$. The family $\mathscr{L}$ is partially ordered by the inclusion relation $\subseteq$ between subsets of X. For each $A \in \mathscr{L}_0$ the set

$$U_A = \{x \in X: x \text{ is an upper bound for } A \text{ and } x \notin A\}$$

is non-empty because, if x is an upper bound for A and $y \in X$ is such that $x \leq y$ and $x \neq y$, then $y \in U_A$. Let $U_\phi = \{x_0\}$, where x_0 is an arbitrary element of X. Let g be a mapping with domain $\{U_A\}: A \in \mathscr{L}\}$ such that $g(U_A) \in U_A$ for all $A \in \mathscr{L}$. For each $A \in \mathscr{L}$ let $f(A) = A \cup \{g(U_A)\}$. By definition of g and U_A we have $a \leq g(U_A)$ for all $a \in A$ and all $A \in \mathscr{L}_0$. It is now clear that $f(A) \in \mathscr{L}$ for all $A \in \mathscr{L}$ and hence f maps $\mathscr{L}$ into itself.

We shall prove that $\mathscr{L}$, partially ordered by inclusion, and f satisfy the conditions of Lemma 1.5.1. First we observe that $\emptyset \in \mathscr{L}$ and $\emptyset \subseteq A$ for all $A \in \mathscr{L}$ so $\mathscr{L}$ satisfies condition (a) of Lemma 1.5.1. Next let $\mathscr{M}$ be a non-empty subfamily of $\mathscr{L}$ such that $\mathscr{M}$ is linearly ordered by inclusion and let $A = \bigcup_{B \in \mathscr{M}} B$. Let $a,b \in A$. There are sets $C,D \in \mathscr{M}$ with $a \in C$ and $b \in D$. Since $\mathscr{M}$ is linearly ordered by inclusion either $C \subseteq D$ or $D \subseteq C$ and in either case we see that there is one set in $\mathscr{M}$ which contains both a and b. Since each set in $\mathscr{M}$ is a linearly ordered subset of X it follows that either $a \leq b$ or $b \leq a$. This proves that $A \in \mathscr{L}$ and it is then easy to see that $A = \sup \mathscr{M}$. Thus $\mathscr{L}$ satisfies condition (b) of Lemma 1.5.1. By definition of f we have $A \subseteq f(A)$ so condition (c) of Lemma 1.5.1 is satisfied. Finally, by definition of g and U_A, we have $g(U_A) \notin A$ for all $A \in \mathscr{L}$ and therefore $f(A) \sim A = \{g(U_A)\}$

for all $A \in \mathscr{L}$. It follows immediately that condition (d) of Lemma 1.5.1 is satisfied.

We can now conclude from Lemma 1.5.1 that there is a set $A_0 \in \mathscr{L}$ with $f(A_0) = A_0$. Since $f(A) \sim A \neq \emptyset$ for all $A \in \mathscr{L}$ we have a contradiction. $\square$

It is interesting to observe that Zorn's lemma is equivalent to the Axiom of Choice. The proof that it implies the Axiom of Choice is not nearly so difficult as the proof of the converse implication; it provides a typical illustration of how Zorn's lemma is used so we feel that it is worthwhile to include it here.

Let $\mathscr{A}$ be a non-empty family of non-empty sets and let $\mathscr{F}$ be the set of all mappings f with $\mathrm{dom}(f) \subseteq \mathscr{A}$ and $f(A) \in A$ for all $A \in \mathrm{dom}(f)$. We have to prove that $\mathscr{F}$ contains a mapping whose domain is $\mathscr{A}$.

We observe first that $\mathscr{F}$ is non-empty. Indeed since $\mathscr{A}$ is non-empty there exists $A_0 \in \mathscr{A}$, and since A_0 is non-empty there exists $a_0 \in A_0$. Obviously the mapping $A_0 \to a_0$ with domain $\{A_0\}$ is in $\mathscr{F}$. We define a relation $\leq$ on $\mathscr{F}$ by setting $f \leq g$ if and only if g is an extension of f. It is simple to verify that the relation $\leq$ is a partial order on $\mathscr{F}$. (We have defined the relation $\leq$ in terms of the traditional language—in terms of the definition of a mapping as a set of ordered pairs we have $f \leq g$ if and only if $f \subseteq g$.)

Let $\mathscr{F}_1$ be a non-empty linearly ordered subset of $\mathscr{F}$ and let $\mathscr{A}_1 = \bigcup_{f \in \mathscr{F}_1} \mathrm{dom}(f)$. We remark that if $A_1 \in \mathscr{A}_1$ and $f,g \in \mathscr{F}_1$ are such that $A_1 \in \mathrm{dom}(f)$ and $A_1 \in \mathrm{dom}(g)$ then $f(A_1) = g(A_1)$ because either $f \leq g$ or $g \leq f$. Consequently we can define a mapping f_1 with domain $\mathscr{A}_1$ as follows: for each $A_1 \in \mathscr{A}_1$ let $f_1(A_1) = f(A_1)$ where f is any mapping in $\mathscr{F}_1$ such that $A_1 \in \mathrm{dom}(f)$. It is clear that $f_1 \in \mathscr{F}$ and that $f \leq f_1$ for all $f \in \mathscr{F}_1$ so f_1 is an upper bound for $\mathscr{F}_1$.

We have now verified that $\mathscr{F}$ satisfies the conditions of Zorn's lemma and hence $\mathscr{F}$ has a maximal element, f say. We shall show that $\mathrm{dom}(f) = \mathscr{A}$. Suppose on the contrary that $\mathrm{dom}(f) \neq \mathscr{A}$. Then there exists $A' \in \mathscr{A} \sim \mathrm{dom}(f)$, and since A' is non-empty there exists $a' \in A'$. Let

$$f'(A) = \begin{cases} f(A) \text{ when } A \in \mathrm{dom}(f), \\ a' \text{ when } A = A'. \end{cases}$$

Obviously $f' \in \mathscr{F}$, $f \leq f'$ and $f \neq f'$. This contradicts the maximality of f so we must have $\mathrm{dom}(f) = \mathscr{A}$. This completes the deduction of the Axiom of Choice from Zorn's lemma.

1.6 Real and Complex Numbers

Throughout this book R will denote the real field and C will denote the complex field. We shall assume that R has the following properties:

(a) R is a field,
(b) there is a linear order relation $\leq$ on R such that (b_1) if $x,y,z \in$ R and $x \leq y$ then $x+z \leq y+z$, and (b_2) if $x,y \in$ R, $0 \leq x$ and $0 \leq y$ then $0 \leq xy$,
(c) each non-empty subset of R that is bounded above has a supremum,
(d) the field Q of rational numbers is a subfield of R.

All the familiar properties of real numbers can be deduced from these assumptions (see for example GLEASON [9, Chapters 8 and 9]). We remark that condition (d) is redundant in the sense that a field which satisfies condition (b) contains a subfield Q_0 which is isomorphic to the field Q of rational numbers: that is to say there is a one-to-one mapping θ of Q_0 onto Q with $\theta(x+y) = \theta(x)+\theta(y)$ and $\theta(xy) = \theta(x)\theta(y)$ for all $x,y \in Q_0$. We shall not give a systematic development of the properties of the real number system here but content ourselves with deriving three fundamental results—the Bolzano–Weierstrass theorem, Borel's theorem and Cauchy's general principle of convergence—to which we shall want to refer in later chapters. The significance of Cauchy's general principle of convergence will emerge in Chapter 2 and that of the Bolzano-Weierstrass theorem and Borel's theorem in Chapter 4.

First we must introduce some notation. Given a non-empty subset A of R that is bounded above it will not be true in general that $\sup A$ belongs to A; if $\sup A$ does belong to A we shall often write $\max A$ instead of $\sup A$. Similarly when A is bounded below and contains $\inf A$ we shall often write $\min A$ instead of $\inf A$. In particular when $A = \{a_1, a_2, \ldots, a_n\}$ is a finite subset of R both $\sup A$ and $\inf A$ belong to A and we shall always write $\max\{a_1, a_2, \ldots, a_n\}$ and $\min\{a_1, a_2, \ldots, a_n\}$ instead of $\sup A$ and $\inf A$, respectively.

There are certain subsets of R which occur so often that it is convenient to have a special notation for them. If $a,b \in$ R with $a \leq b$ we shall write

$$(a,b) = \{x \in \mathsf{R} : a < x < b\}, \tag{1}$$
$$[a,b] = \{x \in \mathsf{R} : a \leq x \leq b\}, \tag{2}$$
$$(a,b] = \{x \in \mathsf{R} : a < x \leq b\}, \tag{3}$$
$$[a,b) = \{x \in \mathsf{R} : a \leq x < b\}. \tag{4}$$

We remark that the sets (a,b), $(a,b]$ and $[a,b)$ are non-empty if and only if $a < b$. The set $[a,b]$ is always non-empty. If $a \in$ R we shall write

$$(a,\infty) = \{x \in \mathsf{R} : x > a\}, \tag{5}$$
$$[a,\infty) = \{x \in \mathsf{R} : x \geq a\}, \tag{6}$$
$$(-\infty,a) = \{x \in \mathsf{R} : x < a\}, \tag{7}$$
$$(-\infty,a] = \{x \in \mathsf{R} : x \leq a\}. \tag{8}$$

A subset of R is called an *interval* if and only if it is of one of the above forms or is equal to R. Intervals of the form (1), (2), (3) and (4) are called *bounded intervals*; they are obviously bounded subsets of R. Intervals of the form (5), (6), (7) and (8) and the set R itself are called *unbounded intervals*; they are obviously not bounded subsets of R. An *open interval* is an interval which is of the form (1), (5) or (7) or is equal to R; a *closed interval* is an interval which is of the form (2), (6) or (8) or is equal to R (the reasons for calling R both an open and a closed interval will appear in Chapter 2); and a *half-open* (or *half-closed*) interval is an interval which is of the form (3) or (4).

Next let us recall some elementary definitions. We say that a sequence (x_n) in R is *increasing* if and only if $x_n \leq x_{n+1}$ for $n = 1, 2, \ldots$; *decreasing* if and only if $x_{n+1} \leq x_n$ for $n = 1, 2, \ldots$; *monotonic* if and only if it is either increasing or decreasing; and *bounded* if and only if the set $\{x_n : n = 1, 2, \ldots\}$ is bounded. Also we say that a sequence (x_n) in R *converges to a limit* $x \in \mathsf{R}$ if and only if, given any $\varepsilon > 0$, there is a positive integer N such that $| x_n - x | < \varepsilon$ whenever $n \geq N$. It is easy to see that a sequence in R can converge to at most one limit; if (x_n) converges to the limit x we shall write $\lim_{n\to\infty} x_n = x$.

We are now in a position to make some deductions from assumptions (a), (b), (c) and (d).

1.6.1 LEMMA *Let $a,b \in \mathsf{R}$ with $a > 0$ and $b \geq 0$. Then there is a positive integer n such that $na \geq b$.*

PROOF Suppose that the lemma is false. Then $n \leq b/a$ for $n = 1, 2, \ldots$, and hence the set N of all positive integers is bounded above. Let $k = \sup \mathsf{N}$. Then $k-1$ is not an upper bound for N and so there exists $m \in \mathsf{N}$ with $k-1 < m$. But then $k < m+1$ which contradicts the definition of k. □

Any field which satisfies condition (b) above and the condition of Lemma 1.6.1 is called an *Archimedean ordered field.* Using Lemma 1.6.1 we can calculate the limits of the sequences (n^{-1}) and (2^{-n}). Let $\varepsilon > 0$. Setting $a = \varepsilon$ and $b = 2$ in Lemma 1.6.1 we obtain a positive integer N with $N\varepsilon \geq 2 > 1$. Thus $0 < n^{-1} < \varepsilon$ for $n \geq N$ and hence $\lim_{n\to\infty} n^{-1} = 0$. Also, by induction, $2^n \geq n+1$ for $n = 1, 2, \ldots$ so $0 < 2^{-n} < \varepsilon$ for $n \geq N + 1$ and hence $\lim_{n\to\infty} 2^{-n} = 0$. We shall use below the fact that $\lim_{n\to\infty} 2^{-n} = 0$ together with the following two elementary results: if (x_n) and (y_n) are convergent sequences in R then (x_n+y_n) also converges and $\lim_{n\to\infty} (x_n+y_n) = \lim_{n\to\infty} x_n + \lim_{n\to\infty} y_n$; and if (x_n) is a convergent sequence in R with $a \leq x_n \leq b$ for $n = 1, 2, \ldots$ then $a \leq \lim_{n\to\infty} x_n \leq b$.

1.6.2 THEOREM *Each bounded monotonic sequence in* R *is convergent.*

PROOF Let (x_n) be a bounded increasing sequence in R and let $x = \sup\{x_n: n = 1, 2, \ldots\}$. Let $\varepsilon > 0$. Then $x - \varepsilon$ is not an upper bound for the set $\{x_n: n=1, 2, \ldots\}$ and so there is a positive integer N with $x_N > x - \varepsilon$. Therefore $x - \varepsilon < x_N \leq x_n \leq x < x + \varepsilon$ for $n \geq N$. This proves that (x_n) converges to x. The proof that a bounded decreasing sequence converges is similar. □

1.6.3 THEOREM (Bolzano-Weierstrass) *Each bounded sequence in* R *has a convergent subsequence.*

PROOF By Theorem 1.6.2 it is sufficient to prove that each sequence in R has a monotonic subsequence. We shall say that y_m is a *maximum* of a sequence (y_n) in R if and only if $y_m \geq y_n$ for $n = 1, 2, \ldots$. Suppose that (y_n) is a sequence in R with no maximum. We shall prove that (y_n) has an increasing subsequence. Let $n_1 = 1$ and suppose that, for some integer $k \geq 2$, integers $n_2, n_3, \ldots, n_k$ have been chosen so that $n_j < n_{j+1}$ and $y_{n_j} \leq y_{n_{j+1}}$ for $j = 1, 2, \ldots, k-1$. There is at least one integer $n \geq n_k + 1$ such that $y_n > \max\{y_{n_1}, y_{n_2}, \ldots, y_{n_k}\}$ (for otherwise $\max\{y_1, y_2, \ldots, y_{n_k}\}$ would be a maximum of (y_n)); let n_{k+1} be the least such integer. Clearly $n_{k+1} > n_k$ and $y_{n_{k+1}} > y_{n_k}$. This defines inductively a sequence (n_k) of positive integers. Obviously (y_{n_k}) is an increasing subsequence of (y_n).

Suppose now that (x_n) is a sequence in R which has no increasing subsequence. We shall prove that (x_n) has a decreasing subsequence. For each integer $k \geq 0$ the sequence $(x_{k+n})_{n \geq 1}$ is a subsequence of (x_n) and so can have no increasing subsequence. It follows therefore from the first part of the proof that for each integer $k \geq 0$ the sequence $(x_{k+n})_{n \geq 1}$ has a maximum. Let n_1 be the least positive integer such that x_{n_1} is a maximum of (x_n) and suppose that, for some integer $k \geq 2$, integers $n_2, n_3, \ldots, n_k$ have been chosen so that $n_j < n_{j+1}$ and $x_{n_{j+1}}$ is a maximum of the sequence $(x_{n_j+m})_{m \geq 1}$ for $j = 1, 2, \ldots, k-1$. Let n_{k+1} be the least integer $\geq n_k + 1$ such that $x_{n_{k+1}}$ is a maximum of the sequence $(x_{n_k+m})_{m \geq 1}$. This defines inductively a subsequence (x_{n_k}) of (x_n) which is obviously decreasing. □

1.6.4 THEOREM (Borel) *Let I be a bounded closed interval and let $\mathscr{J}$ be a family of open intervals with $I \subseteq \bigcup_{J \in \mathscr{J}} J$. Then there is a finite subfamily of $\mathscr{J}$, $\{J_1, J_2, \ldots, J_n\}$ say, such that $I \subseteq \bigcup_{k=1}^{n} J_k$.*

PROOF We shall say that a subfamily $\mathscr{J}_0$ of $\mathscr{J}$ *covers* a subinterval I_0 of I if and only if $I_0 \subseteq \bigcup_{J \in \mathscr{J}_0} J$. Suppose that the theorem is false. We shall define inductively a sequence (I_k) of subintervals of $I = [a,b]$ with the following properties: for $k = 1, 2,...$

(a) $I_k = [a_k, b_k]$ and $b_k - a_k = (b-a)\, 2^{-k+1}$,
(b) $I_{k+1} \subseteq I_k$,
(c) I_k cannot be covered by a finite subfamily of $\mathscr{J}$.

Let $I_1 = I$ and suppose that, for some integer $k \geq 2$, subintervals $I_2, I_3, ..., I_k$ of I have been chosen to satisfy conditions (a), (b) and (c). Let $I'_{k+1} = [a_k, \frac{1}{2}(a_k + b_k)]$ and $I''_{k+1} = [\frac{1}{2}(a_k + b_k), b_k]$. Since I_k satisfies condition (c) so must at least one of I'_{k+1} and I''_{k+1}. If I'_{k+1} satisfies (c) let $I_{k+1} = I'_{k+1}$; otherwise let $I_{k+1} = I''_{k+1}$. It is obvious that I_{k+1} satisfies conditions (a), (b) and (c). This completes the definition of the sequence (I_k).

It follows from condition (b) that (a_k) is a bounded increasing sequence and so, by Theorem 1.6.2, it converges. Since

$$b_k = a_k + (b - a)\, 2^{-k+1}$$

the sequence (b_k) also converges and $\lim_{k\to\infty} b_k = \lim_{k\to\infty} a_k$ (see the remarks on p. 18). Let $c = \lim_{k\to\infty} a_k$. We have $a_1 \leq a_k \leq b_1$ for $k = 1, 2,...$ because $I_k \subseteq I_1$, consequently $a_1 \leq c \leq b_1$ (again see the remarks on p. 18). Thus $c \in [a_1,b_1] = I$ and since $\mathscr{J}$ covers I we have $c \in J_0$ for some $J_0 \in \mathscr{J}$. Let $J_0 = (p,q)$. Then $p < c < q$ and hence, since $c = \lim_{k\to\infty} a_k = \lim_{k\to\infty} b_k$, there is a positive integer k_0 such that $p < a_k < q$ and $p < b_k < q$ for $k \geq k_0$. It follows that $I_{k_0} = [a_{k_0}, b_{k_0}] \subseteq (p,q) = J_0$. This contradicts (c). □

The method used in this proof of Borel's theorem is usually called the method of repeated bisection.

1.6.5 THEOREM (Cauchy's general principle of convergence) *A sequence (x_n) in R converges if and only if for each $\varepsilon > 0$ there is a positive integer N such that $|x_m - x_n| < \varepsilon$ whenever $m \geq N$ and $n \geq N$.*

PROOF Suppose that (x_n) is a convergent sequence in R. Let $\lim_{n\to\infty} x_n = x$ and $\varepsilon > 0$. Then there is a positive integer N such that $|x - x_n| < \frac{1}{2}\varepsilon$ when $n \geq N$. Let $m,n \geq N$. Then

$$|x_m - x_n| \leq |x_m - x| + |x - x_n| < \varepsilon.$$

Suppose conversely that (x_n) is a sequence in R and that for each $\varepsilon > 0$ there is a positive integer N such that $|x_m - x_n| < \varepsilon$ when

$m,n \geq N$. In particular taking $\varepsilon = 1$ we obtain an integer M such that $|x_m - x_n| < 1$ when $m,n \geq M$. Thus if $n \geq M$ we have

$$|x_n| \leq |x_n - x_M| + |x_M| < 1 + |x_M|$$

and hence, for $n = 1,2,...,$

$$|x_n| \leq \max\{|x_1|, |x_2|,..., |x_{M-1}|, 1 + |x_M|\}.$$

This proves that (x_n) is a bounded sequence and consequently, by Theorem 1.6.3, (x_n) has a convergent subsequence, (x_{n_k}) say, with limit x. Let $\varepsilon > 0$. Then there is an integer k_0 such that $|x_{n_k} - x| < \frac{1}{2}\varepsilon$ for $k \geq k_0$ and there is an integer N such that $|x_m - x_n| < \frac{1}{2}\varepsilon$ for $m,n \geq N$. Since (x_{n_k}) is a subsequence of (x_n) we have $n_k \geq N$ for $k \geq N$. Let $K = \max\{k_0, N\}$. Then if $n \geq N$ we have

$$|x_n - x| \leq |x_n - x_{n_K}| + |x_{n_K} - x| < \varepsilon.$$

This proves that the sequence (x_n) converges. □

The next two theorems give information about the field of rational numbers.

1.6.6 THEOREM (a) *The set* Q *of all rational numbers is countable.*
(b) *The set* R *of all real numbers is uncountable.*

PROOF (a) Let Q^+ and Q^- be the sets of all positive and negative rational numbers, respectively. Then $Q = Q^+ \cup \{0\} \cup Q^-$. Since $r \to -r$ is a one-to-one mapping of Q^+ onto Q^- Theorem 1.4.5 shows that it is sufficient to prove that Q^+ is countable. That Q^+ is countable follows from Theorems 1.4.2 and 1.4.3 because $(m,n) \to m/n$ is a mapping of $N \times N$ onto Q^+.

(b) We shall use the following property of real numbers: each $x \in [0,1)$ has a unique triadic expansion in the form

$$x = \sum_{m=1}^{\infty} \alpha_m 3^{-m}, \tag{9}$$

where $\alpha_m = 0$, 1 or 2 and $\{m : \alpha_m \neq 2\}$ is infinite (cf. Exercise 2.2(6)). For brevity we shall call (9) the *standard* triadic expansion of x.

By Theorem 1.4.1 it is sufficient to prove that $[0,1)$ is uncountable. Suppose, on the contrary, that $[0,1)$ is countable and let $n \to x_n$ be an enumeration of $[0,1)$. For each positive integer n let

$$x_n = \sum_{m=1}^{\infty} \alpha_{mn} 3^{-m}$$

be the standard triadic expansion of x_n. Let

$$x = \sum_{m=1}^{\infty} \beta_m 3^{-m}, \tag{10}$$

where $\beta_m = 0$ when $\alpha_{mm} \neq 0$ and $\beta_m = 1$ when $\alpha_{mm} = 0$. Obviously $x \in [0,1)$, and (10) is the standard triadic expansion of x. Since the standard triadic expansion of a number in [0,1) is unique we must have, by definition of β_n, $x \neq x_n$ for $n = 1, 2, \ldots$. This contradicts the assumption that $n \to x_n$ is an enumeration of [0,1).□

The proof of Theorem 1.6.6 is due to G. Cantor (1829–1920) who was the founder of modern set theory. The next theorem shows that although, as we have just seen, there are 'far fewer' rational numbers than real numbers, the rationals are 'thickly and evenly distributed' among the reals.

1.6.7 THEOREM *Let $a,b \in \mathsf{R}$ and $a < b$. Then there is a rational number r with $a < r < b$.*

PROOF By Lemma 1.6.1 there is a positive integer n with $(b-a)n > 1$. Suppose first that $b \geq 0$. Then by Lemma 1.6.1 again there is a positive integer p with $p/n \geq b$; let m be the least such integer. Then we must have

$$\frac{m-1}{n} < b \leq \frac{m}{n}.$$

If $(m-1)/n \leq a$ then

$$b-a \leq \frac{m}{n} - \frac{m-1}{n} = \frac{1}{n}$$

which contradicts the choice of n; therefore

$$a < \frac{m-1}{n} < b.$$

This proves the theorem when $b \geq 0$.

Suppose now that $b \leq 0$. Then $-a > -b \geq 0$ and so, by the first part of the proof, there is a rational number r with $-b < r < -a$. The number $-r$ is rational and $a < -r < b$. This proves the theorem when $b \leq 0$.□

This completes our discussion of the real number system.

We shall assume that the complex number system C has the following properties:

(a) C is a field,
(b) R is a subfield of C,
(c) there is an element $i \in \mathsf{C}$ with $i^2 = -1$,
(d) each element $z \in \mathsf{C}$ can be written in the form $z = x + iy$ with $x,y \in \mathsf{R}$.

We assume also that the reader is already familiar with the following elementary properties of complex numbers. Each element $z \in \mathsf{C}$ has exactly one representation in the form $z = x + iy$ with $x,y \in \mathsf{R}$; we shall write $x = \operatorname{Re} z$ and $y = \operatorname{Im} z$. Then there is a one-to-one mapping $z \to (\operatorname{Re} z, \operatorname{Im} z)$ of C onto $\mathsf{R} \times \mathsf{R}$. This representation of complex numbers by ordered pairs of real numbers is usually called the *Argand diagram*. (It is not difficult to construct a field with the properties (a), (b), (c) and (d) by defining suitable operations of addition and multiplication on $\mathsf{R} \times \mathsf{R}$: see, for example, GLEASON [9, Chapter 10]).

If $z \in \mathsf{C}$ and $z = x + iy$ with $x,y \in \mathsf{R}$ we shall write $\bar{z} = x - iy$ and $|z| = (x^2 + y^2)^{1/2}$; the complex number $\bar{z}$ is called the *complex conjugate* of z and the non-negative real number $|z|$ is called the *absolute value* (or *modulus*) of z. We remark that $z = 0$ if and only if $|z| = 0$. Also if $z \neq 0$ there is a unique real number θ, with $0 \leq \theta < 2\pi$, such that $z = |z| (\cos \theta + i \sin \theta)$. For brevity we shall write

$$e^{i\theta} = \cos \theta + i \sin \theta$$

for all $\theta \in \mathsf{R}$.

We end this section with some remarks about mappings whose ranges are subsets of R or C. We shall adopt the traditional terminology and call a mapping of a set X into R a *real-valued function on X*. We shall use the symbol 1 to denote both the real number 1 and the constant function on X whose value at each element of X is the real number 1. It will always be clear from the context whether 1 denotes a real number or a function.

Given two real-valued functions f and g on a set X we shall write $f \leq g$ if and only if $f(x) \leq g(x)$ for all $x \in X$. It is easy to verify that the relation $\leq$ is a partial order on the set of all real-valued functions on X.

A real-valued function $j \to x_j$ on the set $\{1, 2, \ldots, n\}$ is called an (*ordered*) *n-tuple of real numbers* and is denoted by $(x_1, x_2, \ldots, x_n)$. The set of all n-tuples of real numbers will be denoted by R^n. We shall usually refer to the elements of R^n as 'the points of R^n'. We remark that two points $(x_1, x_2, \ldots, x_n)$ and $(y_1, y_2, \ldots, y_n)$ of R^n are equal if and only if $x_j = y_j$ for $j = 1, 2, \ldots, n$. In particular we see that 2-tuples have the characteristic property of ordered pairs of real numbers; we shall therefore write R^2 interchangeably with $\mathsf{R} \times \mathsf{R}$. Complex-valued functions on a set X and the set C^n of all n-tuples of complex numbers are defined analogously.

CHAPTER 2

Metric Spaces

Limit processes are fundamental to analysis; the reader will be familiar with the importance of the elementary concepts of convergence of a sequence of real numbers, and continuity and differentiability of real-valued functions of a real variable. In this chapter we shall study limit processes in a general setting. Let us recall some definitions. A sequence (ξ_n) of real numbers converges to the limit ξ if and only if for each $\varepsilon > 0$ there is an integer N such that $|\xi_n - \xi| < \varepsilon$ whenever $n \geq N$. A real-valued function of a real variable is continuous at a point ξ if and only if for each $\varepsilon > 0$ there exists $\delta > 0$ such that $|f(\xi) - f(\eta)| < \varepsilon$ whenever $|\xi - \eta| < \delta$. We recall these definitions in order to point out that they can be formulated entirely in terms of the distance $|\xi - \eta|$ between the two points ξ and η on the real line. Many of the properties of convergent sequences and continuous functions depend only on properties of this distance, and not directly on the algebraic nature of the real number system.

Metric spaces are sets on which there is defined a notion of 'distance between pairs of points', and they provide the general setting in which we study convergence and continuity. In the context of metric spaces, the meaning and content of many of the concepts of classical analysis are illuminated, and many of the theorems of classical analysis can be given their simplest and most transparent proofs.

This chapter is devoted to the elementary theory of metric spaces. We shall be concerned mainly with establishing terminology, and most of the theorems are simple deductions from the definitions. The most important concepts introduced in this chapter are those of convergence, completeness and continuity. In the final section we shall give our first example of how results about metric spaces can be used to solve problems in classical analysis.

2.1 Definition and Examples

The concept of a metric space was formulated in 1906 by M. Fréchet.

Much of the early development of the theory is due to M. Fréchet and F. Hausdorff.

2.1.1 DEFINITION Let X be a non-empty set. A *metric on* X is a mapping d of $X \times X$ into R, that satisfies the following conditions:

(a) $d(x,y) \geq 0$ for all $x,y \in X$,
(b) $d(x,y) = 0$ if and only if $x = y$,
(c) $d(x,y) = d(y,x)$ for all $x,y \in X$,
(d) $d(x,z) \leq d(x,y) + d(y,z)$ for all $x,y,z \in X$.

A *metric space* is a pair (X,d) in which X is a non-empty set and d is a metric on X.

A metric is also called a *distance function*. Condition (c) expresses the fact that d is symmetric in x and y; inequality (d) is usually called the *triangle inequality*. Conditions (a)–(d) will sometimes be referred to as the metric space axioms.

It is customary to omit all mention of the metric d, and to write 'the metric space X' instead of 'the metric space (X,d)'. Normally this convention will give rise to no confusion. However two distinct metrics d_1 and d_2 may occur naturally, and in the same context, on a single set X. In such a situation the distinction between the two metric spaces (X,d_1) and (X,d_2) must be carefully preserved. We shall usually call the elements of a metric space X 'the points of X'.

The rest of this section is devoted to examples of metric spaces, some of which are of considerable importance in classical analysis and will be studied in detail in later chapters. For the present, the examples will serve to indicate the scope of the definition of a metric space and to illustrate the concepts introduced later in the chapter. In all the examples it will be obvious that the conditions (a), (b) and (c) of Definition 2.1.1 are satisfied; the verification of condition (d) is sometimes more difficult.

2.1.2 EXAMPLE The function d defined by $d(\xi,\eta) = |\xi-\eta|$ for all $\xi,\eta \in \mathsf{R}$ is a metric on the set R of real numbers.

To see that (d) is satisfied, let $\xi,\eta,\zeta \in \mathsf{R}$. Then

$$\begin{aligned} d(\xi,\zeta) &= |\xi-\zeta| \\ &= |(\xi-\eta)+(\eta-\zeta)| \\ &\leq |\xi-\eta| + |\eta-\zeta| \\ &= d(\xi,\eta) + d(\eta,\zeta). \end{aligned}$$

Whenever R is considered as a metric space we shall assume that the metric is the one just defined, unless the contrary is explicitly stated.

The number $d(\xi,\eta)$ is, of course, the usual 'distance' between the points ξ and η on the real line.

2.1.3 EXAMPLE The function d defined by $d(\xi,\eta) = |\xi-\eta|$ for all $\xi,\eta \in \mathsf{C}$ is a metric on the set C of complex numbers. To prove that (d) is satisfied it is sufficient to prove that $|\xi+\eta| \leq |\xi| + |\eta|$ for all $\xi,\eta \in \mathsf{C}$ (see Example 2.1.2); this follows from the inequalities

$$\begin{aligned} |\xi+\eta|^2 &= (\xi+\eta)(\bar{\xi}+\bar{\eta}) \\ &= \xi\bar{\xi} + \xi\bar{\eta} + \eta\bar{\xi} + \eta\bar{\eta} \\ &= |\xi|^2 + 2\mathrm{Re}(\xi\bar{\eta}) + |\eta|^2 \\ &\leq |\xi|^2 + 2|\xi|\,|\eta| + |\eta|^2 \\ &= (\,|\xi| + |\eta|\,)^2. \end{aligned}$$

This argument is generalized in Example 2.1.5. Whenever C is considered as a metric space we shall assume that the metric is the one just defined, unless the contrary is explicitly stated. The real number $d(\xi,\eta)$ is the usual 'distance' between the points ξ and η in the complex plane.

2.1.4 EXAMPLE Let

$$d(x,y) = \left(\sum_{r=1}^{n} (\xi_r - \eta_r)^2\right)^{1/2}$$

for all $x = (\xi_1, \xi_2, \ldots, \xi_n)$ and $y = (\eta_1, \eta_2, \ldots, \eta_n)$ in R^n. Then d is a metric on R^n and is called the *Euclidean metric*. The metric space (R^n, d) is called *n-dimensional Euclidean space.*

We shall deal with the verification of the triangle inequality in the next example. We remark that in the space R^2 the number

$$d(x,y) = ((\xi_1-\eta_1)^2 + (\xi_2-\eta_2)^2)^{1/2}$$

is the usual 'distance' between the points $x = (\xi_1,\xi_2)$ and $y = (\eta_1,\eta_2)$. In the plane, the triangle inequality reduces to a well-known geometrical theorem: the length of one side of a triangle is not greater than the sum of the lengths of the other two sides.

Since complex numbers can be represented as points in the Argand diagram (or complex plane), there is a natural one-to-one correspondence between C and R^2 that preserves distances. (This remark will be made more precise in Theorem 2.9.2.)

2.1.5 EXAMPLE Let

$$d(x,y) = \left(\sum_{r=1}^{n} |\xi_r - \eta_r|^2\right)^{1/2}$$

for all $x = (\xi_1, \xi_2, \ldots, \xi_n)$ and $y = (\eta_1, \eta_2, \ldots, \eta_n)$ in C^n. Then d is a metric on C^n.

We shall verify the triangle inequality for d. (This includes the triangle inequality for Example 2.1.4.)

The triangle inequality is trivially satisfied if any two of the points concerned coincide. Suppose then that the points $x = (\xi_1, \xi_2, \ldots, \xi_n)$, $y = (\eta_1, \eta_2, \ldots, \eta_n)$, and $z = (\zeta_1, \zeta_2, \ldots, \zeta_n)$ are all distinct. We have to prove that

$$\left(\sum_{r=1}^{n} |\xi_r - \zeta_r|^2\right)^{1/2} \leq \left(\sum_{r=1}^{n} |\xi_r - \eta_r|^2\right)^{1/2} + \left(\sum_{r=1}^{n} |\eta_r - \zeta_r|^2\right)^{1/2}. \quad (1)$$

Let $\xi_r - \eta_r = \alpha_r$ and $\eta_r - \zeta_r = \beta_r$. Then (1) is equivalent to

$$\left(\sum_{r=1}^{n} |\alpha_r + \beta_r|^2\right)^{1/2} \leq \left(\sum_{r=1}^{n} |\alpha_r|^2\right)^{1/2} + \left(\sum_{r=1}^{n} |\beta_r|^2\right)^{1/2}. \quad (2)$$

Squaring both sides of (2) and using the equality

$$|a+b|^2 = |a|^2 + 2\mathrm{Re}\, a\bar{b} + |b|^2,$$

we see that (2) is equivalent to

$$\mathrm{Re} \sum_{r=1}^{n} \alpha_r \bar{\beta}_r \leq \left(\sum_{r=1}^{n} |\alpha_r|^2\right)^{1/2} \left(\sum_{r=1}^{n} |\beta_r|^2\right)^{1/2}. \quad (3)$$

Since x, y and z are all distinct, $\sum_{r=1}^{n} |\alpha_r|^2 \neq 0$ and $\sum_{r=1}^{n} |\beta_r|^2 \neq 0$.

Let $\lambda_r = \alpha_r \Big/ \left(\sum_{s=1}^{n} |\alpha_s|^2\right)^{1/2}$ and $\mu_r = \beta_r \Big/ \left(\sum_{s=1}^{n} |\beta_s|^2\right)^{1/2}$. Then (3) is equivalent to

$$\mathrm{Re} \sum_{r=1}^{n} \lambda_r \bar{\mu}_r \leq 1, \quad (4)$$

where $\sum_{r=1}^{n} |\lambda_r|^2 = \sum_{r=1}^{n} |\mu_r|^2 = 1$. But, for $r = 1, 2, \ldots, n$,

$$|\lambda_r|^2 - 2\,\mathrm{Re}\, \lambda_r \bar{\mu}_r + |\mu_r|^2 = |\lambda_r - \mu_r|^2 \geq 0$$

and consequently

$$2 - 2\,\mathrm{Re} \sum_{r=1}^{n} \lambda_r \bar{\mu}_r \geq 0,$$

when $\sum_{r=1}^{n} |\lambda_r|^2 = \sum_{r=1}^{n} |\mu_r|^2 = 1$. This proves (4) and hence (1). The inequality (2) is a special case of Minkowski's inequality (Lemma 3.2.8).

In situations where the same considerations apply equally to real numbers and complex numbers it is convenient to treat both cases simultaneously, and throughout the rest of this book we shall use the symbol K to stand for either R or C.

2.1.6 EXAMPLE In certain contexts metrics other than those of the last two examples occur naturally on K^n. Two such metrics are given by

$$d_0(x,y) = \max\{|\xi_r - \eta_r| : r = 1, 2, \ldots, n\}$$

and

$$d_1(x,y) = \sum_{r=1}^{n} |\xi_r - \eta_r|$$

for all $x = (\xi_1, \xi_2, \ldots, \xi_n)$ and $y = (\eta_1, \eta_2, \ldots, \eta_n)$ in K^n.

Verification of the triangle inequality for d_0 and d_1 is left to the reader.

The next metric space is rather a trivial one, but it often provides a simple counter-example to a plausible conjecture; it also shows that a metric can be defined on any non-empty set.

2.1.7 EXAMPLE Let X be a non-empty set. The function d defined by

$$d(x,y) = \begin{cases} 0 & \text{when } x = y, \\ 1 & \text{when } x \neq y \end{cases}$$

is a metric on X. The metric space (X,d) is the simplest example of a *discrete* metric space. The general definition of a discrete metric space will be given later (Definition 2.2.14).

The next example is one of the most important metric spaces that occurs in questions concerning classical analysis. We need a definition.

2.1.8 DEFINITION Let E be a non-empty set. A mapping f of E into K is said to be *bounded* if and only if $\{|f(t)| : t \in E\}$ is a bounded set of real numbers. Thus f is bounded if and only if there is a real number M such that $|f(t)| \leq M$ for all $t \in E$.

2.1.9 EXAMPLE Let E be a non-empty set and let $B_{\mathsf{K}}(E)$ be the set of all bounded mappings of E into K. If $f,g \in B_{\mathsf{K}}(E)$ then

$$|f(t) - g(t)| \leq |f(t)| + |g(t)|$$

and the set $\{|f(t) - g(t)| : t \in E\}$ is a bounded set of non-negative real numbers. Let

$$d(f,g) = \sup\{|f(t) - g(t)| : t \in E\}$$

for all $f,g \in B_{\mathsf{K}}(E)$. Then d is a metric on $B_{\mathsf{K}}(E)$. Let us verify the triangle inequality for d. Given $f,g,h \in B_{\mathsf{K}}(E)$ and $t \in E$,

$$\begin{aligned} |f(t) - h(t)| &= |(f(t) - g(t)) + (g(t) - h(t))| \\ &\leq |f(t) - g(t)| + |g(t) - h(t)| \\ &\leq d(f,g) + d(g,h), \end{aligned}$$

and consequently

$$d(f,h) = \sup\{|f(t) - h(t)| : t \in E\} \leq d(f,g) + d(g,h).$$

The metric d is called the *uniform metric* on $B_K(E)$. The metric d_0 of Example 2.1.6 is the uniform metric on $B_K(E)$, where E is the set $\{1, 2,..., n\}$. The next example is also a special case of Example 2.1.9.

2.1.10 Example Let $E = \{n : n = 1, 2,...\}$. Then $B_K(E)$ is the set of all bounded sequences in K and is usually denoted† by m. The metric on m is given by

$$d(x,y) = \sup\{ | \xi_n - \eta_n | : n = 1, 2,...\},$$

where $x = (\xi_n)$ and $y = (\eta_n)$ are in m.

The next example defines further metric spaces whose elements are sequences.

2.1.11 Example Let ℓ denote† the set of all sequences (ξ_n) in K such that the series $\sum_n |\xi_n|$ converges. Let

$$d(x,y) = \sum_{n=1}^{\infty} | \xi_n - \eta_n |$$

for all $x = (\xi_n)$ and $y = (\eta_n)$ in ℓ.

Since $| \xi_n - \eta_n | \leq |\xi_n| + |\eta_n|$ and the series $\sum_n |\xi_n|$ and $\sum_n |\eta_n|$ converge, the series defining d converges. We leave the verification of the triangle inequality to the reader.

Our final examples are metric spaces whose elements are continuous functions. Such metric spaces occur frequently in analysis.

2.1.12 Example Let $C_R([a,b])$ denote the set of all continuous real-valued functions on the closed interval $[a,b]$ of real numbers. It is well known that the functions in $C_R([a,b])$ are bounded (Corollary 4.3.3), and thus as in Example 2.1.9 we can define the *uniform metric* d on $C_R([a,b])$ by

$$d(f,g) = \sup\{ | f(t) - g(t) | : a \leq t \leq b\}.$$

† It would be more precise to write ℓ_K and m_K instead of ℓ and m but it is customary to omit the subscript K. It will always be clear from the context whether ℓ is to be interpreted as ℓ_K, ℓ_R, or ℓ_C. A similar remark applies to m.

2.1.13 EXAMPLE Let $C^1_{\mathsf{R}}([a,b])$ denote the set of all real-valued functions on the closed interval $[a,b]$ that have continuous first derivatives on $[a,b]$. More precisely, $f \in C^1_{\mathsf{R}}([a,b])$ if and only if f is differentiable at each point of the open interval (a,b), differentiable from the right and left at a and b respectively, and the derivative f' is continuous on $[a,b]$ (where $f'(a)$ and $f'(b)$ are the right-hand and left-hand derivatives at a and b respectively, see Definition 3.9.1). We can define a metric d on $C^1_{\mathsf{R}}([a,b])$ by

$$d(f,g) = \sup\{\,|f(t) - g(t)| : a \leq t \leq b\} + \sup\{\,|f'(t) - g'(t)| : a \leq t \leq b\},$$

for all $f,g \in C^1_{\mathsf{R}}([a,b])$.

EXERCISES

(1) Prove that if (X,d) is a metric space then

$$|\,d(x,y) - d(y,z)\,| \leq d(x,z)$$

for all $x,y,z \in X$.

(2) Let (X,d) be a metric space and $x_1, x_2, \ldots, x_n$ points of X. Prove that

$$d(x_1,x_n) \leq d(x_1,x_2) + d(x_2,x_3) + \ldots + d(x_{n-1},x_n).$$

(3) Mappings d_1 and d_2 of $\mathsf{R} \times \mathsf{R}$ into R are defined by

$$d_1(x,y) = \exp|\,x - y\,|,$$

and

$$d_2(x,y) = \max\{x - y, 0\}.$$

Is either d_1 or d_2 a metric on R?

(4) A mapping d of $X \times X$ into R satisfies the conditions

(a) $d(x,y) = 0$ if and only if $x = y$, and

(b) $d(x,z) \leq d(y,x) + d(y,z)$ for all $x,y,z \in X$.

Prove that d is a metric on X.

(5) Show that the conditions

(a) $e(x,y) = 0$ if and only if $x = y$, and

(b) $e(x,z) \leq e(x,y) + e(y,z)$ for all $x,y,z \in X$,

are not sufficient to ensure that the mapping e of $X \times X$ into R is a metric on X.

(6) Let X be a non-empty set. A *pseudo-metric on* X is a mapping d of $X \times X$ into R that satisfies the conditions:

(a) $d(x,y) \geq 0$ for all $x,y \in X$,

(b) $d(x,x) = 0$ for all $x \in X$,

(c) $d(x,y) = d(y,x)$ for all $x,y \in X$,

(d) $d(x,z) \leq d(x,y) + d(y,z)$ for all $x,y,z \in X$.

Let d be a pseudo-metric on X and define a relation $\sim$ on X by $x \sim y$ if and only if $d(x,y) = 0$. Show that $\sim$ is an equivalence relation on X and that the set of equivalence classes is a metric space with respect to a metric induced in a natural way by d.

For $x = (\xi_1,\xi_2)$ and $y = (\eta_1,\eta_2)$ in R^2 a mapping d is defined by $d(x,y) = | \xi_1 - \eta_1 |$. Show that d is a pseudo-metric on R^2. Describe the related equivalence classes and induced metric.

2.2 Open and Closed Sets

The terminology and concepts introduced in this section and the next are inspired by two- and three-dimensional Euclidean geometry. Throughout this section X will denote a metric space with metric d.

2.2.1 DEFINITION Let x be a point of X and r a non-negative real number. The set

$$B(x,r) = \{y \in X: d(x,y) < r\}$$

is called the *open ball with centre x and radius r*, and the set

$$B'(x,r) = \{y \in X: d(x,y) \leq r\}$$

is called the *closed ball with centre x and radius r*. The set

$$S(x,r) = \{y \in X: d(x,y) = r\}$$

is called the *sphere with centre x and radius r*.

By condition (b) of Definition 2.1.1, if $r > 0$, then $x \in B(x,r)$. By conditions (a) and (b) of Definition 2.1.1 we have $B(x,0) = \emptyset$ and $B'(x,0) = \{x\}$. Obviously for all $r \geq 0$,

$$B'(x,r) = B(x,r) \cup S(x,r),$$

and for $0 \leq r_1 < r_2$,

$$B'(x,r_1) \subseteq B(x,r_2) \subseteq B'(x,r_2).$$

In R^3 with the Euclidean metric, the terms open and closed ball and sphere have their usual meaning. In R^2 with the Euclidean metric, $B(x,r)$ is the open disc with centre x and radius r, and $S(x,r)$ is the circle with centre x and radius r. In R^2 with the metric d_0 (Example 2.1.6), $B(x,r)$ is the interior of a square with centre x and side of length $2r$, and $S(x,r)$ is the perimeter of this square. In R the open ball $B(x,r)$ is the open interval $(x-r, x+r)$ and $S(x,r)$ is the set consisting of the two points $x-r$ and $x+r$.

If X is the discrete metric space of Example 2.1.7 then for $0 < r < 1$ and all $x \in X$ we have $B(x,r) = \{x\}$, while for $r \geq 1$ and all $x \in X$ we have $B(x,r) = X$. Figure 2.1 illustrates the closed ball $B'(f,r)$ in the space $C_{\mathsf{R}}([a,b])$ with the uniform metric (Example 2.1.12); it consists of all the functions $g \in C_{\mathbf{R}}([a,b])$ whose graphs lie within the band which has vertical width $2r$ and is centred on the graph of f.

The motion of 'nearness to a point' is very important in metric spaces. The following definition formalizes this idea.

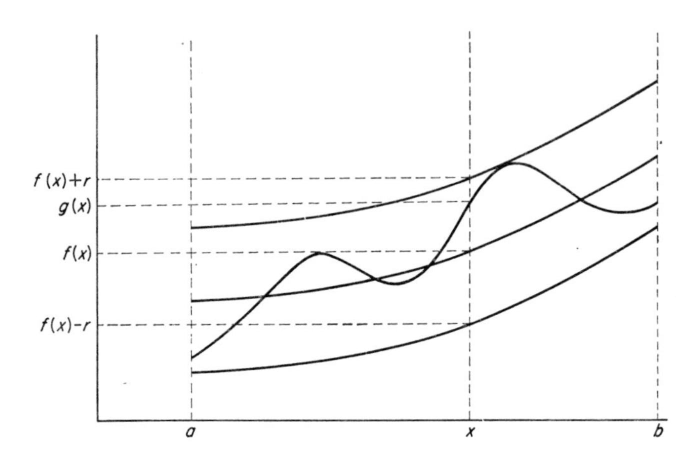

FIG. 2.1. A function g in the closed ball $B'(f,r)$

2.2.2 DEFINITION A subset A of X is said to be *open in the metric space* X if and only if, for each $x \in A$, there is a positive real number r, which depends on x, such that $B(x,r) \subseteq A$.

When it is clear which metric space is being considered, we shall speak simply of open sets or open subsets of X. Intuitively a set A is open when it contains all points of X that are 'sufficiently near to a point of A'. The following properties of open sets are important.

2.2.3 THEOREM (a) *The set X and the empty set $\emptyset$ are open.*
(b) *The union of an arbitrary family of open sets is open.*
(c) *The intersection of a finite family of open sets is open.*

PROOF It is obvious that X is open. Since the set $\emptyset$ has no points, Definition 2.2.2 is vacuously satisfied and so $\emptyset$ is open. (Explicitly, a subset A of X is not open if and only if there is a point $x \in A$ such that the open ball $B(x,r)$ is not contained in A for any $r > 0$. The set $\emptyset$ has no points and so this condition cannot be satisfied.) This proves (a).

Let $\mathscr{A}$ be a family of open subsets of X and let $x \in \bigcup_{A \in \mathscr{A}} A$. Then $x \in A_0$ for some $A_0 \in \mathscr{A}$, and therefore there exists $r > 0$ such that $B(x,r) \subseteq A_0$. But $A_0 \subseteq \bigcup_{A \in \mathscr{A}} A$, so $B(x,r) \subseteq \bigcup_{A \in \mathscr{A}} A$, which shows that $\bigcup_{A \in \mathscr{A}} A$ is open.

Finally, let $A_1, A_2,..., A_n$ be open subsets of X. If $\bigcap_{k=1}^{n} A_k = \emptyset$, (a) shows that $\bigcap_{k=1}^{n} A_k$ is open. Suppose that $\bigcap_{k=1}^{n} A_k \neq \emptyset$ and let $x \in \bigcap_{k=1}^{n} A_k$. For $k = 1, 2,..., n$ we can find $r_k > 0$ such that $B(x,r_k) \subseteq A_k$. Let $r = \min\{r_1, r_2,..., r_n\}$. Then $r > 0$ and $B(x,r) \subseteq B(x,r_k)$ for $k = 1, 2,..., n$, so $B(x,r) \subseteq \bigcap_{k=1}^{n} A_k$. This proves (c). □

An examination of the proof of the above theorem will reveal that we have used neither the symmetry of the metric nor the triangle inequality, which is the most important and restrictive of the metric space axioms. Both symmetry and the triangle inequality, however, play an important part in the next theorem, which establishes the existence of a large class of open sets in any metric space and reconciles our use of 'open' in the terms 'open ball' and 'open set'.

2.2.4 THEOREM *The sets $B(x,r)$ and $X \sim B'(x,r)$ are open for each $x \in X$ and each non-negative real number r.*

PROOF Let $x \in X$ and $r \geq 0$. We shall prove first that $B(x,r)$ is open. Since $B(x,0) = \emptyset$ and $\emptyset$ is open (Theorem 2.2.3), we may suppose that $r > 0$. Let $y \in B(x,r)$. To show that $B(x,r)$ is open we have to find $s > 0$ such that $B(y,s) \subseteq B(x,r)$. (See Fig. 2.2 which illustrates† the proof in the Euclidean space R^2.)

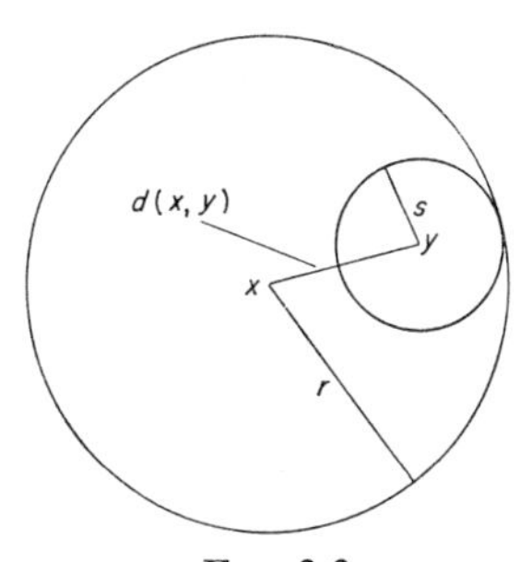

FIG. 2.2

Let $s = r - d(x,y)$. Since $y \in B(x,r)$, $d(x,y) < r$ and so $s > 0$. Let $z \in B(y,s)$. Then $d(y,z) < s$ and the triangle inequality gives

$$\begin{aligned} d(x,z) &\leq d(x,y) + d(y,z) \\ &< d(x,y) + s \\ &= r, \end{aligned}$$

† Diagrams can often help by suggesting methods of proof, but they must always be interpreted with care, and they can be misleading.

which shows that $z \in B(x,r)$. Hence $B(y,s) \subseteq B(x,r)$, and this shows that $B(x,r)$ is open.

We shall prove now that $X \sim B'(x,r)$ is open. If $B'(x,r) = X$, then $X \sim B'(x,r) = \emptyset$ and so $X \sim B'(x,r)$ is open. Suppose, then, that $B'(x,r) \neq X$, and let $y \in X \sim B'(x,r)$. Let $s = d(x,y) - r$. Since $y \notin B'(x,r)$, $d(x,y) > r$, and hence $s > 0$. We shall show that

$$B(y,s) \subseteq X \sim B'(x,r).$$

(See Fig. 2.3.) Let $z \in B(y,s)$. Then $d(y,z) < s$ and by the triangle inequality and the symmetry of d

$$\begin{aligned} d(x,y) &\leq d(x,z) + d(z,y) \\ &= d(x,z) + d(y,z) \\ &< d(x,z) + s, \end{aligned}$$

from which it follows that

$$d(x,z) > d(x,y) - s = r.$$

Thus $z \notin B'(x,r)$ and so $B(y,s) \subseteq X \sim B'(x,r)$. This proves that $X \sim B'(x,r)$ is open. $\square$

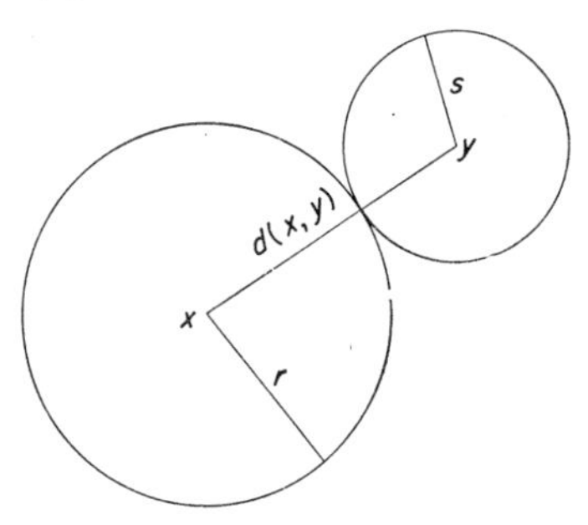

FIG. 2.3.

It follows from the last theorem that the bounded open intervals (a,b) are open subsets of R. Since

$$(a,\infty) = \bigcup_{n>a} (a,n) \quad \text{and} \quad (-\infty,a) = \bigcup_{n<a} (n,a),$$

it follows from Theorem 2.2.3 that the unbounded open intervals (a,∞) and $(-\infty,a)$ are also open subsets of R. It is of course easy to verify directly from the definitions of an open interval and an open set that the open intervals of R are open sets. It is also easy to verify that the half-open intervals and the closed intervals (other than R itself) are not open subsets of R.

We saw in Theorem 2.2.3 that the intersection of a finite family of open sets is open. The following example shows that the intersection of an arbitrary family of open sets may fail to be open. The sets

$(-1/n, 1/n)$, $n = 1, 2,\ldots$, are open in R, but $\bigcap_{n=1}^{\infty} (-1/n, 1/n) = \{0\}$ and the set $\{0\}$ is not open in R.

2.2.5 THEOREM *A non-empty subset of X is open if and only if it is the union of a family of open balls.*

PROOF Let A be a non-empty subset of X. By Theorems 2.2.3 and 2.2.4, if A is the union of a family of open balls then A is open. Suppose conversely that A is open. Then for each point $x \in A$ there exists $r_x > 0$ such that $B(x,r_x) \subseteq A$. Since $x \in B(x,r_x)$ it follows that $A = \bigcup_{x \in A} B(x,r_x)$. □

For open subsets of R Theorem 2.2.5 can be improved. We need two lemmas.

2.2.6 LEMMA *Let A be a non-empty subset of R with the property that $[a,b] \subseteq A$ whenever $a,b \in A$. Then A is an interval.*

PROOF Suppose first that A is bounded and let $M = \sup A$ and $m = \inf A$. Obviously $A \subseteq [m,M]$. Let $x \in (m,M)$. Then there are points $a,b \in A$ with $m \leq a < x$ and $x < b \leq M$. Therefore, by hypothesis, $x \in A$. This shows that $(m,M) \subseteq A \subseteq [m,M]$ and it follows that exactly one of the following holds: $A = (m,M)$, $A = [m,M)$, $A = (m,M]$, or $A = [m,M]$.

We leave the reader to verify that if A is bounded above but unbounded below then either $A = (-\infty,M)$ or $A = (-\infty,M]$, where $M = \sup A$; if A is unbounded above but bounded below then either $A = (m,\infty)$ or $A = [m,\infty)$, where $m = \inf A$; and if A is unbounded above and below then $A = \mathsf{R}$. □

2.2.7 LEMMA *Let $\mathscr{I}$ be a family of open intervals in R, no two of which are disjoint. Then $\bigcup_{I \in \mathscr{I}} I$ is an open interval.*

PROOF Let $I_0 = \bigcup_{I \in \mathscr{I}} I$, and let $a,b \in I_0$ and $a < c < b$. Then $a \in I_1$ and $b \in I_2$ for some $I_1,I_2 \in \mathscr{I}$. Let $I_1 = (a_1,b_1)$ and $I_2 = (a_2,b_2)$. Then $a_1 < a < c < b < b_2$. If $b_1 \leq a_2$ then $I_1 \cap I_2 = \emptyset$ which is impossible, so $a_2 < b_1$. Then either $c < b_1$ in which case $c \in I_1$ or $c \geq b_1$, in which case $c \in I_2$ and consequently in either case $c \in I_0$. It now follows from Lemma 2.2.6 that I_0 is an interval. The interval I_0 must be an open interval because it is an open set (Theorem 2.2.3), and closed ($\neq \mathsf{R}$) and half-open intervals are not open sets. □

2.2.8 THEOREM *Each non-empty open subset of* R *is the union of a countable family of pairwise disjoint open intervals.*

PROOF Let A be a non-empty open subset of R. For each $x \in A$ let I_x be the union of all the open intervals that contain x and are contained in A. (Such intervals exist because A is open.) By Lemma 2.2.7, I_x is an open interval. Obviously $A = \bigcup_{x \in A} I_x$. We shall prove that, for each pair $x,y \in A$, either $I_x = I_y$ or $I_x \cap I_y = \emptyset$. Let $x,y \in A$ and suppose that $I_x \cap I_y \neq \emptyset$. By Lemma 2.2.7 the set $I_x \cup I_y$ is an open interval which contains both x and y. Consequently $I_x \cup I_y \subseteq I_x$ and $I_x \cup I_y \subseteq I_y$, which shows that $I_x = I_y$. Let $\mathscr{F}$ be the family of all distinct sets of the form I_x with $x \in A$. It is obvious that the union of the family $\mathscr{F}$ is A and that the sets in $\mathscr{F}$ are pairwise disjoint. It remains only to show that $\mathscr{F}$ is countable. By Theorems 1.6.6 and 1.4.1 the set $\mathsf{Q} \cap A$ of all rational numbers in A is countable. We define a mapping f of $\mathsf{Q} \cap A$ into $\mathscr{F}$ as follows: given $r \in \mathsf{Q} \cap A$, let $f(r)$ be the unique set in $\mathscr{F}$ that contains r. (The set $f(r)$ is unique because the sets in $\mathscr{F}$ are pairwise disjoint.) By Theorem 1.6.7 each open interval in R contains a rational number, therefore f maps $\mathsf{Q} \cap A$ onto $\mathscr{F}$ and hence $\mathscr{F}$ is countable. □

2.2.9 DEFINITION A subset A of X is said to be a *neighbourhood of a point* $x \in X$ if and only if there is a positive real number r such that $B(x,r) \subseteq A$.

Since $x \in B(x, r)$ for $r > 0$, if A is a neighbourhood of x then $x \in A$. Intuitively if A is a neighbourhood of x then A contains all points of X that are 'sufficiently near to x'. It is easy to see that a set is open if and only if it is a neighbourhood of each of its points. It is sometimes convenient to use neighbourhoods of a point that are not open sets. For example in the Euclidean space R^n the closed ball $B'(x,r)$ is a neighbourhood of x which is not open in R^n if $r > 0$.

2.2.10 LEMMA *Let* $x \in X$. *Then*

(a) *if* A *is a neighbourhood of* x *and* $B \supseteq A$, B *is a neighbourhood of* x, *and* (b) *if* A *and* B *are neighbourhoods of* x, $A \cap B$ *is a neighbourhood of* x.

PROOF Statement (a) is obvious and the proof of (b) is similar to that of Theorem 2.2.3 (c). □

2.2.11 DEFINITION A subset A of X is said to be *closed in the metric space* X if and only if $X \sim A$ is open in X.

As with open sets we shall usually speak of closed sets or closed subsets of X. Obviously to each statement about open sets there corresponds a statement about closed sets, and vice versa.

2.2.12 THEOREM (a) *The set X and the empty set $\varnothing$ are closed.*
(b) *The intersection of an arbitrary family of closed sets is closed.*
(c) *The union of a finite family of closed sets is closed.*

PROOF The proof follows directly from Theorem 2.2.3 and the following set-theoretical identities:

$$X \sim \varnothing = X, X \sim X = \varnothing,$$
$$X \sim \bigcap_{A \in \mathscr{A}} A = \bigcup_{A \in \mathscr{A}} (X \sim A),$$

and
$$X \sim \bigcup_{j=1}^{n} A_j = \bigcap_{j=1}^{n} (X \sim A_j). \square$$

The union of an arbitrary family of closed sets may not be closed (Exercise (3)).

2.2.13 THEOREM *The set $B'(x,r)$ is closed for each $x \in X$ and each non-negative real number r.*

PROOF This theorem follows directly from Theorem 2.2.4. □

Since $B'(x,0) = \{x\}$, a set consisting of a single point is closed, and therefore, by Theorem 2.2.12, all finite subsets of X are closed.

The simplest closed subsets of R are the bounded, closed intervals $[a,b]$, which are the closed balls in R, and the unbounded, closed intervals $(-\infty,a]$ and $[a,\infty)$, which are the complements in R of the open intervals (a,∞) and $(-\infty,a)$, respectively, and the unbounded closed interval R. The bounded half-open intervals $(a,b]$ and $[a,b)$ are neither open nor closed.

In any metric space the whole space and the empty set are both open and closed. In Section 2.10 we shall take up the question of the existence of sets, other than the whole space and the empty set, that are both open and closed. In the metric space of Example 2.1.7 the answer to this question is trivial: every subset is both open and closed (Exercise (1)).

2.2.14 DEFINITION A metric space in which every subset is both open and closed is said to be *discrete*.

We shall see in Section 2.10 that R has no subsets that are both open and closed, other than R and $\varnothing$.

It follows from Theorem 2.2.8 that each proper closed subset of R

is the complement in R of the union of a countable family of pairwise disjoint open intervals. The following example shows how complicated such a set may be.

2.2.15 Example (The Cantor set) We shall describe the construction of the Cantor set C leaving the reader to supply the details. The set C is obtained from the closed interval $[0,1]$ by removing first the open central third $(1/3,2/3)$, then the open central thirds $(1/9,2/9)$ and $(7/9,8/9)$ of each of the two remaining closed intervals, then the open central thirds of each of the four remaining closed intervals, and so on, at each stage removing the open central thirds of the closed intervals remaining at the preceding stage. Thus at the nth stage the subset C_n of $[0,1]$ which remains is the union of 2^n pairwise disjoint closed intervals each of length 3^{-n}. It is not difficult to prove (Exercise (7)) that the left-hand end points of these intervals are the 2^n rational numbers of the form

$$\frac{\alpha_1}{3} + \frac{\alpha_2}{3^2} + \ldots + \frac{\alpha_n}{3^n}$$

with each α_m equal to 0 or 2. The Cantor set C is then the intersection $\bigcap_{n=1}^{\infty} C_n$ and so is a closed subset of $[0,1]$.

All the end points of the intervals which make up the sets C_n (forming a countable set) belong to C. However, these points do not exhaust C. Indeed, there is a one-to-one mapping of the interval $(0,1]$ into C and therefore, by Theorem 1.6.6, C is not countable (Exercise (8)).

The Cantor set is of considerable theoretical interest. It is named after the German mathematician G. Cantor who, in the last quarter of the nineteenth century, introduced the concepts of open and closed subsets of R^n and made a penetrating study of their structure.

Exercises

In these Exercises the metric spaces are those of Examples 2.1.2–2.1.13, *the metric for the space* R^n *is the Euclidean metric of Example* 2.1.4.

(1) Show that every subset of the metric space of Example 2.1.7 is both open and closed.

(2) Determine whether the following subsets of the metric spaces indicated are open, closed, open and closed, or neither open nor closed:

(a) $\{(\xi_1, \xi_2, \ldots, \xi_n) \in \mathsf{R}^n : \xi_i > 0 \text{ for } i = 1, 2, \ldots, n\}$,

(b) $\{(\xi_1, \xi_2, \ldots, \xi_n) \in \mathsf{R}^n : \xi_i \text{ rational for } i = 1, 2, \ldots, n\}$,

(c) $\{(\xi_n) \in \ell : \xi_n < 1/n \text{ for } n = 1, 2, \ldots\}$,

(d) $\{f \in B_{\mathsf{K}}(E): f(t) = 0 \text{ for } t \in A\}$ where A is a non-empty subset of E,

(e) $\{f \in C^1_{\mathsf{R}}([a,b]): f(a) + f'(a) = 0\}$.

(3) Give an example of a countable family of closed subsets of R, whose union is not closed.

(4) Prove that an open subset of R^n can be expressed as the union of a countable family of open balls in R^n.

(5) Show that if $n \geq 2$ there are open subsets of R^n that cannot be expressed as the union of a countable family of pairwise disjoint open balls in R^n.

The three exercises which follow concern the Cantor set. Further exercises on the Cantor set follow Sections 2.3 *and* 2.7.

(6) Show that if b is an integer ≥ 2 each $x \in [0,1]$ has a *b-adic expansion* $x = \sum_{m=1}^{\infty} \alpha_m b^{-m}$ in which α_m is an integer with $0 \leq \alpha_m \leq b-1$ for $m = 1, 2, \ldots$. Show further that if $x \in [0,1]$ has more than one b-adic expansion then it has precisely two, one and only one of which contains only a finite number of non-zero terms. (Hint: For $x \in [0,1)$ define α_m inductively by $\alpha_1 = [bx]$, $\alpha_{m+1} = [b^{m+1}(x - \sum_{k=1}^{m} \alpha_k b^{-k})]$ where the symbol $[y]$ denotes the greatest integer not greater than y.)

(7) Prove that the left-hand end points of the intervals whose union is C_n are the rational numbers of the form

$$\frac{\alpha_1}{3} + \frac{\alpha_2}{3^2} + \ldots + \frac{\alpha_n}{3^n}$$

with each α_m equal to 0 or 2, and that $x \in C_n$ if and only if x has a triadic expansion $x = \sum_{m=1}^{\infty} \alpha_m 3^{-m}$ with $\alpha_m \neq 1$ for $m = 1, 2, \ldots, n$. Hence prove that $x \in C$ if and only if x has a triadic expansion $x = \sum_{m=1}^{\infty} \alpha_m 3^{-m}$ with $\alpha_m \neq 1$ for all $m = 1, 2, \ldots$.

(8) Each $x \in (0,1]$ has a unique dyadic expansion $x = \sum_{m=1}^{\infty} \alpha_m 2^{-m}$ in which $\alpha_m = 1$ for an infinite set of integers m. Show that

$$f(x) = \sum_{m=1}^{\infty} 2\alpha_m 3^{-m}$$

then defines a one-to-one increasing mapping of $(0,1]$ into the Cantor set C. Determine the complement in C of the range of f.

2.3 Closure, Frontier and Interior

Throughout this section X will denote a metric space with metric d.

2.3.1 Definition The *closure in the metric space* X of a subset A of X is the intersection of the family of all closed subsets of X that contain A. (This family is non-empty because X is closed and contains A.) The closure of A in X will be denoted by $\bar{A}$ or by A^-.

2.3.2 Theorem *Let A be a subset of X. Then $\bar{A}$ is closed and $A \subseteq \bar{A}$. Further A is closed if and only if $A = \bar{A}$.*

Proof It is obvious that $A \subseteq \bar{A}$, and it follows from Theorem 2.2.12 that $\bar{A}$ is closed. If A is closed, then, by definition of $\bar{A}$, $\bar{A} \subseteq A$ and consequently $A = \bar{A}$. □

The closure $\bar{A}$ of A is the smallest closed subset of X that contains A in the sense that, if B is closed and $A \subseteq B$, then $\bar{A} \subseteq B$.

2.3.3 Theorem $(A \cup B)^- = \bar{A} \cup \bar{B}$ *for all subsets A and B of X.*

Proof $(A \cup B)^-$ is closed and $A \subseteq A \cup B \subseteq (A \cup B)^-$, so by the above remark $\bar{A} \subseteq (A \cup B)^-$. Similarly $\bar{B} \subseteq (A \cup B)^-$ and thus

$$\bar{A} \cup \bar{B} \subseteq (A \cup B)^-.$$

Also, $\bar{A} \cup \bar{B}$ is closed (Theorems 2.3.2 and 2.2.12) and

$$A \cup B \subseteq \bar{A} \cup \bar{B},$$

so $(A \cup B)^- \subseteq \bar{A} \cup \bar{B}$. □

The reader should note that, in general, $(A \cap B)^- \neq \bar{A} \cap \bar{B}$ (Exercise (2)).

The following theorem is one of the most useful characterizations of the closure of a set.

2.3.4 Theorem *Let A be a non-empty subset of X and $x \in X$. Then $x \in \bar{A}$ if and only if every non-empty open ball with centre x contains at least one point of A.*

Proof Suppose that $x \in \bar{A}$ and that $B(x,r) \cap A = \emptyset$ for some $r > 0$. Then $A \subseteq X \sim B(x,r)$ and consequently, since $X \sim B(x,r)$ is closed (Theorem 2.2.4), $\bar{A} \subseteq X \sim B(x,r)$. This is a contradiction because $x \in \bar{A}$ and $x \in B(x,r)$.

Suppose, conversely, that $B(x,r) \cap A \neq \emptyset$ for all $r > 0$ and $x \notin \bar{A}$. Since $X \sim \bar{A}$ is open and $x \in X \sim \bar{A}$, there exists $r_0 > 0$ such that $B(x,r_0) \subseteq X \sim \bar{A}$, which is a contradiction. □

The intuitive content of the last theorem is that a point x of X is in the closure of A if and only if there are points of A 'arbitrarily close to x'.

Since $B(x,r) \subseteq B'(x,r)$ and $B'(x,r)$ is closed (Theorem 2.2.13), we have $B(x,r)^- \subseteq B'(x,r)$ for all $x \in X$ and all $r \geq 0$. It can happen that $B(x,r)^- \neq B'(x,r)$ (cf. Exercise (6)(a)); however it is an easy consequence of Theorem 2.3.4 that $B(x,r)^- = B'(x,r)$ in the Euclidean space R^n.

Let Q be the set of rational numbers. By Theorem 1.6.7, there is a point of Q between any two real numbers, and therefore, by Theorem 2.3.4, $\bar{\mathsf{Q}} = \mathsf{R}$.

2.3.5 DEFINITION A subset of X whose closure is X is said to be *dense in* X.

We have just seen that Q is dense in R. It is often an important, and sometimes a difficult question to determine whether or not a given subset of a metric space is dense in the space. It follows at once from Theorem 2.3.4 that a subset A of X is dense in X if and only if A has non-empty intersection with each non-empty open ball in X, or, equivalently, if and only if A has non-empty intersection with each non-empty open subset of X.

2.3.6 DEFINITION The metric space X is said to be *separable* if and only if there is a countable subset of X that is dense in X.

Since the set of all rational numbers is countable (Theorem 1.6.6) and dense in R, the metric space R is separable, and it follows easily that the Euclidean space R^n is separable (Exercise (11)).

In some problems great simplifications can be made when it is known that the metric space concerned is separable. It is therefore important to know which of the metric spaces that occur frequently in analysis are separable. In general the metric space $B_{\mathsf{K}}(E)$ (Example 2.1.9) is not separable (see Exercise 3.4(5)); the space $C_{\mathsf{R}}([a,b])$ is separable (Theorem 4.5.12).

It is useful to distinguish two classes of points in the closure of a set.

2.3.7 DEFINITION Let A be a subset of X. A point $x \in X$ is said to be a *point of accumulation of* A if and only if, for each positive real number r, the set $B(x,r) \sim \{x\}$ contains at least one point of A. The set of all accumulation points of A is called the *derived set of* A and is usually denoted by A'. A point of A that is not an accumulation point of A is called an *isolated point of* A.

A point of accumulation of A is also called a *limit point of* A. It is not difficult to prove that the derived set A' is closed (Exercise (5)). A point of accumulation of A need not belong to A. It follows from Theorem 2.3.4 that $A' \subseteq \bar{A}$ and $\bar{A} \sim A'$ is precisely the set of isolated points of A. Consequently $\bar{A} = A \cup A'$ and, by Theorem 2.3.2, A is closed if and only if $A' \subseteq A$. It is easy to see that a point $x \in A$ is an isolated point of A if and only if there exists $r > 0$ such that

$$A \cap B(x,r) = \{x\}.$$

It is also easy to see that a point $x \in X$ is an accumulation point of A if and only if each open set containing x (or even each neighbourhood of x) contains a point of A which is different from x.

2.3.8 Lemma *If x is an accumulation point of a set A, then every non-empty open ball with centre x contains infinitely many points of A.*

Proof Suppose that for some $r > 0$ the set $B(x,r) \sim \{x\}$ contains only a finite number of points of A, say $\{x_1, x_2, \ldots, x_n\}$. Let

$$r_0 = \min\{d(x,x_1), \ldots, d(x,x_n)\}.$$

Then $0 < r_0 \leq r$, and it is obvious that $B(x,r_0) \sim \{x\}$ contains no point of A, which is a contradiction. □

In the duality between open sets and closed sets the concept corresponding to closure is interior.

2.3.9 Definition The *interior in the metric space X* of a subset A of X is the union of the family of all open subsets of X that are contained in A (this family is non-empty since $\emptyset$ is open and contained in A). The interior of A is denoted by A°.

There are analogues of Theorems 2.3.2 and 2.3.3 for A°. We shall be content to state (without proof) the analogue of Theorem 2.3.2.

2.3.10 Theorem *Let A be a subset of X. Then A° is open and $A^\circ \subseteq A$. Further A is open if and only if $A = A^\circ$.*

The interior A° of A is precisely the set of all points of which A is a neighbourhood, and it is the largest open subset of X that is contained in A in the sense that if B is open and $B \subseteq A$, then $B \subseteq A^\circ$. It follows that $x \in A^\circ$ if and only if there exists $r > 0$ such that $B(x,r) \subseteq A$.

We shall now formulate explicitly the relationship between closure and interior.

2.3.11 Theorem *Let A be a subset of X. Then*

$$A^\circ = X \sim (X \sim A)^-$$

and

$$\bar{A} = X \sim (X \sim A)^\circ.$$

Proof Since $X \sim A \subseteq (X \sim A)^-$ we have $X \sim (X \sim A)^- \subseteq A$. But $X \sim (X \sim A)^-$ is open (Theorem 2.3.2), so $X \sim (X \sim A)^- \subseteq A^\circ$.

On the other hand, $X \sim A^\circ$ is closed (Theorem 2.3.10) and $X \sim A \subseteq X \sim A^\circ$, so $(X \sim A)^- \subseteq X \sim A^\circ$ and hence

$$A^\circ \subseteq X \sim (X \sim A)^-.$$

This shows that $A^\circ = X \sim (X \sim A)^-$, and the other relation follows from this one if we replace A by $X \sim A$ and take complements. □

2.3.12 DEFINITION Let A be a subset of X. The set $\bar{A} \sim A^\circ$ is called the *frontier in the metric space* X of A and is denoted by $\mathrm{Fr}\, A$.

The frontier of A is also called the *boundary* of A. The relations of Theorem 2.3.11 give

$$\mathrm{Fr}\, A = \bar{A} \cap (X \sim A)^- = X \sim (A^\circ \cup (X \sim A)^\circ),$$

which shows that $\mathrm{Fr}\, A$ is a closed set and that $\mathrm{Fr}\, A = \mathrm{Fr}\,(X \sim A)$. It is also clear that

$$\bar{A} = A^\circ \cup \mathrm{Fr}\, A$$

and

$$X = A^\circ \cup (X \sim A)^\circ \cup \mathrm{Fr}\, A.$$

Theorem 2.3.4 and the remarks after Theorem 2.3.10 show that $x \in \mathrm{Fr}\, A$ if and only if each non-empty open ball with centre x contains at least one point of A and at least one point of $X \sim A$. In particular each isolated point of A is in $\mathrm{Fr}\, A$.

Intuitively $\mathrm{Fr}\, A$ is the set of points of X that can be 'approached arbitrarily closely by points of A and by points of $X \sim A$'. In the Euclidean space R^n the frontier of a non-empty open ball is the closed sphere with the same centre and radius (Exercise (6)), which is what our geometrical intuition leads us to expect. Geometrical intuition, however, can be misleading: for example, the frontier in R of the set Q of rational numbers is R (Exercise (6)).

EXERCISES

(1) Prove that if $\mathscr{A}$ is a family of subsets of a metric space then

$$\bigcup_{A \in \mathscr{A}} \bar{A} \subseteq \Big(\bigcup_{A \in \mathscr{A}} A\Big)^- \text{ and } \Big(\bigcap_{A \in \mathscr{A}} A\Big)^- \subseteq \bigcap_{A \in \mathscr{A}} \bar{A}.$$

Give an example of a countable family $\{A_n : n = 1, 2, \ldots\}$ of subsets of R with $\bigcup_{n=1}^{\infty} \bar{A}_n \neq \Big(\bigcup_{n=1}^{\infty} A_n\Big)^-$

(2) Prove that if A,B are subsets of a metric space with

$$\mathrm{Fr}\, A \cap \mathrm{Fr}\, B = \varnothing$$

then $(A \cap B)^- = \bar{A} \cap \bar{B}$ and $(A \cup B)^\circ = A^\circ \cup B^\circ$. Give an example of subsets A,B of R with $(A \cap B)^- \neq \bar{A} \cap \bar{B}$.

(3) Show that if A is a subset of a metric space X then $\mathrm{Fr}\, \bar{A} \subseteq \mathrm{Fr}\, A$.

For which metric spaces X are $\operatorname{Fr} \bar{A}$ and $\operatorname{Fr} A$ necessarily equal for all subsets A of X?

(4) Show that a metric space is discrete if and only if every point of the space is isolated.

(5) Let X be a metric space and A a subset of X. Prove that the derived set A' of A is closed.

(6) Find the closures, the interiors and the frontiers of
(a) a subset A of a discrete metric space X,
(b) the set of all rational numbers in R,
(c) an open ball in the Euclidean space R^n.

(7) A subset A of R is said to be an additive subgroup of R if and only if $-\xi \in A$ and $\xi + \eta \in A$ whenever $\xi,\eta \in A$. Prove that an additive subgroup of R either is dense in R or is the set of all integral multiples of some real number.

(8) Prove that the Cantor set of Example 2.2.15 has neither isolated points nor interior points.

(9) Let A be the subset

$$\{(m/n,\ 1/n): n = 1, 2,\ldots;\ m = 0,\ \pm 1,\ \pm 2,\ldots\}$$

of the Euclidean space R^2. Prove that $\bar{A}$ is the union of A and the set $\{(\xi,0): \xi \in \mathsf{R}\}$.

(10) Find the frontier of the subset $\{(\xi_1,\xi_2): \xi_2 = 0\}$ of the Euclidean space R^2.

(11) Show that the metric spaces R^n and C^n of Examples 2.1.4 and 2.1.5 are separable.

(12) Show that a discrete metric space is separable if and only if it is countable.

2.4 Subspaces

Throughout this section X will denote a metric space with metric d, and Y will denote a non-empty subset of X. There is a natural metric on Y. It is easy to see that the restriction, d_1 say, of d to $Y \times Y$ is a metric on Y.

2.4.1 DEFINITION The metric space (Y,d_1) is called a *subspace* of the metric space (X,d).

We shall usually say simply that Y is a subspace of X, and we shall use the same symbol d to denote both the metric on X and the metric on Y. The idea of a subspace may seem trivial, but the use of subspaces may considerably simplify an argument.

There may be subsets of Y that are open in the subspace Y but are not open in the metric space X (Exercise (1)). The following theorem characterizes sets that are open or closed in the subspace Y in terms of sets that are open or closed in the space X.

2.4.2 THEOREM *Let A be a subset of Y.*

(a) *A is open in the subspace Y if and only if there is a subset A_1 of X that is open in the metric space X and such that $A = Y \cap A_1$.*

(b) *A is closed in the subspace Y if and only if there is a subset A_1 of X that is closed in the metric space X and such that $A = Y \cap A_1$.*

(c) *Let $\mathrm{cl}_X A$ and $\mathrm{cl}_Y A$ denote, respectively, the closures of A in the metric space X and in the subspace Y. Then $\mathrm{cl}_Y A = Y \cap \mathrm{cl}_X A$.*

PROOF For each $x \in X$, $y \in Y$ and $r > 0$, let

$$B_X(x,r) = \{z \in X: d(x,z) < r\}$$

and

$$B_Y(y,r) = \{z \in Y: d(y,z) < r\}.$$

Then $B_X(x,r)$ and $B_Y(y,r)$ are the open balls with centres x and y and radius r in the metric space X and the subspace Y, respectively. Clearly

$$B_Y(y,r) = Y \cap B_X(y,r)$$

for all $y \in Y$ and all $r > 0$.

Suppose that A is open in the subspace Y, and let

$$\Lambda = \{(y,r): y \in A,\ r > 0, \text{ and } B_Y(y,r) \subseteq A\}.$$

Since A is open in the subspace Y, the set Λ is non-empty and

$$A = \bigcup_{(y,r)\in\Lambda} B_Y(y,r).$$

Let

$$A_1 = \bigcup_{(y,r)\in\Lambda} B_X(y,r).$$

Then

$$\begin{aligned} Y \cap A_1 &= Y \cap \bigcup_{(y,r)\in\Lambda} B_X(y,r) \\ &= \bigcup_{(y,r)\in\Lambda} [Y \cap B_X(y,r)] \\ &= \bigcup_{(y,r)\in\Lambda} B_Y(y,r) \\ &= A, \end{aligned}$$

while, by Theorems 2.2.3 and 2.2.4, the set A_1 is open in X.

Suppose, conversely, that $A = Y \cap A_1$ where A_1 is open in X. Let $y \in A$. Then $B_X(y,r) \subseteq A_1$ for some $r > 0$ and consequently

$$B_Y(y,r) = Y \cap B_X(y,r) \subseteq Y \cap A_1 = A.$$

This shows that A is open in the subspace Y. This completes the proof of (a).

Statement (b) follows easily from (a) and the definition of a closed set when we observe that, if $A = Y \cap A_1$, then $Y \sim A = Y \cap (X \sim A_1)$.

Finally we shall prove (c). Obviously $A \subseteq Y \cap \mathrm{cl}_X A$. By (b) $Y \cap \mathrm{cl}_X A$ is closed in the subspace Y, and consequently $\mathrm{cl}_Y A \subseteq Y \cap \mathrm{cl}_X A$. On the other hand (b) shows that $\mathrm{cl}_Y A = Y \cap A_1$ for some subset A_1 of X that is closed in X. But then $A \subseteq \mathrm{cl}_Y A \subseteq A_1$ and hence $\mathrm{cl}_X A \subseteq A_1$. It follows that $\mathrm{cl}_Y A = Y \cap A_1 \supseteq Y \cap \mathrm{cl}_X A$. This completes the proof of (c). □

For brevity we shall often write 'A is open in Y' and 'A is closed in Y' instead of 'A is open in the subspace Y' and 'A is closed in the subspace Y'.

2.4.3 THEOREM *A subspace of a separable metric space is separable.*

PROOF Let Y be a subspace of a separable metric space X and let $A = \{x_n : n = 1, 2, \ldots\}$ be a countable subset of X that is dense in X. Since $\bar{A} = X$ it follows from Theorem 2.3.4 that, for each $y \in Y$ and each positive integer m, there is an integer n such that $x_n \in B(y,1/m)$. Let

$$\Lambda = \{(n,m) : Y \cap B(x_n,1/m) \neq \emptyset\}.$$

We have just seen that Λ is non-empty. For each $(n,m) \in \Lambda$, choose a point $y_{nm} \in Y \cap B(x_n,1/m)$. The set $B = \{y_{nm} : (n,m) \in \Lambda\}$ is countable by Theorems 1.4.3 and 1.4.1. We shall prove that B is dense in Y. Let $y \in Y$ and $r > 0$, and choose a positive integer m with $1/m \leq r/2$. As above there is an integer n such that $x_n \in B(y,1/m)$. Then $(n,m) \in \Lambda$ and

$$\begin{aligned} d(y,y_{nm}) &\leq d(y,x_n) + d(x_n,y_{nm}) \\ &< 1/m + 1/m \\ &\leq r \end{aligned}$$

which shows that $y_{nm} \in B(y,r)$. By Theorem 2.3.4 the point y is in the closure of B in the subspace Y, and hence B is dense in Y. □

EXERCISES

(1) Let Y be a subspace of a metric space X. Prove that every subset of Y open in Y is also open in X if and only if Y is open in X. Prove also the statement obtained by replacing 'open' by 'closed'.

(2) Let Y be a subspace of a metric space X and A a subset of Y. The interiors of A in the spaces Y and X are denoted by $\mathrm{int}_Y A$ and $\mathrm{int}_X A$ respectively. Prove that

$$\mathrm{int}_X A \subseteq \mathrm{int}_Y A \subseteq \mathrm{int}_X A \cup \mathrm{Fr}\ Y.$$

(3) Let X be a separable metric space and A a subset of X with no

accumulation points. Prove that A is countable. Deduce that if B is a subset of X with a countable set of accumulation points then B itself is countable.

2.5 Sequences

The ideas introduced in this section will bring us closer to the subject matter of classical analysis. Throughout the section X will denote a metric space with metric d.

2.5.1 DEFINITION A sequence (x_n) in X is said to *converge in the metric space X to the point $x \in X$* if and only if, for each positive real number ε, there exists a positive integer N, which depends on ε, such that $d(x,x_n) < \varepsilon$ for all $n \geq N$. A sequence is said to *converge in X* if and only if it converges in X to some point of X. If a sequence converges in X to the point x, then x is called a *limit* of the sequence.

Except when it is necessary to specify which metric space is being considered we shall say simply that the sequence (x_n) converges to the limit x. Intuitively (x_n) converges to x if and only if 'x_n is arbitrarily close to x for all sufficiently large integers n'.

2.5.2 LEMMA *A convergent sequence has exactly one limit.*

PROOF Suppose that (x_n) converges in X and has two distinct limits, x and y. Then $d(x,y) > 0$ and consequently there are integers N_1 and N_2 such that $d(x,x_n) < \frac{1}{2}d(x,y)$ for all $n \geq N_1$ and $d(y,x_n) < \frac{1}{2}d(x,y)$ for all $n \geq N_2$. Let $N = \max\{N_1,N_2\}$, then

$$d(x,y) \leq d(x,x_N) + d(x_N,y) = d(x,x_N) + d(y,x_N) < d(x,y),$$

which is a contradiction. □

We shall use the notation $\lim_{n\to\infty} x_n = x$ to mean that the sequence (x_n) converges and has the limit x.

In the metric spaces R and C the notion of a convergent sequence introduced above coincides with the classical notion of a convergent sequence of real or complex numbers, and Definition 2.5.1 was, of course, inspired by these classical models.

The definition of convergence in an arbitrary metric space can be reformulated in terms of convergence of a sequence of real numbers. In fact, it is clear that a sequence (x_n) in X converges to the limit $x \in X$ if and only if the sequence $(d(x,x_n))$ of real numbers converges to zero: in symbols $\lim_{n\to\infty} x_n = x$ if and only if $\lim_{n\to\infty} d(x,x_n) = 0$.

On the other hand we can remove all explicit mention of the metric from the definition of convergence: $\lim_{n\to\infty} x_n = x$ if and only if, for each neighbourhood V of x, the point $x_n \in V$ for all but a finite number of positive integers n.

2.5.3 THEOREM *Let (x_k) be a sequence in K^n with $x_k = (\xi_{1k}, \xi_{2k},\dots, \xi_{nk})$ for $k = 1, 2,\dots$, and let d be the metric on K^n defined in Examples 2.1.4 or 2.1.5. Then the sequence (x_k) converges in (K^n, d) to the limit*

$$x = (\xi_1, \xi_2,\dots, \xi_n)$$

if and only if $\lim_{k\to\infty} \xi_{mk} = \xi_m$ for $m = 1, 2,\dots, n$.

PROOF Let $\varepsilon > 0$ and suppose first that $\lim_{k\to\infty} x_k = x$. Then there is an integer K such that $d(x,x_k) < \varepsilon$ for all $k \geq K$, and hence, for $m = 1, 2,\dots, n$ and $k \geq K$,

$$|\xi_{mk} - \xi_m| \leq \left(\sum_{j=1}^{n} |\xi_{jk} - \xi_j|^2\right)^{1/2} = d(x_k,x) < \varepsilon.$$

This proves that $\lim_{k\to\infty} \xi_{mk} = \xi_m$.

Suppose now that $\lim_{k\to\infty} \xi_{mk} = \xi_m$ for $m = 1, 2,\dots, n$. Then there are integers $K_1, K_2,\dots, K_n$ such that, for $m = 1, 2,\dots, n$, $|\xi_{mk} - \xi_m| < n^{-\frac{1}{2}}\varepsilon$ for all $k \geq K_m$. Let $K = \max\{K_1, K_2,\dots, K_n\}$. Then, if $k \geq K$,

$$\begin{aligned} d(x,x_k) &= \left(\sum_{m=1}^{n} |\xi_m - \xi_{mk}|^2\right)^{1/2} \\ &\leq n^{1/2} \max\{|\xi_m - \xi_{mk}| : m = 1, 2,\dots, n\} \\ &< \varepsilon. \end{aligned}$$

This shows that $\lim_{k\to\infty} x_k = x$. □

Let E be a non-empty set. A sequence (f_n) in the metric space $B_{\mathsf{K}}(E)$ (Example 2.1.9) that converges in that space to the point $f \in B_{\mathsf{K}}(E)$ is said to *converge uniformly to f on E*. Suppose that (f_n) converges uniformly to f on E and let $\varepsilon > 0$. Then there is an integer N such that $d(f, f_n) < \varepsilon$ for all $n \geq N$, and consequently, by definition of the metric on $B_{\mathsf{K}}(E)$,

$$\sup\{|f(t) - f_n(t)| : t \in E\} < \varepsilon \tag{1}$$

for all $n \geq N$. It follows from (1) that

$$|f(t) - f_n(t)| < \varepsilon \tag{2}$$

for all $n \geq N$ and all $t \in E$. Inequality (2) shows that the sequence $(f_n(t))$ converges in K for each $t \in E$ and $\lim_{n\to\infty} f_n(t) = f(t)$ for each $t \in E$.

Let (g_n) be a sequence of real or complex-valued functions on E such that the sequence of $(g_n(t))$ converges for each $t \in E$ and let

$$g(t) = \lim_{n\to\infty} g_n(t)$$

for all $t \in E$. Then the sequence (g_n) is said to *converge pointwise to g on E*. Thus, if (f_n) is a sequence in $B_K(E)$ that converges uniformly to f on E, then (f_n) also converges pointwise to f on E. We shall see below that the converse is false.

The reader who is familiar with the concept of uniform convergence in classical analysis will see from (1) or (2) that convergence in the metric space $B_K([a,b])$ is just uniform convergence on $[a,b]$ in the classical sense. The metric space $B_K([a,b])$ provides a natural context in which to study the classical concept of uniform convergence. This concept was formulated independently by G. C. Stokes, Ph. L. Seidel and A. L. Cauchy around 1850. It was introduced in order to answer the question of when the pointwise limit of a sequence of continuous functions is continuous. (Cf. Theorem 2.8.3 and Section 8.3.)

2.5.4 Example Let $E = [0,1]$,

$$f_n(t) = \begin{cases} 1 \text{ for } 0 \leq t \leq 1/n, \\ 0 \text{ for } 1/n < t \leq 1, \end{cases}$$

and

$$f(t) = \begin{cases} 1 \text{ for } t = 0, \\ 0 \text{ for } 0 < t \leq 1, \end{cases}$$

obviously $f_n \in B_R(E)$ and $f \in B_R(E)$. Let $0 < t \leq 1$ and let $N(t)$ be the least integer greater than $1/t$. Then, if $n \geq N(t)$, $|f_n(t) - f(t)| = 0$. Also, for $n = 1, 2, \ldots$, $|f_n(0) - f(0)| = 0$. Therefore (f_n) converges to f pointwise on E. On the other hand, for $n = 1, 2, \ldots$,

$$d(f, f_n) = \sup \{ |f(t) - f_n(t)| : 0 \leq t \leq 1 \} = 1.$$

Therefore (f_n) does not converge to f uniformly on E.

The distinction between pointwise and uniform convergence is sufficiently important to merit further discussion. Let (f_n) be a sequence in $B_K(E)$ that converges pointwise on E to a function f in $B_K(E)$, and let $\varepsilon > 0$. Then, given $t \in E$, there is a positive integer $N(t)$, which will depend on t, such that

$$|f(t) - f_n(t)| < \varepsilon \tag{3}$$

for all $n \geq N(t)$.

The difference between (2) and (3) is obvious: in (2) the integer N is independent of the particular choice of the point $t \in E$, and in (3) the integer $N(t)$ may vary with the point t. Example 2.5.4 shows that (3) can be satisfied for each $t \in E$ without (2) (or equivalently (1)) being satisfied.

We now return to the general theory of convergence. Perhaps the most useful characterization of the closure of a set is in terms of sequences.

2.5.5 THEOREM *Let A be a non-empty subset of X and $x \in X$. Then $x \in \bar{A}$ if and only if there is a sequence (x_n) in A such that* $\lim_{n\to\infty} x_n = x$.

PROOF Suppose that $x \in \bar{A}$. By Theorem 2.3.4 for each positive integer n the open ball $B(x,1/n)$ contains at least one point of A. For $n = 1, 2,\ldots$ choose $x_n \in A \cap B(x,1/n)$. It is obvious that

$$\lim_{n\to\infty} x_n = x.$$

Suppose, on the other hand, that there is a sequence (x_n) in A such that $\lim_{n\to\infty} x_n = x$. Then, for each $r > 0$, there is an integer N_r such that $d(x,x_n) < r$ for all $n \geq N_r$. In particular $x_{N_r} \in A \cap B(x,r)$ and therefore, by Theorem 2.3.4, $x \in \bar{A}$. □

Theorems 2.5.5 and 2.3.2 have the following corollary.

2.5.6 COROLLARY *A non-empty subset A of X is closed if and only if the limit of each convergent sequence of points of A belongs to A.*

Let (x_n) be a sequence in X with $\lim_{n\to\infty} x_n = x$ and let (x_{n_k}) be a subsequence of (x_n) (see Section 1.2). Since (n_k) is a strictly increasing sequence of positive integers we must have $n_k \geq k$ for $k = 1, 2,\ldots$, from which it follows that $\lim_{k\to\infty} x_{n_k} = x$. Thus every subsequence of a convergent sequence converges to the limit of the original sequence. A sequence that does not converge may still have a convergent subsequence. For example, let $x_{2n} = n$ and $x_{2n+1} = 1/n$ for $n = 1, 2,\ldots$, then (x_n) does not converge but the subsequence (x_{2n+1}) converges to 0.

The concept of cluster point of a sequence is closely related to that of a subsequence.

2.5.7 DEFINITION Let (x_n) be a sequence in X. A point $x \in X$ is said to be a *cluster point* of the sequence (x_n) if and only if, for each $r > 0$, the set $\{n\colon x_n \in B(x,r)\}$ is infinite.

Obviously x is a cluster point of (x_n) if and only if, for each neighbourhood V of x, the set $\{n\colon x_n \in V\}$ is infinite.

2.5.8 LEMMA *Let (x_n) be a sequence in X and $x \in X$. Then x is a cluster point of (x_n) if and only if there is a subsequence (x_{n_k}) of (x_n) such that* $\lim_{k\to\infty} x_{n_k} = x$.

PROOF It is obvious that if there is a subsequence (x_{n_k}) of (x_n) with $\lim_{k\to\infty} x_{n_k} = x$, then x is a cluster point of (x_n). Suppose, conversely, that x is a cluster point of (x_n). We shall define by induction a strictly increasing sequence, (n_k), of positive integers. Let n_1 be the least integer in the set $\{n: x_n \in B(x,1)\}$, which is non-empty because x is a cluster point of (x_n). Suppose that $n_1, n_2,..., n_{k-1}$ have been defined with $n_1 < n_2 < \ldots < n_{k-1}$. Since x is a cluster point of (x_n), the set $\{n: n > n_{k-1}$ and $x_n \in B(x,1/k)\}$ is non-empty; let n_k be its least member. This completes the definition of the sequence (n_k). Since $x_{n_k} \in B(x,1/k)$ we have $\lim_{k\to\infty} x_{n_k} = x$. □

It is clear that a convergent sequence has a unique cluster point, which is the limit of the sequence. Exercise (7) illustrates various possibilities for the cluster points of a non-convergent sequence. The following lemma will be used in Chapter 4.

2.5.9 LEMMA *Let (x_n) be a sequence in X, let*

$$E_n = \{x_m: m = n, n+1,...\}$$

for $n = 1, 2,...,$ and let E be the set of cluster points of (x_n). Then

$$E = \bigcap_{n=1}^{\infty} \bar{E}_n.$$

PROOF Let $x \in E$. For each $r > 0$ the set $\{n: x_n \in B(x,r)\}$ is infinite, so $E_n \cap B(x,r) \neq \emptyset$ for $n = 1, 2,....$ Therefore, by Theorem 2.3.4, we have $x \in \bar{E}_n$ for $n = 1, 2,....$ Hence $x \in \bigcap_{n=1}^{\infty} \bar{E}_n$. This shows that

$$E \subseteq \bigcap_{n=1}^{\infty} \bar{E}_n.$$

Suppose, on the other hand, that $x \in \bigcap_{n=1}^{\infty} \bar{E}_n$ and let $r > 0$. By Theorem 2.3.4, $B(x,r) \cap E_n \neq \emptyset$ for $n = 1, 2,....$ It follows easily that the set $\{n: x_n \in B(x,r)\}$ must be infinite. Thus $x \in E$ and $\bigcap_{n=1}^{\infty} \bar{E}_n \subseteq E$. □

We end this section with a simple, but useful, lemma on convergence in subspaces.

2.5.10 LEMMA *Let Y be a subspace of X and let (y_n) be a sequence in Y. Then*

(a) *if (y_n) converges in Y to the limit y, then (y_n) converges in X to the limit y;*
(b) *if (y_n) converges in X to the limit y and if $y \in Y$, then (y_n) converges in Y to the limit y.*

PROOF The proofs of (a) and (b) follow directly from the definition of convergence and the fact that the metric on Y is the restriction to Y of the metric on X. □

EXERCISES

(1) Show that '$< \varepsilon$' can be replaced by '$\leq \varepsilon$' in Definition 2.5.1.

(2) Let X be a metric space and A a subset of X. Prove that A is open if and only if every sequence (x_n) which is convergent to a point of A has $x_n \in A$ for all sufficiently large n.

(3) Prove that x is an accumulation point of a set A if and only if x is the limit of a sequence in $A \sim \{x\}$.

(4) Let A,B be the subsets of R given by

$$A = \left\{\frac{1}{2^n}: n = 1, 2,\ldots\right\},$$

$$B = \left\{\frac{1}{2^n} + \frac{1}{3^m}: n = 1, 2,\ldots; m = 1, 2,\ldots\right\}.$$

Prove that $A' = \{0\}$, and

$$B' = \{0\} \cup \left\{\frac{1}{2^n}: n = 1, 2,\ldots\right\} \cup \left\{\frac{1}{3^n}: n = 1, 2,\ldots\right\}.$$

(5) Characterize the convergent sequences of a discrete metric space.

(6) Show that in the statement of Theorem 2.5.3 the metric d can be replaced by either of the metrics d_0 or d_1 defined in Example 2.1.6.

Deduce that if a subset of K^n is closed or open with respect to one of the metrics d, d_0, d_1 then it is closed or open, respectively, with respect to them all.

(7) Find the cluster points of the sequence (x_n) in R when

(a) $x_n = n(1 + (-1)^n)$,

(b) $x_n = (-1)^n$,

(c) $n \to x_n$ is an enumeration of the rationals.

(8) Show that the set of cluster points of a sequence in a metric space is closed.

(9) Let (x_n) be a sequence in a metric space, let $A = \{x_n: n = 1, 2,\ldots\}$ and let B be the set of cluster points of (x_n). Prove that $\bar{A} = A \cup B$.

(10) A real sequence (x_n) is bounded and the sequence (y_n) is defined by $y_n = \sup \{x_m: m = n, n+1,\ldots\}$. Prove that (y_n) is a bounded and decreasing sequence and that the limit of (y_n) is the greatest cluster point of the sequence (x_n).

The greatest and least cluster points of a bounded real sequence (x_n) are customarily denoted by lim sup x_n and lim inf x_n or by $\overline{\lim}\, x_n$ and $\underline{\lim}\, x_n$.

2.6 Cauchy Sequences

Throughout this section X will denote a metric space with metric d. The property of being convergent in X is not an intrinsic property of a sequence, i.e. we cannot, by considering the sequence (x_n) alone, determine whether or not a sequence (x_n) in X converges. However, one of the most important properties of the real number system is Cauchy's general principle of convergence (Theorem 1.6.5) which provides an intrinsic characterization of the convergent sequences in R. The concepts introduced below arise naturally in any attempt to generalize Theorem 1.6.5.

2.6.1 Definition A sequence (x_n) is said to be a *Cauchy sequence* if and only if, for each positive real number ε, there exists a positive integer N, which depends on ε, such that $d(x_m,x_n) < \varepsilon$ for all $m,n \geq N$.

2.6.2 Lemma *Each convergent sequence in X is a Cauchy sequence.*

Proof Let (x_n) be a sequence in X with $\lim_{n\to\infty} x_n = x$, and let $\varepsilon > 0$. Then there is an integer N such that $d(x,x_n) < \frac{1}{2}\varepsilon$ for all $n \geq N$, and consequently, if $n \geq N$ and $m \geq N$

$$d(x_m,x_n) \leq d(x_m,x) + d(x,x_n) < \varepsilon.$$

Thus (x_n) is a Cauchy sequence. □

The general principle of convergence asserts that a sequence in R converges if and only if it is a Cauchy sequence and we have just seen that one half of this assertion is true—almost trivially—in any metric space. However it is easy to give an example of a metric space in which a Cauchy sequence need not converge.

2.6.3 Example Let X be the subspace $(0,1)$ of R. Then

$$d(s,t) = | s - t |$$

for all $s,t \in X$. We shall show that the sequence $(1/n)$ is a Cauchy sequence in X that does not converge in X. First let $\varepsilon > 0$ and let N be the least integer $> \varepsilon^{-1}$. If $m,n \geq N$ then

$$d\left(\frac{1}{m},\frac{1}{n}\right) = \left|\frac{1}{m} - \frac{1}{n}\right| \leq \max\left\{\frac{1}{m},\frac{1}{n}\right\} \leq \frac{1}{N} < \varepsilon.$$

This shows that $(1/n)$ is a Cauchy sequence.

Now let $t \in X$ and let N be the least integer $> 2t^{-1}$. Then, if $n \geq N$,

$$d\left(t,\frac{1}{n}\right) = \left|t - \frac{1}{n}\right| \geq t - \frac{1}{n} \geq t - \frac{1}{N} > t - \frac{t}{2} = \frac{t}{2}.$$

This shows that $(1/n)$ does not converge to t and, therefore, that $(1/n)$ does not converge in X.

2.6.4 DEFINITION A metric space X is said to be *complete* if and only if every Cauchy sequence in X converges in X.

It follows from Lemma 2.6.2 that a complete metric space is one with the property that a sequence in it converges if and only if it is a Cauchy sequence. The most important example of a complete metric space is the real line R. The concept of completeness for metric spaces was introduced by M. Fréchet.

2.6.5 THEOREM *The Euclidean space* R^n *is complete.*

PROOF Let (x_k) be a Cauchy sequence in R^n and let

$$x_k = (\xi_{1k}, \xi_{2k}, \ldots, \xi_{nk})$$

for $k = 1, 2, \ldots$. The inequality

$$|\xi_{mk} - \xi_{mj}| \leq \left(\sum_{l=1}^{n} |\xi_{lk} - \xi_{lj}|^2\right)^{1/2} = d(x_k, x_j)$$

shows that $(\xi_{mk})_{k\geq 1}$ is a Cauchy sequence in R for $m = 1, 2, \ldots, n$. By Theorem 1.6.5 the sequence $(\xi_{mk})_{k\geq 1}$ converges; let $\xi_m = \lim_{k\to\infty} \xi_{mk}$ for $m = 1, 2, \ldots, n$, and let $x = (\xi_1, \xi_2, \ldots, \xi_n)$. By Theorem 2.5.3 $\lim_{k\to\infty} x_k = x$. This proves that R^n is complete. □

2.6.6 THEOREM *The metric space* (C^n, d), *where* d *is the metric defined in Example* 2.1.5, *is complete.*

PROOF We shall prove first that C is complete. Let (z_n) be a Cauchy sequence in C and let $z_n = x_n + iy_n$, where $x_n, y_n \in \mathsf{R}$. Since

$$|x_m - x_n| \leq |z_m - z_n| \text{ and } |y_m - y_n| \leq |z_m - z_n|,$$

both (x_n) and (y_n) are Cauchy sequences in R and therefore converge —to the limits x and y say. Let $z = x + iy$. Then

$$|z - z_n| \leq |x - x_n| + |y - y_n|$$

and so $\lim_{n\to\infty} z_n = z$. Thus C is complete.

The completeness of (C^n, d) follows from that of C exactly as the completeness of R^n follows from that of R. □

We remark that Theorem 2.6.6 can be obtained as an immediate consequence of Theorems 2.6.5 and 2.9.2 and Lemma 2.9.4.

2.6.7 THEOREM *Let* E *be a non-empty set. The metric space* $B_{\mathsf{K}}(E)$ (*Example* 2.1.9) *is complete.*

PROOF Let (f_n) be a Cauchy sequence in $B_{\mathsf{K}}(E)$ and $\varepsilon > 0$. Then there is an integer N such that $d(f_m, f_n) < \frac{1}{2}\varepsilon$ for all $m \geq N$ and all $n \geq N$. Since $d(f_m, f_n) = \sup\{\,|f_m(t) - f_n(t)| : t \in E\}$ we have

$$|f_m(t) - f_n(t)| < \tfrac{1}{2}\varepsilon \tag{1}$$

for all $m \geq N$, $n \geq N$ and $t \in E$. By (1), $(f_n(t))$ is a Cauchy sequence in K, and hence converges in K, for each $t \in E$.

We define a mapping f of E into K by $f(t) = \lim_{n\to\infty} f_n(t)$ for all $t \in E$. We shall prove that $f \in B_{\mathsf{K}}(E)$ and $\lim_{n\to\infty} f_n = f$. Inequality (1) gives

$$|f(t) - f_n(t)| = \lim_{m\to\infty} |f_m(t) - f_n(t)| \leq \tfrac{1}{2}\varepsilon \tag{2}$$

for all $n \geq N$ and all $t \in E$, and thus

$$|f(t)| \leq |f(t) - f_N(t)| + |f_N(t)| \leq \tfrac{1}{2}\varepsilon + |f_N(t)|$$

for all $t \in E$. Since f_N is bounded this shows that f is bounded and so f belongs to $B_{\mathsf{K}}(E)$. Now (2) gives

$$d(f, f_n) = \sup\{\,|f(t) - f_n(t)| : t \in E\} \leq \tfrac{1}{2}\varepsilon < \varepsilon$$

for all $n \geq N$. This proves that $\lim_{n\to\infty} f_n = f$, and hence that $B_{\mathsf{K}}(E)$ is complete. □

The following result on the completeness of subspaces is often useful.

2.6.8 THEOREM *Let Y be a subspace of X. Then*

(a) *if Y is complete, then Y is a closed subset of X;*
(b) *if X is complete and Y is a closed subset of X, then Y is complete.*

PROOF (a) Let (y_n) be a sequence in Y that converges in X—to the limit y, say. By Lemma 2.6.2 (y_n) is a Cauchy sequence and therefore converges in Y. But then it follows from Lemma 2.5.10 that the limit of (y_n) in Y must be y and hence $y \in Y$. Corollary 2.5.6 now shows that Y is a closed subset of X.

(b) Let (y_n) be a Cauchy sequence in Y. Then (y_n) converges in X—to the limit y, say. Since Y is closed, Corollary 2.5.6 shows that $y \in Y$, and then Lemma 2.5.10 shows that (y_n) converges to y in Y. Thus Y is complete. □

The following lemma on Cauchy sequences in an arbitrary metric space is sometimes useful.

2.6.9 LEMMA *A Cauchy sequence which has a convergent subsequence is itself convergent.*

PROOF Let (x_n) be a Cauchy sequence in X that has a convergent subsequence, (x_{n_k}) say. Let $x = \lim_{k\to\infty} x_{n_k}$ and $\varepsilon > 0$. Then there are integers N and K such that $d(x_m,x_n) < \frac{1}{2}\varepsilon$ for all $m,n \geq N$, and $d(x,x_{n_k}) < \frac{1}{2}\varepsilon$ for all $k \geq K$. Let $M = \max\{N,K\}$. If $m \geq M$ then, since $n_m \geq m$,

$$d(x,x_m) \leq d(x,x_{n_m}) + d(x_{n_m},x_m) < \varepsilon.$$

This proves that $\lim_{n\to\infty} x_n = x$. □

EXERCISES

(1) Prove that the metric spaces (K^n,d_0) and (K^n,d_1) of Example 2.1.6 are complete.

(2) Prove that the metric space ℓ of Example 2.1.11 is complete.

2.7 Continuous Mappings

In this section we shall generalize the classical notion of continuity to mappings of one metric space into another.

2.7.1 DEFINITION Let X and Y be metric spaces with metrics d_X and d_Y respectively. A mapping f of X into Y is said to be *continuous at a point* $x_0 \in X$ if and only if, for each positive real number ε, there exists a positive real number δ, which depends on ε and x_0, such that $d_Y(f(x_0),f(x)) < \varepsilon$ whenever $d_X(x_0,x) < \delta$. A mapping of X into Y is said to be *continuous on* X if and only if it is continuous at each point of X.

Intuitively f is continuous at x_0 if and only if '$f(x)$ is arbitrarily close to $f(x_0)$ whenever x is sufficiently close to x_0'. For mappings of an interval into R, continuity as defined above obviously coincides with the usual notion of continuity in classical analysis. It will be useful to have a reformulation of the definition of continuity at a point in terms of neighbourhoods.

2.7.2 LEMMA *A mapping f of a metric space X into a metric space Y is continuous at a point $x_0 \in X$ if and only if $f^{-1}(V)$ is a neighbourhood of x_0 whenever V is a neighbourhood of $f(x_0)$.*

PROOF Suppose that f is continuous at x_0 and let V be a neighbourhood of $f(x_0)$. There exists $r > 0$ such that $B(f(x_0),r) \subseteq V$ and there exists $\delta > 0$ such that $d_Y(f(x_0),f(x)) < r$ whenever $d_X(x_0,x) < \delta$. This means that if $x \in B(x_0,\delta)$, then $f(x) \in B(f(x_0),r)$, and consequently

$$B(x_0,\delta) \subseteq f^{-1}(B(f(x_0),r)) \subseteq f^{-1}(V).$$

This proves that $f^{-1}(V)$ is a neighbourhood of x_0.

Suppose conversely that $f^{-1}(V)$ is a neighbourhood of x_0 for each neighbourhood V of $f(x_0)$ and let $\varepsilon > 0$. Since $B(f(x_0),\varepsilon)$ is a neighbourhood of $f(x_0)$ the set $f^{-1}(B(f(x_0),\varepsilon))$ is a neighbourhood of x_0 and therefore there exists $\delta > 0$ such that

$$B(x_0,\delta) \subseteq f^{-1}(B(f(x_0),\varepsilon)).$$

It follows that if $d_X(x_0,x) < \delta$ then $d_Y(f(x_0),f(x)) < \varepsilon$. This proves that f is continuous at x_0. □

2.7.3 THEOREM *A mapping f of a metric space X into a metric space Y is continuous on X if and only if f^{-1} (A) is open in X for all open subsets A of Y.*

PROOF Suppose that f is continuous on X and let A be an open subset of Y. We have to show that $f^{-1}(A)$ is open. Since $\emptyset$ is open in X, we may suppose that $f^{-1}(A) \neq \emptyset$. Let $x \in f^{-1}(A)$. Then $f(x) \in A$ and so A is a neighbourhood of $f(x)$. By Lemma 2.7.2 the set $f^{-1}(A)$ is a neighbourhood of x. Thus $f^{-1}(A)$ is a neighbourhood of each of its points and so is open.

Suppose conversely that $f^{-1}(A)$ is open in X for all open subsets A of Y. Let $x \in X$. For each $\varepsilon > 0$ the set $B(f(x),\varepsilon)$ is open and so $f^{-1}(B(f(x),\varepsilon))$ is open. Therefore $f^{-1}(B(f(x),\varepsilon))$ is a neighbourhood of x, and it follows, as in the last part of the proof of Lemma 2.7.2, that f is continuous at x. This shows that f is continuous on X. □

The proof of the next theorem illustrates the brevity and elegance that can be obtained by using the characterization of continuity in terms of open sets given by Theorem 2.7.3.

2.7.4 THEOREM *Let X, Y and Z be metric spaces and let f and g be continuous mappings of X into Y and Y into Z, respectively. Then $g \circ f$ is a continuous mapping of X into Z.*

PROOF Let A be an open subset of Z. By Theorem 2.7.3 the set $g^{-1}(A)$ is an open subset of Y and thus, by Theorem 2.7.3 again, $f^{-1}(g^{-1}(A))$ is an open subset of X. Since $(g \circ f)^{-1}(A) = f^{-1}(g^{-1}(A))$, it follows from Theorem 2.7.3 that $g \circ f$ is continuous on X. □

Naturally there is an analogue of Theorem 2.7.3 for closed sets.

2.7.5 THEOREM *A mapping f of a metric space X into a metric space Y is continuous on X if and only if $f^{-1}(A)$ is closed in X for all closed subsets A of Y.*

PROOF The proof follows from Theorem 2.7.3, the definition of a closed set, and the observation that $f^{-1}(Y \sim A) = X \sim f^{-1}(A)$ for all subsets A of Y. □

Next we turn to the relations between continuity and closure, and continuity and convergence.

2.7.6 THEOREM *Let f be a mapping of a metric space X into a metric space Y. Then the following three statements are all equivalent:*

(a) *f is continuous on X,*
(b) *$f^{-1}(B)^- \subseteq f^{-1}(\bar{B})$ for all subsets B of Y,*
(c) *$f(\bar{A}) \subseteq f(A)^-$ for all subsets A of X.*

PROOF We shall prove first that (a) and (b) are equivalent and then that (b) and (c) are equivalent. Assume (a), and let B be a subset of Y. By Theorem 2.7.5 the set $f^{-1}(\bar{B})$ is closed and thus, since $f^{-1}(B) \subseteq f^{-1}(\bar{B})$ we have $f^{-1}(B)^- \subseteq f^{-1}(\bar{B})$. Conversely assume (b) and let B be a closed subset of Y. Since $B = \bar{B}$ we have $f^{-1}(B)^- \subseteq f^{-1}(\bar{B}) = f^{-1}(B)$ and therefore $f^{-1}(B) = f^{-1}(B)^-$. This shows that $f^{-1}(B)$ is closed. It follows from Theorem 2.7.5 that f is continuous on X.

Now assume (b) and let A be a subset of X. Then if $B = f(A)$ we have $A \subseteq f^{-1}(B)$ and $\bar{A} \subseteq f^{-1}(B)^- \subseteq f^{-1}(\bar{B})$. Thus $f(\bar{A}) \subseteq \bar{B} = f(A)^-$. Finally, assume (c) and let B be a subset of Y. Then if $A = f^{-1}(B)$ we have $f(\bar{A}) \subseteq f(A)^-$ and so $\bar{A} \subseteq f^{-1}(f(A)^-)$. But $B \supseteq f(A)$ so

$$f^{-1}(B)^- = \bar{A} \subseteq f^{-1}(f(A)^-) \subseteq f^{-1}(\bar{B}). \ \square$$

2.7.7 THEOREM *A mapping f of a metric space X into a metric space Y is continuous at a point $x_0 \in X$ if and only if $\lim_{n\to\infty} f(x_n) = f(x_0)$ for all sequences (x_n) in X with $\lim_{n\to\infty} x_n = x_0$.*

PROOF Suppose that f is continuous at x_0 and that (x_n) is a sequence in X with $\lim_{n\to\infty} x_n = x_0$. Given $\varepsilon > 0$, there exists $\delta > 0$ such that $d_Y(f(x_0),f(x)) < \varepsilon$ whenever $d_X(x_0,x) < \delta$ and there exists an integer N such that $d_X(x_0,x_n) < \delta$ for all $n \geq N$. Thus, if $n \geq N$ we have $d_Y(f(x_0),f(x_n)) < \varepsilon$. This shows that $\lim_{n\to\infty} f(x_n) = f(x_0)$.

Suppose conversely that $\lim_{n\to\infty} f(x_n) = f(x_0)$ for all sequences (x_n) in X with $\lim_{n\to\infty} x_n = x_0$, but that f is not continuous at x_0. Then there exists $\varepsilon > 0$ with the following property: for each $\delta > 0$ there exists $x \in X$ such that $d_X(x_0,x) < \delta$ and $d_Y(f(x_0),f(x)) \geq \varepsilon$. In particular for each positive integer n there exists $x_n \in X$ such that $d_X(x_0,x_n) < 1/n$ and $d_Y(f(x_0),f(x_n)) \geq \varepsilon$. Clearly $\lim_{n\to\infty} x_n = x_0$ but the sequence $(f(x_n))$ does not converge to $f(x_0)$. This is a contradiction. □

We end this section with some observations on continuity and subspaces. Let f be a mapping of a metric space X into a metric space Y and suppose that Y_1 is a subset of Y such that $f(X) \subseteq Y_1$. It follows immediately from the definitions of continuity and subspace that f is a continuous mapping of X into Y if and only if it is a continuous mapping of X into the subspace Y_1. Now let X_1 be a non-empty subset of X. It is obvious that the restriction to X_1 of a continuous mapping of X into Y is a continuous mapping of the subspace X_1 into Y.

Exercises

(1) Show that '$< \varepsilon$' can be replaced by '$\leq \varepsilon$' in Definition 2.7.1.

(2) Show that a constant mapping of one metric space into another is continuous.

(3) Show that if X and Y are metric spaces and X is discrete then any mapping of X into Y is continuous.

(4) Prove that if a mapping of K^n into a metric space is continuous with respect to one of the metrics d, d_0, d_1 (see Examples 2.1.4–2.1.6) then it is continuous with respect to them all.

(5) Let X, Y and Z be metric spaces, let f be a mapping of X into Y that is continuous at $x_0 \in X$, and let g be a mapping of Y into Z that is continuous at $f(x_0)$. Show that $g \circ f$ is continuous at x_0.

(6) Let X and Z be metric spaces and f a mapping of X into Z. Subsets Y_1 and Y_2 of X are closed, $X = Y_1 \cup Y_2$ and the restrictions $f|Y_1, f|Y_2$ of f are continuous on Y_1 and Y_2 respectively. Prove that f is continuous on X. Do the same problem with 'closed' replaced by 'open'.

(7) Let f and g be continuous mappings of a metric space X into a metric space Y. Show that if $f(x) = g(x)$ for all x in a dense subset of X then $f = g$.

(8) A continuous mapping f of R into R has $f(x + y) = f(x) + f(y)$ for all $x, y \in \mathsf{R}$. Prove that $f(x) = f(1)x$ for all $x \in \mathsf{R}$.

(9) Show that if f is a continuous mapping of X into R with $f(x) \geq 0$ for all $x \in Y$ then $f(x) \geq 0$ for all $x \in \bar{Y}$.

(10) Give an example of a continuous mapping f of R into R such that $\{x: f(x) \geq 0\}$ is not the closure of $\{x: f(x) > 0\}$. (Cf. Theorem 2.7.6.)

(11) If f is a mapping of a metric space X into R with the property that the sets $\{x: f(x) < t\}$ and $\{x: f(x) > t\}$ are open for every rational number t prove that f is continuous on X.

Give an example of a mapping f of $[0, 1]$ into R which is not continuous on $[0,1]$ but which has $\{x: f(x) < t\}$ open for every $t \in \mathsf{R}$.

(12) Show that an increasing mapping g_1 of the Cantor set onto $[0,1]$ is defined by $g_1(x) = \sum_{m=1}^{\infty} \frac{1}{2}\alpha_m 2^{-m}$ when $x = \sum_{m=1}^{\infty} \alpha_m 3^{-m}$ with $\alpha_m = 0$ or 2 for $m = 1, 2, \ldots$. (See Exercises 2.2(7), (8).) Show further that the function g defined by $g(x) = \sup \{g_1(t): t \in C, t \leq x\}$ is a continuous and increasing mapping of $[0,1]$ onto $[0,1]$ which is constant on each of the intervals complementary to the Cantor set. Sketch the graph of g.

(13) Let X be a closed subset of R and let f be a continuous mapping of X into R. Prove that $\{(x,f(x)): x \in X\}$, the graph of f, is a closed subset of the Euclidean space R^2.

(14) Find the closure in the Euclidean space R^2 of the set

$$\left\{(\xi,\eta): \eta = \sin\frac{1}{\xi} \text{ and } 0 < \xi \leq 1\right\}.$$

2.8 Real and Complex-Valued Continuous Functions

In this section we shall derive some elementary properties of real or complex-valued continuous functions on a metric space. We begin by defining some elementary operations on such functions.

2.8.1 DEFINITION Let E be a non-empty set. Given mappings f and g of E into K and $\alpha \in \mathsf{K}$ we define the mappings $f + g$, αf, fg, and $|f|$ of E into K as follows:

$$(f+g)(t) = f(t) + g(t),$$
$$(\alpha f)(t) = \alpha f(t),$$
$$(fg)(t) = f(t)g(t),$$

and

$$|f|(t) = |f(t)|$$

for all $t \in E$. Further, if $f(t) \neq 0$ for all $t \in E$, we define the mapping $1/f$ of E into K by

$$\left(\frac{1}{f}\right)(t) = \frac{1}{f(t)}$$

for all $t \in E$. Suppose now that f and g are mappings of E into R. Then we can define two further mappings $f \vee g$ and $f \wedge g$ of E into R by

$$(f \vee g)(t) = \max\{f(t), g(t)\}$$

and

$$(f \wedge g)(t) = \min\{f(t), g(t)\}$$

for all $t \in E$. It is easy to verify that

$$f \vee g = \tfrac{1}{2}(f + g + |f - g|)$$

and

$$f \wedge g = \tfrac{1}{2}(f + g - |f - g|). \tag{1}$$

We shall discuss the algebraic significance of these operations in

Chapters 3 and 4 (see especially Section 4.5); for the moment we are concerned only to show that they preserve continuity. By definition of the metrics on R and C, a mapping f of a metric space X into K is continuous at a point $x_0 \in X$ if and only if, for each $\varepsilon > 0$, there exists $\delta > 0$ such that $|f(x_0) - f(x)| < \varepsilon$ whenever $d(x_0,x) < \delta$. The proofs of the assertions of the following theorem are direct generalizations of the familiar proofs in the case when X is the real line.

2.8.2 THEOREM *Let f and g be continuous mappings of a metric space X into K and let $\alpha \in$ K. Then the mappings $f + g$, αf, fg and $|f|$ are continuous on X, and so is the mapping $1/f$ if it is defined. If f and g are continuous mappings of X into R, then the mappings $f \vee g$ and $f \wedge g$ are continuous on X.*

PROOF Let x_0 be an arbitrary point of X. We shall prove that all the mappings are continuous at x_0. First consider $f + g$. Given $\varepsilon > 0$ there exists $\delta_1 > 0$ such that $|f(x_0) - f(x)| < \frac{1}{2}\varepsilon$ whenever $d(x_0,x) < \delta_1$ and there exists $\delta_2 > 0$ such that $|g(x_0) - g(x)| < \frac{1}{2}\varepsilon$ whenever $d(x_0,x) < \delta_2$. Let $\delta = \min\{\delta_1, \delta_2\}$. Then, if $d(x_0,x) < \delta$,

$$\begin{aligned} |(f+g)(x_0) - (f+g)(x)| &= |(f(x_0) - f(x)) + (g(x_0) - g(x))| \\ &\le |f(x_0) - f(x)| + |g(x_0) - g(x)| \\ &< \varepsilon. \end{aligned}$$

This proves that $f + g$ is continuous at x_0.

It is trivial to prove that αf is continuous at x_0, so consider next fg. As above, given η with $0 < \eta \le 1$, there exists $\delta > 0$ such that

$$|f(x_0) - f(x)| < \eta \text{ and } |g(x_0) - g(x)| < \eta$$

whenever $d(x_0,x) < \delta$. Then, if $d(x_0,x) < \delta$,

$$\begin{aligned} |f(x)| &\le |f(x) - f(x_0)| + |f(x_0)| \\ &< \eta + |f(x_0)| \\ &\le 1 + |f(x_0)| \end{aligned}$$

and

$$\begin{aligned} |(fg)(x_0) - (fg)(x)| &= |f(x_0)g(x_0) - f(x)g(x)| \\ &= |g(x_0)(f(x_0) - f(x)) + f(x)(g(x_0) - g(x))| \\ &\le |g(x_0)|\,|f(x_0) - f(x)| + |f(x)|\,|g(x_0) - g(x)| \\ &< \eta(|g(x_0)| + |f(x_0)| + 1). \end{aligned}$$

Given $\varepsilon > 0$, let $\eta = \min\{1, \varepsilon(|g(x_0)| + |f(x_0)| + 1)^{-1}\}$. Then, by what we have just proved, there exists $\delta > 0$ such that

$$|(fg)(x_0) - (fg)(x)| < \varepsilon$$

whenever $d(x_0,x) < \delta$ and therefore fg is continuous at x_0.

Suppose now that $f(x) \ne 0$ for all $x \in X$. Given η with

$$0 < \eta \le \tfrac{1}{2}|f(x_0)|,$$

there exists $\delta > 0$ such that $|f(x_0) - f(x)| < \eta$ whenever $d(x_0,x) < \delta$. Thus, if $d(x_0,x) < \delta$,

$$\begin{aligned} |f(x)| &= |f(x_0) - (f(x_0) - f(x))| \\ &\geq |f(x_0)| - |f(x_0) - f(x)| \\ &\geq \tfrac{1}{2}|f(x_0)| \end{aligned}$$

and

$$\left|\frac{1}{f(x_0)} - \frac{1}{f(x)}\right| = \frac{|f(x) - f(x_0)|}{|f(x)|\,|f(x_0)|} < \frac{2\eta}{|f(x_0)|^2}.$$

Given $\varepsilon > 0$, let $\eta = \min\{\frac{1}{2}|f(x_0)|, \frac{1}{2}\varepsilon|f(x_0)|^2\}$. Then, by what we have just proved, there exists $\delta > 0$ such that

$$\left|\left(\frac{1}{f}\right)(x_0) - \left(\frac{1}{f}\right)(x)\right| < \varepsilon$$

whenever $d(x_0,x) < \delta$. This shows that $1/f$ is continuous at x_0.

Finally it follows at once from the inequality

$$||f|(x_0) - |f|(x)| = ||f(x_0)| - |f(x)|| \leq |f(x_0) - f(x)|$$

that $|f|$ is continuous at x_0.

If f and g are continuous real-valued functions on X then the results proved above, together with the relations (1) show that $f \vee g$ and $f \wedge g$ are continuous. □

We have already seen (Example 2.1.12) that the set $C_{\mathsf{R}}([a,b])$ of all continuous real-valued functions on the interval $[a,b]$ is a metric space. The metric on $C_{\mathsf{R}}([a,b])$ is defined by

$$d(f,g) = \sup\{|f(t) - g(t)| : a \leq t \leq b\}. \tag{2}$$

This metric space plays an important role in the applications of the theory of metric spaces to classical analysis.

Before we can use (2) to define a metric on $C_{\mathsf{R}}([a,b])$ we need to know that the functions in $C_{\mathsf{R}}([a,b])$ are bounded. Unfortunately a continuous real-valued function on an arbitrary metric space X may fail to be bounded on X (for example, the function f given by $f(t) = t$ for all $t \in \mathsf{R}$ is continuous and unbounded on R) so we cannot define a metric on the set of all continuous real-valued functions on X by analogy with (2). We must restrict ourselves to considering bounded continuous functions on X.

Let X be a metric space and let $C_{\mathsf{K}}(X)$ denote the set of all bounded continuous mappings of X into K. Then $C_{\mathsf{K}}(X)$ is a subset of $B_{\mathsf{K}}(X)$ and is therefore, as a subspace of $B_{\mathsf{K}}(X)$, a metric space. The metric on $C_{\mathsf{K}}(X)$ is called the *uniform metric*; it is given by

$$d(f,g) = \sup\{|f(x) - g(x)| : x \in X\}$$

for all $f,g \in C_{\mathsf{K}}(X)$.

2.8.3 THEOREM *The metric space $C_K(X)$ is complete.*

PROOF Since $C_K(X)$ is a subspace of $B_K(X)$, which is complete (Theorem 2.6.7), Theorem 2.6.8 shows that, to prove that $C_K(X)$ is complete, it is sufficient to prove that $C_K(X)$ is a closed subset of $B_K(X)$. To prove that $C_K(X)$ is closed we shall use the characterization of closed sets given in Corollary 2.5.6.

Let (f_n) be a sequence in $C_K(X)$ with $\lim_{n\to\infty} f_n = f$. We have to prove that $f \in C_K(X)$. Let $x_0 \in X$ and $\varepsilon > 0$. There is an integer N such that $d(f,f_n) < \frac{1}{3}\varepsilon$ for all $n \geq N$, and thus

$$|f(x) - f_n(x)| \leq d(f,f_n) < \tfrac{1}{3}\varepsilon \tag{3}$$

for all $x \in X$ and all $n \geq N$. The function f_N is continuous at x_0 so there exists $\delta > 0$ such that

$$|f_N(x_0) - f_N(x)| < \tfrac{1}{3}\varepsilon \tag{4}$$

whenever $d(x_0,x) < \delta$. If $d(x_0,x) < \delta$ then (3) and (4) give

$$\begin{aligned} |f(x_0) - f(x)| &\leq |f(x_0) - f_N(x_0)| + |f_N(x_0) - f_N(x)| \\ &\qquad + |f_N(x) - f(x)| \\ &< \varepsilon. \end{aligned}$$

This shows that f is continuous at x_0 and hence that $f \in C_K(X)$. It follows now from Corollary 2.5.6 that $C_K(X)$ is closed. □

The essential content of the proof of the last theorem is the fact that the uniform limit of a sequence of bounded continuous functions is a bounded continuous function. This result was first proved by G. C. Stokes in the middle of the last century.

So far, we have no reason to suppose that, on an arbitrary metric space, there are any continuous real-valued functions other than the constant functions. We shall now show that $C_K(X)$ has an abundance of non-constant functions.

2.8.4 LEMMA *Let A be a non-empty subset of a metric space X, and for each $x \in X$ let*

$$d(x,A) = \inf\{d(x,y) : y \in A\}.$$

Then (a) $d(x,A) = 0$ *if and only if* $x \in \bar{A}$,

(b) $|d(x,A) - d(y,A)| \leq d(x,y)$ *for all* $x,y \in X$.

PROOF By Theorem 2.3.4 the point $x \in \bar{A}$ if and only if, for each $r > 0$, there exists $y \in A$ such that $d(x,y) < r$. This proves (a).

For each $x,y \in X$ and $z \in A$ we have

$$d(x,A) \leq d(x,z) \leq d(x,y) + d(y,z)$$

and hence

$$d(x,A) \le d(x,y) + d(y,A).$$

Similarly

$$d(y,A) \le d(x,y) + d(x,A).$$

This proves (b). □

The number $d(x,A)$ is called the *distance* of the point x from the set A.

2.8.5 THEOREM *Let A and B be disjoint closed subsets of a metric space X. Then there is a continuous real-valued function f on X such that $f(x) = 0$ for all $x \in A$, $f(x) = 1$ for all $x \in B$, and $0 \le f(x) \le 1$ for all $x \in X$.*

PROOF Lemma 2.8.4(b) shows that the mappings $x \to d(x,A)$ and $x \to d(x,B)$ are continuous on X. Since A and B are closed and $A \cap B = \emptyset$, Lemma 2.8.4(a) shows that $d(x,A) + d(x,B) > 0$ for all $x \in X$. Let

$$f(x) = \frac{d(x,A)}{d(x,A) + d(x,B)}$$

for all $x \in X$. By Theorem 2.8.2 and the above remarks, f is continuous on X. It is trivial to verify that f satisfies the other conditions of the theorem. □

EXERCISES

(1) Let λ be a real number and X the set of all continuous real-valued functions ϕ defined on $[0, \infty)$ which are such that the set

$$\{| \phi(x) | e^{-\lambda x}: x \ge 0\}$$

is bounded. Show that X is a complete metric space with respect to the metric d defined by

$$d(\phi,\psi) = \sup \{ | \phi(x) - \psi(x) | e^{-\lambda x}: x \ge 0\}.$$

(2) Show that

$$d(f,g) = \int_0^1 | f(t) - g(t) | \, dt$$

defines a metric d on $C_{\mathsf{R}}([0,1])$.

For each positive integer n define f_n by

$$f_n(t) = \begin{cases} 1 & \text{for } 0 \le t < \frac{1}{2}, \\ 1 - 2^n(t - \frac{1}{2}) & \text{for } \frac{1}{2} \le t < \frac{1}{2} + 2^{-n}, \\ 0 & \text{for } \frac{1}{2} + 2^{-n} \le t \le 1. \end{cases}$$

Show that $f_n \in C_{\mathsf{R}}([0,1])$ and that, in the metric space $(C_{\mathsf{R}}([0,1]),d)$, the sequence (f_n) is a Cauchy sequence which does not converge.

(3) Give an example of a sequence (f_n) in $B_{\mathsf{R}}([0,1])$ with $(f_n(t))$ convergent for each $t \in [0,1]$ which is such that the function f, defined by $f(t) = \lim_{n\to\infty} f_n(t)$ for $t \in [0,1]$, is not bounded.

(4) Let X be an infinite set and d a metric on X. Prove that there exists a continuous mapping g of X into $[0,1]$ such that $g(X)$ is an infinite set.

2.9 Homeomorphism, Equivalence and Isometry

In Examples 2.1.4, 2.1.5 and 2.1.6 we have defined three different metrics d, d_0 and d_1 on the set K^n and we have seen (Exercises 2.5(6), 2.6(1) and 2.7(4)) that the metric spaces (K^n,d), (K^n,d_0) and (K^n,d_1) have the same class of open sets, the same class of convergent sequences, the same class of Cauchy sequences, and the same class of continuous mappings. The following definitions will help to clarify this situation.

2.9.1 DEFINITION Let f be a one-to-one mapping of a metric space X onto a metric space Y. The mapping f is said to be a *homeomorphism* if and only if the mappings f and f^{-1} are continuous on X and Y respectively. The mapping f is said to be a *uniform equivalence*† if and only if there are positive real numbers k and m such that

$$d_X(x_1,x_2) \leq k\, d_Y(f(x_1),f(x_2))$$

and

$$d_Y(f(x_1),f(x_2)) \leq m\, d_X(x_1,x_2)$$

for all $x_1,x_2 \in X$. (Here d_X and d_Y denote the metrics on X and Y respectively.) Finally the mapping f is said to be an *isometry* if and only if

$$d_X(x_1,x_2) = d_Y(f(x_1),f(x_2))$$

for all $x_1,x_2 \in X$. It is obvious that an isometry is a uniform equivalence and a uniform equivalence is a homeomorphism. We shall give examples below which show that the three concepts are distinct.

It is obvious that f is a homeomorphism of X onto Y if and only if f^{-1} is a homeomorphism of Y onto X. We shall say that two metric spaces X and Y are *homeomorphic* if and only if there exists a homeomorphism of X onto Y. It follows from Theorem 2.7.4 that the composition of two homeomorphisms is again a homeomorphism, and it is then not difficult to see that the relation of being homeomorphic is an

†The reader is warned that some authors use the term 'uniform equivalence' in a less restrictive sense.

equivalence relation on the class of all metric spaces. Analogously we shall say that two metric spaces X and Y are *uniformly equivalent* if and only if there is a uniform equivalence of X onto Y, and that X and Y are *isometric* if and only if there is an isometry of X onto Y. The relations of being uniformly equivalent and isometric are obviously equivalence relations. Two metric spaces that are isometric are indistinguishable in so far as their properties as metric spaces are concerned.

2.9.2 THEOREM *The metric spaces* (C^n,d) *(Example* 2.1.5) *and* (R^{2n},d) *(Example* 2.1.4) *are isometric.*

PROOF It is easy to verify that the mapping

$$(\xi_1 + i\eta_1, \xi_2 + i\eta_2, \ldots, \xi_n + i\eta_n) \rightarrow (\xi_1, \xi_2, \ldots, \xi_n, \eta_1, \eta_2, \ldots, \eta_n)$$

is an isometry of C^n onto R^{2n}. □

If f is a homeomorphism of a metric space X onto a metric space Y then by Theorem 2.7.3 there is a one-to-one correspondence $A \rightarrow f(A)$ between the open subsets of X and the open subsets of Y. The family of open subsets of a metric space is called the *topology of the space*, and any property of a metric space that can be formulated entirely in terms of the open subsets of the space is called a *topological property* of the space. Consequently homeomorphic metric spaces have the same topological properties and are sometimes said to be *topologically equivalent*. Examples of topological properties of a metric space are openness and closedness of subsets, convergence of sequences and continuity of mappings. The following example shows that completeness is not a topological property.

2.9.3 EXAMPLE Let

$$X = \{n : n = 1, 2, \ldots\} \text{ and } Y = \{1/n : n = 1, 2, \ldots\}.$$

It is easy to see that both X and Y regarded as subspaces of R are discrete metric spaces. It follows therefore from Theorem 2.7.3 that the mapping $n \rightarrow 1/n$ is a homeomorphism of X onto Y. Further it follows from Theorem 2.3.4 that X is a closed subset of R. Since R is complete Theorem 2.6.8 shows that the subspace X is also complete. On the other hand the sequence $(1/n)$ is obviously a Cauchy sequence in Y that does not converge in Y, so Y is not complete.

2.9.4 LEMMA *Any metric space uniformly equivalent to a complete metric space is complete.*

PROOF Let X be a metric space and suppose that there is a uniform equivalence f of X onto a complete metric space Y. By definition there exists $k > 0$ such that $d_Y(f(x),f(x')) \leq k\, d_X(x,x')$ for all $x,x' \in X$. Let (x_n) be a Cauchy sequence in X. Since $d_Y(f(x_m),f(x_n)) \leq k\, d_X(x_m,x_n)$, it follows that $(f(x_n))$ is a Cauchy sequence in Y and hence converges. But $x_n = f^{-1}(f(x_n))$ and f^{-1} is continuous on Y so (x_n) converges by Theorem 2.7.7. □

2.9.5 EXAMPLE The metric spaces (K^n,d), (K^n,d_0) and (K^n,d_1) of Examples 2.1.4, 2.1.5 and 2.1.6 are uniformly equivalent to each other. We recall that

$$d(x,y) = \left(\sum_{r=1}^{n} |\, \xi_r - \eta_r \,|^2\right)^{1/2},$$

$$d_0(x,y) = \max\{\, |\, \xi_r - \eta_r \,| : r = 1, 2, \ldots, n\},$$

and

$$d_1(x,y) = \sum_{r=1}^{n} |\, \xi_r - \eta_r \,|,$$

where $x = (\xi_1, \xi_2, \ldots, \xi_n)$ and $y = (\eta_1, \eta_2, \ldots, \eta_n)$.

The inequalities

$$d_0(x,y) \leq d(x,y) \leq n^{1/2}\, d_0(x,y)$$

and

$$d_0(x,y) \leq d_1(x,y) \leq n\, d_0(x,y),$$

which are easily verified, show that the identity mapping $x \to x$ is a uniform equivalence of any one of (K^n,d), (K^n,d_0) and (K^n,d_1) onto any other. None of these uniform equivalences is an isometry.

Our last example is of a one-to-one continuous mapping of one metric space onto another that is not a homeomorphism.

2.9.6 EXAMPLE Let

$$d(f,g) = \sup\{\, |f(t) - g(t)| : 0 \leq t \leq 1\}$$

and

$$d_1(f,g) = \sup\{\, |f(t) - g(t)| : 0 \leq t \leq 1\} + \sup\{\, |f'(t) - g'(t)| : 0 \leq t \leq 1\},$$

for all $f,g \in C^1_{\mathsf{R}}([0,1])$ (see Example 2.1.13). For brevity we write C^1 for $C^1_{\mathsf{R}}([0,1])$. Both d and d_1 are metrics on C^1 and, since $d(f,g) \leq d_1(f,g)$ for all $f,g \in C^1$, the identity mapping $f \to f$ is a one-to-one continuous mapping of (C^1,d_1) onto (C^1,d).

Let $f_n(t) = n^{-1} \sin nt$ for $0 \leq t \leq 1$ and $n = 1, 2, \ldots$. Then $f_n \in C^1$ and we have $d(0,f_n) = n^{-1}$ and $d_1(0,f_n) = 1 + n^{-1}$. This shows that (f_n) converges to the limit 0 in (C^1,d) but does not converge in (C^1,d_1). It follows then from Theorem 2.7.7 that the identity mapping $f \to f$ is not a continuous mapping of (C^1,d) into (C^1,d_1).

EXERCISES

(1) Prove that homeomorphism, uniform equivalence and isometry are equivalence relations.

(2) Show that

$$f(x) = a + x(b - a)$$

defines a homeomorphism of $[0,1]$ onto $[a,b]$.

(3) Show that the mapping f of $(0,1)$ into R defined by

$$f(x) = \frac{1}{1-x} - \frac{1}{x}$$

is a homeomorphism of $(0,1)$ onto R and that restriction of f gives homeomorphisms of $(\frac{1}{2},1)$ onto $(0,\infty)$ and $[\frac{1}{2},1)$ onto $[0,\infty)$.

(4) Prove that the subspaces (a,b), $(-\infty,a)$ and (a,∞) of R (where $b > a$) and the space R itself are homeomorphic and that the subspaces $(a,b]$, $[a,b)$, $(-\infty,a]$ and $[a,\infty)$ of R are homeomorphic.

(5) A real-valued function f is defined on $[0,\infty)$, is non-decreasing, $f(0) = 0$, $f(u) > 0$ if $u > 0$, and $f(u + v) \leq f(u) + f(v)$ for all $u \geq 0$ and $v \geq 0$. Let d be a metric on a set X and let d_1 be defined by $d_1(x,y) = f(d(x,y))$ for all $x,y \in X$. Show that d_1 is also a metric on X and that the identity mapping on X is a homeomorphism of the spaces (X,d) and (X,d_1) if and only if either (X,d) is a discrete space or f is continuous at 0. Show further that if f is continuous at 0 then (X,d) is complete if and only if (X,d_1) is complete.

(6) Let d be defined by

$$d(\xi,\eta) = \frac{|\xi - \eta|}{1 + |\xi - \eta|}$$

for all $\xi,\eta \in \mathsf{R}$. Show that d is a metric on R and that (R,d) is complete.

(7) Let d be defined by

$$d(x,y) = \sum_{n=1}^{\infty} 2^{-n} \frac{|\xi_n - \eta_n|}{1 + |\xi_n - \eta_n|}$$

for all sequences of real numbers $x = (\xi_n)$ and $y = (\eta_n)$. Show that the set of all real sequences is a complete metric space with respect to d.

(8) Let (X,d) be a metric space. Prove that d_1 defined by

$$d_1(x,y) = \min\{d(x,y), 1\}$$

for all $x,y \in X$ is also a metric on X and that the identity mapping on X is a homeomorphism between (X,d) and (X,d_1).

The Completion of a Metric Space

(9) Let (X,d) be a metric space. Prove that there exists a complete metric space $\tilde{X}$ (a *completion* of X) with the property that there is an

isometric mapping of X onto a dense subset of $\tilde{X}$. (Hint: If (x_n) and (y_n) are two Cauchy sequences in X then the sequence $(d(x_n,y_n))$ is a Cauchy sequence in R. A pseudo-metric d_1 (see Exercise 2.1(6)) is defined on the set of all Cauchy sequences in X by

$$d_1((x_n),(y_n)) = \lim_{n\to\infty} d(x_n,y_n).$$

Use the result of Exercise 2.1(6) and consider the natural mapping of X onto the set of constant sequences in X.)

(10) Show that the completion of a metric space is unique in the sense that if f_1 and f_2 are isometries of X onto dense subsets of the complete metric spaces Y_1 and Y_2, respectively, then there is an isometry g of Y_1 onto Y_2 with $f_2 = g \circ f_1$.

(11) Let X be a metric space and x_0 a point of X. Show that

$$f_x(y) = d(y,x) - d(y,x_0)$$

for $x,y \in X$ defines an isometry $x \to f_x$ of X into $C_{\mathsf{R}}(X)$. Hence give an alternative proof that every metric space has a completion.

2.10 Connected Sets

Throughout this section X will denote a metric space.

2.10.1 Definition The metric space X is said to be *connected* if and only if $\emptyset$ and X are the only subsets of X that are both open and closed in X. A non-empty subset Y of X is said to be *connected* if and only if the subspace Y is a connected metric space.

It is not obvious from the above definition that the connected subsets of the Euclidean space R^2 are those subsets of R^2 that we would intuitively call connected. However, experience has shown that Definition 2.10.1 is the most suitable definition of connectedness in the context of a general metric space, and we shall see in Theorem 2.10.9 that connectedness in this sense does coincide with the intuitive notion of connectedness, at least for open sets.

The connectedness of a subset Y of X is an intrinsic property of Y: that is it depends only on the metric of the subspace Y. Thus if Z is a subset of X such that $Y \subseteq Z$, then Y is a connected subset of the metric space X if and only if Y is a connected subset of the subspace Z. This remark will enable us to simplify some proofs.

For reference we set out explicitly a condition that a metric space be not connected: X is not connected if and only if there exist non-empty disjoint open subsets A and B of X such that $X = A \cup B$, or, equivalently, if and only if there exist non-empty disjoint closed subsets A and B of X such that $X = A \cup B$.

It is useful to have a criterion for the connectedness of a subset of X in terms of the closure operation in X.

2.10.2 DEFINITION Two non-empty subsets A and B of X are said to be *separated* if and only if $\bar{A} \cap B = \emptyset$ and $A \cap \bar{B} = \emptyset$. Obviously separated subsets of X are disjoint.

2.10.3 LEMMA *A subset of X is connected if and only if it cannot be expressed as the union of two separated subsets of X.*

PROOF Let Y be a subset of X. Suppose first that $Y = A \cup B$, where A and B are separated subsets of X. Then

$$\bar{A} \cap Y = \bar{A} \cap (A \cup B) = (\bar{A} \cap A) \cup (\bar{A} \cap B) = A$$

and similarly $\bar{B} \cap Y = B$. It follows then from Theorem 2.4.2 that A and B are closed in Y, and thus Y is not connected.

Suppose, conversely, that Y is not connected. Then there exist non-empty disjoint subsets A and B of Y that are closed in Y and such that $Y = A \cup B$. Since A and B are closed in Y, Theorem 2.4.2 gives $\bar{A} \cap Y = A$ and $\bar{B} \cap Y = B$ and consequently

$$\bar{A} \cap B = \bar{A} \cap (\bar{B} \cap Y) = (\bar{A} \cap Y) \cap (\bar{B} \cap Y) = A \cap B = \emptyset.$$

Similarly $A \cap \bar{B} = \emptyset$. This proves that A and B are separated subsets of X. □

2.10.4 THEOREM *A non-empty subset of R is connected if and only if it is an interval.*

PROOF Let A be a non-empty subset of R that is not an interval. By Lemma 2.2.6 there exist $a,b,c \in \mathsf{R}$ such that $a,b \in A$, $c \notin A$ and $a < c < b$. The sets $A \cap (-\infty,c)$ and $A \cap (c,\infty)$ are non-empty and disjoint. Also $A = (A \cap (-\infty,c)) \cup (A \cap (c,\infty))$ and by Theorem 2.4.2 the sets $A \cap (-\infty,c)$ and $A \cap (c,\infty)$ are open in A. Consequently A is not connected.

On the other hand let I be an interval and suppose that I is not connected. By Lemma 2.10.3 there are separated subsets A and B of R such that $I = A \cup B$. Choose $a \in A$ and $b \in B$. We may suppose, without loss of generality, that $a < b$. Then $[a, b] \subseteq I$ and $A \cap [a,b] \neq \emptyset$. Let $c = \sup (A \cap [a,b])$. Then $a \leq c \leq b$ and so $c \in I$. Since $I = A \cup B$ either $c \in A$ or $c \in B$. Suppose first that $c \in A$. Then $c \neq b$ and, by definition of c, if $c < t < b$ then $t \notin A$ and consequently $t \in B$. It follows now from Theorem 2.3.4 that $c \in \bar{B}$, which is a contradiction because A and B are separated. Suppose now that $c \in B$. Since $c = \sup (A \cap [a,b])$ it follows easily from Theorem 2.3.4 that

$c \in (A \cap [a,b])^- \subseteq \bar{A}$, which is again a contradiction because A and B are separated. These contradictions prove that I must be connected. □

Next we show that the property of connectedness is preserved under continuous mappings.

2.10.5 THEOREM *Let f be a continuous mapping of a connected metric space X into a metric space Y. Then $f(X)$ is a connected subset of Y.*

PROOF By the remarks after Definition 2.10.1 we may suppose, without loss in generality, that $f(X) = Y$. Suppose that Y is not connected, and let A be a non-empty subset of Y that is both open and closed in Y and different from Y. Then $f^{-1}(A)$ is non-empty and different from X and, by Theorems 2.7.3 and 2.7.5, is both open and closed in X. This is a contradiction because X is connected. Therefore Y must be connected. □

As a corollary of the last two theorems we obtain a famous theorem of Darboux on continuous real-valued functions of a real variable.

2.10.6 COROLLARY *Let f be a real-valued continuous function on an interval I and let $t_1,t_2 \in I$. Then for each real number s between $f(t_1)$ and $f(t_2)$ there exists $t \in I$ such that $s = f(t)$.*

PROOF By Theorem 2.10.4 the interval I is connected and therefore by Theorem 2.10.5 $f(I)$ is a connected subset of R. But then Theorem 2.10.4 shows that $f(I)$ is an interval. Consequently if s is between $f(t_1)$ and $f(t_2)$ then $s \in f(I)$. □

Let us turn now to the problem of identifying the connected subsets of R^n and C^n. We need a lemma and a definition.

2.10.7 LEMMA *Let A and B be separated subsets of X and let Y be a connected subset of X such that $Y \subseteq A \cup B$. Then either $Y \subseteq A$ or $Y \subseteq B$.*

PROOF In the first part of the proof of Lemma 2.10.3 we saw that A and B are closed in the subspace $A \cup B$. Since $A \cap B = \emptyset$ we have $A = (A \cup B) \sim B$ and $B = (A \cup B) \sim A$. Hence A and B are open in the subspace $A \cup B$. But Y is a connected subset of the subspace $A \cup B$ and, by Theorem 2.4.2 the sets $Y \cap A$ and $Y \cap B$ are both open and closed in Y. Therefore either $Y \cap A = Y$ in which case $Y \subseteq A$, or $Y \cap B = Y$ in which case $Y \subseteq B$. □

Given $x = (\xi_1, \xi_2,\ldots, \xi_n)$ and $y = (\eta_1, \eta_2,\ldots, \eta_n)$ in K^n and $s,t \in \mathsf{K}$ we shall write

$$sx + ty = (s\xi_1 + t\eta_1, s\xi_2 + t\eta_2,\ldots, s\xi_n + t\eta_n).$$

(See Example 3.2.5.)

2.10.8 Definition Let $x,y \in \mathsf{K}^n$. The set $\{(1-t)x+ty: 0 \leq t \leq 1\}$ is called the *closed line segment joining* x *to* y and will be denoted by xy. More generally, let $x_1, x_2, \ldots, x_m$ be m points of K^n and let

$$f(t) = (k-t)x_k + (t-k+1)x_{k+1} \tag{1}$$

for $k-1 \leq t \leq k$ and $k = 1, 2, \ldots, m-1$. The set

$$\{f(t): 0 \leq t \leq m-1\}$$

is called a *polygon* in K^n and is denoted by $x_1x_2 \ldots x_m$. The point x_1 is called the *initial point*, and the point x_m the *final point* of the polygon $x_1x_2 \ldots x_m$, and the polygon $x_1x_2 \ldots x_m$ is said *to join* x_1 *to* x_m.

The polygon $x_1x_2 \ldots x_m$ is the union of the line segments $x_1x_2, x_2x_3, \ldots, x_{m-1}x_m$. On K^n, let d be the metric defined in Examples 2.1.4 and 2.1.5. It is easy to verify that the mapping f defined by (1) is a continuous mapping of $[0, m-1]$ into (K^n,d) and therefore, by Theorems 2.10.4 and 2.10.5, a polygon is a connected subset of (K^n,d).

2.10.9 Theorem *A non-empty open subset A of (K^n,d) is connected if and only if each pair of points of A can be joined by a polygon that lies wholly in A. In particular (K^n,d) is connected.*

Proof Let A be a non-empty subset of K^n such that each pair of points in A can be joined by a polygon that lies wholly in A. Suppose that A is not connected. Then, by Lemma 2.10.3, there are separated subsets B and C of K^n with $A = B \cup C$. Choose points $b \in B$ and $c \in C$. Since a polygon is a connected subset of K^n, Lemma 2.10.7 shows that b and c cannot be joined by a polygon that lies wholly in A, which is a contradiction. Therefore A is connected. (Notice that we have not had to assume that A is open for this part of the theorem.)

Now let A be a non-empty open connected subset of K^n. Choose a point $x_0 \in A$. Let A_1 be the set of all points $x \in A$ with the property that there exists a polygon that lies wholly in A and joins x_0 to x and let $A_2 = A \sim A_1$. We shall prove first that A_1 and A_2 are open in K^n. Since $x_0 \in A_1$ the set A_1 is not empty. Let $x \in A_1$. The set A is open and $x \in A$ so there exists $r > 0$ such that $B(x,r) \subseteq A$. Let $y \in B(x,r)$. It is easy to verify that the line segment xy lies in $B(x,r)$ and, *a fortiori*, in A. But $x \in A_1$ so there is a polygon, $x_0x_1x_2 \ldots x_kx$ say, that lies in A and joins x_0 to x. Clearly the polygon $x_0x_1x_2 \ldots x_kxy$ joins x_0 to y and lies in A, and consequently $y \in A_1$. Thus $B(x,r) \subseteq A_1$. This shows that A_1 is open in K^n.

If A_2 is empty then it is open. Suppose that A_2 is not empty and let $x \in A_2$. Again there exists $r > 0$ such that $B(x,r) \subseteq A$. We shall show that $B(x,r) \subseteq A_2$. If this is not so then $A_1 \cap B(x,r) \neq \emptyset$. Choose

$y \in A_1 \cap B(x,r)$. There is a polygon $x_0x_1x_2 \ldots x_ky$ that lies in A and therefore, as before, the polygon $x_0x_1x_2 \ldots x_kyx$ lies in A, which is a contradiction because $x \in A_2$. Thus $B(x,r) \subseteq A_2$. This shows that A_2 is open in K^n.

It follows now from Theorem 2.4.2 that A_1 and A_2 are open in A. But A is connected, A_1 and A_2 are disjoint and $A_1 \neq \varnothing$. So we must have $A_2 = \varnothing$. Thus any point of A can be joined to x_0 by a polygon that lies wholly in A and it follows at once that any pair of points of A can be joined by a polygon that lies wholly in A. □

We end this section by showing that a metric space that is not connected can be decomposed into connected subsets that are 'as large as possible' (Theorem 2.10.13).

2.10.10 Lemma *Let $\mathscr{A}$ be a family of connected subsets of X such that $\bigcap_{A \in \mathscr{A}} A \neq \varnothing$. Then $\bigcup_{A \in \mathscr{A}} A$ is connected.*

Proof Suppose, on the contrary, that $\bigcup_{A \in \mathscr{A}} A$ is not connected. By Lemma 2.10.3 there are separated subsets B and C of X such that $\bigcup_{A \in \mathscr{A}} A = B \cup C$. By Lemma 2.10.7, for each $A \in \mathscr{A}$ either $A \subseteq B$ or $A \subseteq C$. Choose a point $x_0 \in \bigcap_{A \in \mathscr{A}} A$. Either $x_0 \in B$ or $x_0 \in C$. Suppose, for instance, that $x_0 \in B$. Then since $x_0 \in A$ for all $A \in \mathscr{A}$ we have $A \subseteq B$ for all $A \in \mathscr{A}$, and consequently $\bigcup_{A \in \mathscr{A}} A \subseteq B$. This is a contradiction. □

2.10.11 Lemma *Let A be a connected subset of X. Then each subset B of X with $A \subseteq B \subseteq \bar{A}$ is connected. In particular $\bar{A}$ is connected.*

Proof Suppose that $A \subseteq B \subseteq \bar{A}$ and that B is not connected. By Lemma 2.10.3 there exist separated subsets C and D of X such that $B = C \cup D$. Since $A \subseteq C \cup D$, Lemma 2.10.7 shows that either $A \subseteq C$ or $A \subseteq D$. Suppose, for instance, that $A \subseteq C$. Then $\bar{A} \subseteq \bar{C}$ and consequently $\bar{A} \cap D = \varnothing$, because C and D are separated. But then $D = B \cap D = \varnothing$, which is a contradiction. □

2.10.12 Definition Let A be a non-empty subset of X. A subset B of A is said to be a *component* of A if and only if B is connected and there is no connected subset C of A, different from B, such that $B \subseteq C$.

It is not obvious from the definition that components exist: we need Lemmas 2.10.10 and 2.10.11 to establish their existence.

2.10.13 THEOREM *Let A be a non-empty subset of X. Each point of A is contained in exactly one component of A and each connected subset of A is contained in exactly one component of A. Further the components of A are closed in A.*

PROOF We may, without loss of generality, suppose that $A = X$. Let $x \in X$. The set $\{x\}$ is connected, so the family of all connected subsets of X that contain x is not empty; let A_x be the union of this family. Lemma 2.10.10 shows that A_x is connected and it is now obvious that A_x is a component of X.

Let C be a non-empty connected subset of X and choose $x \in C$. Then $C \subseteq A_x$ by construction of A_x. Let B be any component of X that contains x. Since B is connected $B \subseteq A_x$. But B is a component of X so $B = A_x$. This proves that each connected subset of X (including the sets consisting of single points) is contained in a unique component of X. It follows at once from the definition of a component and Lemma 2.10.11 that each component is closed in X. □

The components of a set need not be open in the set: for example $\{0\}$ is a component of the set Q of all rational numbers but is not open in Q.

EXERCISES

(1) Prove that a metric space X is connected if and only if every non-empty proper subset of X has a non-empty frontier.

(2) Prove that a metric space X is not connected if and only if there exists a continuous mapping of X onto a metric space with two points.

(3) A subset Y of a metric space X is connected and Y has non-empty intersections with a set A and its complement $X \sim A$. Show that Y has a non-empty intersection with Fr A.

(4) Prove that a countable subset of a metric space cannot be connected unless it consists of a single point.

(5) For $n = 1, 2, \ldots$ the subset A_n of a metric space X is connected and $A_n \cap A_{n+1} \neq \emptyset$. Prove that $\bigcup_{n=1}^{\infty} A_n$ is connected.

(6) Prove that the components of Q, the set of all rational numbers, consist of single points.

In Exercises (7)–(10) *the metrics are those of Example* 2.1.4.

(7) Prove that the components of an open subset of R^n are open in R^n.

(8) Let A be the set of all those points in R^2 neither of whose co-

ordinates is rational, and B the set of all those points in R^2 at least one of whose coordinates is rational. Determine whether or not A and B are connected.

(9) Let I be the subset $\{(0,\eta)\colon -1 \leq \eta \leq 1\}$ of R^2. Prove that $\left\{\left(\xi, \sin\frac{1}{\xi}\right)\colon \xi > 0\right\} \cup I$ is a connected subset of R^2.

(10) Prove that the unit circle $\{(s,t)\colon s^2 + t^2 = 1\}$ is a connected subset of R^2.

(11) The most direct way of showing that two given spaces are homeomorphic is to construct a homeomorphism. (See Exercises 2.9(2), (3), (4).) To show that two spaces are not homeomorphic is quite a different problem, one which can be extremely difficult. It is solved if one finds a topological property possessed by one space but not by the other, for then the two spaces cannot be homeomorphic. Connectedness is a topological property (Theorem 2.10.5) and so we can conclude that, for example, R and $\mathsf{R} \sim \{0\}$ are not homeomorphic.

Prove that no two of the intervals (a,b), $(a,b]$ and $[a,b]$, where $b > a$, are homeomorphic.

(12) Prove that a continuous one-to-one mapping of $(0,1)$ into R is monotonic.

(13) (a) Prove that if g is a continuous one-to-one mapping of (a,b), where $a < b$, into R then the range of g is an open interval.

(b) Prove that if f is a one-to-one mapping of $(0,1)$ onto $[0,1]$ then f is discontinuous at an infinite set of points.

(c) Give an example of a one-to-one mapping of $(0,1)$ onto $[0,1]$ that is continuous on the complement of a countable set.

2.11 The Contraction Mapping Theorem

We end this chapter with a simple result about complete metric spaces that has interesting and important applications in classical analysis.

2.11.1 Definition Let X be a metric space with metric d. A mapping T of X into itself is said to be a *contraction mapping* if and only if there is a real number α, with $0 < \alpha < 1$, such that

$$d(T(x),T(y)) \leq \alpha d(x,y)$$

for all $x,y \in X$.

To simplify our notation we shall write Tx for $T(x)$. It is obvious that a contraction mapping is continuous. The following property of contraction mappings was discovered by the Polish mathematician

S. Banach in 1922. The idea of the proof goes back to E. Picard's proof of Theorem 2.11.9.

2.11.2 THEOREM (The contraction mapping theorem) *Let T be a contraction mapping of a complete metric space X into itself. Then there is a unique point $u \in X$ such that $Tu = u$.*

PROOF By hypothesis there is a real number α with $0 < \alpha < 1$ such that $d(Tx,Ty) \leq \alpha d(x,y)$ for all $x,y \in X$. Choose any point $x_0 \in X$ and let (x_n) be the sequence defined inductively by $x_{n+1} = Tx_n$ for $n = 0, 1, 2,\ldots$. We shall prove that (x_n) is a Cauchy sequence. For $k = 1, 2\ldots$, we have

$$d(x_{k+1},x_k) = d(Tx_k,Tx_{k-1}) \leq \alpha d(x_k,x_{k-1}).$$

Repeated application of this inequality gives

$$\begin{aligned} d(x_{k+1},x_k) &\leq \alpha d(x_k,x_{k-1}) \\ &\leq \alpha^2 d(x_{k-1},x_{k-2}) \\ &\cdots\cdots \\ &\cdots\cdots \\ &\leq \alpha^k d(x_1,x_0). \end{aligned} \tag{1}$$

Now let m, n be positive integers with $m > n$. By the triangle inequality and (1)

$$\begin{aligned} d(x_m,x_n) &= d(x_m,x_{m-1}) + d(x_{m-1},x_{m-2}) + \ldots + d(x_{n+1},x_n) \\ &\leq (\alpha^{m-1} + \alpha^{m-2} + \ldots + \alpha^n)d(x_1,x_0) \\ &= \alpha^n(\alpha^{m-n-1} + \alpha^{m-n-2} + \ldots + 1)d(x_1,x_0) \\ &< \frac{\alpha^n}{1-\alpha}d(x_1,x_0). \end{aligned}$$

But $\lim_{n\to\infty} \alpha^n d(x_1,x_0) = 0$ so given $\varepsilon > 0$ there exists an integer N such that $d(x_m,x_n) < \varepsilon$ for $m > n \geq N$. This proves that (x_n) is a Cauchy sequence. Since X is complete (x_n) converges. Let $\lim_{n\to\infty} x_n = u$. Since T is continuous Theorem 2.7.7 gives $\lim_{n\to\infty} Tx_n = Tu$. On the other hand $Tx_n = x_{n+1}$, so $\lim_{n\to\infty} Tx_n = \lim_{n\to\infty} x_{n+1}$, which shows that $Tu = u$.

Let $v \in X$ be such that $Tv = v$. Then $d(u,v) = d(Tu,Tv) \leq \alpha d(u,v)$. But $d(u,v) \geq 0$ and $0 < \alpha < 1$ so $d(u,v) = 0$, and hence $u = v$. □

A point $u \in X$ such that $Tu = u$ is called a *fixed point of T*.

We shall use the contraction mapping theorem to prove a theorem of Picard on the existence and uniqueness of solutions of a certain class of first-order ordinary differential equations. First we must say what we mean by the solution of a differential equation.

2.11.3 DEFINITION Let f be a real-valued function on a non-empty subset D of the Euclidean space R^2. A real-valued function ϕ on an interval I is said to be *a solution of the differential equation*

$$\frac{dy}{dx} = f(x,y) \tag{2}$$

on the interval I if and only if $(t,\phi(t)) \in D$ for all $t \in I$, ϕ is differentiable† on I, and $\phi'(t) = f(t,\phi(t))$ for all $t \in I$.

The above differential equation can be transformed into an 'integral equation' of the form

$$y(x) = y_0 + \int_{x_0}^{x} f(t,y(t))\,dt \tag{3}$$

where y is a real-valued function on I which has to be determined.

We shall have to assume that the reader is familiar with some form of integral defined on continuous functions. In Sections 3.7, 3.8 and 3.9 we shall give a self-contained treatment of an elementary integral that is adequate for the purposes of this book.

2.11.4 DEFINITION Let f be a continuous real-valued function on a non-empty subset D of the Euclidean space R^2. A real-valued function ϕ on an interval I containing x_0 is said to be a *solution of the integral equation* (3) *on* I if and only if $(t,\phi(t)) \in D$ for all $t \in I$, ϕ is continuous on I and

$$\phi(x) = y_0 + \int_{x_0}^{x} f(t,\phi(t))\,dt$$

for all $x \in I$. It is clear that the function $t \to f(t,\phi(t))$ is continuous on I so the integral $\int_{x_0}^{x} f(t,\phi(t))\,dt$ is defined for each $x \in I$.

2.11.5 LEMMA *Let f be a continuous real-valued function on a non-empty subset D of the Euclidean space R^2, let I be an interval, let $x_0 \in I$ and let ϕ be a real valued function on I. Then ϕ is a solution of the differential equation* (2) *on I and $\phi(x_0) = y_0$ if and only if ϕ is a solution of the integral equation* (3) *on I.*

PROOF Suppose that ϕ is a solution of (2) on I and $\phi(x_0) = y_0$. Then ϕ is differentiable on I and hence continuous on I. Also, since $\phi'(t) = f(t,\phi(t))$ for all $t \in I$,

$$\phi(x) - y_0 = \phi(x) - \phi(x_0) = \int_{x_0}^{x} \phi'(t)\,dt = \int_{x_0}^{x} f(t,\phi(t))\,dt$$

for all $x \in I$. (See Theorem 3.9.9.) Thus ϕ is a solution of (3) on I.

† If I contains one or both of its end points, differentiability at such a point means differentiability from the right or the left. (See Definition 3.9.1.)

Conversely suppose that ϕ is a solution of (3) on I. Then ϕ is continuous on I and

$$\phi(x) = y_0 + \int_{x_0}^{x} f(t,\phi(t))\, dt$$

for all $x \in I$. Consequently ϕ is differentiable on I and

$$\phi'(x) = f(x,\phi(x))$$

for all $x \in I$ (Theorem 3.9.7). This shows that ϕ is a solution of (2) on I Obviously $\phi(x_0) = y_0$. □

We shall use the contraction mapping theorem to show that, under suitable conditions, the integral equation (3) has a unique solution on some interval containing x_0. We shall need two lemmas.

2.11.6 Lemma *Let X be the set of all continuous mappings of the closed interval $[a,b]$ into the closed interval $[c,d]$. Then the subspace X of $C_{\mathsf{R}}([a,b])$ is complete.*

Proof The metric space $C_{\mathsf{R}}([a,b])$ is complete (Theorem 2.8.3) so by Theorem 2.6.8 we have only to prove that X is closed. Let (f_n) be a sequence in X with $\lim_{n\to\infty} f_n = f$. For each $t \in [a,b]$ we have

$$| f_n(t) - f(t) | \leq d(f_n, f)$$

and hence $\lim_{n\to\infty} f_n(t) = f(t)$. But, for each $t \in [a,b]$, we have $c \leq f_n(t) \leq d$ for $n = 1, 2,\ldots$, so $c \leq f(t) \leq d$. This shows that $f \in X$ and therefore, by Corollary 2.5.6, X is closed. □

2.11.7 Lemma *Let f be a continuous real-valued function on a non-empty subset D of the Euclidean space R^2, and let $\alpha > 0$ and $K > 0$ be such that the rectangle*

$$\{(x,y) : | x - x_0 | \leq \alpha, | y - y_0 | \leq \alpha K\}$$

is contained in D and

$$| f(x,y) | < K$$

for $| x - x_0 | \leq \alpha$ and $| y - y_0 | \leq \alpha K$. If ϕ is a solution of (3) on the interval $[x - \alpha, x_0 + \alpha]$ then

$$| \phi(x) - y_0 | \leq K | x - x_0 |$$

for all $x \in [x_0 - \alpha, x_0 + \alpha]$.

Proof Since ϕ is continuous on $[x_0 - \alpha, x_0 + \alpha]$ and $\phi(x_0) = y_0$ there exists $\delta > 0$ such that $| \phi(x) - y_0 | \leq \alpha K$ for $| x - x_0 | \leq \delta$. Consequently, for $x_0 \leq x \leq x_0 + \delta$,

$$\begin{aligned} | \phi(x) - y_0 | &= \left| \int_{x_0}^{x} f(t,\phi(t))\, dt \right| \\ &\leq | x - x_0 | \sup \{ | f(t,\phi(t)) | : x_0 \leq t \leq x\} \\ &\leq K | x - x_0 |. \end{aligned}$$

Let J be the set of real numbers x such that $x_0 \leq x \leq x_0 + \alpha$ and $| \phi(x) - y_0 | > K(x - x_0)$, and suppose that $J \neq \emptyset$. Let $x_1 = \inf J$. By what we have already proved $x_1 > x_0$. Since ϕ is continuous it is easy to see that $| \phi(x_1) - y_0 | = K(x_1 - x_0)$. By the classical mean value theorem there exists x_2 such that $x_0 < x_2 < x_1$ and

$$\phi(x_1) - y_0 = (x_1 - x_0)\phi'(x_2).$$

Thus $| \phi'(x_2) | = K$. On the other hand $x_2 \notin J$ so

$$| \phi(x_2) - y_0 | \leq K(x_2 - x_0) \leq \alpha K$$

and hence $| \phi'(x_2) | = | f(x_2,\phi(x_2)) | < K$, which is a contradiction. This proves that $| \phi(x) - y_0 | \leq K(x - x_0)$ for $x_0 \leq x \leq x_0 + \alpha$. Similarly we can prove that

$$| \phi(x) - y_0 | \leq K | x - x_0 | \text{ for } x_0 - \alpha \leq x \leq x_0. \ \square$$

A real-valued function f on a subset D of R^2 is said to satisfy a *Lipschitz condition on D* (with respect to the second variable) if and only if there exists $M > 0$ such that

$$| f(x,y_1) - f(x,y_2) | \leq M | y_1 - y_2 | \tag{4}$$

for all (x,y_1), $(x,y_2) \in D$.

2.11.8 Theorem *Let D be a non-empty open subset of the Euclidean space R^2, let f be a continuous real-valued function on D that satisfies the Lipschitz condition* (4) *and let $(x_0,y_0) \in D$. Then there is a positive real number α such that the integral equation* (3) *has a unique solution on the interval* $[x_0 - \alpha, x_0 + \alpha]$.

Proof Since f is continuous at (x_0,y_0) there is an open disc U with centre (x_0,y_0) and positive radius such that $| f(x,y) - f(x_0,y_0) | < 1$ for all $(x,y) \in U \cap D$. The set D is open so we may suppose, by decreasing the radius of U if necessary, that $U \subseteq D$. Let $K = 1 + | f(x_0,y_0) |$. Then, for all $(x,y) \in U$,

$$| f(x,y) | \leq | f(x,y) - f(x_0,y_0) | + | f(x_0,y_0) | < K. \tag{5}$$

Now choose $\alpha > 0$ so that $\alpha M < 1$ and

$$\{(x,y) : | x - x_0 | \leq \alpha, | y - y_0 | \leq \alpha K\} \subseteq U.$$

Let X be the set of all continuous mappings of $I = [x_0 - \alpha, x_0 + \alpha]$ into $[y_0 - \alpha K, y_0 + \alpha K]$. It follows from Lemma 2.11.6 that X is a complete subspace of $C_{\mathrm{R}}(I)$. For each $\phi \in X$ the function $t \to f(t,\phi(t))$ is defined and continuous on I. Let

$$(T\phi)(x) = y_0 + \int_{x_0}^{x} f(t,\phi(t))\, dt \tag{6}$$

for all $x \in I$ and all $\phi \in X$. If $x_1,x_2 \in I$, then

$$\begin{aligned} |(T\phi)(x_1) - (T\phi)(x_2)| &= \left| \int_{x_0}^{x_1} f(t,\phi(t))\, dt - \int_{x_0}^{x_2} f(t,\phi(t))\, dt \right| \\ &= \left| \int_{x_2}^{x_1} f(t,\phi(t))\, dt \right| \\ &\leq |x_1 - x_2| \sup \{ |f(t,\phi(t))| : t \in I\} \\ &\leq K |x_1 - x_2|. \end{aligned}$$

This shows that $T\phi$ is continuous on I. Similarly if $x \in I$

$$\begin{aligned} |(T\phi)(x) - y_0| &= \left| \int_{x_0}^{x} f(t,\phi(t))\, dt \right| \\ &\leq K|x - x_0| \\ &\leq \alpha K. \end{aligned}$$

Therefore $T\phi \in X$.

We shall prove that $\phi \to T\phi$ is a contraction mapping on X. Let $\phi_1,\phi_2 \in X$. Then by (6), (5) and (4) we have for all $x \in I$,

$$\begin{aligned} |(T\phi_1)(x) - (T\phi_2)(x)| &= \left| \int_{x_0}^{x} (f(t,\phi_1(t)) - f(t,\phi_2(t)))\, dt \right| \\ &\leq |x - x_0| \sup\{ |f(t,\phi_1(t)) - f(t,\phi_2(t))| : t \in I\} \\ &\leq M\alpha \sup \{ |\phi_1(t) - \phi_2(t)| : t \in I\} \\ &= M\alpha d(\phi_1,\phi_2). \end{aligned}$$

Consequently

$$\begin{aligned} d(T\phi_1,T\phi_2) &= \sup\{ |(T\phi_1)(x) - (T\phi_2)(x)| : x \in I\} \\ &\leq M\alpha d(\phi_1,\phi_2). \end{aligned}$$

This shows that $\phi \to T\phi$ is a contraction mapping because $M\alpha < 1$. Since X is complete Theorem 2.11.1 shows that there is a unique function $\phi \in X$ such that $T\phi = \phi$. By (6)

$$\phi(x) = y_0 + \int_{x_0}^{x} f(t,\phi(t))\, dt$$

for all $x \in I$, so ϕ is a solution of (3) on I.

Let ψ be any solution of (3) on I. It follows from Lemma 2.11.7 that $\psi \in X$ and therefore by (6) $\psi = T\psi$. But ϕ is the unique fixed point of T, so $\psi = \phi$. □

From Theorem 2.11.8 and Lemma 2.11.5 we obtain Picard's theorem.

2.11.9 THEOREM (Picard) *Let D, f and (x_0,y_0) satisfy the conditions of Theorem 2.11.8. Then there is a positive real number α such that the differential equation (2) has a unique solution ϕ in the interval $[x_0-\alpha, x_0+\alpha]$ which satisfies the boundary condition $\phi(x_0) = y_0$.*

The proof of Picard's theorem is constructive in the sense that it yields explicitly a sequence (ϕ_n) of continuous functions that converges uniformly on $[x_0-\alpha, x_0+\alpha]$ to the solution ϕ. To see this we have only to observe that ϕ is the fixed point of the contraction mapping T given by (6). The proof of the contraction mapping theorem then shows that, if ψ is any function in X and (ϕ_n) is defined inductively by $\phi_1 = \psi$ and $\phi_{n+1} = T\phi_n$ for $n = 1, 2,...,$ then (ϕ_n) converges to ϕ in X, that is (ϕ_n) converges uniformly to ϕ on $[x_0-\alpha, x_0+\alpha]$.

EXERCISES

(1) Let T be a mapping of a complete metric space X into itself. Define T^n inductively by $T^1 = T$ and $T^{n+1} = T \circ T^n$. Prove that if T^n is a contraction mapping for some positive integer n then T has a unique fixed point.

(2) Let (X,d) be a complete metric space and S a contraction mapping on X satisfying

$$d(Sx,Sy) \leq \alpha d(x,y) \qquad (*)$$

for all $x,y \in X$, where $0 \leq \alpha < 1$. Prove that if u is the unique fixed point of S then $d(u,x) \leq (1-\alpha)^{-1}d(Sx,x)$ for all $x \in X$.

Mappings T, T_1, T_2,... of X into itself satisfy the condition (*) and $Tx = \lim_{n\to\infty} T_n x$ for all $x \in X$. Prove that if u, u_1, u_2,... are the fixed points of T, T_1, T_2,... then $u = \lim_{n\to\infty} u_n$.

(3) Let α_{ij}, for $i,j = 1, 2,..., n$, be complex numbers with

$$\sum_{j=1}^{n} |\alpha_{ij}| \leq \alpha$$

for $i = 1, 2,..., n$, where $0 \leq \alpha < 1$, and let $\beta_1, \beta_2,..., \beta_n$ be any complex numbers. Prove, by considering the metric space (C^n, d_0) of Example 2.1.6, that the system of equations

$$\xi_i - \sum_{j=1}^{n} \alpha_{ij}\xi_j = \beta_i, \qquad i = 1, 2,..., n,$$

has exactly one solution.

(4) Let f be a continuous real-valued function defined on the subset $\{(x,y) : x \geq 0\}$ of the Euclidean space R^2 and such that

$$|f(x,y_1) - f(x,y_2)| \leq M|y_1 - y_2|, \quad \text{and} \quad |f(x,0)| \leq M$$

for all $x \geq 0$ and all $y_1, y_2 \in \mathsf{R}$, where $M > 0$. Use the contraction mapping theorem to prove that if $\lambda > M$ there is a function ϕ in the metric space X of Exercise 2.8(1) which satisfies $\phi(0) = 0$ and

$$\phi'(t) = f(t, \phi(t))$$

for all $t \geq 0$.

CHAPTER 3

Normed Linear Spaces

The set R^n of all ordered n-tuples of real numbers is a linear space over the real field. In Examples 2.1.4 and 2.1.6 we defined three metrics on R^n. The reader will easily verify that, if d is any one of these metrics, then

$$d(x,y) = d(x-y,0) \quad \text{and} \quad d(\alpha x,0) = |\alpha|\, d(x,0) \qquad (*)$$

for all $x,y \in \mathsf{R}^n$ and all $\alpha \in \mathsf{R}$. Thus these metric space structures on R^n are closely related to the linear space structure of R^n. Most of the other examples of metric spaces given in Section 2.1 are also linear spaces and the relations (*) are satisfied for these spaces.

Linear spaces that are also metric spaces in which the metric and linear space structures are related by (*) are called *normed linear spaces.* These spaces were introduced in the early years of this century, and many of their deeper properties were discovered by the Polish mathematician S. Banach, whose book *Théorie des opérations linéaires* [3] published in 1932, is still one of the best sources of information about them. The theory of normed linear spaces lies at the heart of the subject called *functional analysis.*

Except for Chapter 4 and part of Chapter 8, where we shall return to the study of metric spaces, the rest of this book is devoted to the study of normed linear spaces and their applications in classical analysis. In this chapter we shall develop the elementary theory of normed linear spaces. The most important results are those concerning bounded linear mappings obtained in Section 3.5. In Section 3.7 we shall apply these results to develop an elementary theory of integration.

3.1 Linear Spaces

In this section we shall review some of the elementary properties of linear spaces. We are interested only in linear spaces over the real or complex fields, and as usual K will denote either R or C.

3.1.1 DEFINITION A *linear space over* K is a quadruple $(E,\mathsf{K},+,.)$ where E is a non-empty set, $+$ is a mapping $(x,y) \to x+y$ of $E \times E$ into E, and $.$ is a mapping $(\alpha,x) \to \alpha.x$ of $\mathsf{K} \times E$ into E, and the following conditions are satisfied:

(a) $x + (y + z) = (x + y) + z$ for all $x,y,z \in E$,
(b) $x + y = y + x$ for all $x,y \in E$,
(c) there exists $\theta \in E$ such that $x + \theta = x$ for all $x \in E$,
(d) for each $x \in E$ there exists $-x \in E$ such that $x + (-x) = \theta$,
(e) $(\alpha+\beta).x = \alpha.x + \beta.x$ for all $x \in E$ and $\alpha,\beta \in \mathsf{K}$,
(f) $\alpha.(x+y) = \alpha.x + \alpha.y$ for all $x,y \in E$ and $\alpha \in \mathsf{K}$,
(g) $\alpha.(\beta.x) = (\alpha\beta).x$ for all $x \in E$ and $\alpha,\beta \in \mathsf{K}$,
(h) $1.x = x$ for all $x \in E$.

We shall usually speak of the linear space E instead of the linear space $(E,\mathsf{K},+,.)$. When we wish to emphasize which field is being considered we shall speak of the linear space (E,K). A linear space over R is often called a *real linear space* and a linear space over C is often called a *complex linear space*. A linear space over K is also called a *vector space over* K.

It is important to notice that every complex linear space is also a real linear space. We shall make this assertion more precise. Let $(E,\mathsf{C},+,.)$ be a complex linear space. It is obvious that we obtain a real linear space $(E,\mathsf{R},+,.)$ when we restrict the mapping $(\alpha,x) \to \alpha.x$ of $\mathsf{C} \times E$ into E to $\mathsf{R} \times E$. $(E,\mathsf{R},+,.)$ is called the *real linear space associated with the complex linear space* $(E,\mathsf{C},+,.)$.

Let E be a linear space over K. It is easy to verify that the elements θ and $-x$ whose existence is postulated in conditions (c) and (d) of Definition 3.1.1 are unique. The element θ is called the *zero of the linear space E*. It is also easy to verify that $(-1).x = -x$ for all $x \in E$. We shall write $x-y$ for $x+(-y)$ and αx for $\alpha.x$. It is easy to see that $0x = \theta$ for all $x \in E$ and $\alpha\theta = \theta$ for all $\alpha \in \mathsf{K}$. It is convenient and will cause no confusion to use the same symbol 0 to denote both the zero of K and the zero of the linear space E.

Given $x_1, x_2,\ldots, x_n$ in E and $\alpha_1, \alpha_2,\ldots, \alpha_n$ in K, the element $\alpha_1x_1 + \alpha_2x_2 + \ldots + \alpha_nx_n$ of E is called a *linear combination of* $x_1, x_2,\ldots, x_n$. A subset A of E is said to *span* E if and only if each element of E can be expressed as a linear combination of elements of A. Elements $x_1, x_2,\ldots, x_n$ of E are said to be *linearly dependent* if and only if there exist elements $\alpha_1, \alpha_2,\ldots, \alpha_n$ of K, not all zero, such that $\alpha_1x_1 + \alpha_2x_2 + \ldots + \alpha_nx_n = 0$. A non-empty subset A of E is said to be *linearly independent* if and only if there is no finite subset of A whose elements are linearly dependent. A linearly independent subset of E

that spans E is called a *basis for E*. It is clear that a non-empty subset A of E is a basis for E if and only if each element of E can be expressed uniquely as a linear combination of elements of A. The space E is said to be *finite dimensional* if and only if E either is $\{0\}$ or has a finite basis. If E is not finite dimensional it is said to be *infinite dimensional.* It can be proved that if E is finite dimensional and $E \neq \{0\}$ then all bases for E are finite and have the same number of elements (see HALMOS [12, p. 13]). For our purposes this is the central property of finite dimensional linear spaces. If E is finite dimensional and $E \neq \{0\}$ the number of elements in any basis for E is called the *dimension of E*. We shall adopt the convention that the linear space $\{0\}$ has dimension 0.

The reader who is not already familiar with the theory of linear spaces will find an excellent introduction to this theory in HALMOS [12, §1 – §22].

3.2 Normed Linear Spaces

In the introductory remarks to this chapter we said that a normed linear space was a linear space on which there was defined a metric satisfying the condition (*). It is, however, more convenient to take as fundamental the concept of a norm on a linear space, and define the metric of a normed linear space in terms of its norm.

3.2.1 DEFINITION Let E be a linear space over K. A mapping $x \to \| x \|$ of E into R is called a *norm on E* if and only if it satisfies the following conditions:

(a) $\| x \| \geq 0$ for all $x \in E$,
(b) if $x \in E$ and $\| x \| = 0$ then $x = 0$,
(c) $\| \alpha x \| = |\alpha| \, \| x \|$ for all $x \in E$ and $\alpha \in \mathsf{K}$,
(d) $\| x+y \| \leq \| x \| + \| y \|$ for all $x,y \in E$.

A *normed linear space* over K is a pair $(E, \| \,.\, \|)$, where E is a linear space over K and $\| \,.\, \|$ is a norm on E.

We shall usually speak of the normed linear space E rather than of the normed linear space $(E, \| \,.\, \|)$. Since $0x = 0$ for all $x \in E$, condition (c) gives $\| 0 \| = 0$. Therefore we may replace (b) by the condition

(b)′ $\| x \| = 0$ if and only if $x = 0$.

It follows by induction from (d) that, if $x_1, x_2, \ldots, x_n$ are points of E, then

$$\| x_1 + x_2 + \ldots + x_n \| \leq \| x_1 \| + \| x_2 \| + \ldots + \| x_n \|.$$

The following inequality is often useful:

$$\big| \| x \| - \| y \| \big| \leq \| x - y \| \tag{1}$$

for all $x, y \in E$. To prove (1) we have only to observe that

$$\begin{aligned} \| x \| &= \| (x-y) + y \| \\ &\leq \| x - y \| + \| y \| \end{aligned}$$

and

$$\begin{aligned} \| y \| &= \| (y-x) + x \| \\ &\leq \| y - x \| + \| x \| \\ &= \| (-1)(x-y) \| + \| x \| \\ &= |-1| \, \| x - y \| + \| x \| \\ &= \| x - y \| + \| x \|. \end{aligned}$$

Many of the metric spaces in the Examples of Section 2.1 are also normed linear spaces. We shall list them again below, together with some new examples. These examples will serve to illustrate the concepts introduced in this chapter.

3.2.2 EXAMPLE The real line R is a real linear space and $x \to |x|$ is a norm on R.

3.2.3 EXAMPLE The complex plane C is a complex linear space and $x \to |x|$ is a norm on C. Whenever R and C are considered as normed linear spaces the norms will be those defined here.

3.2.4 EXAMPLE The complex plane C is also a real linear space (C, R) and $x \to |x|$ is a norm on (C, R).

More generally, let (E, C) be a complex linear space and let $x \to \| x \|$ be a norm on (E, C). It is obvious that $x \to \| x \|$ is also a norm on the real linear space (E, R) associated with (E, C) (see Section 3.1). The space $(E, \mathsf{R}, \| \cdot \|)$ will be called the *real normed linear space associated with the complex normed linear space* $(E, \mathsf{C}, \| \cdot \|)$.

3.2.5 EXAMPLE The set K^n of all n-tuples of elements of K is a linear space. The linear space operations on K^n are defined by

$$x+y = (\xi_1+\eta_1, \xi_2+\eta_2, \ldots, \xi_n+\eta_n)$$

and

$$\alpha x = (\alpha\xi_1, \alpha\xi_2, \ldots, \alpha\xi_n),$$

for all $\alpha \in \mathsf{K}$ and all $x = (\xi_1, \xi_2, \ldots, \xi_n)$ and $y = (\eta_1, \eta_2, \ldots, \eta_n)$ in K^n.

Let

$$\| x \| = \left(\sum_{r=1}^{n} | \xi_r |^2 \right)^{1/2}$$

for all $x = (\xi_1, \xi_2, \ldots, \xi_n) \in \mathsf{K}^n$. Then $\| \cdot \|$ is a norm on K^n. In fact all the norm conditions are satisfied obviously except (d), which follows from inequality (2) of Section 2.1.

3.2.6 Example Let p be a real number ≥ 1 and let ℓ^p denote† the set of all sequences (ξ_n) in K such that the infinite series $\sum_n |\xi_n|^p$ converges. The sequence $(\xi_n+\eta_n) \in \ell^p$ whenever $(\xi_n) \in \ell^p$ and $(\eta_n) \in \ell^p$, because

$$|\xi_n+\eta_n|^p \leq 2^p \max\{|\xi_n|^p, |\eta_n|^p\}$$
$$\leq 2^p(|\xi_n|^p + |\eta_n|^p),$$

and it is easy to verify that ℓ^p is a linear space with respect to the linear space operations defined by $x+y = (\xi_n+\eta_n)$ and $\alpha x = (\alpha\xi_n)$ for all $x = (\xi_n)$ and $y = (\eta_n)$ in ℓ^p and all $\alpha \in$ K.

Let

$$\| x \| = \left(\sum_{n=1}^{\infty} |\xi_n|^p\right)^{1/p}$$

for all $x = (\xi_n) \in \ell^p$. We shall prove that $\| \,.\, \|$ is a norm on ℓ^p. This is trivial if $p = 1$; if $p > 1$, the only difficulty lies in verification of (*d*) which in this case is a famous inequality due to Minkowski. The proof of Minkowski's inequality depends on another inequality due to Hölder.

3.2.7 Lemma (Hölder) *Let $p > 1$ and let $q = p/(p-1)$. Then, if $(\xi_n) \in \ell^p$ and $(\eta_n) \in \ell^q$, the series $\sum_n \xi_n\eta_n$ converges absolutely and*

$$\sum_{n=1}^{\infty} |\xi_n\eta_n| \leq \left(\sum_{n=1}^{\infty} |\xi_n|^p\right)^{1/p}\left(\sum_{n=1}^{\infty} |\eta_n|^q\right)^{1/q}. \tag{2}$$

Proof We shall prove first that, if $a \geq 0$ and $b \geq 0$, then

$$a^{1/p}b^{1/q} \leq \frac{a}{p} + \frac{b}{q}. \tag{3}$$

This is obvious if either $a = 0$ or $b = 0$ so we may suppose that $a > 0$ and $b > 0$. If $0 \leq m \leq 1$, the function $f(x) = m(x-1) - (x^m - 1)$ has non-negative derivative for all $x \geq 1$, and hence

$$x^m - 1 \leq m(x-1) \tag{4}$$

for $x \geq 1$. Inequality (3) follows from (4) if we set $x = b/a$, and $m = 1/q$ if $a \leq b$, and $x = a/b$ and $m = 1/p$ if $a > b$.

Inequality (2) is obvious if at least one of the terms on the right-hand side is zero so we may suppose that $\sum_{n=1}^{\infty} |\xi_n|^p \neq 0$ and $\sum_{n=1}^{\infty} |\eta_n|^q \neq 0$.

† See the footnote on p. 29. We shall write ℓ instead of ℓ^1 (see Example 2.1.11).

Then, for $n = 1, 2, \ldots$, (3) gives

$$\frac{|\xi_n|}{\left(\sum_{m=1}^{\infty} |\xi_m|^p\right)^{1/p}} \cdot \frac{|\eta_n|}{\left(\sum_{m=1}^{\infty} |\eta_m|^q\right)^{1/q}} \leq \frac{1}{p} \frac{|\xi_n|^p}{\sum_{m=1}^{\infty} |\xi_m|^p} + \frac{1}{q} \frac{|\eta_n|^q}{\sum_{m=1}^{\infty} |\eta_m|^q}$$

from which it follows that $\sum_n |\xi_n \eta_n|$ converges and

$$\frac{\sum_{n=1}^{\infty} |\xi_n \eta_n|}{\left(\sum_{m=1}^{\infty} |\xi_m|^p\right)^{1/p} \left(\sum_{m=1}^{\infty} |\eta_m|^q\right)^{1/q}} \leq \frac{1}{p} + \frac{1}{q} = 1.$$

This proves (2). □

3.2.8 Lemma (Minkowski) *Let $p > 1$. Then for all (ξ_n) and (η_n) in ℓ^p*

$$\left(\sum_{n=1}^{\infty} |\xi_n + \eta_n|^p\right)^{1/p} \leq \left(\sum_{n=1}^{\infty} |\xi_n|^p\right)^{1/p} + \left(\sum_{n=1}^{\infty} |\eta_n|^p\right)^{1/p}. \tag{5}$$

Proof. We have already observed that the series on the left-hand side of (5) converges. Let $q = p/(p-1)$. Since $(\xi_n + \eta_n) \in \ell^p$ we have $((\xi_n + \eta_n)^{p-1})_{n \geq 1} \in \ell^q$ and hence, by Lemma 3.2.7, the series $\sum_n |\xi_n + \eta_n|^{p-1} |\xi_n|$ and $\sum_n |\xi_n + \eta_n|^{p-1} |\eta_n|$ converge and

$$\begin{aligned}
\sum_{n=1}^{\infty} |\xi_n + \eta_n|^p &\leq \sum_{n=1}^{\infty} |\xi_n + \eta_n|^{p-1} |\xi_n| + \sum_{n=1}^{\infty} |\xi_n + \eta_n|^{p-1} |\eta_n| \\
&\leq \left(\sum_{n=1}^{\infty} |\xi_n|^p\right)^{1/p} \left(\sum_{n=1}^{\infty} |\xi_n + \eta_n|^{(p-1)q}\right)^{1/q} \\
&\quad + \left(\sum_{n=1}^{\infty} |\eta_n|^p\right)^{1/p} \left(\sum_{n=1}^{\infty} |\xi_n + \eta_n|^{(p-1)q}\right)^{1/q} \\
&= \left(\sum_{n=1}^{\infty} |\xi_n + \eta_n|^p\right)^{1/q} \left(\left(\sum_{n=1}^{\infty} |\xi_n|^p\right)^{1/p} + \left(\sum_{n=1}^{\infty} |\eta_n|^p\right)^{1/p}\right).
\end{aligned}$$

If $\sum_{n=1}^{\infty} |\xi_n + \eta_n|^p \neq 0$ then (5) follows directly from the last inequality; and if $\sum_{n=1}^{\infty} |\xi_n + \eta_n|^p = 0$ then (5) is obvious. □

3.2.9 Example Given $f, g \in B_{\mathrm{K}}(E)$ and $\alpha \in \mathrm{K}$, let

$$(f+g)(t) = f(t) + g(t) \text{ and } (\alpha f)(t) = af(t)$$

for all $t \in E$. Then $f+g, \alpha f \in B_{\mathsf{K}}(E)$ and it is easy to verify that $B_{\mathsf{K}}(E)$ is a linear space. It is also easy to verify that if

$$\| f \| = \sup \{ | f(t) | : t \in E\}$$

for all $f \in B_{\mathsf{K}}(E)$ then $\| \, . \, \|$ is a norm on $B_{\mathsf{K}}(E)$. The above norm is called the *uniform norm on* $B_{\mathsf{K}}(E)$ (cf. Example 2.1.9).

3.2.10 EXAMPLE If $E = \{1, 2,..., n\}$ then $B_{\mathsf{K}}(E)$ is just K^n. The uniform norm on K^n is given by

$$\| x \| = \max \{ |\xi_1|, |\xi_2|,..., |\xi_n| \}$$

for all $x = (\xi_1, \xi_2,..., \xi_n) \in \mathsf{K}^n$.

3.2.11 EXAMPLE If $E = \{n : n = 1, 2,...\}$ then $B_{\mathsf{K}}(E)$ is the set m of all bounded sequences in K. The uniform norm on m is given by

$$\| x \| = \sup \{ |\xi_n| : n = 1, 2,...\}$$

for all $x = (\xi_n) \in m$.

Our last example is perhaps the most important.

3.2.12 EXAMPLE Recall that $C_{\mathsf{K}}(X)$ is the set of all bounded continuous mappings of a metric space X into K. Theorem 2.8.2 shows that if $f,g \in C_{\mathsf{K}}(X)$ and $\alpha \in \mathsf{K}$, then $f+g \in C_{\mathsf{K}}(X)$ and $\alpha f \in C_{\mathsf{K}}(X)$, and it is easy to verify that $C_{\mathsf{K}}(X)$ is a linear space. Also, if

$$\| f \| = \sup \{ | f(x) | : x \in X\}$$

for all $f \in C_{\mathsf{K}}(X)$ then $\| \, . \, \|$ is a norm on $C_{\mathsf{K}}(X)$, and is called the *uniform norm*.

EXERCISES

(1) Show that

$$\| f \| = \left(\int_0^1 | f(t) |^2 \, dt \right)^{1/2}$$

defines a norm on $C_{\mathsf{R}}([0,1])$.

(2) For $p > 0$ let ϕ_p be defined on R^2 by

$$\phi_p(x) = (|\xi_1|^p + |\xi_2|^p)^{1/p}$$

where $x = (\xi_1,\xi_2)$. Show that ϕ_p is not a norm on R^2 if $p < 1$.

(3) Sketch the closed unit balls in R^2 which correspond to the norms defined, for $x = (\xi_1,\xi_2) \in \mathsf{R}^2$, by

(a) $\| x \| = \max \{ |\xi_1|,|\xi_2| \}$,
(b) $\| x \| = |\xi_1| + |\xi_2|$,
(c) $\| x \| = (\xi_1^2 + \xi_2^2)^{1/2}$.

(4) Let X be a metric space. Prove that the linear space $C_{\mathsf{R}}(X)$ is finite dimensional if and only if X is a finite set.

3.3 The Metric on a Normed Linear Space

Let E be a normed linear space with norm $\|.\|$ and let $d(x,y) = \| x - y \|$ for all $x,y \in E$. We shall show that d is a metric on E. It is obvious that $d(x,y) \geq 0$ for all $x,y \in E$ and that $d(x,y) = 0$ if and only if $x = y$. If $x,y \in E$ then

$$d(x,y) = \| (-1)(y - x) \| = |-1| \, \| y - x \| = d(y,x)$$

and if $x,y,z \in E$ then

$$\begin{aligned} d(x,z) &= \| (x - y) + (y - z) \| \\ &\leq \| x - y \| + \| y - z \| \\ &= d(x,y) + d(y,z). \end{aligned}$$

Whenever we consider a normed linear space E as a metric space we shall assume that the metric on E is the one just defined. The ball $\{x \in E : \| x \| < 1\}$, with centre 0, is called the *open unit ball in E* and the ball $\{x \in E : \| x \| \leq 1\}$ is called the *closed unit ball in E*. We shall always work directly in terms of the norm of E rather than use the metric d. If the reader keeps in mind the definition of d, this should cause him no trouble. We remark that if (E, C) is a complex normed linear space and (E, R) is the associated real normed linear space then the metrics defined on E by the norms of (E, C) and (E, R) coincide.

Many of the normed linear spaces defined in the Examples of the last section were also cited as examples of metric spaces in Section 2.1, so we now have two definitions of metrics on them: the metric defined in terms of the norm of Section 3.2, and the metric defined directly in Section 2.1. Let us note here, for future use, the relations between these definitions.

The metrics on the normed linear spaces R and C of Examples 3.2.2 and 3.2.3 are the metrics defined in Examples 2.1.2 and 2.1.3. The metric on K^n derived from the norm of Example 3.2.5 is the metric defined in Examples 2.1.4 and 2.1.5, and the metric on K^n derived from the norm of Example 3.2.10 is the metric d_0 of Example 2.1.6. The metric on $B_{\mathsf{K}}(E)$ derived from the uniform norm (Example 3.2.9) is the metric defined in Example 2.1.9, and the metric on $C_{\mathsf{K}}(X)$ derived from the uniform norm (Example 3.2.12) is the uniform metric defined in Section 2.8.

We shall now obtain some elementary properties of sequences in a normed linear space, that generalize familiar properties of sequences of real numbers.

3.3.1 LEMMA *Let (x_n) and (y_n) be sequences in a normed linear space E over K with $\lim_{n\to\infty} x_n = x$ and $\lim_{n\to\infty} y_n = y$, and let (α_n) be a sequence in K with $\lim_{n\to\infty} \alpha_n = \alpha$. Then $\lim_{n\to\infty} (x_n + y_n) = x + y$, $\lim_{n\to\infty} \alpha_n x_n = \alpha x$, and $\lim_{n\to\infty} \| x_n \| = \| x \|$.*

PROOF Let $\varepsilon > 0$. There are integers N_1 and N_2 such that $\| x - x_n \| < \frac{1}{2}\varepsilon$ for all $n \geq N_1$ and $\| y - y_n \| < \frac{1}{2}\varepsilon$ for all $n \geq N_2$. Let $N = \max\{N_1, N_2\}$. For all $n \geq N$,

$$\begin{aligned} \| (x + y) - (x_n + y_n) \| &= \| (x - x_n) + (y - y_n) \| \\ &\leq \| x - x_n \| + \| y - y_n \| \\ &< \varepsilon. \end{aligned}$$

This proves that $\lim_{n\to\infty} (x_n + y_n) = x + y$.

Next let $1 \geq \eta > 0$. Then, as above, there is an integer N_0 such that $\| x - x_n \| < \eta$ and $| \alpha - \alpha_n | < \eta$ for all $n \geq N_0$. Thus, for all $n \geq N_0$,

$$\| x_n \| \leq \| x_n - x \| + \| x \| \leq 1 + \| x \|,$$

and

$$\begin{aligned} \| \alpha x - \alpha_n x_n \| &= \| \alpha(x - x_n) + (\alpha - \alpha_n)x_n \| \\ &\leq \| \alpha(x - x_n) \| + \| (\alpha - \alpha_n)x_n \| \\ &= |\alpha| \, \| x - x_n \| + | \alpha - \alpha_n | \, \| x_n \| \\ &< (|\alpha| + \| x \| + 1)\eta. \end{aligned}$$

Given $\varepsilon > 0$, let $\eta = \min\{1, \varepsilon(|\alpha| + \| x \| + 1)^{-1}\}$. Then, by what we have just proved, there is an integer N_0 such that $\| \alpha x - \alpha_n x_n \| < \varepsilon$ for all $n \geq N_0$. This proves that $\lim_{n\to\infty} \alpha_n x_n = \alpha x$.

Finally, by inequality (1) of Section 3.2 we have

$$| \| x \| - \| x_n \| | \leq \| x - x_n \|$$

and hence $\lim_{n\to\infty} \| x_n \| = \| x \|$. □

3.3.2 DEFINITION A complete normed linear space is called a *Banach space.*

All the normed linear spaces in the Examples of Section 3.2 are Banach spaces. For the space K^n (Example 3.2.5) this was proved in Theorems 2.6.5 and 2.6.6, and for the spaces $B_{\mathsf{K}}(E)$ and $C_{\mathsf{K}}(X)$ this was proved in Theorems 2.6.7 and 2.8.3.

3.3.3 EXAMPLE We shall prove that, for $p \geq 1$, the space ℓ^p is a Banach space. Let (x_n) be a Cauchy sequence in ℓ^p and let $x_n = (\xi_{jn})_{j\geq 1}$ for $n = 1, 2, \ldots$. Given $\varepsilon > 0$, there is an integer N with $\| x_m - x_n \| < \varepsilon$

for all $m,n \geq N$ and therefore

$$\sum_{j=1}^{\infty} | \xi_{jm} - \xi_{jn} |^p < \varepsilon^p \tag{1}$$

for all $m,n \geq N$. Thus, for each positive integer k,

$$| \xi_{km} - \xi_{kn} |^p \leq \sum_{j=1}^{\infty} | \xi_{jm} - \xi_{jn} |^p < \varepsilon^p$$

for all $m,n \geq N$. This shows that $(\xi_{kn})_{n\geq 1}$ is a Cauchy sequence in K. Hence $(\xi_{kn})_{n\geq 1}$ converges. Let $\xi_k = \lim_{n\to\infty} \xi_{kn}$ for $k = 1, 2,...,$ and let J be a positive integer. If $m,n \geq N$, then, by (1),

$$\sum_{j=1}^{J} | \xi_{jm} - \xi_{jn} |^p \leq \sum_{j=1}^{\infty} | \xi_{jm} - \xi_{jn} |^p < \varepsilon^p,$$

and consequently, for all $n \geq N$,

$$\sum_{j=1}^{J} | \xi_j - \xi_{jn} |^p = \lim_{m\to\infty} \sum_{j=1}^{J} | \xi_{jm} - \xi_{jn} |^p \leq \varepsilon^p.$$

Since J is an arbitrary positive integer this proves that the series $\sum_j | \xi_j - \xi_{jn} |^p$ converges for all $n \geq N$ and that

$$\sum_{j=1}^{\infty} | \xi_j - \xi_{jn} |^p = \lim_{J\to\infty} \sum_{j=1}^{J} | \xi_j - \xi_{jn} |^p \leq \varepsilon^p \tag{2}$$

for all $n \geq N$. Therefore the sequence $(\xi_j - \xi_{jN})_{j\geq 1}$ is in ℓ^p, and, since $(\xi_{jN})_{j\geq 1}$ is in ℓ^p it follows that (ξ_j) is in ℓ^p. Let $x = (\xi_j)$. Then (2) gives $\| x - x_n \| \leq \varepsilon$ for all $n \geq N$. This proves that $\lim_{n\to\infty} x_n = x$. Thus ℓ^p is complete.

In a normed linear space we can introduce the notion of an infinite series.

3.3.4 DEFINITION Let E be a normed linear space. A pair of sequences (x_n) and (s_n) in E such that $s_n = x_1 + x_2 + \ldots + x_n$ for $n = 1, 2,...$ is called an *infinite series* and is denoted by $\sum_n x_n$. The point s_n is called the *nth partial sum* of the series $\sum_n x_n$.

3.3.5 DEFINITION Let $\sum_n x_n$ be an infinite series in a normed linear space E and let s_n be the nth partial sum of the series. The series $\sum_n x_n$ is said to *converge* if and only if the sequence of partial sums (s_n) converges. If $\sum_n x_n$ converges and $s = \lim_{n\to\infty} s_n$ then s is called the *sum of the series* $\sum_n x_n$ and we write $s = \sum_{n=1}^{\infty} x_n$.

The reader is warned that the symbol $\sum_{n=1}^{\infty} x_n$, which we shall use to denote the sum of the convergent series $\sum_n x_n$, is often used to denote the series $\sum_n x_n$ itself. We shall preserve the distinction between the series $\sum_n x_n$ and its sum $\sum_{n=1}^{\infty} x_n$.†

It is obvious that the convergent series in the normed linear space R are just the convergent series of real numbers in the usual sense of elementary analysis. Many properties of series of real numbers generalize to series in normed linear spaces or Banach spaces. We shall consider only those generalizations that we shall need later.

3.3.6 LEMMA *Let $\sum_n x_n$ be an infinite series in a Banach space E. The series $\sum_n x_n$ converges if and only if, for each $\varepsilon > 0$, there exists an integer N such that*

$$\| x_{n+1} + x_{n+2} + \ldots + x_{n+k} \| < \varepsilon$$

for all $n \geq N$ and $k = 1, 2, \ldots$.

PROOF Let $s_n = x_1 + x_2 + \ldots + x_n$ for $n = 1, 2, \ldots$. Then

$$s_{n+k} - s_n = x_{n+1} + x_{n+2} + \ldots + x_{n+k},$$

so the condition of the lemma is precisely the condition that (s_n) be a Cauchy sequence. □

3.3.7 DEFINITION An infinite series $\sum_n x_n$ in a normed linear space E is said to be *absolutely convergent* if and only if the series of real numbers $\sum_n \| x_n \|$ is convergent.

Clearly the absolutely convergent series in the normed linear space R are just the absolutely convergent series of real numbers in the usual sense of elementary analysis. The following lemma generalizes a well-known property of absolutely convergent series of real numbers.

3.3.8 LEMMA *Let $\sum_n x_n$ be an absolutely convergent series in a Banach space E. Then the series $\sum_n x_n$ converges and*

$$\left\| \sum_{n=1}^{\infty} x_n \right\| \leq \sum_{n=1}^{\infty} \| x_n \|.$$

† These notational conventions have already been used in Examples 2.1.11, 3.2.6 and 3.3.3.

PROOF Let $\varepsilon > 0$. Applying Lemma 3.3.6 to the convergent series $\sum_n \| x_n \|$ we obtain an integer N such that

$$\| x_{n+1} \| + \| x_{n+2} \| + \ldots + \| x_{n+k} \| < \varepsilon$$

for all $n \geq N$ and $k = 1, 2, \ldots$. By the remark on p. 85 we now have $\| x_{n+1} + x_{n+2} + \ldots + x_{n+k} \| \leq \| x_{n+1} \| + \| x_{n+2} \| + \ldots + \| x_{n+k} \| < \varepsilon$ for all $n \geq N$ and $k = 1, 2, \ldots$ and it follows from Lemma 3.3.6 that $\sum_n x_n$ converges.

Finally, for $m = 1, 2, \ldots$, we obtain as above

$$\left\| \sum_{n=1}^{m} x_n \right\| \leq \sum_{n=1}^{m} \| x_n \|$$

and therefore Lemma 3.3.1 gives

$$\left\| \sum_{n=1}^{\infty} x_n \right\| = \lim_{m \to \infty} \left\| \sum_{n=1}^{m} x_n \right\| \leq \lim_{m \to \infty} \sum_{n=1}^{m} \| x_n \| = \sum_{n=1}^{\infty} \| x_n \|. \square$$

EXERCISES

(1) Let E be a linear space and d a metric on E with $d(x,y) = d(x-y, 0)$ and $d(\alpha x, 0) = |\alpha|\, d(x,0)$ for all $x,y \in E$ and $\alpha \in \mathsf{K}$. Verify that $x \to d(x,0)$ is a norm on E and that d is the metric defined by the norm.

(2) Let E be a normed linear space over K, $x_0 \in E$ with $x_0 \neq 0$ and $\alpha_0 \in \mathsf{K}$ with $\alpha_0 \neq 0$. Prove that the mapping $\alpha \to \alpha x_0$ of K into E is a homeomorphism and that the mappings $x \to \alpha_0 x$ and $x \to x_0 + x$ are homeomorphisms of E onto itself.

(3) Let E be a normed linear space and let $r > 0$. Prove that E and its subset $B(0,r)$ are homeomorphic.

(4) For a subset of A of a normed linear space E over K and $\alpha \in \mathsf{K}$ let $\alpha A = \{\alpha x\colon\ x \in A\}$. Prove that, for $\alpha \neq 0$, $(\alpha A)^- = \alpha(A^-)$ and $(\alpha A)^\circ = \alpha(A^\circ)$.

(5) Prove that in a normed linear space the closure of the open unit ball is the closed unit ball.

(6) Prove that if the series $\sum_n x_n$ and $\sum_n y_n$ in a normed linear space over K converge and if $\alpha, \beta \in \mathsf{K}$ then the series $\sum_n (\alpha x_n + \beta y_n)$ converges and

$$\sum_{n=1}^{\infty} (\alpha x_n + \beta y_n) = \alpha \sum_{n=1}^{\infty} x_n + \beta \sum_{n=1}^{\infty} y_n.$$

(7) If π is a one-to-one mapping of the set of all positive integers onto itself then the series $\sum_n x_{\pi(n)}$ is said to be a *rearrangement* of the series

$\sum_n x_n$. Prove that if $\sum_n x_n$ is an absolutely convergent series in a Banach space then $\sum_n x_{\pi(n)}$ is absolutely convergent and $\sum_{n=1}^{\infty} x_{\pi(n)} = \sum_{n=1}^{\infty} x_n$.

(8) A subset A of a linear space E is said to be *convex* if

$$\alpha x + (1-\alpha)y \in A$$

whenever $x,y \in A$ and $0 \leq \alpha \leq 1$. Prove that the open balls and closed balls in a normed linear space are convex.

(9) Prove that a convex subset of a normed linear space is connected.

(10) Extend Theorem 2.10.9 to a general normed linear space.

3.4 Linear Subspaces

We begin with some algebraic definitions.

3.4.1 DEFINITION A non-empty subset F of a linear space E over K is said to be a *linear subspace of E* if and only if $x+y \in F$ and $\alpha x \in F$ for all $x,y \in F$ and all $\alpha \in \mathsf{K}$.

A linear subspace is also called a *linear manifold.* Some writers use the term 'subspace' alone to mean 'linear subspace' (or even to mean only 'closed linear subspace of a normed linear space'); we shall not adopt this convention. It is obvious that a linear subspace of a linear space is itself a linear space. Trivially E is a linear subspace of itself.

It is easy to verify that the intersection of a family of linear subspaces is a linear subspace (the intersection is non-empty because every linear subspace contains 0).

3.4.2 DEFINITION Let A be a non-empty subset of a linear space E. The intersection of the family of all the linear subspaces of E that contain A is called the *linear hull of A*.

Thus the linear hull, F say, of A is a linear subspace of E that contains A. Clearly F is the smallest linear subspace of E that contains A in the sense that, if G is a linear subspace of E and $A \subseteq G$, then $F \subseteq G$. In the next lemma we characterize F explicitly in terms of A.

3.4.3 LEMMA *The linear hull of a non-empty subset A of a linear space E is the set of all finite linear combinations of elements of A.*

PROOF Let F be the linear hull of A and let G be the set of all finite linear combinations of elements of A. Obviously G is a linear subspace of E and $A \subseteq G$, therefore $F \subseteq G$. On the other hand, since F is a linear subspace of E and $A \subseteq F$, all finite linear combinations of elements of A are in F. Thus $G \subseteq F$. □

The linear hull of A is often called the *linear subspace of E spanned by A* (*or generated by A*) and is often defined to be the set of all finite linear combinations of elements of A. We have chosen Definition 3.4.2 because it has an analogue for closed linear subspaces.

Suppose now that F is a linear subspace of a normed linear space E. The restriction to F of the norm on E is obviously a norm on F, so F is a normed linear space. Whenever we consider a linear subspace of a normed linear space as a normed linear space we shall assume that the norm on the linear subspace has been obtained by restriction from the norm of the linear space. It is clear that the metric defined on F by its norm coincides with the restriction to F of the metric defined on E by its norm and therefore F is a subspace of the metric space E in the sense of Definition 2.4.1. In particular, by Theorem 2.6.8, a closed linear subspace of a Banach space is a Banach space.

3.4.4 THEOREM *The closure of a linear subspace of a normed linear space is a linear subspace.*

PROOF Let F be a linear subspace of a normed linear space E, let $x,y \in \bar{F}$, and let $\alpha \in \mathsf{K}$. By Theorem 2.5.5 there are sequences (x_n) and (y_n) in F with $\lim_{n\to\infty} x_n = x$ and $\lim_{n\to\infty} y_n = y$. By Lemma 3.3.1 we have $x+y = \lim_{n\to\infty} (x_n+y_n)$ and $\alpha x = \lim_{n\to\infty} \alpha x_n$. But $x_n+y_n \in F$ and $\alpha x_n \in F$, because F is a linear subspace; consequently, by Theorem 2.5.5, we have $x+y \in \bar{F}$ and $\alpha x \in \bar{F}$. This proves that $\bar{F}$ is a linear subspace. □

3.4.5 DEFINITION Let A be a non-empty subset of a normed linear space E. The intersection of the family of all closed linear subspaces of E that contain A is called the *closed linear hull of A*.

The closed linear hull, F say, of A is obviously a linear subspace of E, and it is a closed subset of E because the intersection of any family of closed sets is closed (Theorem 2.2.12). Clearly F is the smallest closed linear subspace of E that contains A in the sense that, if G is a closed linear subspace of E and $A \subseteq G$, then $F \subseteq G$.

3.4.6 LEMMA *The closed linear hull of a non-empty subset A of a normed linear space E is the closure of the linear hull of A.*

PROOF Let F be the closed linear hull of A and let G be the linear hull of A. Since F is a linear subspace and $A \subseteq F$ we have $G \subseteq F$ and therefore $\bar{G} \subseteq F$. On the other hand $A \subseteq G \subseteq \bar{G}$ and, by Theorem 3.4.4, $\bar{G}$ is a linear subspace of E so $F \subseteq \bar{G}$. □

We recall that a basis for a linear space E is a linearly independent subset of E whose linear hull is E. The following weaker concept is more useful in normed linear spaces. We recall that a subset of a metric space X is dense in X if and only if its closure is X.

3.4.7 DEFINITION A non-empty subset A of a normed linear space E is said to be *fundamental in* E if and only if the linear hull of A is dense in E.

It follows directly from Lemma 3.4.6 that A is fundamental in E if and only if the closed linear hull of A is E. We recall that a metric space X is separable if and only if there is a countable subset of X that is dense in X.

3.4.8 THEOREM *A normed linear space E is separable if and only if there exists a countable subset of E that is fundamental in E.*

PROOF Suppose that E is separable, and let A be a countable subset of E that is dense in E. Then the linear hull of A, which contains A, must also be dense in E, so A is fundamental in E.

The converse is not quite so trivial. Suppose that there exists a countable subset of A of E that is fundamental in E. Choose a countable subset K_0 of K that is dense in K. (If $\mathsf{K} = \mathsf{R}$ take K_0 to be the set of all rational numbers, and if $\mathsf{K} = \mathsf{C}$ take K_0 to be the set of all complex numbers with rational real and imaginary parts, which is countable by Theorem 1.4.4.) Let B be the set of all finite linear combinations of the form $\beta_1 x_1 + \beta_2 x_2 + \ldots + \beta_n x_n$, where n is an arbitrary positive integer, $\beta_1, \beta_2, \ldots, \beta_n$ are arbitrary points of K_0 and $x_1, x_2, \ldots, x_n$ are arbitrary points of A. We shall prove that B is countable and dense in E.

First we prove that B is countable. By Theorem 1.4.4 the set $\mathsf{K}_0 \times A$ is countable, and hence by Theorem 1.4.6 the family $\mathscr{A}$ of all finite subsets of $\mathsf{K}_0 \times A$ is also countable. The mapping

$$\{(\beta_1, x_1), (\beta_2, x_2), \ldots, (\beta_n, x_n)\} \rightarrow \beta_1 x_1 + \beta_2 x_2 + \ldots + \beta_n x_n$$

maps $\mathscr{A}$ onto B and consequently, by Theorem 1.4.2, B is countable.

Now we prove that B is dense in E. Let $x \in E$ and $\varepsilon > 0$. The linear hull of A is dense in E and consequently, by Lemma 3.4.3 and Theorem 2.3.4, there are points $x_1, x_2, \ldots, x_n$ in A, and $\alpha_1, \alpha_2, \ldots, \alpha_n$ in K, such that

$$\| x - (\alpha_1 x_1 + \alpha_2 x_2 + \ldots + \alpha_n x_n) \| < \tfrac{1}{2}\varepsilon.$$

But K_0 is dense in K so there are points $\beta_1, \beta_2, \ldots, \beta_n$ in K_0 such that $| \alpha_m - \beta_m | < \frac{1}{2}\varepsilon n^{-1}(1 + \| x_n \|)^{-1}$ for $m = 1, 2, \ldots, n$. Then

$$\beta_1 x_1 + \beta_2 x_2 + \ldots + \beta_n x_n \in B$$

and

$$\begin{aligned}
&\| x - (\beta_1 x_1 + \beta_1 x_2 + \ldots + \beta_n x_n) \| \\
&\quad \leq \| x - (\alpha_1 x_1 + \alpha_2 x_2 + \ldots + \alpha_n x_n) \| \\
&\qquad + \| (\alpha_1 - \beta_1) x_1 + (\alpha_2 - \beta_2) x_2 + \ldots + (\alpha_n - \beta_n) x_n \| \\
&\quad < \tfrac{1}{2}\varepsilon + | \alpha_1 - \beta_1 | \, \| x_1 \| + | \alpha_2 - \beta_2 | \, \| x_2 \| + \ldots + | \alpha_n - \beta_n | \, \| x_n \| \\
&\quad < \varepsilon.
\end{aligned}$$

This proves that B is dense in E (Theorem 2.3.4). □

3.4.9 EXAMPLE We shall prove that the Banach space ℓ^p is separable. Let $e_n = (\delta_{mn})_{m \geq 1}$ for $n = 1, 2, \ldots$, where $\delta_{mn} = 0$ if $m \neq n$ and $\delta_{mm} = 1$. Obviously $e_n \in \ell^p$. Let $x = (\xi_n) \in \ell^p$ and $\varepsilon > 0$. There is an integer N with $\sum_{m=n+1}^{\infty} | \xi_m |^p < \varepsilon^p$ for all $n \geq N$. If $n \geq N$ then

$$\| x - (\xi_1 e_1 + \xi_2 e_2 + \ldots + \xi_n e_n) \| = \left(\sum_{m=n+1}^{\infty} | \xi_m |^p \right)^{1/p} < \varepsilon.$$

This shows that the set $\{e_n : n = 1, 2, \ldots\}$ is fundamental in ℓ^p.

3.4.10 EXAMPLE We shall prove that m is not separable. Let A be the subset of m consisting of all sequences (ξ_n) with the property that, for each positive integer n, either $\xi_n = 0$ or $\xi_n = 1$. The set A is uncountable. To see this suppose, on the contrary, that A is countable and let $n \to (\xi_{nm})_{m \geq 1}$ be an enumeration of A. Let $x_0 = (\xi_n)$ where $\xi_n = 0$ if $\xi_{nn} = 1$ and $\xi_n = 1$ if $\xi_{nn} = 0$. Obviously $x_0 \in A$ and $x_0 \neq x_n$ for $n = 1, 2, \ldots$, so we have a contradiction. This shows that A is uncountable.

Suppose now that m is separable and let B be a countable subset of m that is dense in m. We shall obtain a contradiction by constructing a one-to-one mapping of A onto a subset of B. For each $x \in m$ the open ball $B(x, \frac{1}{2})$ contains at least one point of B so corresponding to each point $x \in A$ we can choose a point $z_x \in B \cap B(x, \frac{1}{2})$. If $x, y \in A$ and $x \neq y$ then $\| x - y \| = 1$ and therefore

$$\begin{aligned}
\| y - z_x \| &\geq | \, \| y - x \| - \| x - z_x \| \, | \\
&\geq \| y - x \| - \| x - z_x \| \\
&> 1/2.
\end{aligned}$$

This shows that $x \to z_x$ is a one-to-one mapping of A into B. Since B is countable so is the set $\{z_x : x \in A\}$ (Theorem 1.4.1) and hence, since the mapping $x \to z_x$ is one-to-one, A is countable. This contradiction shows that m cannot be separable.

Let us now return to some simple algebraic considerations. Given linear subspaces F and G of a linear space E we shall write

$$F+G = \{x+y : x \in F, y \in G\}.$$

It is easy to verify that $F+G$ is a linear subspace of E. The case when $F+G = E$ and $F \cap G = \{0\}$ is particularly interesting.

3.4.11 DEFINITION Let F and G be linear subspaces of a linear space E. The space E is said to be the *direct sum of F and G* if and only if $E = F+G$ and $F \cap G = \{0\}$. If E is the direct sum of F and G we write $E = F \oplus G$.

It is clear that $E = F \oplus G$ if and only if each element $x \in E$ has a unique representation in the form $x = y+z$ where $y \in F$ and $z \in G$. It can be proved that, given a linear subspace F of E, there exists at least one linear subspace G of E such that $E = F \oplus G$; any such linear subspace G is said to be *complementary to F in E*. (The reader will find a proof of this result for finite dimensional linear spaces in HALMOS [12, p. 31]. We shall not need the more general result for infinite dimensional spaces but we shall give an indication of its proof in Exercise 5.3(10).)

Suppose now that E is a normed linear space, and F is a closed linear subspace of E. By the remarks above there exist linear subspaces of E complementary to F in E; however these linear subspaces may not be closed. In 1937 F. J. Murray gave an example of a normed linear space E and a closed linear subspace F of E with the property that there are no *closed* linear subspaces complementary to F in E. In Chapter 8 we shall return to the question of the existence of complementary closed linear subspaces (see Lemma 8.8.8, Theorem 8.8.10 and also Lemma 8.7.3).

The existence of complementary linear subspaces plays an important role in the theory of finite dimensional linear spaces. The quotient spaces introduced below can sometimes be used as effective substitutes for complementary linear subspaces.

We begin with the purely algebraic situation. Let F be a linear subspace of a linear space E, let $x+F = \{x+y : y \in F\}$ for all $x \in E$, and let $E/F = \{x+F : x \in E\}$. The sets $x+F$ are called the *cosets of F in E*. We observe that $F = 0+F$. Obviously $x_1+F = x_2+F$ if and only if $x_1-x_2 \in F$ and consequently, for each pair $x_1,x_2 \in E$, either $(x_1+F) \cap (x_2+F) = \emptyset$ or $x_1+F = x_2+F$. Also, if $x_1, x_2, y_1, y_2 \in E$ and $x_1-x_2, y_1-y_2 \in F$, then $(x_1+y_1)-(x_2+y_2) \in F$ and $\alpha x_1 - \alpha x_2 \in F$ for all $\alpha \in \mathsf{K}$, because F is a linear subspace. It follows that we can define mappings

$$(x+F, y+F) \to (x+F) + (y+F)$$

of $E/F \times E/F$ into E/F and $(\alpha, x+F) \to \alpha(x+F)$ of $\mathsf{K} \times E/F$ into E/F unambiguously by

$$(x + F) + (y + F) = (x + y) + F$$

and

$$\alpha(x + F) = \alpha x + F$$

for all $x,y \in E$ and all $\alpha \in \mathsf{K}$.

It is easy to verify that with the operations defined above E/F is a linear space over K. We remark that F is the zero of E/F.

3.4.12 DEFINITION Let F be a linear subspace of a linear space E. The linear space E/F defined above is called the *quotient space of E modulo F*. The mapping $x \to x + F$ of E onto E/F is called the *canonical mapping* of E onto E/F.

The relationship between the quotient space E/F and linear subspaces of E complementary to F is simple. Let G be a linear subspace of E such that $E = F \oplus G$. It is easy to verify that the restriction to G of the canonical mapping $x \to x+F$ is a one-to-one mapping of G onto E/F which preserves the linear space operations. Such a mapping between two linear spaces is called an isomorphism. Thus any linear subspace of E that is complementary to F is 'isomorphic' to E/F.

3.4.13 THEOREM *Let F be a closed linear subspace of a normed linear space E, and let*

$$\| x+F \| = \inf\{ \| x+y \| : y \in F\}$$

for all $x \in F$. The mapping $x+F \to \| x+F \|$ is a norm on E/F, and further, if E is a Banach space, so is E/F.

PROOF It is obvious that $\| x+F \| \geq 0$ for all $x \in E$. Suppose that $\| x+F \| = 0$ for some $x \in E$. Then $\inf\{ \| x-y \| : y \in F\} = 0$, because $y \in F$ if and only if $-y \in F$, and therefore by Lemma 2.8.4 we have $x \in F$. Thus $x+F = F$ which is the zero of E/F.

Next let $x,y \in F$. Then, since F is a linear subspace,

$$\begin{aligned}
\| (x+F) + (y+F) \| &= \| (x + y)+F \| \\
&= \inf\{ \| x + y + z \| : z \in F\} \\
&= \inf\{ \| (x + z_1) + (y + z_2) \| : z_1,z_2 \in F\} \\
&\leq \inf\{ \| x + z_1 \| + \| y + z_2 \| : z_1,z_2 \in F\} \\
&= \inf\{ \| x + z_1 \| : z_1 \in F\} \\
&\qquad + \inf\{ \| y + z_2 \| : z_2 \in F\} \\
&= \| x+F \| + \| y+F \|.
\end{aligned}$$

Finally let $x \in E$ and let $\alpha \in \mathbb{K}$ with $\alpha \neq 0$. Then

$$\begin{aligned} \| \alpha(x + F) \| &= \| \alpha x + F \| \\ &= \inf\{ \| \alpha x + y \| : y \in F\} \\ &= \inf\{ \| \alpha(x + (1/\alpha)y) \| : y \in F\} \\ &= |\alpha| \inf\{ \| x + (1/\alpha)y \| : y \in F\} \\ &= |\alpha| \, \| x+F \|, \end{aligned}$$

because F is a linear subspace. This proves that $x+F \to \| x+F \|$ is a norm.

Suppose now that E is a Banach space. Let (x_n+F) be a Cauchy sequence in E/F. We shall define inductively a subsequence (x_{n_k}) of (x_n) such that

$$\| x_{n_k} - x_{n_{k+1}} + F \| < 2^{-k} \tag{1}$$

for $k = 1, 2, \ldots$. Since (x_n+F) is a Cauchy sequence, for each positive integer k there is a positive integer N_k such that

$$\| x_n - x_m + F \| < 2^{-k} \tag{2}$$

for all $m,n \geq N_k$. Let $n_1 = N_1$ and suppose that $n_2, n_3, \ldots, n_k$ have been defined so that $n_1 < n_2 < \ldots < n_k$ and $N_j \leq n_j$ for $j = 1, 2, \ldots, k$. Let $n_{k+1} = \max\{N_{k+1}, n_k+1\}$. This defines a strictly increasing sequence (n_k), and (1) follows from (2). To simplify our notation let $y_k = x_{n_k}$. We shall define inductively a sequence (z_k) in E such that $z_k \in y_k+F$ and $\| z_k - z_{k+1} \| < 2^{-k+1}$ for $k = 1, 2, \ldots$. Choose $z_1 \in y_1+F$ and suppose that $z_2, z_3, \ldots, z_k$ have been chosen to satisfy the above conditions. Then $y_k+F = z_k+F$ and, by (1), we have $\| z_k - y_{k+1} + F \| < 2^{-k}$. By definition of the norm in E/F there exists $z_{k+1} \in y_{k+1} + F$ such that $\|z_k - z_{k+1}\| \leq \|z_k - y_{k+1} + F \| + 2^{-k}$. Then $\|z_k - z_{k+1}\| < 2^{-k+1}$ as required.

The series $\sum_k \| z_k - z_{k+1} \|$ converges and therefore, by Lemma 3.3.8, the series $\sum_k (z_k - z_{k+1})$ converges. But

$$(z_1 - z_2) + (z_2 - z_3) + \ldots + (z_m - z_{m+1}) = z_1 - z_{m+1}$$

so the sequence (z_m) also converges. Let $z = \lim_{k\to\infty} z_k$. Since $z_k \in y_k+F$,

$$\| (z+F) - (y_k+F) \| = \| z - y_k + F \| \leq \| z - z_k \|.$$

This shows that $\lim_{k\to\infty} (y_k+F) = z+F$. Thus the Cauchy sequence (x_n+F) has a convergent subsequence $(x_{n_k}+F)$ and consequently, by Lemma 2.6.9, must converge. This proves that E/F is complete. □

We remark that the normed linear space E/F is always defined irrespective of whether there exists a closed linear subspace of E complementary to F.

EXERCISES

In Exercises (1)–(4) *F is a linear subspace of a normed linear space E.*

(1) Prove that if $F \neq E$ then the interior of F is empty.

(2) Let F be closed. Prove that the canonical mapping $x \to x+F$ of E onto the quotient space E/F is continuous and that it maps open subsets of E onto open subsets of E/F.

(3) Let F be closed and let X be a closed subset of E such that $X+F = X$. Prove that the image of X under the canonical mapping $x \to x+F$ is a closed subset of E/F.

(4) Let F be closed. Prove that if F and E/F are both complete then E is complete. Deduce that every finite dimensional normed linear space is complete.

(5) Prove that the space $B_{\mathsf{R}}(E)$ of bounded mappings of a non-empty set E into R (see Example 3.2.9) is separable if and only if E is a finite set.

(6) Let X be a metric space and x_0 a point of X. Prove that the set of functions $f \in B_{\mathsf{K}}(X)$ which are continuous at x_0 is a closed linear subspace of $B_{\mathsf{K}}(X)$.

(7) Determine the points in (0,1) at which the function f defined by

$$f(x) = \sum_{n=1}^{\infty} \frac{(nx)}{n^2}$$

is continuous. (The symbol (t) is written in place of $t - [t]$ where $[t]$ denotes the greatest integer not greater than t.)

(8) Let c denote the linear subspace of m which consists of all the convergent sequences and let c_0 denote the linear subspace of m which consists of all the sequences that converge to zero. Prove that c is a separable Banach space and that c_0 is a closed linear subspace of c.

(9) Let e_n, $n = 1, 2, \ldots$, be the sequences defined in Example 3.4.9.

(a) Prove that if $x = (\xi_n) \in \ell$ then $\sum_n \xi_n e_n$ is absolutely convergent in ℓ and $x = \sum_{n=1}^{\infty} \xi_n e_n$.

(b) Prove that if $x = (\xi_n) \in c_0$ then $\sum_n \xi_n e_n$ is convergent in c_0 and $x = \sum_{n=1}^{\infty} \xi_n e_n$. Show that $\sum_n \xi_n e_n$ may not be absolutely convergent.

(10) A point x of a subset A of a linear space E is said to be an *extreme point* of A if it cannot be expressed in the form $x = \alpha y + (1 - \alpha)z$ with $y,z \in A$, $y \neq z$ and $0 < \alpha < 1$.

Show that the extreme points of a subset of a normed linear space are frontier points of the set.

Determine all the extreme points of the closed unit balls of the spaces ℓ, c, c_0 and $C_{\mathsf{R}}([0,1])$.

3.5 Bounded Linear Transformations

Many problems of classical analysis are essentially problems about continuous linear mappings of one normed linear space into another. We shall take up the systematic study of continuous linear mappings in Chapter 6, where the reader will also find a discussion of several examples. In this section we shall obtain some elementary properties of these mappings.

3.5.1 Definition Let E and F be linear spaces over K and let T be a mapping of E into F. Then T is said to be a *linear mapping*† *of the linear space* (E, K) *into the linear space* (F, K) if and only if

$$T(\alpha x + \beta y) = \alpha Tx + \beta Ty$$

for all $x,y \in E$ and all $\alpha,\beta \in \mathsf{K}$. A linear mapping of E into F is often called a *linear transformation of E into F* and sometimes a *linear operator*. We shall generally reserve the term linear operator to denote a linear transformation of a linear space into itself. It is easy to see that if T is a linear transformation of E into F then $T0 = 0$ and $T(-x) = -Tx$ for all $x \in E$.

3.5.2 Lemma *Let T be a linear transformation of a normed linear space E into a normed linear space F. If T is continuous at one point of E then T is continuous on E.*

Proof Suppose that T is continuous at x_0. Given $\varepsilon > 0$ there exists $\delta > 0$ such that $\| Tx_0 - Tx \| < \varepsilon$ whenever‡ $\| x_0 - x \| < \delta$. Let $x_1 \in E$. Then, for all $x \in E$ with $\| x_1 - x \| < \delta$,

$$\| x_0 - (x - x_1 + x_0) \| = \| x_1 - x \| < \delta,$$

and so $\quad \| Tx_1 - Tx \| = \| Tx_0 - T(x - x_1 + x_0) \| < \varepsilon.$

This proves that T is continuous at x_1. □

3.5.3 Definition A linear transformation T of a normed linear space E into a normed linear space F is said to be *bounded* if and only if $\{ \| Tx \| : \| x \| \leq 1\}$ is a bounded set of real numbers. Thus T is bounded

† For linear mappings it is customary to denote the value of T at x by Tx rather than $T(x)$.

‡ It will simplify our notation and cause no confusion if we use the symbol $\|.\|$ to denote both the norm on E and the norm on F.

if and only if there is a real number M such that $\| Tx \| \leq M$ whenever $\| x \| \leq 1$.

It is important to notice that (in the case $F = \mathsf{K}$) a bounded linear transformation T of E into K is *not* a bounded mapping of E into K in the sense of Definition 2.1.8 unless $Tx = 0$ for all $x \in E$.

3.5.4 THEOREM *A linear transformation T of a normed linear space E into a normed linear space F is continuous on E if and only if it is bounded.*

PROOF Suppose that T is continuous on E. In particular T is continuous at 0 so, since $T0 = 0$, there exists $\delta > 0$ such that $\| Tx \| < 1$ whenever $\| x \| < \delta$. Let $\| x \| \leq 1$. Then $\| \frac{1}{2}\delta x \| = \frac{1}{2}\delta \| x \| < \delta$ and hence $\| T(\frac{1}{2}\delta x) \| < 1$. But $T(\frac{1}{2}\delta x) = \frac{1}{2}\delta Tx$ and thus

$$\| Tx \| = \| \frac{2}{\delta} T(\tfrac{1}{2}\delta x) \| = \frac{2}{\delta} \| T(\tfrac{1}{2}\delta x) \| < \frac{2}{\delta}.$$

This shows that T is bounded.

Suppose conversely that T is bounded. To prove that T is continuous on E it is, by Lemma 3.5.2, sufficient to prove that T is continuous at 0. Let $\varepsilon > 0$ and let M be such that $\| Tx \| \leq M$ whenever $\| x \| \leq 1$. Choose $\delta > 0$ so that $\delta M < \varepsilon$. If $\| x \| < \delta$ then $\| \delta^{-1} x \| = \delta^{-1} \| x \| < 1$ and hence $\| T(\delta^{-1}x) \| \leq M$. Therefore, for $\| x \| < \delta$,

$$\| Tx \| = \| \delta T(\delta^{-1}x) \| = \delta \| T(\delta^{-1}x) \| \leq \delta M < \varepsilon.$$

This proves that T is continuous at 0. □

The equivalence of boundedness and continuity is fundamental to the theory of linear transformations.

3.5.5 DEFINITION Let T be a bounded linear transformation of a normed linear space E into a normed linear space F. The non-negative real number $\| T \|$ defined by

$$\| T \| = \sup \{ \| Tx \| : \| x \| \leq 1 \} \tag{1}$$

is called the *norm* or *bound* of T.

The use of the word norm will be justified in Theorem 3.5.7 below. The following alternative expressions for the norm of T will be useful.

$$\| T \| = \sup \{ \| Tx \| : \| x \| = 1 \}, \tag{2}$$

$$\| T \| = \sup \left\{ \frac{\| Tx \|}{\| x \|} : x \in E \text{ and } x \neq 0 \right\}, \tag{3}$$

$$\| T \| = \inf \{ M \geq 0 : \| Tx \| \leq M \| x \| \text{ for all } x \in E \}. \tag{4}$$

In (2) and (3) we are assuming, of course, that E contains non-zero points.

Let M_2, M_3 and M_4 denote the numbers on the right-hand sides of (2), (3) and (4), respectively. The truth of the equalities (2), (3) and (4) will follow from the inequalities

$$\| T \| \geq M_2 = M_3 \geq M_4 \geq \| T \|.$$

The inequality $\| T \| \geq M_2$ is obvious. Since T is linear we can write

$$M_3 = \sup \{ \| T(\| x \|^{-1}x) \| : x \in E \text{ and } x \neq 0\}.$$

As x varies over the set of non-zero points of E, the point $\| x \|^{-1}x$ varies over the set $\{y : \| y \| = 1\}$ and consequently $M_2 = M_3$. It follows from the definition of M_3 that $\| Tx \| \leq M_3 \| x \|$ for all $x \in E$, from which it follows that $M_4 \leq M_3$. Finally, if $\| Tx \| \leq M \| x \|$ for all $x \in E$, then $\| Tx \| \leq M$ for all $x \in E$ with $\| x \| \leq 1$, and therefore $\| T \| \leq M$. It follows that $\| T \| \leq M_4$. This completes the proof of the equalities (2), (3) and (4).

From (3) we obtain the important inequality

$$\| Tx \| \leq \| T \| \, \| x \| \tag{5}$$

for all $x \in E$.

3.5.6 DEFINITION Let S and T be linear transformations of a linear space E into a linear space F and let $\alpha \in K$. We define the mappings $S+T$ and αT of E into F by

$$(S+T)x = Sx + Tx$$

and

$$(\alpha T)x = \alpha(Tx)$$

for all $x \in E$.

It is easy to verify that the mappings $S+T$ and αT are linear and that, with these operations, the set of all linear transformations of E into F is a linear space (see HALMOS [12, pp. 56–7]). The zero of this linear space is the linear transformation 0 defined by $0x = 0$ for all $x \in E$.

3.5.7 THEOREM *The set $L(E,F)$ of all bounded linear transformations of a normed linear space E into a normed linear space F is a linear space and the mapping $T \to \| T \|$ is a norm on $L(E,F)$. Further, if F is a Banach space, then so is $L(E,F)$.*

PROOF To prove that $L(E,F)$ is a linear space it will be sufficient to prove that $S+T \in L(E,F)$ and $\alpha T \in L(E,F)$ for all $S,T \in L(E,F)$ and all $\alpha \in K$. By (5), if $\| x \| \leq 1$ then

$$\begin{aligned} \| (S+T)x \| &= \| Sx + Tx \| \\ &\leq \| Sx \| + \| Tx \| \\ &\leq \| S \| \, \| x \| + \| T \| \, \| x \| \\ &\leq \| S \| + \| T \|, \end{aligned}$$

which shows that $S+T \in L(E,F)$ and $\| S+T \| \leq \| S \| + \| T \|$. Since $\| (\alpha T)x \| = \| \alpha(Tx) \| = |\alpha| \, \| Tx \|$, it follows directly from (1) that $\alpha T \in L(E,F)$ and $\| \alpha T \| = |\alpha| \, \| T \|$. Suppose finally that $T \in L(E,F)$ and $\| T \| = 0$. Then, by (5), $\| Tx \| = 0$ and hence $Tx = 0$ for all $x \in E$. This shows that $T = 0$. This completes the proof that $L(E,F)$ is a linear space and that $T \to \| T \|$ is a norm on $L(E,F)$.

Suppose now that F is a Banach space. Let (T_n) be a Cauchy sequence in $L(E,F)$ and let $\varepsilon > 0$. There is an integer N such that $\| T_m - T_n \| < \varepsilon$ for all $m,n \geq N$. Let $x \in E$. Then, if $m,n \geq N$,

$$\begin{aligned} \| T_m x - T_n x \| &= \| (T_m - T_n)x \| \\ &\leq \| T_m - T_n \| \, \| x \| \\ &< \varepsilon \| x \|. \end{aligned} \tag{6}$$

This shows that $(T_n x)$ is a Cauchy sequence in F. Since F is complete $(T_n x)$ converges. Let

$$Tx = \lim_{n \to \infty} T_n x. \tag{7}$$

Since $x \in E$ is arbitrary this defines a mapping T of E into F. We shall prove that $T \in L(E,F)$ and $\lim\limits_{n \to \infty} T_n = T$.

Let $x,y \in E$ and $\alpha,\beta \in \mathsf{K}$. Then by (7), the linearity of T_n and Lemma 3.3.1

$$\begin{aligned} T(\alpha x + \beta y) &= \lim_{n \to \infty} T_n(\alpha x + \beta y) \\ &= \lim_{n \to \infty} (\alpha T_n x + \beta T_n y) \\ &= \alpha \lim_{n \to \infty} T_n x + \beta \lim_{n \to \infty} T_n y \\ &= \alpha Tx + \beta Ty. \end{aligned}$$

This shows that T is linear.

Finally, for all $x \in E$ and all positive integers n, (7) and Lemma 3.3.1 give

$$\begin{aligned} Tx - T_n x &= (\lim_{m \to \infty} T_m x) - T_n x \\ &= \lim_{m \to \infty} (T_m x - T_n x) \end{aligned}$$

and therefore, again by Lemma 3.3.1,

$$\| Tx - T_n x \| = \lim_{m \to \infty} \| T_m x - T_n x \|.$$

But then, by (6),

$$\| Tx - T_n x \| \leq \varepsilon \| x \| \tag{8}$$

for all $x \in E$ and all $n \geq N$. Inequalities (5) and (8) give, for all $x \in E$,

$$\begin{aligned} \| Tx \| &\leq \| Tx - T_N x \| + \| T_N x \| \\ &\leq \varepsilon \| x \| + \| T_N \| \, \| x \| \\ &= (\varepsilon + \| T_N \|) \, \| x \|. \end{aligned}$$

This shows that $T \in L(E,F)$. Inequality (8) now gives

$$\| T - T_n \| = \sup \{ \| Tx - T_n x \| : \| x \| \leq 1\} \leq \varepsilon$$

for all $n \geq N$. This shows that $\lim_{n \to \infty} T_n = T$. This completes the proof that $L(E,F)$ is a Banach space. □

If E and F are normed linear spaces over the same field we shall denote by $L(E,F)$ the normed linear space of bounded linear transformations of E into F. We shall need the following lemma on the composition of two bounded linear transformations.

3.5.8 LEMMA *Let E, F and G be normed linear spaces, $T \in L(E,F)$, and $S \in L(F,G)$. Then $S \circ T \in L(E,G)$ and*

$$\| S \circ T \| \leq \| S \| \| T \|.$$

PROOF It is easy to verify that $S \circ T$ is a linear transformation. For all $x \in E$, inequality (5) gives

$$\| (S \circ T)x \| = \| S(Tx) \| \leq \| S \| \| Tx \| \leq \| S \| \| T \| \| x \|,$$

which shows that $S \circ T \in L(E,F)$ and $\| S \circ T \| \leq \| S \| \| T \|$. □

It is easy to verify that if T is a one-to-one linear transformation of a linear space E onto a linear space F then the inverse mapping T^{-1} is also linear. There is a simple criterion for the continuity of T^{-1}.

3.5.9 LEMMA *Let T be a linear transformation of a non-zero normed linear space E onto a normed linear space F. Then T is one-to-one and T^{-1} is bounded if and only if there is a positive real number m such that $\| Tx \| \geq m$ for all $x \in E$ with $\| x \| = 1$.*

PROOF Suppose that T is one-to-one and that $T^{-1} \in L(F,E)$. Since $E \neq \{0\}$ we have $T^{-1} \neq 0$ and therefore $\| T^{-1} \| > 0$. Let $x \in E$ with $\| x \| = 1$. Then

$$1 = \| x \| = \| T^{-1}(Tx) \| \leq \| T^{-1} \| \| Tx \|$$

and hence $\| Tx \| \geq \| T^{-1} \|^{-1}$.

Conversely suppose that there exists $m > 0$ such that $\| Tx \| \geq m$ for $\| x \| = 1$. Then $\| Tx \| \geq m \| x \|$ for all $x \in E$ (because if $x \neq 0$ then $\| \| x \|^{-1} x \| = 1$). It follows that T is one-to-one. Let $y \in F$ and $x = T^{-1}y$. Then $y = Tx$ and

$$\| T^{-1}y \| = \| x \| \leq \frac{1}{m} \| Tx \| = \frac{1}{m} \| y \|.$$

This shows that $T^{-1} \in L(F,E)$. □

Our final theorem has some interesting applications, and is the basis of the elementary theory of integration developed in Section 3.7. We recall (Theorem 3.4.4) that the closure of a linear subspace is a linear subspace.

3.5.10 THEOREM *Let E_0 be a linear subspace of a normed linear space E and let T_0 be a bounded linear transformation of E_0 into a Banach space F. Then there is a unique bounded linear transformation T of $\bar{E}_0$ into F such that $Tx = T_0x$ for all $x \in E_0$. Further $\| T \| = \| T_0 \|$.*†

PROOF Let $x \in \bar{E}_0$. By Theorem 2.5.5 there is a sequence (x_n) in E_0 such that $\lim_{n\to\infty} x_n = x$. By Lemma 2.6.2 the sequence (x_n) is a Cauchy sequence. Since

$$\| T_0x_m - T_0x_n \| = \| T_0(x_m - x_n) \| \leq \| T_0 \| \, \| x_m - x_n \|$$

(T_0x_n) is also a Cauchy sequence and, therefore, (T_0x_n) converges because F is a Banach space. Let (y_n) be any other sequence in E_0 with $\lim_{n\to\infty} y_n = x$. Then (T_0y_n) also converges. We shall prove that $\lim_{n\to\infty} T_0x_n = \lim_{n\to\infty} T_0y_n$. We have $\| T_0x_n - T_0y_n \| \leq \| T_0 \| \, \| x_n - y_n \|$ from which it follows that $\lim_{n\to\infty} \| T_0x_n - T_0y_n \| = 0$ because $\lim_{n\to\infty} \| x_n - y_n \| = 0$. Therefore, by Lemma 3.3.1,

$$\lim_{n\to\infty} T_0x_n = \lim_{n\to\infty} T_0y_n.$$

We can now define a mapping T of $\bar{E}_0$ into F without ambiguity by $Tx = \lim_{n\to\infty} T_0x_n$ where $x \in \bar{E}_0$ and (x_n) is any sequence in E_0 with $\lim_{n\to\infty} x_n = x$. We shall prove that T satisfies the conditions of the theorem.

First let $x \in E_0$. If $x_n = x$ for $n = 1, 2, \ldots$, then $\lim_{n\to\infty} x_n = x$ and, since $T_0x_n = T_0x$ for $n = 1, 2, \ldots$, we have $Tx = \lim_{n\to\infty} T_0x_n = T_0x$.

Next let $x, y \in \bar{E}_0$ and $\alpha, \beta \in \mathsf{K}$. If (x_n) and (y_n) are sequences in E_0 with $\lim_{n\to\infty} x_n = x$ and $\lim_{n\to\infty} y_n = y$ then, by Lemma 3.3.1,

$$\lim_{n\to\infty} (\alpha x_n + \beta y_n) = \alpha x + \beta y.$$

But $\alpha x_n + \beta y_n \in E_0$ and therefore, again by Lemma 3.3.1,

$$\begin{aligned} T(\alpha x + \beta y) &= \lim_{n\to\infty} T_0(\alpha x_n + \beta y_n) \\ &= \lim_{n\to\infty} (\alpha T_0x_n + \beta T_0y_n) \\ &= \alpha Tx + \beta Ty. \end{aligned}$$

This proves that T is linear.

† Notice that $\| T_0 \| = \sup \{ \| T_0x \| : x \in E_0 \text{ and } \| x \| \leq 1 \}$ while $\| T \| = \sup \{ \| Tx \| : x \in \bar{E}_0 \text{ and } \| x \| \leq 1 \}$.

Finally let $x \in \bar{E}_0$ with $\| x \| \leq 1$. If (x_n) is a sequence in E_0 with $\lim_{n\to\infty} x_n = x$ then, by Lemma 3.3.1, $\lim_{n\to\infty} \| x_n \| = \| x \| \leq 1$. Also, by (5), we have $\| T_0 x_n \| \leq \| T_0 \| \| x_n \|$ for $n = 1, 2, \ldots$, so, by Lemma 3.3.1 again,

$$\| Tx \| = \lim_{n\to\infty} \| T_0 x_n \| \leq \lim_{n\to\infty} \| T_0 \| \| x_n \| \leq \| T_0 \|.$$

This proves that T is bounded and $\| T \| \leq \| T_0 \|$. On the other hand

$$\begin{aligned} \| T_0 \| &= \sup \{ \| T_0 x \| : x \in E_0 \text{ and } \| x \| \leq 1 \} \\ &= \sup \{ \| Tx \| : x \in E_0 \text{ and } \| x \| \leq 1 \} \\ &\leq \sup \{ \| Tx \| : x \in \bar{E}_0 \text{ and } \| x \| \leq 1 \} \\ &= \| T \| \end{aligned}$$

so $\| T_0 \| = \| T \|$.

It remains only to prove that T is unique. This follows from Exercise 2.7(7), but we give a proof here for completeness. Suppose that T_1 is a bounded linear transformation of $\bar{E}_0$ into F such that $T_1 x = T_0 x$ for all $x \in E_0$. Let $x \in \bar{E}_0$ and let (x_n) be a sequence in E_0 with $\lim_{n\to\infty} x_n = x$. Since T_1 is continuous in $\bar{E}_0$ (Theorem 3.5.4) Theorem 2.7.7 gives $\lim_{n\to\infty} T_1 x_n = T_1 x$. By definition of T we have $Tx = \lim_{n\to\infty} T_0 x_n$ and by hypothesis $T_0 x_n = T_1 x_n$ for $n = 1, 2, \ldots$, consequently $Tx = T_1 x$. This proves that $T_1 = T$. □

The mapping T is an extension of the mapping T_0 (see Section 1.2); it is usual to say that T_0 has been *extended by continuity* to a bounded linear transformation T on $\bar{E}_0$.

EXERCISES

(1) Let T be a linear transformation of a normed linear space E into a normed linear space F with the property that the set

$$\{ \| Tx_n \| : n = 1, 2, \ldots \}$$

is bounded whenever (x_n) is a sequence convergent to 0 in E. Prove that T is a bounded linear transformation.

(2) Let M be a positive real number, let y_n, $n = 1, 2, \ldots$, be points in ℓ with $\| y_n \| \leq M$ for $n = 1, 2, \ldots$ and let e_n, $n = 1, 2, \ldots$, be the points of ℓ defined in Example 3.4.9. Prove that there is a unique bounded linear transformation T on ℓ with $Te_n = y_n$ for $n = 1, 2, \ldots$. Prove further that $Tx = \sum_{n=1}^{\infty} \xi_n y_n$ for $x = (\xi_n) \in \ell$.

3.6 Linear Homeomorphisms

In this short section we consider homeomorphisms, uniform equivalences and isometries that are also linear mappings.

3.6.1 Definition Let T be a one-to-one linear mapping of a normed linear space E onto a normed linear space F. The mapping T is said to be a *linear homeomorphism* if and only if T and T^{-1} are continuous on E and F, respectively. The mapping T is said to be a *linear isometry* if and only if $\| Tx \| = \| x \|$ for all $x \in E$. It is clear that a linear isometry is an isometry in the sense of Definition 2.9.1.

It is obvious that T is a linear homeomorphism of E onto F if and only if T^{-1} is a linear homeomorphism of F onto E. We shall say that two normed linear spaces E and F are *linearly homeomorphic* if and only if there exists a linear homeomorphism of E onto F. It is not difficult to see that the relation of being linearly homeomorphic is an equivalence relation on the class of all normed linear spaces. Analogously we shall say that two normed linear spaces E and F are *linearly isometric* if and only if there exists a linear isometry of E onto F. Two normed linear spaces that are linearly isometric are indistinguishable in so far as their properties as normed linear spaces are concerned.

Let T be a linear homeomorphism of a normed linear space E onto a normed linear space F. By Theorem 3.5.4 and Lemma 3.5.9 there exist $M > 0$ and $m > 0$ such that $m \leq \| Tx \| \leq M$ for all $x \in E$ with $\| x \| = 1$, and consequently

$$m\| x \| \leq \| Tx \| \leq M\| x \| \tag{1}$$

for all $x \in E$. It follows directly from (1) that T is a uniform equivalence (see Definition 2.9.1). Linearly homeomorphic normed linear spaces are often said to be *equivalent*. Lemma 2.9.4 shows that any normed linear space that is linearly homeomorphic to a Banach space is itself a Banach space.

3.6.2 Definition Two norms $\|.\|_1$ and $\|.\|_2$ on a linear space E are said to be *equivalent* if and only if the identity mapping $x \to x$ is a linear homeomorphism of the normed linear space $(E, \|.\|_1)$ onto the normed linear space $(E, \|.\|_2)$.

It follows from (1) that $\|.\|_1$ and $\|.\|_2$ are equivalent norms if and only if there exist $M > 0$ and $m > 0$ such that $m\| x \|_1 \leq \| x \|_2 \leq M\| x \|_1$ for all $x \in E$.

Exercises

(1) Prove that each normed linear space is linearly isometric to a dense linear subspace of a Banach space (cf. Exercises 2.9(9) and (10)).

(2) Verify that linear homeomorphism and linear isometry are equivalence relations.

(3) Let E and F be normed linear spaces over the same field and let T be a one-to-one linear mapping of E onto F with the property that (Tx_n) is a Cauchy sequence in F if and only if (x_n) is a Cauchy sequence in E. Prove that T is a linear homeomorphism.

(4) Use the results of Exercise 3.4(10) to show that no two of the real normed linear spaces ℓ, c, c_0 and $C_{\mathsf{R}}([0,1])$ are linearly isometric.

3.7 An Elementary Integral

In this section we shall develop the theory of an elementary integral. The reader will find an alternative treatment of this integral in the book of DIEUDONNÉ [7, §8.7]. The definition of the integral is a straightforward application of Theorem 3.5.10. The set of functions to which the theory applies includes all continuous functions but is smaller than the set of Riemann integrable functions. However, this integral is adequate for the purposes of this book and, in compensation for the restricted domain of definition, the development is simpler than that of the Riemann integral. Furthermore, the integral can be defined for mappings of an interval into a Banach space without introducing any new difficulties.

Throughout this section E will denote a real or complex Banach space and $I = [a,b]$ will denote a bounded closed interval of real numbers with $a < b$. We shall denote by $B_E(I)$ the set of all mappings f of I into E such that $\{\|f(t)\| : t \in I\}$ is a bounded set of real numbers. Given $f,g \in B_E(I)$ and $\alpha \in \mathsf{K}$, let $(f+g)(t) = f(t) + g(t)$ and $(\alpha f)(t) = \alpha f(t)$ for all $t \in I$, and

$$\|f\| = \sup\{\|f(t)\| : t \in I\}. \tag{1}$$

For all $t \in I$ we have

$$\|f(t) + g(t)\| \leq \|f(t)\| + \|g(t)\| \text{ and } \|\alpha f(t)\| = |\alpha| \, \|f(t)\|,$$

from which it follows that $f+g$, $\alpha f \in B_E(I)$ and that

$$\|f+g\| \leq \|f\| + \|g\| \quad \text{and} \quad \|\alpha f\| = |\alpha| \, \|f\|.$$

It is now easy to verify that $B_E(I)$ is a linear space and $f \to \|f\|$ is a norm on $B_E(I)$.

The following lemma is a direct generalization of Theorem 2.6.7.

3.7.1 LEMMA *$B_E(I)$ is a Banach space.*

PROOF Let (f_n) be a Cauchy sequence in $B_E(I)$ and let $\varepsilon > 0$. There is a positive integer N such that $\|f_m - f_n\| < \varepsilon$ for all $m,n \geq N$. By (1), for all $t \in I$ and all $m,n \geq N$ we have

$$\|f_m(t) - f_n(t)\| \leq \|f_m - f_n\| < \varepsilon. \tag{2}$$

This shows that $(f_n(t))$ is a Cauchy sequence in E. Since E is a Banach space $(f_n(t))$ converges. Let $f(t) = \lim_{n\to\infty} f_n(t)$ for all $t \in I$. Inequality (2) and Lemma 3.3.1 give

$$\| f(t) - f_n(t) \| = \lim_{m\to\infty} \| f_m(t) - f_n(t) \| \leq \varepsilon$$

for all $t \in I$ and $n \geq N$, which shows that $f - f_n \in B_E(I)$ and $\| f - f_n \| \leq \varepsilon$ for all $n \geq N$. But $f = (f - f_N) + f_N$ so $f \in B_E(I)$ and $\lim_{n\to\infty} f_n = f$. □

Our integral is defined initially on the class of all 'stepped-mappings'.

3.7.2 Definition A mapping f of I into E is called a *stepped-mapping* if and only if there is a finite subset A of I, containing a and b, and such that the restrictions of f to each of the open intervals (a_{k-1}, a_k), $k = 1, 2,..., n$, are constant, where $k \to a_k$ is the enumeration of A for which $a = a_0 < a_1 < \ldots < a_n = b$. Given a stepped-mapping f, any such set A is called an *f-partition*. When A is an f-partition and we write $A = \{a_0, a_1,..., a_n\}$ it will be understood that

$$a = a_0 < a_1 < \ldots < a_n = b.$$

A stepped mapping f has many different f-partitions; in particular any finite subset of I that contains an f-partition is obviously an f-partition. We shall denote the set of all stepped-mappings of I into E by $S_E(I)$. The term stepped-mapping is used here because, when $E = \mathsf{R}$, such a mapping is always called a *step-function*.

3.7.3 Lemma *The set $S_E(I)$ is a linear subspace of $B_E(I)$.*

Proof Let $f \in S_E(I)$ and let $\{a_0, a_1,..., a_n\}$ be an f-partition. If $y_1, y_2,..., y_n$ are such that $f(t) = y_k$ for $a_{k-1} < t < a_k$ and $k = 1, 2,..., n$ then

$$\{f(t) : t \in I\} = \{y_1, y_2,..., y_n, f(a_0), f(a_1),..., f(a_n)\}.$$

It follows that $f \in B_E(I)$. Thus $S_E(I) \subseteq B_E(I)$.

Now let $f,g \in S_E(I)$ and $\alpha,\beta \in \mathsf{K}$. If P and Q are f- and g-partitions, respectively, then obviously $P \cup Q$ is an $(\alpha f + \beta g)$-partition. Thus $\alpha f + \beta g \in S_E(I)$. This shows that $S_E(I)$ is a linear subspace of $B_E(I)$. □

Let $f \in S_E(I)$. With each f-partition P we associate a point $S(f;P)$ of E as follows. Let $P = \{a_0, a_1,..., a_n\}$ be an f-partition and let $y_1, y_2,..., y_n \in E$ be chosen as in the proof of Lemma 3.7.3. We define

$$S(f;P) = \sum_{k=1}^{n} (a_k - a_{k-1}) y_k.$$

When it is obvious which stepped-mapping f is being considered we shall write $S(P)$ instead of $S(f;P)$.

3.7.4 LEMMA *Let $f \in S_E(I)$. Then $S(P) = S(Q)$ for all f-partitions P and Q.*

PROOF Let $P = \{a_0, a_1, \ldots, a_n\}$ be an f-partition and let $y_1, y_2, \ldots, y_n \in E$ be chosen again as in the proof of Lemma 3.7.3. Let P' be an f-partition that can be obtained from P by adding one extra point, say

$$P' = \{a_0, a_1, \ldots, a_{r-1}, a', a_{r+1}, \ldots, a_n\}.$$

Then

$$\begin{aligned} S(P') &= \sum_{k=1}^{r} (a_k - a_{k-1}) y_k + (a' - a_r) y_{r+1} \\ &\qquad + (a_{r+1} - a') y_{r+1} + \sum_{k=r+2}^{n} (a_k - a_{k-1}) y_k \\ &= \sum_{k=1}^{n} (a_k - a_{k-1}) y_k \\ &= S(P). \end{aligned}$$

Now let Q be an f-partition with $P \subseteq Q$. There is a finite number of f-partitions, $P_0, P_1, \ldots, P_m$ say, such that $P = P_0 \subseteq P_1 \subseteq \ldots \subseteq P_m = Q$ and P_r can be obtained from P_{r-1} by adding one extra point, for $r = 1, 2, \ldots, m$. By what we have already proved

$$S(P) = S(P_0) = S(P_1) = \ldots = S(P_m) = S(Q).$$

Finally let Q be an arbitrary f-partition. Then $Q' = P \cup Q$ is an f-partition, $P \subseteq Q'$ and $Q \subseteq Q'$. Thus $S(P) = S(Q') = S(Q)$. □

3.7.5 DEFINITION Let $f \in S_E(I)$. We shall write

$$\int_a^b f = \int_a^b f(t)\, dt = S(f;P),$$

where P is any f-partition.

Lemma 3.7.4 shows that the definition of $\int_a^b f$ is free from ambiguity. The point $\int_a^b f$ of E is called the *integral* of f. For a step-function f it is clear that $\int_a^b f$ is just the area under the graph $y = f(x)$ between the lines $x = a$ and $x = b$.

3.7.6 LEMMA *The mapping $f \to \int_a^b f$ is a bounded linear transformation of $S_E(I)$ into E and*

$$\left\| \int_a^b f \right\| \leq (b - a) \, \| f \|$$

for all $f \in S_E(I)$.

PROOF Let $f,g \in S_E(I)$, $\alpha,\beta \in K$, let P be an f-partition and Q a g-partition. Let $P \cup Q = \{a_0, a_1,\ldots, a_n\}$. Then $P \cup Q$ is an $(\alpha f + \beta g)$-partition, an f-partition and a g-partition. Let $y_1, y_2,\ldots, y_n$ and $z_1, z_2,\ldots, z_n$ in E be such that $f(t) = y_k$ and $g(t) = z_k$ for $a_{k-1} < t < a_k$ and $k = 1, 2,\ldots, n$. Then

$$(\alpha f + \beta g)(t) = \alpha f(t) + \beta g(t) = \alpha y_k + \beta z_k$$

for $a_{k-1} < t < a_k$ and $k = 1, 2,\ldots, n$ and hence

$$\begin{aligned}\int_a^b (\alpha f + \beta g) &= \sum_{k=1}^{n} (a_k - a_{k-1})(\alpha y_k + \beta z_k) \\ &= \alpha \sum_{k=1}^{n} (a_k - a_{k-1}) y_k + \beta \sum_{k=1}^{n} (a_k - a_{k-1}) z_k \\ &= \alpha \int_a^b f + \beta \int_a^b g.\end{aligned}$$

This shows that the mapping $f \to \int_a^b f$ is linear. By definition of $\| f \|$ we have $\| y_k \| \leq \| f \|$ for $k = 1, 2,\ldots, n$ and therefore

$$\begin{aligned}\left\| \int_a^b f \right\| &= \left\| \sum_{k=1}^{n} (a_k - a_{k-1}) y_k \right\| \\ &\leq \sum_{k=1}^{n} (a_k - a_{k-1}) \| y_k \| \\ &\leq (b-a) \| f \|.\end{aligned}$$

This shows that the linear transformation $f \to \int_a^b f$ is bounded. □

3.7.7 THEOREM *There is a unique bounded linear transformation T of $S_E(I)^-$ into E such that*

$$Tf = \int_a^b f$$

for all $f \in S_E(I)$.

This linear transformation T also satisfies the condition

$$\| Tf \| \leq (b-a) \| f \|$$

for all $f \in S_E(I)^-$.

PROOF Let $T_0 f = \int_a^b f$ for all $f \in S_E(I)$. We saw in Lemma 3.7.6 that T_0 is a bounded linear transformation of $S_E(I)$ into E and consequently by Theorem 3.5.10 there is a unique bounded linear transformation T

of $S_E(I)^-$ into E such that $Tf = T_0 f$ for all $f \in S_E(I)$. Also $\| T \| = \| T_0 \|$. Lemma 3.7.6 shows that $\| T_0 \| \leq b-a$, and hence

$$\| Tf \| \leq \| T \| \, \| f \| \leq (b-a) \, \| f \|$$

for all $f \in S_E(I)^-$. □

3.7.8 Definition Let $f \in S_E(I)^-$. We shall write

$$\int_a^b f = \int_a^b f(t) \, dt = Tf,$$

where T is the bounded linear transformation of $S_E(I)^-$ into E whose existence was established in Theorem 3.7.7. The point $\int_a^b f$ of E is called the *integral* of f.

For mappings $f \in S_E(I)$ we have now two definitions of the integral $\int_a^b f$ (Definitions 3.7.5 and 3.7.8); Theorem 3.7.7 shows that these definitions coincide. Further $f \to \int_a^b f$ is a bounded linear transformation of $S_E(I)^-$ into E and

$$\left\| \int_a^b f \right\| \leq (b-a) \, \| f \| \tag{3}$$

for all $f \in S_E(I)^-$.

3.7.9 Lemma *Let (f_n) be a sequence in $S_E(I)^-$ with $\lim_{n\to\infty} f_n = f$. Then*

$$\lim_{n\to\infty} \int_a^b f_n = \int_a^b f.$$

Proof Theorem 2.5.6 shows that $f \in S_E(I)^-$. Since a bounded linear transformation is continuous (Theorem 3.5.4) Theorem 2.7.7 shows that $\lim_{n\to\infty} \int_a^b f_n = \int_a^b f$. □

We shall need the following three lemmas later. The first is a familiar property of integrals. In order to simplify our notation we shall adopt the following convention.

Let $a < c < b$ and let f be a mapping of $I = [a,b]$ into E. If $f \mid [a, c] \in S_E([a, c])$ then we shall write $f \in S_E([a, c])$. The expressions $f \in S_E([a,c])^-$ and $\int_a^c f$ will be used similarly.

3.7.10 Lemma *Let $f \in S_E(I)^-$ and let $a < c < b$. Then $f \in S_E([a,c])^-$, $f \in S_E([c,b])^-$. Also*

$$\int_a^b f = \int_a^c f + \int_c^b f.$$

PROOF Let $I_1 = [a,c]$ and $I_2 = [c,b]$. Suppose first that $f \in S_E(I)$ and let P be an f-partition of I. Let $P' = P \cup \{c\}$ and let $P_1 = P' \cap I_1$ and $P_2 = P' \cap I_2$. Obviously P_1 and P_2 are f-partitions of I_1 and I_2, respectively, so $f \in S_E(I_1)$ and $f \in S_E(I_2)$. It is also obvious that

$$\int_a^b f = S(P') = S(P_1) + S(P_2) = \int_a^c f + \int_c^b f.$$

This proves the lemma for stepped-mappings.

Suppose now that $f \in S_E(I)^-$. Let $\|.\|$, $\|.\|_1$ and $\|.\|_2$ denote the norms on $B_E(I)$, $B_E(I_1)$ and $B_E(I_2)$, respectively. Clearly $\| g \|_1 \leq \| g \|$ and $\| g \|_2 \leq \| g \|$ for all $g \in B_E(I)$. By Theorem 2.5.5 there is a sequence (f_n) in $S_E(I)$ with $\lim_{n\to\infty} \| f_n - f \| = 0$. By the first part of the proof $f_n \in S_E(I_1)$ and $f_n \in S_E(I_2)$. Also, by the remarks above, $\lim_{n\to\infty} \| f_n - f \|_1 = 0$ and $\lim_{n\to\infty} \| f_n - f \|_2 = 0$. This proves that $f \in S_E(I_1)^-$ and $f \in S_E(I_2)^-$. Lemma 3.7.9 gives

$$\int_a^b f = \lim_{n\to\infty} \int_a^b f_n, \quad \int_a^c f = \lim_{n\to\infty} \int_a^c f_n \quad \text{and} \quad \int_c^b f = \lim_{n\to\infty} \int_c^b f_n.$$

Now the first part of the proof and Lemma 3.3.1 give

$$\begin{aligned} \int_a^b f = \lim_{n\to\infty} \int_a^b f_n &= \lim_{n\to\infty} \left(\int_a^c f_n + \int_c^b f_n \right) \\ &= \lim_{n\to\infty} \int_a^c f_n + \lim_{n\to\infty} \int_c^b f_n = \int_a^c f + \int_c^b f. \ \square \end{aligned}$$

3.7.11 LEMMA *Let* $f \in S_E(I)^-$ *and let* $g(t) = \| f(t) \|$ *for all* $t \in I$. *Then* $g \in S_{\mathsf{R}}(I)^-$ *and*

$$\left\| \int_a^b f \right\| \leq \int_a^b g.$$

PROOF Suppose first that $f \in S_E(I)$. It is obvious that $g \in S_{\mathsf{R}}(I)$ and the last part of the proof of Lemma 3.7.6 shows that $\left\| \int_a^b f \right\| \leq \int_a^b g$. This proves the lemma for stepped-mappings.

Suppose now that $f \in S_E(I)^-$, and let (f_n) be a sequence in $S_E(I)$ with $\lim_{n\to\infty} f_n = f$. Let $g_n(t) = \| f_n(t) \|$. Then for all $t \in I$,

$$\begin{aligned} | g_n(t) - g(t) | &= | \| f_n(t) \| - \| f(t) \| | \\ &\leq \| f_n(t) - f(t) \| \\ &\leq \| f_n - f \|. \end{aligned}$$

This shows that $\| g_n - g \| \leq \| f_n - f \|$ and hence that $\lim_{n\to\infty} g_n = g$.

Thus $g \in S_R(I)^-$. By the first part of the proof

$$\left\| \int_a^b f_n \right\| \leq \int_a^b g_n$$

for $n = 1, 2, \ldots$ and therefore Lemmas 3.7.9 and 3.3.1 give

$$\left\| \int_a^b f \right\| = \lim_{n\to\infty} \left\| \int_a^b f_n \right\| \leq \lim_{n\to\infty} \int_a^b g_n = \int_a^b g. \square$$

Exercises

(1) Let (f_n) be a sequence in $S_E(I)^-$ such that the series $\sum_n f_n$ is convergent in $B_E(I)$. Prove that $\sum_{n=1}^{\infty} f_n$ is in $S_E(I)^-$ and that

$$\int_a^b \left(\sum_{n=1}^{\infty} f_n \right) = \sum_{n=1}^{\infty} \int_a^b f_n.$$

(2) Let $f \in S_E(I)^-$ and let T be a bounded linear transformation of E into a Banach space F. Prove that $T \circ f \in S_F(I)^-$ and

$$\int_a^b (T \circ f) = T\left(\int_a^b f \right).$$

(Hint: Consider first the case when $f \in S_E(I)$.) This result can be used to show that it is possible to change the order of integration in certain repeated integrals of functions of two variables (Exercise 4.3(13)).

(3) Let $f \in S_E(I)^-$ and $\varepsilon > 0$. Prove that there exists $\delta > 0$ with the following property: given any subset $\{x_0, x_1, \ldots, x_n\}$ of I with $a = x_0 < x_1 < \ldots < x_n = b$ and $x_k - x_{k-1} < \delta$ for $k = 1, 2, \ldots, n$ and given points $t_1, t_2, \ldots, t_n$ with $x_{k-1} < t_k < x_k$ for $k = 1, 2, \ldots, n$, then

$$\left\| \int_a^b f - \sum_{k=1}^{n} (x_k - x_{k-1}) f(t_k) \right\| < \varepsilon.$$

(Hint: Consider first the case when $f \in S_E(I)$.)

(4) *The Riemann Integral.* If f is a bounded real-valued function on $I = [a,b]$, the *upper integral* and the *lower integral* of f are defined by

$$\overline{\int} f = \inf \left\{ \int_a^b g : g \text{ is a step function and } f \leq g \right\},$$

$$\underline{\int} f = \sup \left\{ \int_a^b g : g \text{ is a step function and } g \leq f \right\}.$$

The function f is said to be *Riemann integrable* if and only if

$$\overline{\int} f = \underline{\int} f.$$

Prove that if $f \in S_R(I)^-$ then f is Riemann integrable and

$$\overline{\int} f = \underline{\int} f = \int_a^b f.$$

(5) The real-valued function f on $I = [0,1]$ is defined by

$$f(t) = \begin{cases} 1 & \text{if } \dfrac{1}{2n} < t \leq \dfrac{1}{2n-1} \text{ and } n = 1, 2,\ldots, \\[2ex] 0 & \text{if } t = 0 \text{ or } \dfrac{1}{2n+1} < t \leq \dfrac{1}{2n} \text{ and } n = 1, 2,\ldots. \end{cases}$$

Prove that f is Riemann integrable but that $f \notin S_R(I)^-$.

3.8 Regulated Mappings

The notation of this section is the same as that of Section 3.7. We are concerned with obtaining an alternative characterization of the mappings in $S_E(I)^-$.

3.8.1 DEFINITION Let f be a mapping of I into E and let $a \leq t < b$. A point $x \in E$ is said to be a *right-hand limit of f at t* if and only if for each $\varepsilon > 0$ there exists $\delta > 0$, with $t+\delta \leq b$, such that $\| f(s) - x \| < \varepsilon$ for all $t < s < t+\delta$.

Now let $a < t \leq b$. A point $x \in E$ is said to be a *left-hand limit of f at t* if and only if for each $\varepsilon > 0$ there exists $\delta > 0$, with $t-\delta \geq a$, such that $\| f(s) - x \| < \varepsilon$ for all $t-\delta < s < t$.

It is easy to see that f can have at most one right-hand limit at each point of $[a,b)$. If f has a right-hand limit at a point $t \in [a,b)$, we shall denote it by $f(t,+)$. Similarly if f has a left-hand limit at a point $t \in (a,b]$ we shall denote it by $f(t,-)$.

3.8.2 DEFINITION A mapping of I into E is said to be *regulated* if and only if it has a right-hand limit at each point of $[a,b)$ and a left-hand limit at each point of $(a,b]$.

It is obvious that a continuous mapping of I into E is regulated and it is easy to verify that a stepped-mapping of I into E is regulated. We shall prove that $S_E(I)^-$ is precisely the set of all regulated mappings of I into E.

3.8.3 LEMMA *A regulated mapping of I into E is in $B_E(I)$.*

PROOF Let f be a regulated mapping of I into E and let J be the set of all $c \in I$ such that the restriction of f to $[a,c]$ is in $B_E([a,c])$. Obviously

$a \in J$. Let $d = \sup J$. We shall prove that $d \in J$. This is obvious if $d = a$. Suppose $d > a$. There exists $\delta > 0$, such that

$$\| f(t) - f(d,-) \| < 1$$

for $d - \delta < t < d$, so

$$\| f(t) \| \leq \| f(t) - f(d,-) \| + \| f(d,-) \| < 1 + \| f(d,-) \| \qquad (1)$$

for $d - \delta < t < d$. There exists $c \in J$ with $d - \delta < c \leq d$ and then, by the definition of J, there is a real number M such that

$$\| f(t) \| \leq M \qquad (2)$$

for $a \leq t \leq c$. Inequalities (1) and (2) give

$$\| f(t) \| \leq \max \{M, 1 + \| f(d,-) \|, \| f(d) \| \} \qquad (3)$$

for $a \leq t \leq d$, which shows that $d \in J$.

Suppose now that $d < b$. Then there exists $\eta > 0$ such that $\| f(d,+) - f(t) \| < 1$ for $d < t < d + \eta$, so

$$\| f(t) \| \leq 1 + \| f(d,+) \| \qquad (4)$$

for $d < t \leq d + \frac{1}{2}\eta$. Inequalities (3) and (4) show that $d + \frac{1}{2}\eta \in J$, which contradicts the definition of d. Thus $d = b$. □

3.8.4 Lemma *Let f be a regulated mapping of I into E. Then $\| f(t,+) \| \leq \| f \|$ for $a \leq t < b$ and $\| f(t,-) \| \leq \| f \|$ for $a < t \leq b$.*

Proof Let $a \leq t < b$ and $\varepsilon > 0$. There exists $\delta > 0$ such that $\| f(t,+) - f(s) \| < \varepsilon$ for $t < s < t + \delta$ and thus

$$\| f(t,+) \| \leq \| f(t,+) - f(t + \tfrac{1}{2}\delta) \| + \| f(t + \tfrac{1}{2}\delta) \| < \varepsilon + \| f \|.$$

Since $\varepsilon > 0$ is arbitrary $\| f(t,+) \| \leq \| f \|$. There is a corresponding proof of the second inequality. □

3.8.5 Lemma *The set of all regulated mappings of I into E is a closed linear subspace of $B_E(I)$.*

Proof Let A be the set of all regulated mappings of I into E. By Lemma 3.8.3 we have $A \subseteq B_E(I)$. Let $f, g \in A$, $a \leq t < b$, and $\varepsilon > 0$. There exists $\delta > 0$ such that

$$\| f(t,+) - f(s) \| < \tfrac{1}{2}\varepsilon \quad \text{and} \quad \| g(t,+) - g(s) \| < \tfrac{1}{2}\varepsilon$$

for $t < s < t + \delta$. Thus, for $t < s < t + \delta$,

$$\begin{aligned} \| (f(t,+) + g(t,+)) - (f + g)(s) \| & \\ &= \| (f(t,+) - f(s)) + (g(t,+) - g(s)) \| \\ &\leq \| f(t,+) - f(s) \| + \| g(t,+) - g(s) \| \\ &< \varepsilon. \end{aligned}$$

This proves that $f+g$ has a right-hand limit at t and that

$$(f+g)(t,+) = f(t,+) + g(t,+).$$

Similarly $f+g$ has a left-hand limit at each point of $(a,b]$ and hence $f+g \in A$. It is easy to verify that $\alpha f \in A$ for all $\alpha \in \mathsf{K}$ so A is a linear subspace of $B_E(I)$.

Now let (f_n) be a sequence in A with $\lim_{n\to\infty} f_n = f$. Let $a \leq t < b$ and let m and n be positive integers. By what we have already proved $f_m - f_n \in A$ and $(f_m - f_n)(t,+) = f_m(t,+) - f_n(t,+)$ so Lemma 3.8.4 gives $\| f_m(t,+) - f_n(t,+) \| = \| (f_m - f_n)(t,+) \| \leq \| f_m - f_n \|$. But (f_n) is a Cauchy sequence so $(f_n(t,+))$ is a Cauchy sequence in E. Since E is a Banach space, $(f_n(t,+))$ converges. Let $x = \lim_{n\to\infty} f_n(t,+)$. Given $\varepsilon > 0$ there is an integer N_1 such that

$$\| f_n(s) - f(s) \| < \tfrac{1}{3}\varepsilon \tag{5}$$

for all $s \in I$ and all $n \geq N_1$, and there is an integer N_2 such that

$$\| f_n(t,+) - x \| < \tfrac{1}{3}\varepsilon \tag{6}$$

for all $n \geq N_2$. Let $N = \max\{N_1, N_2\}$. There exists $\delta > 0$ such that

$$\| f_N(t,+) - f_N(s) \| < \tfrac{1}{3}\varepsilon \tag{7}$$

for $t < s < t+\delta$. Inequalities (5), (6) and (7) give

$$\begin{aligned}&\| x - f(s) \| \\ &\quad \leq \| x - f_N(t,+) \| + \| f_N(t,+) - f_N(s) \| + \| f_N(s) - f(s) \| < \varepsilon\end{aligned}$$

for $t < s < t+\delta$. This shows that f has the right-hand limit x at t. Similarly f has a left-hand limit at each point of $(a,b]$ and thus $f \in A$. By Corollary 2.5.6 the set A is closed. □

3.8.6 THEOREM *The set $S_E(I)^-$ is precisely the set of all regulated mappings of I into E.*

PROOF Let A be the set of all regulated mappings of I into E. We have already seen that A is closed and $S_E(I) \subseteq A$. Thus $S_E(I)^- \subseteq A$.

Let $f \in A$. For $a \leq c \leq b$ let $\|.\|_c$ denote the norm on $B_E([a,c])$, that is $\| g \|_c = \sup\{\| g(t) \| : a \leq t \leq c\}$. Let $\varepsilon > 0$ and let J be the set of all $c \in I$ with the property that there exists $g_c \in S_E([a,c])$ such that† $\| g_c - f \|_c < \varepsilon$. It is obvious that $a \in J$. Let $d = \sup J$. We shall prove that $d \in J$. If $d = a$ then $d \in J$. Suppose that $d > a$. Then there exists $\delta > 0$ such that

$$\| f(d,-) - f(t) \| < \varepsilon \tag{8}$$

† Throughout this proof we shall use the notational convention explained before Lemma 3.7.10.

for $d-\delta < t < d$, and, by definition of d, there exists $c \in J$ with $d-\delta < c \leq d$. By definition of J there exists $g_c \in S_E([a,c])$ such that

$$\| g_c - f \|_c < \varepsilon. \tag{9}$$

If $c = d$ then $d \in J$. If $c < d$, let

$$g_d(t) = \begin{cases} g_c(t) & \text{for } a \leq t \leq c, \\ f(d,-) & \text{for } c < t < d, \\ f(d) & \text{for } t = d. \end{cases}$$

It is obvious that $g_d \in S_E([a,d])$ and it follows from (8) and (9) that $\| g_d - f \|_d < \varepsilon$. Thus $d \in J$.

Suppose now that $d < b$. Then there exists $\eta > 0$ such that

$$\| f(d,+) - f(t) \| < \varepsilon \tag{10}$$

for $d < t < d+\eta$, and, since $d \in J$, there exists $g_d \in S_E([a,d])$ such that

$$\| g_d - f \|_d < \varepsilon. \tag{11}$$

Let

$$g_{d+\frac{1}{2}\eta}(t) = \begin{cases} g_d(t) & \text{for } a \leq t \leq d, \\ f(d,+) & \text{for } d < t \leq d+\frac{1}{2}\eta. \end{cases}$$

It is obvious that $g_{d+\frac{1}{2}\eta} \in S_E([a,d+\frac{1}{2}\eta])$, and it follows from (10) and (11) that $\| g_{d+\frac{1}{2}\eta} - f \|_{d+\frac{1}{2}\eta} < \varepsilon$, which shows that $d + \frac{1}{2}\eta \in J$. This contradicts the definition of d, so $d = b$. It follows now from Theorem 2.3.4 that $f \in S_E(I)^-$ and hence $A \subseteq S_E(I)^-$. □

Finally we shall prove that a regulated function is 'almost' continuous.

3.8.7 THEOREM *Let f be a regulated mapping of I into E. Then there is a countable subset A of I such that f is continuous at each point of the set $I \sim A$.*

PROOF By Theorem 3.8.6 there is a sequence (f_n) in $S_E(I)$ with $\lim_{n\to\infty} f_n = f$. For each positive integer n let A_n be an f_n-partition. Then A_n is a finite subset of E and it is clear that f_n is continuous at each point of the set $I \sim A_n$. Let $A = \bigcup_{n=1}^{\infty} A_n$. Then A is countable (Theorem 1.4.5). Let $t \in I \sim A$ and $\varepsilon > 0$. There is a positive integer N such that

$$\| f_n - f \| < \tfrac{1}{3}\varepsilon \tag{12}$$

for all $n \geq N$. The mapping f_N is continuous at t, because $A_N \subseteq A$, so there exists $\delta > 0$ such that

$$\| f_N(s) - f_N(t) \| < \tfrac{1}{3}\varepsilon \tag{13}$$

for $|s - t| < \delta$. If $|s - t| < \delta$, inequalities (12) and (13) give

$$\begin{aligned} \| f(s) - f(t) \| &\leq \| f(s) - f_N(s) \| + \| f_N(s) - f_N(t) \| + \| f_N(t) - f(t) \| \\ &\leq 2 \| f - f_N \| + \| f_N(s) - f_N(t) \| \\ &< \varepsilon. \end{aligned}$$

This shows that f is continuous at t. □

EXERCISES

(1) Prove that a mapping f of $[a,b]$ into E is continuous if and only if it is regulated, $f(t,+) = f(t,-) = f(t)$ for $a < t < b$, $f(a,+) = f(a)$ and $f(b,-) = f(b)$.

(2) Prove that a monotonic real-valued function on I is regulated on I.

(3) Let f be a regulated mapping of I into E and let g be a continuous mapping of $f(I)^-$ into F. Prove that $g \circ f$ is a regulated mapping of I into F.

(4) Let

$$f(t) = \begin{cases} t \sin \dfrac{1}{t} & \text{for } 0 < t \leq 1, \\ 0 & \text{for } t = 0 \end{cases}$$

and

$$g(t) = \begin{cases} 1 & \text{for } t > 0, \\ 0 & \text{for } t = 0, \\ -1 & \text{for } t < 0. \end{cases}$$

Prove that f is continuous on $[0,1]$, that g is regulated on $[-1,1]$, but that $g \circ f$ is not regulated on $[0,1]$.

(5) Let $f \in S_E(I)^-$ and $\varepsilon > 0$. Prove that there is a continuous mapping g of I into E with

$$\int_a^b \| f(t) - g(t) \| \, dt < \varepsilon.$$

(Hint: Consider first the case when $f \in S_E(I)$.)

3.9 Integration and Differentiation

The notation of this section is the same as that of Section 3.7. In elementary calculus, integration is defined as the operation inverse to differentiation. We shall now define a concept of differentiation for mappings of I into E and show that the operations of differentiation and integration are inverse to each other.

3.9.1 DEFINITION Let f be a mapping of I into E and let $a \leq t < b$. The mapping f is said to be *differentiable from the right at* t if and only if there exists a point $x \in E$ with the following property: for each $\varepsilon > 0$ there exists $\delta > 0$, with $t+\delta \leq b$, such that

$$\left\| \frac{1}{s-t}(f(s) - f(t)) - x \right\| < \varepsilon \tag{1}$$

for $t < s < t+\delta$.

It is easy to verify that if f is differentiable from the right at t, then there is exactly one point $x \in E$ that has the property (1); we shall denote this point of E by $f'_+(t)$ and call it the *right-hand derivative of f at t.*

If $a < t \leq b$, *differentiability of f from the left at t* and the *left-hand derivative of f at t* are defined analogously. The left-hand derivative of f at t is denoted by $f'_-(t)$.

3.9.2 DEFINITION Let f be a mapping of I into E and let $a < t < b$. Then f is said to be *differentiable at t* if and only if f is differentiable from the right and from the left at t and $f'_+(t) = f'_-(t)$. If f is differentiable at t, we shall write $f'(t) = f'_+(t) = f'_-(t)$ and call $f'(t)$ the *derivative of f at t.*

The above definitions are straightforward generalizations of the usual definitions of differentiability from the right and left for real-valued functions of a real variable. In Chapter 7 we shall discuss differentiation in a more general context so we shall be content here to investigate the relationship between integration and differentiation.

The proofs of the following two lemmas are quite elementary and analogous to the proofs of the corresponding results for real-valued functions of a real variable. We leave the reader to supply the details (see Exercise (2)).

3.9.3 LEMMA *Let f and g be mappings of I into E. If f and g are differentiable from the right at a point $t \in [a,b)$ then so is $f+g$ and $(f+g)'_+(t) = f'_+(t) + g'_+(t)$. If f and g are differentiable from the left at a point $t \in (a,b]$ then so is $f+g$ and $(f+g)'_-(t) = f'_-(t) + g'_-(t)$.*

3.9.4 LEMMA *Let f be a mapping of I into E. If f is differentiable from the right at each point of $[a,b)$ and differentiable from the left at each point of $(a,b]$ then f is continuous on I.*

We shall also need two lemmas on integration, the first of which is an immediate consequence of the definition of the integral.

3.9.5 LEMMA *Let $y \in E$ and let $f(t) = y$ for $a \leq t \leq b$. Then f is regulated on I and* $\int_a^b f = (b-a)y$.

3.9.6 LEMMA *Let f be a regulated mapping of I into E. Then*

$$\left\| \int_a^b f \right\| \leq (b-a) \sup \{ \| f(t) \| : a < t < b \}.$$

PROOF Let

$$g(t) = \begin{cases} 0 & \text{for } t = a \text{ and } t = b, \\ f(t) & \text{for } a < t < b. \end{cases}$$

Then $g(t) - f(t) = 0$ for $a < t < b$ so $\{a,b\}$ is a $(g-f)$-partition of I, and $\int_a^b (g-f) = 0$. Since f is regulated and $g = (g-f) + f$, the mapping g is also regulated and

$$\int_a^b g = \int_a^b (g-f) + \int_a^b f = \int_a^b f.$$

Consequently inequality (3) of Section 3.7 gives

$$\begin{aligned} \left\| \int_a^b f \right\| &= \left\| \int_a^b g \right\| \\ &\leq (b-a) \sup \{ \| g(t) \| : a \leq t \leq b \} \\ &= (b-a) \sup \{ \| f(t) \| : a < t < b \}. \square \end{aligned}$$

Before stating our main theorem it is convenient slightly to extend the definition of the integral. Let f be a regulated mapping of I into E and let $a \leq u \leq v \leq b$. We shall write

$$\int_u^u f = 0 \text{ and } \int_v^u f = -\int_u^v f.$$

It follows directly from Lemma 3.7.10 that if $a \leq t_0 \leq b$ and $a \leq t \leq s \leq b$ then

$$\int_{t_0}^s f - \int_{t_0}^t f = \int_t^s f.$$

3.9.7 THEOREM *Let f be a regulated mapping of I into E, let $a \leq t_0 \leq b$ and let*

$$g(t) = \int_{t_0}^t f$$

for $a \leq t \leq b$. Then

(a) *g is continuous on I,*
(b) *g is differentiable from the right at each point $t \in [a,b)$ and $g'_+(t) = f(t,+)$,*
(c) *g is differentiable from the left at each point $t \in (a,b]$ and $g'_-(t) = f(t,-)$.*

Consequently g is differentiable at each point $t \in (a,b)$ at which f is continuous, and $g'(t) = f(t)$ for such points t.

PROOF By Lemma 3.9.4 assertion (a) follows from assertions (b) and (c). Let us prove (b). Let $a \leq t < s \leq b$. The remark immediately preceding the theorem and Lemma 3.9.5 give

$$\frac{1}{s-t}(g(s) - g(t)) - f(t,+) = \left(\frac{1}{s-t}\int_t^s f\right) - f(t,+)$$
$$= \frac{1}{s-t}\int_t^s (f - f(t,+)).$$

Let $\varepsilon > 0$. There exists $\delta > 0$ such that $\| f(u) - f(t,+) \| < \varepsilon$ for $t < u < t + \delta$ and therefore by Lemma 3.9.6 if $t < s < t + \delta$

$$\left\| \frac{1}{s-t}(g(s) - g(t)) - f(t,+) \right\| = \frac{1}{s-t}\left\| \int_t^s (f - f(t,+)) \right\|$$
$$\leq \sup \{ \| f(u) - f(t,+) \| : t < u < s\}$$
$$\leq \varepsilon.$$

This proves (b). The proof of (c) is similar. The final assertion of the theorem follows from (b), (c) and the remark that if f is continuous at a point $t \in (a,b)$ then $f(t,+) = f(t,-) = f(t)$. □

To complete the discussion of the relationship between integration and differentiation, we have to show that if g is any mapping of I into E that satisfies conditions (b) and (c) of Theorem 3.9.7 then $g(t) - g(t_0) = \int_{t_0}^t f$ for $a \leq t \leq b$ and $a \leq t_0 \leq b$. For real-valued functions this follows from the classical mean value theorem. The following lemma is a special case of a general mean value theorem proved in Section 7.3; we have given a separate proof to make this chapter independent of Chapter 7.

3.9.8 LEMMA *Let g be a mapping of I into E that is differentiable from the right at each point of $[a,b)$ and differentiable from the left at each point of $(a,b]$. Suppose that $g'_+(t) = 0$ for $a \leq t < b$ and $g'_-(t) = 0$ for $a < t \leq b$. Then $g(t) = g(a)$ for $a \leq t \leq b$.*

PROOF Let $\varepsilon > 0$ and

$$J = \{c \in I : \| g(t) - g(a) \| \leq \varepsilon(t - a) \text{ for } a \leq t \leq c\}.$$

Obviously $a \in J$. Let $d = \sup J$. We shall prove that $d \in J$. This is obvious if $d = a$, so suppose that $d > a$. Then for each number t with $a \leq t < d$ there exists a number $c_t \in J$ with $t < c_t$ and hence

$$\| g(t) - g(a) \| \leq \varepsilon(t - a). \tag{2}$$

By Lemma 3.9.4 the mapping g is continuous on I, so using (2) we obtain

$$\begin{aligned}\| g(d) - g(a) \| &= \lim_{n\to\infty} \left\| g\left(d - \frac{1}{n}\right) - g(a) \right\| \\ &\leq \lim_{n\to\infty} \varepsilon\left(d - \frac{1}{n} - a\right) \\ &= \varepsilon(d - a). \end{aligned} \tag{3}$$

The inequalities (2) and (3) show that $d \in J$.

Suppose now that $d < b$. Then there exists $\delta > 0$ such that

$$\left\| \frac{1}{t - d}(g(t) - g(d)) \right\| < \varepsilon$$

for $d < t < d + \delta$. Thus, if $d \leq t \leq d + \frac{1}{2}\delta$, we have

$$\| g(t) - g(d) \| \leq \varepsilon(t - d) \tag{4}$$

and from (3) and (4) we obtain

$$\begin{aligned}\| g(t) - g(a) \| &\leq \| g(t) - g(d) \| + \| g(d) - g(a) \| \\ &\leq \varepsilon(t - d) + \varepsilon(d - a) \\ &= \varepsilon(t - a). \end{aligned} \tag{5}$$

The inequalities (5) and (2) show that $d + \frac{1}{2}\delta \in J$, which contradicts the definition of d. Thus $d = b$ and hence $\| g(t) - g(a) \| \leq \varepsilon(t - a)$ for $a \leq t \leq b$. Since $\varepsilon > 0$ is arbitrary it follows that $g(t) = g(a)$ for $a \leq t \leq b$. □

3.9.9 Theorem *Let f be a regulated mapping of I into E and let g be a mapping of I into E that satisfies the conditions*

(a) *g is differentiable from the right at each point $t \in [a,b)$ and $g'_+(t) = f(t,+)$,*
(b) *g is differentiable from the left at each point $t \in (a,b]$ and $g'_-(t) = f(t,-)$.*

Then

$$g(t) = g(t_0) + \int_{t_0}^{t} f$$

for $a \leq t \leq b$ and $a \leq t_0 \leq b$.

Proof Let $a \leq t_0 \leq b$ and let $h(t) = \int_{t_0}^{t} f$ for $a \leq t \leq b$. It follows from Theorem 3.9.7 and Lemma 3.9.3 that $(g - h)'_+(t) = 0$ for $a \leq t < b$ and $(g - h)'_-(t) = 0$ for $a < t \leq b$. Therefore Lemma 3.9.8 gives

$$g(t) - h(t) = (g - h)(t) = (g - h)(t_0) = g(t_0)$$

for $a \leq t \leq b$. □

EXERCISES

(1) (a) Prove that

$$\frac{1}{\alpha}(\| x + \alpha y \| - \| x \|) \leq \frac{1}{\beta}(\| x + \beta y \| - \| x \|)$$

for all $x,y \in E$ and all real numbers α,β with $0 < \alpha \leq \beta$.

(b) Let f be a mapping of I into E that is differentiable from the right at a point $t \in [a,b)$. Let $g(s) = \| f(s) \|$ for all $s \in I$. Prove that g is differentiable from the right at t and that $| g'_+(t) | \leq \| f'_+(t) \|$.

(2) Let f be a mapping of I into E. Prove that if f is differentiable from the right at a point $t \in [a,b)$ then f has a right-hand limit at t and $f(t,+) = f(t)$. Hence prove Lemma 3.9.4.

Exercises (3)–(7) *concern generalizations of classical formulae for a change of variable in an integral and for integration by parts. In the exercises 'f is differentiable on I' means that f is differentiable on* (a,b), *differentiable from the right at a, and differentiable from the left at b.*

(3) Let ϕ and f be regulated mappings of I into K and E, respectively. Prove that the mapping $t \to \phi(t)f(t)$ is regulated on I.

(4) Let ϕ be a mapping of I into R that is differentiable from the right at a point $t \in [a,b)$, let J be an open interval with $\phi(I) \subseteq J$, and let f be a mapping of J into E that is differentiable at the point $\phi(t)$. Prove that the mapping $f \circ \phi$ is differentiable from the right at t and that

$$(f \circ \phi)'_+(t) = \phi'_+(t)f'(\phi(t)).$$

(5) Let ϕ be a mapping of I into R that is differentiable on I and such that the mapping $t \to \phi'(t)$ is regulated on I, let J be a bounded interval with $\phi(I) \subseteq J$ and let f be a continuous mapping of J into E. Prove that the mapping $t \to \phi'(t)f(\phi(t))$ is regulated on I and that

$$\int_a^b \phi'(t)f(\phi(t))\,dt = \int_{\phi(a)}^{\phi(b)} f(t)\,dt.$$

(6) Let ϕ and f be mappings of I into K and E respectively that are differentiable from the right at a point $t \in [a,b)$. Prove that the mapping $t \to \phi(t)f(t)$ is differentiable from the right at t.

(7) Let ϕ and f be mappings of I into K and E, respectively, that are differentiable on I and such that the mappings $t \to \phi'(t)$ and $t \to f'(t)$ are regulated on I. Prove that the mappings $t \to \phi(t)f'(t)$ and $t \to \phi'(t)f(t)$ are regulated on I and that

$$\int_a^b \phi(t)f'(t)\,dt = \phi(b)f(b) - \phi(a)f(a) - \int_a^b \phi'(t)f(t)\,dt.$$

Exercises (8)–(12) *indicate how Picard's theorem* (*Theorem* 2.11.9) *can be extended to systems of first order differential equations and to higher order equations. In these Exercises the norm taken on* R^n *should be that corresponding to the metric* d_0 *of Example* 2.1.6. (*See p.* 90.)

(8) Let $f_1, f_2,\ldots, f_n$ be real-valued functions on I and let f be the mapping of I into R^n defined by $f(t) = (f_1(t), f_2(t),\ldots, f_n(t))$ for all $t \in I$.

(a) Prove that f is continuous on I if and only if $f_1, f_2,\ldots, f_n$ are continuous on I.

(b) Let T_i be the mapping $(x_1, x_2,\ldots, x_n) \to x_i$ of R^n into R. Prove, by applying the result of Exercise 3.7(2) to the mappings T_i, that if f is continuous on I then

$$\int_a^b f = \left(\int_a^b f_1, \int_a^b f_2,\ldots, \int_a^b f_n\right).$$

(c) Prove that f is differentiable on I if and only if $f_1, f_2,\ldots, f_n$ are differentiable on I. Prove also that, if f is differentiable on I, then $f'(t) = (f_1'(t), f_2'(t),\ldots, f_n'(t))$ for all $t \in I$.

(9) Let $f_1, f_2,\ldots, f_n$ be continuous real-valued functions on an open subset D of R^{n+1} and suppose that there exists $M > 0$ with

$$| f_j(x, y_1,\ldots, y_n) - f_j(x, z_1,\ldots, z_n) | \\ \leq M \max \{ | y_1 - z_1 |, | y_2 - z_2 |,\ldots, | y_n - z_n | \}$$

for all points $(x, y_1,\ldots, y_n)$ and $(x, z_1,\ldots, z_n)$ in D and for $j = 1, 2,\ldots, n$. Given a point $(x_0, y_1,\ldots, y_n) \in D$, prove that there exist $\alpha > 0$ and unique real valued functions $\phi_1, \phi_2,\ldots, \phi_n$ on the interval $[x_0 - \alpha, x_0 + \alpha]$ such that

(a) $(x, \phi_1(x),\ldots, \phi_n(x)) \in D$ for $| x - x_0 | \leq \alpha$,

(b) $\phi_j(x_0) = y_j$ for $j = 1, 2,\ldots, n$,

(c) $\phi_1, \phi_2,\ldots, \phi_n$ are differentiable on $[x_0 - \alpha, x_0 + \alpha]$ and

$$\phi_j'(x) = f_j(x, \phi_1(x),\ldots, \phi_n(x))$$

for $| x - x_0 | \leq \alpha$ and $j = 1, 2,\ldots, n$. (Hint: If $x \in \mathsf{R}$ and $z = (z_1,\ldots, z_n) \in \mathsf{R}^n$ are such that $(x, z_1,\ldots, z_n) \in D$, set $f(x,z) = (f_1(x, z_1,\ldots, z_n),\ldots, f_n(x, z_1, \ldots, z_n))$. Adapt the proofs of Lemma 2.11.7 and Theorem 2.11.8 to show that for some $\alpha > 0$ there exists a unique continuous mapping ϕ of $[x_0 - \alpha, x_0 + \alpha]$ into R^n that satisfies

$$\phi(x) = y + \int_{x_0}^{x} f(t,\phi(t))\, dt$$

for $| x - x_0 | \leq \alpha$, where $y = (y_1, y_2,\ldots, y_n)$.)

(10) Let f be a continuous real-valued function on an open set D in R^{n+1} and suppose that there exists $M > 0$ with

$$| f(x, y_1,\ldots, y_n) - f(x, z_1,\ldots, z_n) | \leq M \max\{ | y_1 - z_1 |,\ldots, | y_n - z_n |\}$$

for all points $(x, y_1,..., y_n)$ and $(x, z_1,..., z_n)$ in D. Given a point $(x_0, y_1,..., y_n) \in D$, prove that for some $\alpha > 0$ there exists a unique real-valued function ϕ on $[x_0 - \alpha, x_0 + \alpha]$ such that

(a) ϕ is n-times differentiable on $[x_0 - \alpha, x_0 + \alpha]$,

(b) $(x, \phi(x), \phi'(x),..., \phi^{(n-1)}(x)) \in D$ for $|x - x_0| \leq \alpha$,

(c) $\phi(x_0) = y_1, \phi'(x_0) = y_2,..., \phi^{(n-1)}(x_0) = y_n$,

(d) $\phi^{(n)}(x) = f(x, \phi(x), \phi'(x),..., \phi^{(n-1)}(x))$ for $|x - x_0| \leq \alpha$. (Hint: Reduce the nth order equation to a system of n first order equations.)

(11) Let a_{jk}, $j,k = 1, 2,..., n$, be continuous real-valued functions on an interval $[a,b]$, let $a < x_0 < b$ and let $y_1, y_2,..., y_n$ be real numbers. Prove that for some $\alpha > 0$ there exist unique real-valued functions $\phi_1, \phi_2,..., \phi_n$ on $[x_0 - \alpha, x_0 + \alpha]$ such that

(a) $\phi_j(x_0) = y_j$ for $j = 1, 2,..., n$,

(b) $\phi_1, \phi_2,..., \phi_n$ are differentiable on $[x_0 - \alpha, x_0 + \alpha]$ and

$$\phi_j'(x) = \sum_{k=1}^{n} a_{jk}(x)\, \phi_k(x)$$

for $|x - x_0| \leq \alpha$ and $j = 1, 2,..., n$. (Hint: Show that the conditions of Exercise (9) are satisfied.)

(12) Let $a_1, a_2,..., a_n$ be continuous real-valued functions on an interval $[a,b]$, let $a < x_0 < b$, and let $y_1, y_2,..., y_n$ be real numbers. Prove that there exists $\alpha > 0$ and a unique real-valued function ϕ on $[x_0 - \alpha, x_0 + \alpha]$ such that

(a) ϕ is n-times differentiable on $[x_0 - \alpha, x_0 + \alpha]$,

(b) $\phi(x_0) = y_1, \phi'(x_0) = y_2,..., \phi^{(n-1)}(x_0) = y_n$,

(c) $\phi^{(n)}(x) + a_1(x)\phi^{(n-1)}(x) + \ldots + a_n(x)\phi(x) = 0$ for $|x - x_0| \leq \alpha$.

CHAPTER 4

Compact Metric Spaces

Many concepts in the theory of metric spaces are abstractions of important properties of the real number system; for instance the definition of completeness is motivated by Cauchy's general principle of convergence (Theorem 1.6.5). There are several notions of compactness, each of which is motivated by a property of the real number system. The most important of these notions is obtained by abstraction from Borel's theorem (1.6.4) and is called simply 'compactness'. Our aim in this chapter is to study 'compactness' in metric spaces.

The definition of a compact metric space (Definition 4.1.2) is difficult to grasp intuitively, and it is often very hard to decide whether or not a given metric space is compact. In Section 4.2 we shall obtain several alternative characterizations of compactness which are sometimes more easily verified. In particular we shall show that the notion of compactness is equivalent to the notion of 'sequential compactness'. The definition of 'sequential compactness' (Definition 4.2.1) is inspired by the Bolzano–Weierstrass theorem (1.6.3); it is more elementary and closer in spirit to classical analysis than is the definition of 'compactness'. The concept of sequential compactness is due to M. Fréchet, and that of compactness to the Russian mathematicians P. Alexandrov and P. Urysohn.

The simplest and most familiar examples of compact metric spaces are the bounded closed intervals in the real field. We shall see in Section 4.3 that there are important properties of continuous real-valued functions on a bounded closed interval which generalize to properties of continuous real-valued functions on a compact metric space.

There are many applications of compactness in analysis. In this chapter we use compactness to obtain results about finite-dimensional normed linear spaces, and we prove two results that form the basis for many further applications: the first is M. H. Stone's generalization of the classical Weierstrass polynomial approximation theorem, and the second is the theorem of G. Ascoli and C. Arzelà, on compactness in certain spaces of continuous functions.

4.1 Definition and Elementary Consequences

The following two definitions are motivated by Borel's theorem (1.6.4).

4.1.1 DEFINITION Let Y be a non-empty subset of a metric space X. A family $\mathscr{A}$ of subsets of X is said to be a *cover of* Y (or *to cover* Y) if and only if $Y \subseteq \bigcup_{A \in \mathscr{A}} A$. A subfamily of a cover of Y which is also a cover of Y is called a *subcover of* Y. A finite family of subsets of X which is a cover of Y is called a *finite cover of* Y. A cover of Y each member of which is an open subset of X is called an *open cover of* Y.

4.1.2 DEFINITION A metric space X is said to be *compact* if and only if each open cover of X has a finite subcover. A non-empty subset Y of a metric space is said to be *compact* if and only if the subspace Y is a compact metric space; and it is said to be *relatively compact* if and only if $\bar{Y}$, the closure of Y, is compact.

Let Y be a non-empty subset of a metric space X. By definition Y is compact if and only if each cover of Y, by subsets of Y that are open in Y, has a finite subcover. It follows at once from Theorem 2.4.2 that Y is compact if and only if each open cover of Y has a finite subcover (recall that an open cover of Y is a cover of Y by open subsets of X). This is the most useful criterion for compactness of subsets, and we shall use it below without comment.

It is an easy consequence of Borel's theorem that a non-empty bounded closed subset of R is compact. This criterion for compactness does not extend to an arbitrary metric space (see Exercise (2)); we shall see in Theorem 4.2.11 that it extends to the spaces R^n and C^n. However a compact subset of a metric space must always be bounded and closed.

4.1.3 DEFINITION A non-empty subset A of a metric space X is said to be *bounded* if and only if the set $\{d(x,y) : x,y \in A\}$ of real numbers is bounded. A sequence (x_n) in X is said to be *bounded* if and only if its range, $\{x_n : n = 1, 2, \ldots\}$, is a bounded subset of X.

It is easy to see that A is bounded if and only if there is a point $x_0 \in X$ and a real number $r > 0$ such that $A \subseteq B(x_0,r)$, and it is then clear that when $X = \mathsf{R}$ the concept of boundedness just introduced coincides with the usual concept of boundedness in R.

4.1.4 LEMMA *A compact subset of a metric space is bounded and closed.*

PROOF Let A be a compact subset of a metric space X. Let x_0 be any point of A. The family of open balls $\{B(x_0,n): n = 1, 2,...\}$ is an open cover of A and hence it has a finite subcover of A. If N is the radius of the largest ball in this finite subcover then obviously $A \subseteq B(x_0,N)$. This shows that A is bounded.

Now let $x_0 \in X \sim A$. The family of sets

$$\{X \sim B'(x_0,1/n): n = 1, 2,...\}$$

is an open cover of A and hence it has a finite subcover. If $1/N$ is the radius of the smallest closed ball associated with this finite subcover, then obviously $A \cap B(x_0, 1/N) \subseteq A \cap B'(x_0, 1/N) = \varnothing$. This proves that $x_0 \notin \bar{A}$ (Theorem 2.3.4) and hence that A is closed. □

It follows from Lemma 4.1.4 that a compact subset of a metric space is also relatively compact.

4.1.5 LEMMA *A non-empty closed subset of a compact metric space is compact.*

PROOF Let A be a non-empty closed subset of a compact metric space X and let $\mathscr{B}$ be an open covering of A. The family $\mathscr{C} = \mathscr{B} \cup \{X \sim A\}$ is an open covering of X and therefore has a finite subfamily, $\mathscr{C}_0$ say, which also covers X. Clearly, if $\mathscr{B}_0 = \mathscr{C}_0 \sim \{X \sim A\}$ then $\mathscr{B}_0$ is a finite subfamily of $\mathscr{B}$, and covers A. This proves that A is compact. □

It follows from Lemma 4.1.5 that a non-empty subset of a compact metric space is relatively compact.

4.1.6 LEMMA *A compact metric space is separable.*

PROOF Let X be a compact metric space. For each positive integer n the family $\{B(x,1/n): x \in X\}$ is an open cover of X so it has a finite subcover, $\{B(x_{1n},1/n), B(x_{2n},1/n),..., B(x_{m_n n},1/n)\}$ say. Let

$$A = \{x_{kn}: k = 1, 2,..., m_n, n = 1, 2,...\}.$$

By Theorem 1.4.5 the set A is countable. Let $B(x,r)$ be any non-empty open ball in X and choose a positive integer n so that $1/n < r$. Since $x \in B(x_{kn},1/n)$ for some integer k with $1 \le k \le m_n$ we have

$$x_{kn} \in B(x,1/n) \subseteq B(x,r).$$

This proves that A is dense in X. □

EXERCISES

(1) Prove that a bounded closed subset of R is compact. (See Theorem 1.6.4.)

(2) Give an example to show that it is possible for a metric space to be bounded (in itself) but not compact. (See Exercise 2.9(8).)

(3) Let A be a non-empty subset of R. Prove that if every family of open intervals which covers A has a finite subfamily which also covers A then A is compact.

(4) Prove that if $A_1, A_2, \ldots, A_n$ are compact subsets of a metric space then $\bigcup_{k=1}^{n} A_k$ is compact.

(5) Determine which discrete metric spaces are compact.

(6) Prove that a Cauchy sequence in a metric space is bounded.

(7) A sequence (x_n) in a metric space X is convergent and $x = \lim_{n\to\infty} x_n$. Prove that $\{x\} \cup \{x_n : n = 1, 2, \ldots\}$ is a compact subset of X.

(8) Let f be a mapping of a metric space X into a metric space. Prove that if $f|A$ is continuous on A for all compact subsets A of X then f is continuous on X.

4.2 Conditions Equivalent to Compactness

Our first definition is obviously motivated by the Bolzano–Weierstrass theorem (1.6.3).

4.2.1 DEFINITION A metric space X is said to be *sequentially compact* if and only if each sequence in X has a convergent subsequence. A non-empty subset Y of a metric space is said to be *sequentially compact* if and only if the subspace Y is a sequentially compact metric space; and it is said to be *relatively sequentially compact* if and only if $\bar{Y}$ is sequentially compact.

Lemma 2.5.8 shows that X is sequentially compact if and only if each sequence in X has a cluster point. Lemma 2.5.10 shows that a non-empty subset Y of X is sequentially compact if and only if each sequence in Y has a subsequence which converges in X to a point of Y. It is not difficult to show that a non-empty subset Y of X is relatively sequentially compact if and only if each sequence in Y has a subsequence which converges in X (to a point that need not be in Y).

The Bolzano–Weierstrass theorem shows that a non-empty bounded closed subset of R is sequentially compact.

Our object is to prove that a metric space is compact if and only if it is sequentially compact. The proof is rather long, so we shall break it up into several steps, and in the course of the proof we shall obtain two further useful characterizations of compactness. The first is a simple consequence of the De Morgan formulae.

4.2.2 Definition A family $\mathscr{A}$ of subsets of a given set is said to have the *finite intersection property* if and only if each finite subfamily of $\mathscr{A}$ has non-empty intersection.

4.2.3 Lemma *A metric space X is compact if and only if each family of closed subsets of X with the finite intersection property has non-empty intersection.*

Proof For any family $\mathscr{A}$ of subsets of X we have $\bigcup_{A \in \mathscr{A}} A = X$ if and only if $\bigcap_{A \in \mathscr{A}} (X \sim A) = \emptyset$. Consequently X is compact if and only if each family of closed subsets of X with empty intersection has a finite subfamily which also has empty intersection. The last statement is obviously equivalent to the statement of the lemma. □

It is not difficult now to show that compactness implies sequential compactness.

4.2.4 Lemma *A compact metric space is sequentially compact.*

Proof Let (x_n) be a sequence in a compact metric space X and let $E_n = \{x_m : m = n, n+1, \ldots\}$. If $\{n_1, n_2, \ldots, n_k\}$ is any finite set of positive integers then

$$\bigcap_{j=1}^{k} \bar{E}_{n_j} \supseteq E_N \neq \emptyset,$$

where $N = \max \{n_1, n_2, \ldots, n_k\}$, and hence the family of closed sets $\{\bar{E}_n : n = 1, 2, \ldots\}$ has the finite intersection property. By Lemma 4.2.3 we have $\bigcap_{n=1}^{\infty} \bar{E}_n \neq \emptyset$, and therefore by Lemma 2.5.9 the sequence (x_n) has at least one cluster point. This proves that X is sequentially compact. □

It is rather more difficult to show that a sequentially compact metric space is compact, and in the course of the proof we shall obtain yet another characterization of compactness.

4.2.5 Definition Let X be a metric space and let $\varepsilon > 0$. A non-empty subset A of X is said to be an *ε-net in X* if and only if the family of open balls $\{B(x,\varepsilon) : x \in A\}$ covers X. A finite subset of X that is an ε-net in X is called a *finite ε-net in X*. The metric space X is said to be *totally bounded* if and only if there is a finite ε-net in X for each $\varepsilon > 0$. A non-empty subset of Y of X is said to be *totally bounded* if and only if the subspace Y is a totally bounded metric space.

A totally bounded metric space is sometimes said to be *precompact*. (See Exercise (7).) It is obvious that a totally bounded set is bounded. In terms of coverings we can rephrase the definition of total boundedness as follows: a metric space X is totally bounded if and only if for each $\varepsilon > 0$ there is a finite cover of X consisting of open balls of radius ε. It is now obvious from Definition 4.1.2 that a compact metric space is totally bounded. It is not so obvious that a sequentially compact metric space is totally bounded. That this is in fact so is the next step in our proof of the equivalence of compactness and sequential compactness. For technical reasons it is convenient to prove a slightly more general result.

4.2.6 Lemma *A relatively sequentially compact subset of a metric space is totally bounded.*

Proof Let Y be a relatively sequentially compact subset of a metric space X and suppose that Y is not totally bounded. Then for some $\varepsilon > 0$ there is no finite ε-net in Y. We shall define inductively a sequence (x_n) in Y that has no convergent subsequence. Let x_1 be any point of Y and suppose that points $x_2, x_3, \ldots, x_n$ in Y have been chosen so that $d(x_j,x_k) \geq \varepsilon$ for $1 \leq j < k \leq n$. Since there is no finite ε-net in Y, we have $\bigcup_{m=1}^{n} B_Y(x_m,\varepsilon) \neq Y$, where

$$B_Y(x_m,\varepsilon) = \{y \in Y: d(x_m,y) < \varepsilon\}.$$

Choose a point $x_{n+1} \in Y \sim \bigcup_{m=1}^{n} B_Y(x_m,\varepsilon)$. Clearly $d(x_{n+1},x_m) \geq \varepsilon$ for $m = 1, 2, \ldots, n$. This completes the definition of the sequence (x_n). It is obvious that no subsequence of (x_n) can be a Cauchy sequence, and therefore by Lemma 2.6.2 no subsequence of (x_n) can converge. Since (x_n) is a sequence in Y and Y is relatively sequentially compact, we have a contradiction. □

We need two more lemmas.

4.2.7 Lemma *A sequentially compact metric space is complete.*

Proof Let X be a sequentially compact metric space. Each sequence in X, and in particular each Cauchy sequence in X, has a convergent subsequence, therefore by Lemma 2.6.9 each Cauchy sequence in X converges. □

4.2.8 LEMMA *A non-empty subset of a totally bounded set is totally bounded.*

PROOF Let Y be a non-empty subset of a totally bounded set X and let $\varepsilon > 0$. There is a finite $\frac{1}{2}\varepsilon$-net in X, say $\{x_1, x_2, \ldots, x_n\}$. Since Y is non-empty $Y \cap B(x_j, \frac{1}{2}\varepsilon) \neq \emptyset$ for at least one integer j with $1 \leq j \leq n$. We may, by reindexing the points x_j if necessary, suppose that $Y \cap B(x_j, \frac{1}{2}\varepsilon) \neq \emptyset$ if and only if $j = 1, 2, \ldots, m$. Choose $y_j \in Y \cap B(x_j, \frac{1}{2}\varepsilon)$ for $j = 1, 2, \ldots, m$. Then if $y \in Y$ we have $y \in B(x_j, \frac{1}{2}\varepsilon)$ for some j with $1 \leq j \leq m$ and hence

$$d(y,y_j) \leq d(y,x_j) + d(x_j,y_j) < \varepsilon.$$

This shows that $\{y_1, y_2, \ldots, y_m\}$ is a finite ε-net in Y. □

The equivalence of (b) and (c) in the following theorem was first proved by M. Fréchet.

4.2.9 THEOREM *Let X be a metric space. The following three conditions are all equivalent:*

(a) *X is compact,*

(b) *X is sequentially compact,*

(c) *X is totally bounded and complete.*

PROOF We have already seen in Lemma 4.2.4 that condition (a) implies condition (b), and in Lemmas 4.2.7 and 4.2.8 that condition (b) implies condition (c). It remains therefore to show that condition (c) implies condition (a).

Suppose on the contrary that X is totally bounded and complete but not compact. We shall obtain a contradiction by a construction analogous to the process of repeated bisection that was used in the proof of Borel's theorem. The possibility of 'repeated subdivision' in our general context is a consequence of the hypothesis of total boundedness.

Since X is not compact there is an open cover, $\mathscr{A}$ say, of X which has no finite subcover of X. We shall define inductively a sequence (x_n) in X with the following properties: for $n = 1, 2, \ldots$

(1) $d(x_n,x_{n+1}) \leq 2^{-n+1}$,

(2) the open ball $B(x_n,2^{-n+1})$ cannot be covered by a finite subfamily of $\mathscr{A}$.

By hypothesis there is a finite 1-net in X, say $\{y_1, y_2, \ldots, y_m\}$. Since $X \subseteq \bigcup_{j=1}^{m} B(y_j,1)$ there is at least one integer j, with $1 \leq j \leq m$, such that $B(y_j,1)$ cannot be covered by a finite subfamily of $\mathscr{A}$; let $x_1 = y_j$ where j is the least such integer. Then x_1 satisfies condition (2) (condition (1) is vacuously satisfied).

Suppose that the points $x_2, x_3, \ldots, x_n$ in X have been chosen to satisfy conditions (1) and (2). By Lemma 4.2.8 the open ball $B(x_n, 2^{-n+1})$ is totally bounded and hence there is a finite 2^{-n}-net in $B(x_n, 2^{-n+1})$, say $\{y_1, y_2, \ldots, y_m\}$. Since $B(x_n, 2^{-n+1})$ satisfies condition (2) there is at least one integer j, with $1 \leq j \leq m$, such that $B(y_j, 2^{-n})$ cannot be covered by a finite subfamily of $\mathscr{A}$; let $x_{n+1} = y_j$ where j is the least such integer. Since $x_{n+1} \in B(x_n, 2^{-n+1})$ condition (1) is satisfied; condition (2) is also satisfied. This completes the definition of the sequence (x_n).

If m and n are positive integers with $m > n$ then

$$\begin{aligned} d(x_m, x_n) &\leq d(x_m, x_{m-1}) + d(x_{m-1}, x_{m-2}) + \ldots + d(x_{n+1}, x_n) \\ &\leq \frac{1}{2^{m-2}} + \frac{1}{2^{m-3}} + \ldots + \frac{1}{2^{n-1}} \\ &\leq \frac{1}{2^{n-2}}. \end{aligned}$$

Consequently (x_n) is a Cauchy sequence and converges because, by hypothesis, X is complete. Let $x = \lim_{n \to \infty} x_n$. Since $\mathscr{A}$ covers X we have $x \in A$ for some $A \in \mathscr{A}$ and since A is open $B(x,r) \subseteq A$ for some $r > 0$. There is an integer N such that $d(x_n, x) < \frac{1}{2}r$ and $2^{-n+1} < \frac{1}{2}r$ for $n \geq N$. If $y \in B(x_N, 2^{-N+1})$ then

$$d(y,x) \leq d(y, x_N) + d(x_N, x) < 2^{-N+1} + \tfrac{1}{2}r < r.$$

Thus $B(x_N, 2^{-N+1}) \subseteq B(x,r) \subseteq A$ which contradicts condition (2). This proves that condition (c) implies condition (a). □

4.2.10 COROLLARY (a) *A non-empty subset of a metric space is relatively compact if and only if it is relatively sequentially compact.*

(b) *A non-empty subset of a complete metric space is relatively compact if and only if it is totally bounded.*

PROOF Statement (a) follows immediately from Theorem 4.2.10.

Let Y be a non-empty subset of a metric space X. By (a) and Lemma 4.2.6 if Y is relatively compact then Y is totally bounded. Suppose now that X is complete and Y is totally bounded. Then, by Theorem 2.6.8, the subspace $\bar{Y}$ is complete and so, by Theorem 4.2.9, we have only to prove that $\bar{Y}$ is totally bounded.

Let $\varepsilon > 0$. There is a finite $\frac{1}{2}\varepsilon$-net in Y, say $\{y_1, y_2, \ldots, y_n\}$. Let $x \in \bar{Y}$. By Theorem 2.3.4 there exists $y \in Y$ with $d(x, y) < \frac{1}{2}\varepsilon$. Also $d(y, y_k) < \frac{1}{2}\varepsilon$ for some integer k with $1 \leq k \leq n$ so

$$d(x, y_k) \leq d(x, y) + d(y, y_k) < \varepsilon.$$

This shows that $\{y_1, y_2, \ldots, y_n\}$ is an ε-net in $\bar{Y}$. □

We have already seen that a non-empty subset of R is compact if and only if it is closed and bounded. We can now extend this result to R^n and C^n.

4.2.11 THEOREM (Borel–Lebesgue) *Let* $\| x \| = \left(\sum_{r=1}^{n} | \xi_r |^2 \right)^{1/2}$ *for all* $x = (\xi_1, \xi_2, \ldots, \xi_n)$ *in* K^n. *A non-empty subset of the normed linear space* $(\mathsf{K}^n, \| \cdot \|)$ *is compact if and only if it is bounded and closed.*

PROOF We have seen in Lemma 4.1.4 that a compact subset of any metric space is bounded and closed. Let A be a non-empty bounded closed subset of $(\mathsf{K}^n, \| \cdot \|)$; we have to prove that A is compact. By Theorem 2.9.2 the metric spaces $(\mathsf{C}^n, \| \cdot \|)$ and $(\mathsf{R}^{2n}, \| \cdot \|)$ are isometric so we may suppose that $\mathsf{K} = \mathsf{R}$. By Theorems 2.6.5 and 2.6.8 the subspace A is complete. Therefore by Theorem 4.2.9 it is sufficient to prove that A is totally bounded. Since A is bounded there is a positive integer N such that

$$A \subseteq A_N = \{(\xi_1, \xi_2, \ldots, \xi_n) \in \mathsf{R}^n : | \xi_j | \leq N \text{ for } j = 1, 2, \ldots, n\}.$$

If we can prove that A_N is totally bounded it will follow from Lemma 4.2.8 that A is also totally bounded.

Let $\varepsilon > 0$, let m be the least positive integer with $m\varepsilon \geq \sqrt{n}$, and let E_m be the set of n-tuples $(p_1/m, p_2/m, \ldots, p_n/m)$ where p_j is an integer with $| p_j | \leq mN$ for $j = 1, 2, \ldots, n$. There are exactly $(1 + 2mN)^n$ distinct points in the set E_m. Let $(\xi_1, \xi_2, \ldots, \xi_n) \in A_N$. Then, for $j = 1, 2, \ldots, n$, we have $| \xi_j | \leq N$ and hence we can find an integer p_j with $| p_j | \leq mN$ and $| m\xi_j - p_j | < 1$. Since

$$\| (\xi_1, \xi_2, \ldots, \xi_n) - (p_1/m, p_2/m, \ldots, p_n/m) \| = \left(\sum_{j=1}^{n} (\xi_j - p_j/m)^2 \right)^{1/2} < \frac{\sqrt{n}}{m} \leq \varepsilon$$

E_m is a finite ε-net in A_N. □

The reader may wonder why, in view of the equivalence of compactness and sequential compactness for metric spaces, we have not been content to introduce the simpler concept of sequential compactness only. It is true that, in this book, all the results that depend on compactness can be proved by using sequential compactness only. However the reader who starts to re-prove these results using sequential compactness instead of compactness must soon admit that the proofs based on compactness are shorter and more elegant (and eventually give more insight).

The strongest reason for working with compactness rather than sequential compactness, and for maintaining the distinction between the two notions is to be found in the study of topological spaces rather than in the study of metric spaces. A metric space is a particular kind of topological space. (In a topological space it is the concept of an 'open set' that is the primitive object.) In the context of topological spaces the analogues of compactness and sequential compactness are no longer equivalent, and it is the analogue of compactness that is more natural and more significant. For a discussion of topological spaces we refer the reader to the books of SIMMONS [25, Part 1] and KELLEY [18].

EXERCISES

(1) Prove that if a sequence in a compact metric space has a unique cluster point then it is convergent.

(2) Let Y be a non-empty subset of a metric space X with the property that every sequence in Y has a subsequence which converges in X. Prove that Y is relatively sequentially compact.

(3) (a) Let E be a normed linear space, let X and Y be subsets of E and let X be relatively compact. Show that $(X + Y)^- = \bar{X} + \bar{Y}$.

(b) Give an example of subsets X and Y of R with

$$(X + Y)^- \neq \bar{X} + \bar{Y}.$$

(4) Let f be a continuous mapping of a compact metric space X into a metric space. Prove that $f(\bar{A}) = f(A)^-$ for every subset A of X.

(5) Let X be a totally bounded metric space. Prove that for each $\varepsilon > 0$ there is a largest integer N such that it is possible to find N points $x_1, x_2, \ldots, x_N$ in X with $d(x_i,x_j) \geq \varepsilon$ whenever $i \neq j$.

(6) Let Y be a subset of a metric space X. If for each $\varepsilon > 0$ there is a finite set $\{x_1, x_2, \ldots, x_n\}$ in X with $Y \subseteq \bigcup_{i=1}^{n} B(x_i,\varepsilon)$, prove that Y is totally bounded.

(7) Prove that a metric space is totally bounded if and only if its completion is compact. (See Exercises 2.9(9) and 2.9(10).)

4.3 Compactness and Continuity

The results of this section are generalizations of well-known properties of continuous real-valued functions on bounded closed intervals.

4.3.1 THEOREM *Let f be a continuous mapping of a metric space X into a metric space Y and let A be a compact subset of X. Then $f(A)$ is a compact subset of Y.*

PROOF Let $\mathscr{B}$ be an open cover of $f(A)$. By Theorem 2.7.3 the set $f^{-1}(B)$ is open for each $B \in \mathscr{B}$. Consequently the family $\{f^{-1}(B): B \in \mathscr{B}\}$ is an open cover of the compact set A and hence has a finite subcover of A, say $\{f^{-1}(B_1), f^{-1}(B_2),\ldots, f^{-1}(B_n)\}$. Obviously $\{B_1, B_2,\ldots, B_n\}$ is a finite cover of $f(A)$. This proves that $f(A)$ is compact. □

4.3.2 COROLLARY *Let f be a continuous real-valued function on a compact metric space X. Then there are points x_1 and x_2 in X such that*

$$f(x_1) \leq f(x) \leq f(x_2)$$

for all $x \in X$.

PROOF By Theorem 4.3.1 the set $f(X)$ is a compact subset of R and so by Lemma 4.1.4, $f(X)$ is bounded and closed. Let $\alpha = \inf f(X)$ and $\beta = \sup f(X)$. Since $f(X)$ is closed $\alpha \in f(X)$ and $\beta \in f(X)$ (Theorem 2.3.4), and therefore there are points $x_1,x_2 \in X$ such that $\alpha = f(x_1)$ and $\beta = f(x_2)$. Obviously $f(x_1) \leq f(x) \leq f(x_2)$ for all $x \in X$. □

We can say briefly that a continuous real-valued function on a compact metric space is bounded and attains its bounds. If f is a continuous complex-valued function on a compact metric space X then $|f|$ is a continuous real-valued function on X (Theorem 2.8.2) and hence f is bounded on X (Definition 2.1.8). We recall (Example 3.2.12) that if X is a metric space $C_{\mathsf{K}}(X)$ denotes the set of all bounded continuous mappings of X into K, and that

$$\|f\| = \sup \{|f(x)| : x \in X\}$$

defines a norm on $C_{\mathsf{K}}(X)$. Corollary 4.3.2 shows that if X is a compact metric space then $C_{\mathsf{K}}(X)$ coincides with the set of all continuous mappings of X into K and the norm on $C_{\mathsf{K}}(X)$ is given by

$$\|f\| = \max \{|f(x)| : x \in X\}.$$

4.3.3 COROLLARY *Let f be a continuous real-valued function on a bounded closed interval $[a, b]$. Then f is bounded on $[a, b]$. Further, if $M = \sup \{f(t): a \leq t \leq b\}$ and $m = \inf \{f(t): a \leq t \leq b\}$, then for each y with $m \leq y \leq M$ there exists t with $a \leq t \leq b$ such that $y = f(t)$.*

PROOF Since $[a, b]$ is a compact subset of R the corollary follows directly from Corollaries 4.3.2 and 2.10.6. □

4.3.4 THEOREM *Let f be a one-to-one continuous mapping of a compact metric space X onto a metric space Y. Then f^{-1} is continuous on Y.*

PROOF Since f is one-to-one, $(f^{-1})^{-1}(A) = f(A)$ for all subsets A of X, and therefore, by Theorem 2.7.5, to prove that f^{-1} is continuous on Y

we have only to prove that $f(A)$ is closed in Y for all closed subsets A of X. This is obvious if $A = \emptyset$. If A is closed in X and $A \neq \emptyset$, Lemma 4.1.5 shows that A is compact in X. Theorem 4.3.1 then shows that $f(A)$ is compact in Y, and finally Lemma 4.1.4 shows that $f(A)$ is closed in Y. □

The last theorem has as a corollary a well-known property of continuous real-valued functions of a real variable. We recall that a real-valued function f on the closed interval $[a,b]$ is said to be *strictly increasing on* $[a,b]$ if and only if $f(s) < f(t)$ for all $a \leq s < t \leq b$, and is said to be *strictly decreasing on* $[a,b]$ if and only if $f(s) > f(t)$ for all $a \leq s < t \leq b$.

4.3.5 Corollary *Let f be a continuous real-valued function on the bounded closed interval $[a,b]$ that is either strictly increasing or strictly decreasing on $[a,b]$. Then f is a one-to-one mapping of $[a,b]$ onto the bounded closed interval $f([a,b])$ and f^{-1} is continuous on $f([a,b])$.*

Proof Corollary 4.3.3 shows that $f([a,b])$ is a bounded closed interval, it is obvious that f is one-to-one and the continuity of f^{-1} follows from Theorem 4.3.4. □

In the proof of Corollary 4.3.5 the continuity of f is used twice: once to show that $f([a,b])$ is a bounded closed interval and again to show that f^{-1} is continuous on $f([a,b])$. However it is not difficult to see that, if f is any strictly increasing or strictly decreasing function on $[a,b]$, then f^{-1} is continuous on $f([a,b])$, which need no longer now be an interval (Exercise (1)). Thus the hypothesis of continuity in Corollary 4.3.5 is needed only to ensure that $f([a,b])$ is a bounded closed interval.

We come now to the important concept of uniform continuity which was introduced in 1870 by E. Heine. He proved a special case of Theorem 4.3.8.

4.3.6 Definition Let f be a mapping of a metric space X into a metric space Y. Then f if said to be *uniformly continuous on* X if and only if for each positive real number ε there exists a positive real number δ such that

$$d(f(x_1),f(x_2)) < \varepsilon$$

for all $x_1,x_2 \in X$ with† $d(x_1,x_2) < \delta$.

It is obvious that if a mapping is uniformly continuous on X then it is continuous at each point of X and is therefore continuous on X.

† We have used d to denote both the metric on X and the metric on Y. This should cause no confusion.

The following example shows that a mapping that is continuous on X may fail to be uniformly continuous on X.

4.3.7 EXAMPLE Let $f(t) = t^2$ for all $t \in \mathsf{R}$. Obviously f is continuous on R. Suppose that f is uniformly continuous on R. Then there exists $\delta > 0$ such that $|f(t_1) - f(t_2)| < 1$ for all $t_1,t_2 \in \mathsf{R}$ with $|t_1 - t_2| < \delta$. Therefore

$$\frac{\delta}{2}\left|2t + \frac{\delta}{2}\right| = \left|f\left(t + \frac{\delta}{2}\right) - f(t)\right| < 1$$

for all $t \in \mathsf{R}$ which is clearly impossible. Thus f cannot be uniformly continuous on R.

Uniform continuity is a global condition on the behaviour of a mapping on a set; continuity is a collection of local conditions on the behaviour of a mapping near a point. Precisely, f is continuous on X if and only if for each $\varepsilon > 0$ and for each $x \in X$ there exists $\delta(x) > 0$, which may depend on both x and ε, such that $d(f(x),f(y)) < \varepsilon$ for all $y \in X$ with $d(x,y) < \delta(x)$. The difference between this condition and the condition of Definition 4.3.6 is obvious: in the above condition, to each $x \in X$ there corresponds $\delta(x) > 0$ and $\delta(x)$ will, in general, vary with x; whereas Definition 4.3.6 requires that a single $\delta > 0$ can be chosen, independently of x.

The importance of uniform continuity in analysis rests almost entirely on the following theorem.

4.3.8 THEOREM *Every continuous mapping of a compact metric space X into a metric space Y is uniformly continuous on X.*

PROOF Let f be a continuous mapping of X into Y and let $\varepsilon > 0$. To each point $x \in X$ there corresponds $\delta(x) > 0$ such that

$$d(f(x),f(y)) < \tfrac{1}{2}\varepsilon \tag{1}$$

for all $y \in X$ with $d(x,y) < \delta(x)$. The family of open balls

$$\{B(x,\tfrac{1}{2}\delta(x)) \colon x \in X\}$$

is an open cover of X and therefore has a finite subcover,

$$\{B(x_1,\tfrac{1}{2}\delta(x_1)),\ B(x_2,\tfrac{1}{2}\delta(x_2)),\dots,\ B(x_n,\tfrac{1}{2}\delta(x_n))\}$$

say. Let $\delta = \min\{\tfrac{1}{2}\delta(x_1),\ \tfrac{1}{2}\delta(x_2),\dots,\ \tfrac{1}{2}\delta(x_n)\}$ and let $y_1,y_2 \in X$ with $d(y_1,y_2) < \delta$. For some integer k with $1 \leq k \leq n$, $y_1 \in B(x_k,\tfrac{1}{2}\delta(x_k))$. Thus $d(x_k,y_1) < \tfrac{1}{2}\delta(x_k)$ and

$$d(x_k,y_2) \leq d(x_k,y_1) + d(y_1,y_2) < \tfrac{1}{2}\delta(x_k) + \delta \leq \delta(x_k).$$

Consequently (1) gives

$$d(f(y_1),f(y_2)) \leq d(f(y_1),f(x_k)) + d(f(x_k),f(y_2)) < \varepsilon.$$

This proves that f is uniformly continuous on X. □

Our final theorem is concerned with the convergence of monotonic sequences of continuous functions. It was proved, for functions of a real variable, by U. Dini in 1878. We shall use the terminology established in the discussion on p. 49.

4.3.9 THEOREM (Dini) *Let X be a compact metric space, let (f_n) be a sequence of continuous real-valued functions on X such that $f_n \leq f_{n+1}$ for $n = 1, 2, \ldots$ and suppose that (f_n) converges pointwise on X to a continuous real-valued function f. Then (f_n) converges uniformly on X to f.*

PROOF We have to prove that the sequence (f_n) converges in the Banach space $C_{\mathsf{R}}(X)$. Let $g_n = f - f_n$. Then $g_n \in C_{\mathsf{R}}(X)$ and

$$0 \leq g_{n+1} \leq g_n \tag{2}$$

for $n = 1, 2, \ldots$. Let $\varepsilon > 0$ and $X_n = \{x \in X: g_n(x) < \varepsilon\}$ for $n = 1, 2, \ldots$. Since $X_n = g_n^{-1}((-\infty,\varepsilon))$ Theorem 2.7.3 shows that X_n is open. By hypothesis $\lim_{n\to\infty} g_n(x) = 0$ for all $x \in X$, and consequently $X = \bigcup_{n=1}^{\infty} X_n$. Thus $\{X_n: n = 1, 2, \ldots\}$ is an open cover of X and so has a finite subcover, $\{X_{n_1}, X_{n_2}, \ldots, X_{n_k}\}$ say. It follows from (2) that $X_n \subseteq X_{n+1}$, and hence, for $j = 1, 2, \ldots, k$, $X_{n_j} \subseteq X_N$ where $N = \max\{n_1, n_2, \ldots, n_k\}$. Then

$$X = \bigcup_{j=1}^{k} X_{n_j} = X_N$$

and so $0 \leq g_n(x) < \varepsilon$ for all $x \in X$ and all $n \geq N$. But then for all $n \geq N$,

$$\| f - f_n \| = \max\{\,|f(x) - f_n(x)|: x \in X\} < \varepsilon.$$

This proves that (f_n) converges uniformly on X to f. □

It is clear that Dini's theorem remains true if the condition $f_n \leq f_{n+1}$ is replaced by the condition $f_n \geq f_{n+1}$. The following example shows that the continuity of the limit function f is not a consequence of the other hypotheses of Dini's theorem.

4.3.10 EXAMPLE Let $f_n(t) = 1 - t^n$ for $0 \leq t \leq 1$ and $n = 1, 2, \ldots$. Then $f_n \in C_{\mathsf{R}}([0,1])$ and $f_n \leq f_{n+1}$. However

$$f(t) = \lim_{n\to\infty} f_n(t) = \begin{cases} 1 & \text{if } 0 \leq t < 1, \\ 0 & \text{if } t = 1 \end{cases}$$

so f is not continuous on $[0,1]$.

Since the limit of a uniformly convergent sequence of continuous functions is continuous (Theorem 2.8.3) this shows that the condition that f be continuous cannot be omitted from the hypothesis of Dini's Theorem. The following example shows that the condition $f_n \leq f_{n+1}$ cannot be omitted, either.

4.3.11 EXAMPLE Let

$$g_n(t) = \begin{cases} 2nt & \text{for } 0 \leq t \leq 1/2n, \\ 2\text{–}2nt & \text{for } 1/2n < t \leq 1/n, \\ 0 & \text{for } 1/n < t \leq 1. \end{cases}$$

It is easy to see that $g_n \in C_{\mathsf{R}}(X)$ and $\lim_{n\to\infty} g_n(t) = 0$ for $0 \leq t \leq 1$. But $\| g_n \| = 1$ for $n = 1, 2, \ldots$, so (g_n) cannot converge uniformly to 0 on $[0,1]$.

EXERCISES

(1) Let f be a strictly monotonic real-valued function on $[a,b]$. Prove that f^{-1} is a continuous mapping of the subspace $f([a,b])$ of R onto $[a,b]$. (See the remarks after Corollary 4.3.5.)

(2) Give an example of a one-to-one mapping f of a metric space X onto a metric space Y such that f is continuous but f^{-1} is not continuous.

(3) Let f be a regulated mapping of the interval $I = [a,b]$ into a normed linear space E and let A be a closed subset of I. Prove that $f(A)$ is a relatively compact subset of E. Give an example to show that $f(A)$ may fail to be compact.

(4) A mapping f of $[0,1]$ into R is bounded and $\{(x,f(x)) : x \in [0,1]\}$, the graph of f, is a closed subset of R^2. Prove that f is continuous. Give an example to show that 'bounded' cannot be omitted.

(5) Let X be a metric space which is not compact. Prove that there exists a continuous real-valued function on X which is not bounded.

(6) A mapping f of $[a,b]$ into R is continuous and for each $\xi \in (a,b)$ and $\varepsilon > 0$ there exists $h \in (0,\varepsilon)$ such that $f(\xi + h) > f(\xi)$. Prove that $f(b) > f(a)$.

(7) Prove the following sequence of propositions. Together they constitute an elementary proof of the *fundamental theorem of algebra* which states that every complex polynomial which is not constant has a complex zero, or, in technical language, that the field of complex numbers is algebraically closed.

(a) If a continuous complex valued function p defined on C has the property that for each M there is an r such that $| p(z) | \geq M$ whenever $| z | \geq r$, then there exists $\xi \in \mathsf{C}$ with $| p(\xi) | \leq | p(z) |$ for all $z \in \mathsf{C}$.

(b) If p is a polynomial, that is $p(z) = \alpha_0 + \alpha_1 z + \ldots + \alpha_n z^n$ for all $z \in \mathsf{C}$, where $\alpha_0, \alpha_1, \ldots, \alpha_n \in \mathsf{C}$, and if p is not a constant then p has the property of (a).

(c) If p is a polynomial which is not constant, $\delta > 0$ and $|p(0)| \leq |p(z)|$ whenever $|z| \leq \delta$ then $p(0) = 0$.

(d) If p is a polynomial which is not constant then $p(\xi) = 0$ for some $\xi \in \mathsf{C}$.

(8) Let A be a compact subset of a metric space X and let $x \in X \sim A$. Show that there is a point $y \in A$ with $d(x,y) = d(x,A)$.

(9) Let f be a mapping of a compact metric space X into itself with the property that $d(f(x),f(y)) < d(x,y)$ whenever $x \neq y$. Prove, by considering the mapping $x \to d(x,f(x))$ of X into R, that f has a unique fixed point.

(10) Let X be a compact metric space and f an isometry of X into itself. Prove that $f(X) = X$.

(11) Show that a continuous linear transformation of one normed linear space into another is uniformly continuous.

(12) Let Y be a complete metric space, let A be a dense subset of a metric space X and let f be a uniformly continuous mapping of A into Y. Prove that there exists a uniformly continuous mapping of X into Y which is an extension of f (cf. Theorem 3.5.10).

Is it possible, for every such Y, A and X, to extend every continuous mapping of A into Y to a continuous mapping of X into Y?

(13) *Repeated integrals of a continuous function of two variables.* Let S be the rectangle $\{(s,t): a \leq s \leq b,\ c \leq t \leq d\}$ in R^2 and let k be a continuous complex-valued function on S. Let a mapping g of $[a,b]$ into $C_{\mathsf{C}}([c,d])$ and a mapping h of $[c,d]$ into $C_{\mathsf{C}}([a,b])$ be defined by

$$(g(s))(t) = k(s,t) \text{ and } (h(t))(s) = k(s,t)$$

for $a \leq s \leq b$ and $c \leq t \leq d$. Prove that g and h are continuous.

By considering the mapping $f \to f(s)$ of $C_{\mathsf{C}}([a,b])$ into C and by using Exercise 3.7(2) show that

$$\int_c^d k(s,t)\,dt = \left(\int_c^d h\right)(s)$$

for $a \leq s \leq b$ and so establish the existence of the repeated integrals

$$\int_a^b\left(\int_c^d k(s,t)\,dt\right)ds \text{ and } \int_c^d\left(\int_a^b k(s,t)\,ds\right)dt. \qquad (*)$$

Use Exercise 3.7(2) to prove that

$$\int_a^b\left(\int_c^d h\right) = \int_c^d\left(\int_a^b g\right)$$

and hence show that the two repeated integrals (*) are equal.

(14) Use the contraction mapping theorem in the space $C_{\mathsf{R}}([0,\infty))$ to prove that there is a unique real-valued function f, continuous and bounded on $[0,\infty)$, which satisfies the equation

$$f(x) = 1 + \int_0^x \exp(-t^2) f(xt)\, dt$$

for all $x \geq 0$. (Note that $\int_0^\infty \exp(-t^2)\, dt = \frac{1}{2}\sqrt{\pi}$.)

4.4 Finite-Dimensional Normed Linear Spaces

As an illustration of the use of compactness in analysis we shall establish some important properties of finite-dimensional normed linear spaces, and obtain a characterization of such spaces in terms of compactness.

The Borel–Lebesgue theorem (4.2.11) characterizes the compact subsets of K^n with the norm $\| x \| = \left(\sum_{r=1}^{n} | \xi_r |^2\right)^{1/2}$. This characterization, and Corollary 4.3.2, are the tools used in the proof of the first and principal theorem of this section. This theorem shows that K^n, with the above norm, is, to within equivalence, the typical n-dimensional normed linear space over K.

4.4.1 THEOREM *Any two n-dimensional normed linear spaces over the same field are linearly homeomorphic.*

PROOF It is obviously sufficient to prove that any n-dimensional normed linear space over K is linearly homeomorphic to the space K^n with the norm given by

$$\| x \| = \left(\sum_{r=1}^{n} | \xi_r |^2\right)^{1/2} \tag{1}$$

where $x = (\xi_1, \xi_2, \ldots, \xi_n)$.

Let E be an n-dimensional normed linear space over K with norm $\| \cdot \|$ and let $\{e_1, e_2, \ldots, e_n\}$ be a basis for E. Let

$$Tx = \xi_1 e_1 + \xi_2 e_2 + \ldots + \xi_n e_n$$

for all $x = (\xi_1, \xi_2, \ldots, \xi_n) \in \mathsf{K}^n$. It is well known, and easily proved, that T is a one-to-one linear mapping of K^n onto E. We have to prove that T and T^{-1} are continuous or, equivalently (by Theorem 3.5.4), that T and T^{-1} are bounded.

For all $x = (\xi_1, \xi_2, \ldots, \xi_n) \in \mathsf{K}^n$ we have

$$\begin{aligned}\| Tx \| &= \| \xi_1 e_1 + \xi_2 e_2 + \ldots + \xi_n e_n \| \\ &\leq | \xi_1 | \| e_1 \| + | \xi_2 | \| e_2 \| + \ldots + | \xi_n | \| e_n \| \\ &\leq \| x \| (\| e_1 \| + \| e_2 \| + \ldots + \| e_n \|),\end{aligned}$$

where the last inequality follows from (1). This proves that T is bounded. It is more difficult to prove that T^{-1} is bounded.

Let $A = \{x \in \mathsf{K}^n : \| x \| = 1\}$. Clearly A is bounded and closed so by the Borel–Lebesgue theorem A is compact. Let $f(x) = \| Tx \|$ for all $x \in A$. The mapping $y \to \| y \|$ is continuous on E and f is the composition of the mappings $x \to Tx$ and $y \to \| y \|$ so f is continuous on A (Theorem 2.7.4). Let $m = \inf \{f(x) : x \in A\}$. Obviously $m \geq 0$. Corollary 4.3.2 shows that there exists $x_0 \in A$ with $m = f(x_0)$. Consequently, if $m = 0$, then $\| Tx_0 \| = f(x_0) = 0$, which is impossible because T is one-to-one and $\| x_0 \| = 1$. Thus $m > 0$. Since $\| Tx \| \geq m$ for all $x \in A$ Lemma 3.5.9 shows that T^{-1} is bounded. □

4.4.2 COROLLARY *Each finite-dimensional normed linear space is complete.*

PROOF By Theorem 4.4.1 each finite-dimensional normed linear space over K is linearly homeomorphic to $(\mathsf{K}^n, \| \cdot \|)$, where the norm $\| \cdot \|$ is defined by (1). Now $(\mathsf{K}^n, \| \cdot \|)$ is complete (Theorems 2.6.5 and 2.6.6) and a linear homeomorphism is a uniform equivalence (see p. 110) so it follows from Lemma 2.9.4 that each finite-dimensional normed linear space is complete. □

4.4.3 COROLLARY *Each finite-dimensional linear subspace of a normed linear space is closed.*

PROOF This follows directly from Corollary 4.4.2 and Theorem 2.6.8. □

4.4.4 COROLLARY *A non-empty bounded closed subset of a finite dimensional normed linear space is compact.*

PROOF Let A be a non-empty bounded closed subset of a finite dimensional normed linear space E over K. By Theorem 4.4.1 there is a linear homeomorphism T of E onto $(\mathsf{K}^n, \| \cdot \|)$, where the norm $\| \cdot \|$ is defined by (1). Obviously $T(A)$ is a bounded closed subset of $(\mathsf{K}^n, \| \cdot \|)$ and therefore by the Borel–Lebesgue theorem $T(A)$ is compact. Since $A = T^{-1}(T(A))$ and T^{-1} is continuous Theorem 4.3.1 shows that A is compact. □

It follows easily from Corollary 4.4.4 that a bounded sequence in a finite-dimensional normed linear space has a convergent subsequence. There is a converse of Corollary 4.4.4 which is true and which provides an important characterization of finite-dimensional normed linear spaces. We shall need a lemma.

4.4.5 LEMMA *Let E be a normed linear space and let F be a linear subspace of E that is a proper closed subset of E. For each real number δ with $0 < \delta < 1$ there exists a point $x_0 \in E$ such that $\| x_0 \| = 1$ and $d(x_0,F) \geq \delta$.*

PROOF Choose $x_1 \in E \sim F$ and let $m = d(x_1,F)$. Then $m > 0$, because F is closed and $x_1 \notin F$ (Lemma 2.8.4). Since $m\delta^{-1} > m$ there exists $y_1 \in F$ such that $\| x_1 - y_1 \| \leq m\delta^{-1}$. Let $x_0 = \| x_1 - y_1 \|^{-1}(x_1 - y_1)$. (Note that $x_1 - y_1 \neq 0$ because $x_1 \notin F$.) Then $\| x_0 \| = 1$ and, for all $x \in F$,

$$\begin{aligned} \| x_0 - x \| &= \| \, \| x_1 - y_1 \|^{-1}(x_1 - y_1) - x \, \| \\ &= \| x_1 - y_1 \|^{-1} \, \| x_1 - (y_1 + \| x_1 - y_1 \| x) \| \\ &\geq (m\delta^{-1})^{-1}m \\ &= \delta \end{aligned}$$

because $y_1 + \| x_1 - y_1 \| x \in F$. □

4.4.6 THEOREM *The closed unit ball in a normed linear space E is compact if and only if E is finite-dimensional.*

PROOF If E is finite-dimensional then, by Corollary 4.4.4, the closed unit ball in E is compact. Suppose, conversely, that the closed unit ball, B say, in E is compact. Then B is totally bounded (see p. 135) so there is a finite $\frac{1}{2}$-net, $\{x_1, x_2, \dots, x_n\}$ say, in B. Let F be the linear hull of the set $\{x_1, x_2, \dots, x_n\}$ and suppose that $F \neq E$. The linear subspace F is finite-dimensional and so, by Corollary 4.4.3, F is closed. Now by Lemma 4.4.5 there exists $x_0 \in B$ such that $d(x_0,F) \geq \frac{1}{2}$. In particular $\| x_0 - x_k \| \geq \frac{1}{2}$ for $k = 1, 2, \dots, n$, which is a contradiction because $x_0 \in B$ and $\{x_1, x_2, \dots, x_n\}$ is a $\frac{1}{2}$-net in B. Consequently $F = E$. This proves that E is finite-dimensional. □

Lemma 4.4.5 and Theorem 4.4.6 are essentially due to F. Riesz (1918). Lemma 4.4.5 is used frequently; it is commonly called Riesz's lemma.

EXERCISES

(1) (a) Let F be a finite dimensional subspace of a normed linear space E and let $x \in E \sim F$. Show that there is a $y \in F$ with $\| x - y \| = d(x,F)$.

(b) A normed linear space E is said to be *strictly convex* if the equation $\| x + y \| = \| x \| + \| y \|$ with $x,y \in E$ implies that $\alpha x = \beta y$ for some non-negative real numbers α,β with $\alpha + \beta > 0$.

Prove that there is at most one $y \in F$ with $\| x - y \| = d(x,F)$ for every $x \in E$ and every subspace F of E if and only if E is strictly convex.

(2) There are circumstances in which Lemma 4.4.5 can be improved.

(a) Let E be a normed linear space and let F be a linear subspace of E which is not dense in E. Prove that there is $x_0 \in E$ with $\| x_0 \| = 1$ and $d(x_0,F) = 1$ (that is, it is possible to take $\delta = 1$ in Lemma 4.4.5) if and only if there is $x \notin F$ such that $\| x - y \| = d(x,F)$ for some $y \in F$.

(b) Deduce that if the proper linear subspace F of E is finite-dimensional then it is possible to take $\delta = 1$ in Lemma 4.4.5.

(c) Prove that it is not always possible to take $\delta = 1$ in Lemma 4.4.5 by considering the subspace

$$F = \left\{ f \in C_{\mathsf{R}}([0,1]) : \int_0^{\frac{1}{2}} f(t)\, dt = \int_{\frac{1}{2}}^1 f(t)\, dt \right\}$$

of the space $C_{\mathsf{R}}([0,1])$.

(3) There are alternative proofs of Theorem 4.4.1 and Corollaries 4.4.2, 4.4.3 and 4.4.4, that are independent of considerations of compactness. Exercise 3.4(4) provides such a proof of Corollary 4.4.2, and hence of Corollary 4.4.3; the following five exercises provide such proofs of Theorem 4.4.1 and Corollary 4.4.4.

(a) Let E be an n-dimensional linear space, let F be an $(n - 1)$-dimensional linear subspace of E and let $u \in E \sim F$. Prove that each point $x \in E$ has a unique representation in the form $x = \xi u + y$ with $\xi \in \mathsf{K}$ and $y \in F$.

(b) Let F be a closed linear subspace of a normed linear space E and let $u \in E \sim F$. Prove that there are positive real numbers M and M' (depending on u) such that $| \xi | \leq M \| \xi u + y \|$ and $\| y \| \leq M' \| \xi u + y \|$ for all $\xi \in \mathsf{K}$ and all $y \in F$.

(c) Let E and F be normed linear spaces, and suppose that E is finite-dimensional. Prove that each linear transformation of E into F is bounded. (Hint: Use induction on the dimension of E and the results of (a) and (b). A proof of this result, which uses Theorem 4.4.1, is given in Theorem 6.1.2.)

(d) Use the result of (c) to prove that any two norms on a finite dimensional linear space are equivalent. Deduce that any two n-dimensional normed linear spaces over the same field are linearly homeomorphic.

(e) Use the results of (a) and (b) to prove, by induction on the dimension of the space, that any non-empty bounded closed subset of a finite-dimensional normed linear space is compact.

4.5 The Stone-Weierstrass Theorem

Throughout this section X will denote a compact metric space and $C_{\mathsf{K}}(X)$ will as usual denote the Banach space of all continuous mappings of X into K with the uniform norm.

Our aim is to prove a generalization, due to M. H. Stone, of a classical theorem of Weierstrass on the approximation of continuous real-valued functions by polynomials. Weierstrass's theorem, published in 1885, is one of the most celebrated results of classical analysis; Stone's generalization, published in 1948, is perhaps the most useful known result about $C_{\mathsf{K}}(X)$.

Weierstrass's theorem is not generally included in elementary analysis courses, so we shall begin by proving it. Many proofs are known; the one that we shall give is due to Landau and is similar to, but simpler than, that of Weierstrass.

Let $\alpha_0, \alpha_1, \alpha_2, \ldots, \alpha_n$ be real numbers. The function p defined by

$$p(t) = \alpha_0 + \alpha_1 t + \ldots + \alpha_n t^n$$

for all $t \in \mathsf{R}$ is called a *polynomial function*.

4.5.1 THEOREM (The Weierstrass approximation theorem) *Let f be a continuous real-valued function on a bounded closed interval $[a,b]$ and let $\varepsilon > 0$. Then there exists a polynomial function p such that*

$$|f(t) - p(t)| < \varepsilon$$

for $a \leq t \leq b$.

PROOF If $a = b$ the theorem is trivial so we may suppose that $a < b$. We shall show first that it is sufficient to prove the theorem in the special case $a = 0$ and $b = 1$. The mapping $s \to (b - a)s + a$ is continuous on $[0,1]$, so the function g defined by $g(s) = f((b - a)s + a)$ for $0 \leq s \leq 1$ is continuous on $[0,1]$. If we have proved the theorem in the case $a = 0$ and $b = 1$ then there is a polynomial function q such that $|g(s) - q(s)| < \varepsilon$ for $0 \leq s \leq 1$. Let $p(t) = q((t - a)(b - a)^{-1})$ for all $t \in \mathsf{R}$. Then p is obviously a polynomial function and $|f(t) - p(t)| < \varepsilon$ for $a \leq t \leq b$. Thus we may suppose that $a = 0$ and $b = 1$.

Next we shall show that it is sufficient to prove the theorem in the case $f(0) = f(1) = 0$. Let $g(t) = f(t) - f(0) - (f(1) - f(0))t$ for $0 \leq t \leq 1$. Then g is continuous on $[0,1]$ and $g(0) = g(1) = 0$. If we have proved the theorem for such functions then there is a polynomial function q such that $|g(t) - q(t)| < \varepsilon$ for $0 \leq t \leq 1$. Let

$$p(t) = q(t) + (f(1) - f(0))t + f(0)$$

for all $t \in \mathsf{R}$. Clearly p is a polynomial function and $|f(t) - p(t)| < \varepsilon$ for $0 \leq t \leq 1$. Thus we may suppose that $f(0) = f(1) = 0$.

We can extend f to the whole of R by setting $f(t) = 0$ for $t \in \mathsf{R} \sim [0,1]$. It is clear that the extended function is continuous on R. For $n = 1, 2,\ldots$ let

$$q_n(t) = \begin{cases} \alpha_n(1 - t^2)^n & \text{for } |t| \leq 1, \\ 0 & \text{for } |t| > 1, \end{cases}$$

where

$$\frac{1}{\alpha_n} = \int_{-1}^{1} (1 - t^2)^n \, dt,$$

and let

$$p_n(t) = \int_{-1}^{1} f(t - s) q_n(s) \, ds \tag{1}$$

for all $t \in \mathsf{R}$. Since q_n vanishes outside $[-1, 1]$ and f vanishes outside $[0,1]$, we can also write, for all $t \in \mathsf{R}$

$$p_n(t) = \int_{t-1}^{t} f(t - s) q_n(s) \, ds$$

and a simple change of variable gives

$$p_n(t) = \int_{0}^{1} f(s) q_n(t - s) \, ds,$$

which shows that p_n is a polynomial function.

We need an estimate for the rate of growth of the sequence (α_n). Using the inequality $(1 - t^2)^n \geq 1 - nt^2$, which is valid for $-1 \leq t \leq 1$ and $n = 1, 2,\ldots$, we obtain

$$\begin{aligned} \int_{-1}^{+1} (1 - t^2)^n \, dt &\geq \int_{-n^{-1/2}}^{n^{-1/2}} (1 - t^2)^n \, dt \\ &\geq \int_{-n^{-1/2}}^{n^{-1/2}} (1 - nt^2) \, dt \\ &= \tfrac{4}{3} n^{-\frac{1}{2}}, \end{aligned}$$

and therefore $\alpha_n < n^{1/2}$ for $n = 1, 2,\ldots$. If $0 < \delta \leq 1$, then for $n = 1, 2,\ldots$, $0 \leq q_n(t) \leq \alpha_n(1 - \delta^2)^n$ for $\delta \leq |t| \leq 1$, and consequently

$$0 \leq q_n(t) < n^{1/2}(1 - \delta^2)^n \tag{2}$$

for $\delta \leq |t| \leq 1$. Further it is obvious that, for $n = 1, 2,\ldots$,

$$\int_{-1}^{1} q_n(t) \, dt = 1. \tag{3}$$

Since $[0,1]$ is compact Theorem 4.3.8 shows that f is uniformly continuous on $[0,1]$ and hence there exists $\delta > 0$ such that $|f(s) - f(t)| < \frac{1}{2}\varepsilon$ for all $s,t \in [0,1]$ with $|s - t| < \delta$. Since f vanishes outside $[0,1]$ we have now

$$|f(s) - f(t)| < \tfrac{1}{2}\varepsilon \tag{4}$$

for all $s,t \in \mathsf{R}$ with $|s - t| < \delta$. Obviously we may suppose that $\delta < 1$.

Now let $0 \leq t \leq 1$ and let n be a positive integer. Using (1), (2), (3) and (4) we obtain

$$\begin{aligned}
|p_n(t) - f(t)| &= \left| \int_{-1}^{1} f(t-s)q_n(s)\,ds - f(t) \right| \\
&= \left| \int_{-1}^{1} (f(t-s) - f(t))q_n(s)\,ds \right| \\
&\leq \int_{-1}^{1} |f(t-s) - f(t)|\, q_n(s)\,ds \\
&= \int_{-1}^{-\delta} |f(t-s) - f(t)|\, q_n(s)\,ds \\
&\qquad + \int_{-\delta}^{\delta} |f(t-s) - f(t)|\, q_n(s)\,ds \\
&\qquad + \int_{\delta}^{1} |f(t-s) - f(t)|\, q_n(s)\,ds \\
&< 2\,\|f\| \int_{-1}^{-\delta} q_n(s)\,ds + \tfrac{1}{2}\varepsilon \int_{-\delta}^{\delta} q_n(s)\,ds + 2\,\|f\| \int_{\delta}^{1} q_n(s)\,ds \\
&< 2\,\|f\|\, n^{1/2}(1-\delta^2)^n(1-\delta) + \tfrac{1}{2}\varepsilon \int_{-\delta}^{\delta} q_n(s)\,ds \\
&\qquad + 2\,\|f\|\, n^{1/2}(1-\delta^2)^n(1-\delta) \\
&\leq 4\,\|f\|\, n^{1/2}(1-\delta^2)^n + \tfrac{1}{2}\varepsilon.
\end{aligned}$$

Since† $\lim_{n\to\infty} n^{1/2}(1-\delta^2)^n = 0$, there is a positive integer N such that $4\,\|f\|\, n^{1/2}(1-\delta^2)^n < \frac{1}{2}\varepsilon$ for all $n \geq N$. Thus

$$|p_N(t) - f(t)| < \varepsilon$$

for $0 \leq t \leq 1$. □

The concept of polynomial function is not available on an arbitrary metric space and it is not immediately obvious how to extend the Weierstrass approximation theorem to compact metric spaces. We have to find an alternative statement of the Weierstrass theorem that does not explicitly mention polynomial functions. The alternative formulation makes use of the algebraic structure of $C_{\mathsf{R}}([a,b])$. We know already that

† Since $0 < \delta < 1$ we have $1 - \delta^2 = (1+\beta)^{-1}$ for some $\beta > 0$. Thus $0 \leq n^{\frac{1}{2}}(1-\delta^2)^n = n^{\frac{1}{2}}(1+\beta)^{-n} \leq \beta^{-1}n^{-\frac{1}{2}}$ because $(1+\beta)^n \geq n\beta$.

$C_{\mathsf{R}}([a,b])$ is a Banach space; it is also a linear associative algebra. We begin with some definitions.

4.5.2 DEFINITION A linear space E over K is said to be a *linear associative algebra* if and only if there is a mapping $(x,y) \to xy$ of $E \times E$ into E which satisfies the following conditions:

(a) $x(yz) = (xy)z$ for all $x,y,z \in E$,

(b) $(\alpha x)(\beta y) = (\alpha\beta)(xy)$ for all $x,y \in E$ and $\alpha,\beta \in \mathsf{K}$,

(c) $x(y + z) = xy + xz$ and $(y + z)x = yx + zx$ for all $x,y,z \in E$.

Let E be a linear associative algebra. Then E is said to be *commutative* if and only if

(d) $xy = yx$ for all $x,y \in E$.

An element e of E is called an *identity for E* if and only if

(e) $ex = xe$ for all $x \in E$.

It is easy to see that E can have at most one identity. A non-empty subset F of E is said to be a *subalgebra of E* if and only if F is a linear subspace of E and $xy \in F$ for all $x,y \in F$. The intersection of a family of subalgebras of E is obviously a subalgebra of E. (It is non-empty because every subalgebra contains 0.) Given a non-empty subset A of E the intersection of all the subalgebras of E that contain A is a subalgebra, called the *subalgebra of E generated by A*. Given $x \in E$, the positive integral powers, x^n, of x are defined inductively by $x^n = x(x^{n-1})$. If $x \neq 0$ and E has an identity e we set $x^0 = e$. Suppose that E has an identity e. An expression of the form

$$\alpha_0 e + \alpha_1 x + \alpha_2 x^2 + \ldots + \alpha_n x^n,$$

where $\alpha_0, \alpha_1, \ldots, \alpha_n \in \mathsf{K}$ is called a *polynomial in x*.

We have now enough terminology to reformulate Weierstrass's theorem. First we notice that $C_{\mathsf{K}}(X)$ is a linear associative algebra: we recall that the product fg of two functions $f,g \in C_{\mathsf{K}}(X)$ is defined by $(fg)(x) = f(x)g(x)$ for all $x \in X$. We saw in Theorem 2.8.2 that $fg \in C_{\mathsf{K}}(X)$, and it is easy to verify that the conditions for a linear associative algebra are satisfied. Further $C_{\mathsf{K}}(X)$ is commutative and the function 1 is the identity of $C_{\mathsf{K}}(X)$.

Now take $X = [a,b]$ and let f_0 be defined by $f_0(t) = t$ for $a \leq t \leq b$. If p is the restriction to $[a, b]$ of a polynomial function,

$$p(t) = \alpha_0 + \alpha_1 t + \alpha_2 t^2 + \ldots + \alpha_n t^n$$

say, then, for $a \leq t \leq b$,

$$\begin{aligned} p(t) &= \alpha_0 + \alpha_1 t + \ldots + \alpha_n t^n \\ &= \alpha_0 1 + \alpha_1 f_0(t) + \alpha_2 (f_0(t))^2 + \ldots + \alpha_n (f_0(t))^n \\ &= \alpha_0 1 + \alpha_1 f_0(t) + \alpha_2 f_0^2(t) + \ldots + \alpha_n f_0^n(t) \\ &= (\alpha_0 1 + \alpha_1 f_0 + \alpha_2 f_0^2 + \ldots + \alpha_n f_0^n)(t), \end{aligned}$$

and consequently

$$p = \alpha_0 1 + \alpha_1 f_0 + \alpha_2 f_0^2 + \ldots + \alpha_n f_0^n. \tag{5}$$

Thus every polynomial function can be expressed as a polynomial in the special function f_0. Finally we notice that the set of all functions of the form (5) with $\alpha_0, \alpha_1, \ldots, \alpha_n \in \mathsf{R}$ is precisely the subalgebra of $C_{\mathsf{R}}([a,b])$ generated by the set $\{1,f_0\}$. To see this we have only to remark that the set of all polynomials of the form (5) is obviously a subalgebra of $C_{\mathsf{R}}([a,b])$ that contains the set $\{1,f_0\}$, and that every subalgebra containing this set must contain all such polynomials.

The inequality $|f(t) - p(t)| < \varepsilon$ for $a \leq t \leq b$, which is the conclusion of Weierstrass's theorem, implies that $\|f - p\| < \varepsilon$, so we can restate the theorem as follows: given $f \in C_{\mathsf{R}}([a,b])$ and $\varepsilon > 0$, the subalgebra of $C_{\mathsf{R}}([a,b])$ generated by the set $\{1,f_0\}$ contains a function p such that $\|f - p\| < \varepsilon$. By Theorem 2.3.4 this means that the subalgebra of $C_{\mathsf{R}}([a,b])$ generated by the set $\{1,f_0\}$ is dense in $C_{\mathsf{R}}([a,b])$.

Stone's generalization of the Weierstrass approximation theorem, now universally known as the Stone–Weierstrass theorem, gives sufficient conditions for a subalgebra of $C_{\mathsf{R}}(X)$ to be dense in $C_{\mathsf{R}}(X)$. We shall need several lemmas, the first of which shows that we need consider only closed subalgebras of $C_{\mathsf{R}}(X)$.

4.5.3 LEMMA *The closure of a subalgebra of $C_{\mathsf{K}}(X)$ is a subalgebra of $C_{\mathsf{K}}(X)$.*

PROOF Let A be a subalgebra of $C_{\mathsf{K}}(X)$. By Theorem 3.4.4 the closure $\bar{A}$ of A is a linear subspace of $C_{\mathsf{K}}(X)$. Let $f,g \in \bar{A}$. By Theorem 2.5.5 there are sequences (f_n) and (g_n) in A with $\lim_{n\to\infty} f_n = f$ and $\lim_{n\to\infty} g_n = g$. Let $0 < \eta \leq 1$ and choose an integer N such that $\|f_n - f\| < \eta$ and $\|g_n - g\| < \eta$ for all $n \geq N$. If $n \geq N$, then

$$\|f_n\| \leq \|f_n - f\| + \|f\| < 1 + \|f\|,$$

and

$$\begin{aligned}
\|f_n g_n - fg\| &= \|f_n(g_n - g) + g(f_n - f)\| \\
&\leq \|f_n(g_n - g)\| + \|g(f_n - f)\| \\
&= \max\{|f_n(x)|\,|g_n(x) - g(x)| : x \in X\} \\
&\qquad + \max\{|g(x)|\,|f_n(x) - f(x)| : x \in X\} \\
&\leq \|f_n\|\,\|g_n - g\| + \|g\|\,\|f_n - f\| \\
&< \eta(1 + \|f\| + \|g\|).
\end{aligned}$$

Given $\varepsilon > 0$, let $\eta = \min\{1, \varepsilon(1 + \|f\| + \|g\|)^{-1}\}$ and let N be chosen as above. Then $\|f_n g_n - fg\| < \varepsilon$ for all $n \geq N$. This shows that $\lim_{n\to\infty} f_n g_n = fg$. But $f_n g_n \in A$ so, by Theorem 2.5.5, we have $fg \in \bar{A}$. □

The next lemma is a special case of the Weierstrass approximation theorem. We shall give an alternative proof, and so make the proof of the Stone–Weierstrass theorem independent of Weierstrass's theorem.

4.5.4 LEMMA *For each $\varepsilon > 0$ there exists a polynomial function p such that $p(0) = 0$ and $| \, |t| - p(t) \, | < \varepsilon$ for $-1 \le t \le 1$.*

PROOF Let $-1 \le t \le 1$. We define the sequence $(p_n(t))$ inductively by $p_0(t) = 0$ and

$$p_{n+1}(t) = p_n(t) + \tfrac{1}{2}(t^2 - (p_n(t))^2)$$

for $n = 0, 1, 2, \ldots$. Then $p_1(t) = \frac{1}{2}t^2 \le |t|$. Suppose that $p_n(t) \le |t|$ for some positive integer n. Then

$$\begin{aligned} |t| - p_{n+1}(t) &= (|t| - p_n(t)) - \tfrac{1}{2}(|t|^2 - (p_n(t))^2) \\ &= (|t| - p_n(t))(1 - \tfrac{1}{2}(|t| + p_n(t))) \\ &\ge (|t| - p_n(t))(1 - |t|) \\ &\ge 0 \end{aligned}$$

and hence by induction $p_n(t) \le |t|$ for $n = 1, 2, \ldots$. It is now obvious that $0 \le p_n(t) \le p_{n+1}(t) \le |t|$ for $n = 1, 2, \ldots$ and therefore $(p_n(t))$ converges. Let $q(t) = \lim_{n \to \infty} p_n(t)$. Then $q(t) \ge 0$ and

$$q(t) = q(t) + \tfrac{1}{2}(t^2 - (q(t))^2),$$

so $q(t) = |t|$.

By Dini's theorem (4.3.9) the sequence (p_n) converges uniformly on $[-1,1]$ to q, and therefore there is a positive integer N such that $|q(t) - p_n(t)| < \varepsilon$ for $-1 \le t \le 1$ and all $n \ge N$. Clearly p_N is a polynomial function which satisfies the conditions of the lemma. □

4.5.5 LEMMA *Let A be a closed subalgebra of $C_{\mathbf{R}}(X)$. Then $f \vee g$ and $f \wedge g$ are in A for all f and g in A.*

PROOF Since $f \vee g = \frac{1}{2}(f + g + |f - g|)$
and $f \wedge g = \frac{1}{2}(f + g - |f - g|)$
(see p. 60) it is sufficient to prove that $|f| \in A$ whenever $f \in A$. This is obvious if $f = 0$. Suppose then that $f \in A$ and $f \ne 0$, and let $\varepsilon > 0$. By Lemma 4.5.4 there is a polynomial function p such that $p(0) = 0$ and $| \, |t| - p(t) \, | < \|f\|^{-1}\varepsilon$ for $-1 \le t \le 1$ and therefore since $| \, \|f\|^{-1} f(x) \, | \le 1$ for all $x \in X$,

$$\left| \left| \frac{f(x)}{\|f\|} \right| - p\left(\frac{f(x)}{\|f\|}\right) \right| < \frac{\varepsilon}{\|f\|} \tag{6}$$

for all $x \in X$. Since $p(0) = 0$ we have $p(t) = \alpha_1 t + \alpha_2 t^2 + \ldots + \alpha_n t^n$, where $\alpha_1, \alpha_2, \ldots, \alpha_n \in \mathsf{R}$, and (6) gives

$$\left| |f(x)| - \alpha_1 f(x) - \frac{\alpha_2}{\|f\|}(f(x))^2 - \ldots - \frac{\alpha_n}{\|f\|^{n-1}}(f(x))^n \right| < \varepsilon$$

for all $x \in X$. This shows that

$$\left\| |f| - \alpha_1 f - \frac{\alpha_2}{\|f\|} f^2 - \ldots - \frac{\alpha_n}{\|f\|^{n-1}} f^n \right\| < \varepsilon.$$

The function

$$\alpha_1 f + \frac{\alpha_2}{\|f\|} f^2 + \ldots + \frac{\alpha_n}{\|f\|^{n-1}} f^n$$

is in A because $f \in A$ and A is a subalgebra. Therefore, by Theorem 2.3.4, $|f| \in A$ because A is closed. □

The next lemma gives a sufficient condition for a function in $C_{\mathsf{R}}(X)$ to belong to a given closed subalgebra of $C_{\mathsf{R}}(X)$.

4.5.6 LEMMA *Let A be a closed subalgebra of $C_{\mathsf{R}}(X)$ and $f \in C_{\mathsf{R}}(X)$. Suppose that for each pair x,y of points of X there exists a function $g_{x,y} \in A$ such that $g_{x,y}(x) = f(x)$ and $g_{x,y}(y) = f(y)$. Then $f \in A$.*

PROOF Let $\varepsilon > 0$. Choose a point $x \in X$ which will remain fixed for the moment. For each $y \in X$ let

$$G_{x,y} = \{z \in X : f(z) < g_{x,y}(z) + \varepsilon\} = (f - g_{x,y})^{-1}((-\infty, \varepsilon)).$$

Since $f - g_{x,y}$ is continuous on X Theorem 2.7.3 shows that $G_{x,y}$ is open. Obviously $y \in G_{x,y}$ for all $y \in X$ so $\{G_{x,y} : y \in X\}$ is an open cover of X and therefore has a finite subcover, $\{G_{x,y_1}, G_{x,y_2}, \ldots, G_{x,y_n}\}$ say. Let

$$g_x = g_{x,y_1} \vee g_{x,y_2} \vee \ldots \vee g_{x,y_n}.$$

It follows from Lemma 4.5.5 that $g_x \in A$. Given $z \in X$, there is an integer k with $1 \leq k \leq n$ such that $z \in G_{x,y_k}$ and hence

$$\begin{aligned} f(z) &< g_{x,y_k}(z) + \varepsilon \\ &\leq \max\{g_{x,y_1}(z), g_{x,y_2}(z), \ldots, g_{x,y_n}(z)\} + \varepsilon \\ &= g_x(z) + \varepsilon. \end{aligned} \tag{7}$$

Also $f(x) = g_{x,y}(x)$ for all $y \in X$ so

$$f(x) = g_x(x). \tag{8}$$

We now let x vary over X. For each $x \in X$ let

$$H_x = \{z \in X : g_x(z) - \varepsilon < f(z)\}.$$

As above we see that H_x is open and (8) shows that $x \in H_x$ so $\{H_x : x \in X\}$ is an open cover of X which must have a finite subcover, $\{H_{x_1}, H_{x_2}, \ldots, H_{x_m}\}$ say. Let

$$g = g_{x_1} \wedge g_{x_2} \wedge \ldots \wedge g_{x_m}.$$

It follows from Lemma 4.5.5 that $g \in A$. Given $z \in X$, then $z \in H_{x_k}$ for some k with $1 \leq k \leq m$, and hence

$$\begin{aligned} f(z) &> g_{x_k}(z) - \varepsilon \\ &\geq \min \{g_{x_1}(z), g_{x_2}(z), \ldots, g_{x_m}(z)\} - \varepsilon \\ &= g(z) - \varepsilon. \end{aligned} \tag{9}$$

Also, by (7), we have $f(z) < g_x(z) + \varepsilon$ for all $z \in X$ and all $x \in X$ so

$$f(z) < \min \{g_{x_1}(z), g_{x_2}(z), \ldots, g_{x_m}(z)\} + \varepsilon = g(z) + \varepsilon \tag{10}$$

for all $z \in X$. From (9) and (10) we obtain

$$\| f - g \| = \max \{ | f(z) - g(z) | : z \in X\} < \varepsilon.$$

Since A is closed, it follows from Theorem 2.3.4 that $f \in A$. □

4.5.7 Definition A subset A of $C_{\mathsf{R}}(X)$ is said to *separate the points of X* if and only if for each pair x,y of distinct points of X there exists a function $f \in A$ such that $f(x) \neq f(y)$.

A set consisting of a single point is closed, and therefore Theorem 2.8.5 shows that $C_{\mathsf{R}}(X)$ separates the points of X. Theorem 2.8.5 shows also that for each point of X there is a function in $C_{\mathsf{R}}(X)$ that does not vanish at that point. We shall now see that these two properties are typical of $C_{\mathsf{R}}(X)$.

4.5.8 Theorem (The Stone–Weierstrass theorem) *Let X be a compact metric space and let A be a subalgebra of $C_{\mathsf{R}}(X)$. Suppose that A separates the points of X and that for each point of X there is a function in A that does not vanish at the point. Then A is dense in $C_{\mathsf{R}}(X)$.*

Proof If X has only one point, then $C_{\mathsf{R}}(X)$ is one-dimensional, and it is obvious that $A = C_{\mathsf{R}}(X)$. Therefore we may suppose that X has at least two distinct points. By Lemma 4.5.3 the closure $\bar{A}$ of A is a closed subalgebra of $C_{\mathsf{R}}(X)$. We shall use Lemma 4.5.6 to show that $\bar{A} = C_{\mathsf{R}}(X)$. Let x_1,x_2 be distinct points of X. We shall prove first that there are functions $h_1,h_2 \in A$ with

$$h_1(x_1) = h_2(x_2) = 1$$

and

$$h_1(x_2) = h_2(x_1) = 0.$$

By hypothesis there is a function $g \in A$ such that $g(x_1) \neq g(x_2)$. If $g(x_1) \neq 0$ let

$$h_1 = \frac{g(x_2)g - g^2}{g(x_1)g(x_2) - (g(x_1))^2}.$$

Clearly $h_1 \in A$, $h_1(x_1) = 1$ and $h_1(x_2) = 0$. If $g(x_1) = 0$ then $g(x_2) \neq 0$; let $g_0 \in A$ be such that $g_0(x_1) \neq 0$ and let

$$h_1 = \frac{g_0}{g_0(x_1)} - \frac{g_0(x_2)}{g_0(x_1)g(x_2)}g.$$

Again $h_1 \in A$, $h_1(x_1) = 1$ and $h_1(x_2) = 0$. The function h_2 is constructed similarly.

Now let $f \in C_{\mathsf{R}}(X)$ and let

$$g_{x_1x_2} = f(x_1)h_1 + f(x_2)h_2.$$

Obviously $g_{x_1x_2} \in A$, $g_{x_1x_2}(x_1) = f(x_1)$ and $g_{x_1x_2}(x_2) = f(x_2)$ and therefore Lemma 4.5.6 shows that $f \in \bar{A}$. This proves that $\bar{A} = C_{\mathsf{R}}(X)$. □

A subalgebra of $C_{\mathsf{R}}(X)$ that separates the points of X and contains the constant functions obviously satisfies the conditions of the above theorem, and is therefore dense in $C_{\mathsf{R}}(X)$; it is in this form that the Stone–Weierstrass theorem is often quoted. A comparison of the proofs of the Weierstrass and Stone–Weierstrass theorems is instructive. It shows how the introduction of algebraic techniques into analysis has transformed analysis and changed its emphasis. Classical analysis is mainly concerned with the study of a single function or type of function; functional analysis is concerned with the study of families of functions.

The Weierstrass approximation theorem can be obtained as a corollary of the Stone–Weierstrass theorem. It is easy to verify that the subalgebra of $C_{\mathsf{R}}([a,b])$ generated by the set $\{1,f_0\}$, where $f_0(t) = t$ for $a \leq t \leq b$, contains the constant functions and separates the points of $[a,b]$. It is therefore dense in $C_{\mathsf{R}}([a,b])$, and we have seen that this statement is equivalent to the Weierstrass approximation theorem.

Many other classical theorems on approximation by special classes of functions can be obtained as corollaries of the Stone–Weierstrass theorem. We shall be content to prove another famous theorem of Weierstrass on approximation by trigonometric polynomials. A *trigonometric polynomial* is a function p of the form

$$p(\theta) = \tfrac{1}{2}\alpha_0 + \sum_{r=1}^{n} (\alpha_r \cos r\theta + \beta_r \sin r\theta)$$

for all $\theta \in \mathsf{R}$, where $\alpha_0, \alpha_1, \alpha_2, \ldots, \alpha_n, \beta_1, \beta_2, \ldots, \beta_n \in \mathsf{R}$. A real-valued function f on R is said to be *periodic, with period* 2π, if and only if $f(t + 2\pi) = f(t)$ for all $t \in \mathsf{R}$. A trigonometric polynomial is obviously periodic with period 2π.

4.5.9 THEOREM (Weierstrass) *Let f be a continuous real-valued function on R that is periodic with period 2π and let $\varepsilon > 0$. Then there is a trigonometric polynomial p such that $|f(\theta) - p(\theta)| < \varepsilon$ for all $\theta \in \mathsf{R}$.*

PROOF Let X be the unit circle $\{(\xi_1,\xi_2) : \xi_1^2 + \xi_2^2 = 1\}$ in the Euclidean space R^2. By the Borel–Lebesgue theorem (4.2.11) X is compact. There is a natural one-to-one correspondence $g \to g^*$ between the set of all continuous real-valued periodic functions on R with period 2π and $C_{\mathsf{R}}(X)$; it is defined by $g^*(\cos\theta, \sin\theta) = g(\theta)$ for all $\theta \in \mathsf{R}$. The set, A^* say, of all functions p^* in $C_{\mathsf{R}}(X)$ that correspond to trigonometric polynomials p is easily seen to be a subalgebra of $C_{\mathsf{R}}(X)$ that separates the points of X and contains the constant functions (see Exercise (3)), and therefore, by the Stone–Weierstrass theorem, A^* is dense in $C_{\mathsf{R}}(X)$. This means that there is a trigonometric polynomial p with $\| f^* - p^* \| < \varepsilon$ (Theorem 2.3.4) and therefore, for all $\theta \in \mathsf{R}$,
$| f(\theta) - p(\theta) | = | f^*(\cos\theta, \sin\theta) - p^*(\cos\theta, \sin\theta) | \leq \| f^* - p^* \| < \varepsilon$. □

The following example shows that Theorem 4.5.8 becomes false when we replace $C_{\mathsf{R}}(X)$ by $C_{\mathsf{C}}(X)$.

4.5.10 EXAMPLE† Let D be the closed unit disc $\{z : |z| \leq 1\}$ in the complex plane C and let A be the subset of $C_{\mathsf{C}}(D)$ consisting of those functions in $C_{\mathsf{C}}(D)$ that are analytic on the interior, $\{z : |z| < 1\}$, of D. The Borel–Lebesgue theorem (4.2.11) shows that D is compact, and it is obvious that A is a subalgebra of $C_{\mathsf{C}}(D)$ that separates the points of D and contains the constant functions. However A is not dense in $C_{\mathsf{C}}(D)$. To see this we notice first that $A \neq C_{\mathsf{C}}(D)$ because the function g defined by $g(z) = \bar{z}$ is continuous but not analytic and hence $g \in C_{\mathsf{C}}(D) \sim A$. Thus if we show that A is closed in $C_{\mathsf{C}}(D)$ it will follow that A is not dense in $C_{\mathsf{C}}(D)$. Let (f_n) be a sequence in A with $\lim_{n\to\infty} f_n = f$. Then (f_n) converges to f uniformly on D and, *a fortiori*, uniformly on each compact subset of D. Consequently, by a well-known theorem on analytic functions (due to Weierstrass), $f \in A$ (see AHLFORS [2, p. 174] or HILLE [14, p. 192]). Corollary 2.5.6 now shows that A is closed.

Although in Theorem 4.5.8 we cannot replace $C_{\mathsf{R}}(X)$ by $C_{\mathsf{C}}(X)$ there is a simple extension of the theorem to the algebra $C_{\mathsf{C}}(X)$. Before we can state this extension we must introduce some notation. Let $f \in C_{\mathsf{C}}(X)$. The functions $\bar{f}$, $\operatorname{Re} f$ and $\operatorname{Im} f$ are defined as follows: $\bar{f}(x) = \overline{f(x)}$, $(\operatorname{Re} f)(x) = \operatorname{Re} f(x)$, $(\operatorname{Im} f)(x) = \operatorname{Im} f(x)$ for all $x \in X$. It is clear that $f = \operatorname{Re} f + i \operatorname{Im} f$, $\operatorname{Re} f = 2^{-1}(f + \bar{f})$, and $\operatorname{Im} f = (2i)^{-1}(f - \bar{f})$. For all $x,y \in X$ we have

$$| \bar{f}(x) - \bar{f}(y) | = | \overline{f(x) - f(y)} | = | f(x) - f(y) |$$

so $\bar{f} \in C_{\mathsf{C}}(X)$ and hence $\operatorname{Re} f \in C_{\mathsf{R}}(X)$ and $\operatorname{Im} f \in C_{\mathsf{R}}(X)$.

† The reader who is not familiar with the theory of analytic functions of a complex variable should omit this example.

The subalgebra A of $C_{\mathsf{C}}(D)$ defined in Example 4.5.10 has the property that, if $f \in A$ and f is not constant, then $\bar{f} \notin A$. (This can be shown by an elementary argument using the Cauchy–Riemann equations.) To extend the Stone–Weierstrass theorem to subalgebras of $C_{\mathsf{C}}(X)$ we must restrict our attention to those subalgebras of $C_{\mathsf{C}}(X)$ that contain with each function f the function $\bar{f}$.

4.5.11 THEOREM *Let X be a compact metric space and let A be a subalgebra of $C_{\mathsf{C}}(X)$. Suppose that A separates the points of X, that for each point of X there is a function in A that does not vanish at that point, and that $\bar{f} \in A$ whenever $f \in A$. Then A is dense in $C_{\mathsf{C}}(X)$.*

PROOF By Lemma 4.5.3, the closure $\bar{A}$ of A in $C_{\mathsf{C}}(X)$ is a closed subalgebra of $C_{\mathsf{C}}(X)$, and it follows easily that $B = \bar{A} \cap C_{\mathsf{R}}(X)$ is a closed subalgebra of $C_{\mathsf{R}}(X)$. For each $f \in A$ we have $\bar{f} \in A$ and so

$$\operatorname{Re} f = 2^{-1}(f + \bar{f}) \in A \cap C_{\mathsf{R}}(X) \subseteq B$$

and

$$\operatorname{Im} f = (2i)^{-1}(f - \bar{f}) \in A \cap C_{\mathsf{R}}(X) \subseteq B.$$

Given a pair x,y of distinct points of X there exists $f \in A$ such that $f(x) \neq f(y)$. Then either $(\operatorname{Re} f)(x) \neq (\operatorname{Re} f)(y)$ or $(\operatorname{Im} f)(x) \neq (\operatorname{Im} f)(y)$, which shows that B separates the points of X. Similarly for each point of X there is a function in B that does not vanish there, and consequently, by Theorem 4.5.8 we have $B = C_{\mathsf{R}}(X)$. But then, for each $f \in C_{\mathsf{C}}(X)$ we have $\operatorname{Re} f \in B \subseteq \bar{A}$ and $\operatorname{Im} f \in B \subseteq \bar{A}$ and so

$$f = \operatorname{Re} f + i \operatorname{Im} f \in \bar{A}.$$

This proves that $\bar{A} = C_{\mathsf{C}}(X)$. □

It follows easily from the Weierstrass approximation theorem that $C_{\mathsf{R}}([a,b])$ is separable. Indeed we have seen that the linear hull of the set $\{f_0^n : n = 0, 1, 2 \ldots\}$, where $f_0(t) = t$ for $a \leq t \leq b$, is dense in $C_{\mathsf{R}}([a,b])$ and therefore, by Theorem 3.4.8, $C_{\mathsf{R}}([a,b])$ is separable. A similar, but more complicated, argument based on the Stone–Weierstrass theorem shows that $C_{\mathsf{R}}(X)$ is separable for all compact metric spaces X.

4.5.12 THEOREM *Let X be a compact metric space. Then the Banach space $C_{\mathsf{K}}(X)$ is separable.*

PROOF We shall prove first that $C_{\mathsf{R}}(X)$ is separable. If X has only one point then $C_{\mathsf{R}}(X)$ is one-dimensional and so is separable because R is separable. Thus we may suppose that X has at least two distinct points. The space X is totally bounded (see p. 135) so for each positive integer

n there is a finite $1/n$-net, $\{x_{n1}, x_{n2},\ldots, x_{nm_n}\}$ say, in X. Let $B_{nk} = B(x_{nk},1/n)$ and, for $k = 1, 2,\ldots, m_n$ and $n = 1, 2,\ldots,$ let

$$f_{nk}(x) = d(x,X \sim B_{nk})$$

for all $x \in X$.

Lemma 2.8.4 shows that $f_{nk} \in C_{\mathsf{R}}(X)$. We shall prove that the subalgebra A generated by the set $A_0 = \{f_{nk}: k = 1, 2,\ldots, m_n; n = 1, 2,\ldots\}$ is dense in $C_{\mathsf{R}}(X)$. Let x,y be distinct points of X and let N be the smallest positive integer with $2/N < d(x,y)$. We have $x \in B_{Nk}$ for some k with $1 \leq k \leq m_N$ and therefore

$$\frac{2}{N} < d(x,y) \leq d(x,x_{Nk}) + d(x_{Nk},y) < \frac{1}{N} + d(x_{Nk},y).$$

This shows that $y \in X \sim B_{Nk}$. Since $X \sim B_{Nk}$ is closed, Lemma 2.8.4 now gives $f_{Nk}(x) \neq 0$ and $f_{Nk}(y) = 0$, which shows that A satisfies the conditions of Theorem 4.5.8. Therefore A is dense in $C_{\mathsf{R}}(X)$. It is not difficult to see that A is the linear hull of the set A_1 of all finite products that can be formed from the functions in the set A_0 (cf. p. 154), and consequently A_1 is fundamental in $C_{\mathsf{R}}(X)$. By Theorem 1.4.4 the set $A_0 \times \mathsf{N}$ is countable, and hence, by Theorem 1.4.6, the family $\mathscr{A}$ of all finite subsets of $A_0 \times \mathsf{N}$ is also countable. The mapping

$$\{(g_1,m_1), (g_2,m_2),\ldots, (g_j,m_j)\} \to g_1^{m_1}g_2^{m_2}\cdots g_j^{m_j}$$

maps $\mathscr{A}$ onto A_1 so, by Theorem 1.4.2, the set A_1 is countable. Theorem 3.4.8 now shows that $C_{\mathsf{R}}(X)$ is separable.

It is not difficult now to prove that $C_{\mathsf{C}}(X)$ is also separable. Let $\{g_n: n = 1, 2,\ldots\}$ be a countable subset of $C_{\mathsf{R}}(X)$ that is dense in $C_{\mathsf{R}}(X)$. The subset $\{g_n + ig_m: n,m = 1, 2,\ldots\}$ of $C_{\mathsf{C}}(X)$ is countable (Theorem 1.4.4); we shall show that it is dense in $C_{\mathsf{C}}(X)$. Let $f \in C_{\mathsf{C}}(X)$ and $\varepsilon > 0$. Then $\operatorname{Re} f \in C_{\mathsf{R}}(X)$ and $\operatorname{Im} f \in C_{\mathsf{R}}(X)$ so, by Theorem 2.3.4, there are positive integers n and m with $\| \operatorname{Re} f - g_n \| < \frac{1}{2}\varepsilon$ and $\| \operatorname{Im} f - g_m \| < \frac{1}{2}\varepsilon$. Hence

$$\begin{aligned} \| f - (g_n + ig_m) \| &= \| (\operatorname{Re} f - g_n) + i(\operatorname{Im} f - g_m) \| \\ &\leq \| \operatorname{Re} f - g_n \| + \| \operatorname{Im} f - g_m \| \\ &< \varepsilon. \end{aligned}$$

This shows that $\{g_n + ig_m: n,m = 1, 2,\ldots\}$ is dense in $C_{\mathsf{C}}(X)$. □

Exercises

(1) Give examples to show that if, in the statement of Theorem 4.5.1, the bounded closed interval $[a,b]$ is replaced by the interval $[a,b)$ or by the interval $[a, \infty)$ then the resulting statements are false.

(2) *The Bernstein polynomials of a function.* The Weierstrass approximation theorem naturally leads to the problem of constructing polynomials which are uniform approximations to a given continuous function. In 1912 the Russian mathematician S. N. Bernstein published a proof of the Weierstrass theorem which gives a solution of this problem.

The *Bernstein polynomials* $B_n(f,\cdot)$, $n = 1, 2, \ldots$, of a function f on $[0,1]$ are defined by

$$B_n(f,x) = \sum_{k=0}^{n} \binom{n}{k} x^k (1-x)^{n-k} f\left(\frac{k}{n}\right)$$

for $x \in [0,1]$, where the $\binom{n}{k}$ are the binomial coefficients.

Prove the following statements:

(a)
$$\sum_{k=0}^{n} \binom{n}{k} x^k (1-x)^{n-k} = 1,$$

$$\sum_{k=0}^{n} \frac{k}{n}\binom{n}{k} x^k (1-x)^{n-k} = x,$$

and
$$\sum_{k=0}^{n} \frac{k^2}{n^2}\binom{n}{k} x^k (1-x)^{n-k} = \left(1-\frac{1}{n}\right)x^2 + \frac{1}{n}x;$$

(b) $$f(x) - B_n(f,x) = \sum_{k=0}^{n} \left(f(x) - f\left(\frac{k}{n}\right)\right)\binom{n}{k} x^k (1-x)^{n-k};$$

(c) If $\delta > 0$ and if K_1 and K_2 are the sets of non-negative integers $k \leq n$ with $\left|\frac{k}{n} - x\right| \leq \delta$ and $\left|\frac{k}{n} - x\right| > \delta$, respectively, then

$$\left|\sum_{k\in K_1} \left(f(x) - f\left(\frac{k}{n}\right)\right)\binom{n}{k} x^k (1-x)^{n-k}\right| \leq \sup_{|x-x'|\leq\delta} |f(x) - f(x')|,$$

$$\begin{aligned}\left|\sum_{k\in K_2} \left(f(x) - f\left(\frac{k}{n}\right)\right)\binom{n}{k} x^k (1-x)^{n-k}\right| &\leq \frac{2}{\delta^2}\|f\| \sum_{k=0}^{n}\left(\frac{k}{n} - x\right)^2 \binom{n}{k} x^k (1-x)^{n-k} \\ &\leq \frac{2}{n\delta^2}\|f\|;\end{aligned}$$

(d) If f is a continuous real-valued function on $[0,1]$ then the sequence $(B_n(f,\cdot))$ converges uniformly to f. (A generalization of this result is given in Exercise 5.6(3).)

(3) Show that the set A^* defined in the proof of Theorem 4.5.9 consists of all polynomial functions of the form

$$p(\xi_1,\xi_2) = \sum_{k,l=0}^{n} \alpha_{kl}\xi_1^k\xi_2^l$$

with $\alpha_{kl} \in \mathsf{R}$.

(4) Let $\Gamma = \{z \in \mathsf{C}: |z| = 1\}$. Prove that the set of rational functions p of the form $p(z) = \sum_{k=-n}^{n} \alpha_k z^k$ for $z \in \Gamma$, with $\alpha_k \in \mathsf{C}$, is dense in $C_{\mathsf{C}}(\Gamma)$.

(5) Let functions f_k, $k = 1, 2,...$, be defined by $f_k(x) = \exp(kx)$ for $x \in [a,b]$. Prove that $\{f_k: k = 1, 2,...\}$ is fundamental in $C_{\mathsf{R}}([a,b])$.

(6) Let X be a compact metric space. Prove that the set

$$\{f \in C_{\mathsf{C}}(X): |f(x)| = 1 \text{ for all } x \in X\}$$

is fundamental in $C_{\mathsf{C}}(X)$.

(7) Functions ϕ_n, $n = 1, 2,...$, are defined on the Cantor set C by setting $\phi_n(x) = (-1)^{\alpha_n/2}$ if $x = \sum_{m=1}^{\infty} \alpha_m 3^{-m}$ with $\alpha_m = 0$ or $\alpha_m = 2$ for $m = 1, 2,...$. Prove that the set of all functions of the form $\phi_{n_1}\phi_{n_2}\cdots\phi_{n_k}$, where $n_1, n_2,..., n_k$ and k are positive integers, is a fundamental subset of $C_{\mathsf{R}}(C)$.

(8) Prove that if Y is a closed subset of a compact metric space X then $\{f|Y: f \in C_{\mathsf{R}}(X)\} = C_{\mathsf{R}}(Y)$.

Note that this result can be read as saying that every continuous real-valued function on Y can be extended to a continuous function on X. This is a special case of a topological result known as *Tietze's theorem.*

(9) The separability of $C_{\mathsf{R}}([a,b])$ can be proved by a method which depends upon uniform approximation by step-functions, and which is independent of the Stone–Weierstrass theorem. Let A be the subset of $B_{\mathsf{R}}([a,b])$ consisting of all step-functions f which have rational values and for which there is an f-partition consisting of rational numbers. (See Definition 3.7.2.) Prove that A is countable and that $C_{\mathsf{R}}([a,b]) \subseteq \bar{A}$. Deduce that $C_{\mathsf{R}}([a,b])$ is separable.

In Exercises (10) *and* (11) X *denotes a compact metric space.*

(10) There are extensions of Theorem 4.5.8. Prove by using the method of proof of Theorem 4.5.8, that if A is a subalgebra of $C_{\mathsf{R}}(X)$ and $f \in C_{\mathsf{R}}(X)$ then the following two conditions on f are necessary and sufficient for f to be in the closure of A:

(a) if $g(x) = 0$ for all $g \in A$ then $f(x) = 0$,

(b) if $g(x) = g(y)$ for all $g \in A$ then $f(x) = f(y)$.

(11) A subset I of a linear associative algebra E is an *ideal* of E if it is a linear subspace of E and $fg \in I$, $gf \in I$ for all $f \in I$ and $g \in E$. An ideal

I is *maximal* if $I \neq E$, and if there is no ideal J such that $I \subseteq J$, $J \neq I$, and $J \neq E$. The result of Exercise (10) leads to a characterization of the closed ideals and the maximal ideals of $C_{\mathsf{R}}(X)$. Prove the following statements:

(a) A subset of an algebra which is an ideal of the algebra is also a subalgebra.

(b) The closure of an ideal of $C_{\mathsf{R}}(X)$ is an ideal of $C_{\mathsf{R}}(X)$.

(c) Between the closed ideals I of $C_{\mathsf{R}}(X)$ and the closed subsets Y of X there is a one-to-one correspondence given by

$$I = \{f \in C_{\mathsf{R}}(X) : f(x) = 0 \text{ for all } x \in Y\}.$$

(d) If I is a maximal ideal of $C_{\mathsf{R}}(X)$ then I is closed and, for some $y \in X$, $I = \{f \in C_{\mathsf{R}}(X) : f(y) = 0\}$. (The reader should observe that there is a fairly short direct proof of this result.)

(e) If A is a closed subalgebra of $C_{\mathsf{R}}(X)$ that separates the points of X, then either $A = C_{\mathsf{R}}(X)$ or A is a maximal ideal of $C_{\mathsf{R}}(X)$.

4.6 The Ascoli-Arzelà theorem

The general criteria for compactness obtained in Section 4.2 are seldom easy to verify. In this section we shall establish an important, and often readily verifiable, characterization of the compact subsets of the metric space $C_{\mathsf{K}}(X)$ when X is itself a compact metric space.

4.6.1 Definition Let X be a metric space. A non-empty subset A of $C_{\mathsf{K}}(X)$ is said to be *equicontinuous at a point* $x \in X$ if and only if, for each $\varepsilon > 0$, there exists $\delta > 0$ such that $|f(y) - f(x)| < \varepsilon$ for all $y \in X$ with $d(x,y) < \delta$ and all $f \in A$. The set A is said to be *equicontinuous on* X if and only if it is equicontinuous at each point of X.

Let A be a non-empty subset of $C_{\mathsf{K}}(X)$, $x \in X$ and $\varepsilon > 0$. Since each $f \in A$ is continuous at x there exists $\delta(f) > 0$, which may depend on ε, f and x, such that $|f(y) - f(x)| < \varepsilon$ for all $y \in X$ with $d(x,y) < \delta(f)$. In general $\delta(f)$ will vary with f; the condition that A be equicontinuous at x requires that a single $\delta > 0$ can be chosen independently of $f \in A$. It is easy to see that each finite subset of $C_{\mathsf{K}}(X)$ is equicontinuous on X.

The concept of equicontinuity was introduced independently by G. Ascoli and C. Arzelà in the early 1880's.

4.6.2 Theorem (Ascoli–Arzelà) *Let X be a compact metric space. Then a non-empty subset of $C_{\mathsf{K}}(X)$ is relatively compact if and only if it is bounded and equicontinuous on X.*

PROOF Let A be a non-empty subset of $C_{\mathsf{K}}(X)$. Suppose first that A is relatively compact. Then $\bar{A}$ is compact and hence totally bounded. (See p. 135.) It follows by Lemma 4.2.8 that A is totally bounded and therefore bounded. Let $\varepsilon > 0$ and let $\{f_1, f_2, \ldots, f_n\}$ be a $\frac{1}{3}\varepsilon$-net in A. There are positive real numbers $\delta_1, \delta_2, \ldots, \delta_n$ such that, for $k = 1, 2, \ldots, n$,

$$|f_k(y) - f_k(x)| < \tfrac{1}{3}\varepsilon$$

whenever $d(y,x) < \delta_k$. Let $\delta = \min\{\delta_1, \delta_2, \ldots, \delta_n\}$. Given $f \in A$, then $\|f - f_k\| < \frac{1}{3}\varepsilon$ for some integer k with $1 \le k \le n$, and consequently, if $d(x,y) < \delta$,

$$\begin{aligned}|f(y) - f(x)| &\le |f(y) - f_k(y)| + |f_k(y) - f_k(x)| + |f_k(x) - f(x)| \\ &\le 2\|f - f_k\| + |f_k(y) - f_k(x)| \\ &< \varepsilon.\end{aligned}$$

This proves that A is equicontinuous at x. We remark that, in this part of the proof, we have not used the fact that X is compact.

Suppose now that A is bounded and equicontinuous on X. Since $C_{\mathsf{K}}(X)$ is complete (Theorem 2.8.3), Corollary 4.2.10 shows that, to prove that A is relatively compact, it is sufficient to prove that A is totally bounded. Let $\varepsilon > 0$. By the equicontinuity of A on X, for each $x \in X$ there exists $\delta_x > 0$ such that

$$|f(y) - f(x)| < \tfrac{1}{4}\varepsilon \tag{1}$$

for all $y \in X$ with $d(y,x) < \delta_x$ and all $f \in A$. Since X is compact, the open cover $\{B(x,\delta_x): x \in X\}$ of X has a finite subcover,

$$\{B(x_1,\delta_{x_1}), B(x_2,\delta_{x_2}), \ldots, B(x_n,\delta_{x_n})\}$$

say. Let j be an integer with $1 \le j \le n$. Since A is a bounded subset of $C_{\mathsf{K}}(X)$, the set $A[x_j] = \{f(x_j): f \in A\}$ is a bounded subset of K and so is totally bounded (cf. the proof of Theorem 4.2.11). Choose a $\frac{1}{4}\varepsilon$-net, $\{\xi_{j1}, \xi_{j2}, \ldots, \xi_{jk_j}\}$ say, in $A[x_j]$.

For each n-tuple $(m_1, m_2, \ldots, m_n)$ of integers, with $1 \le m_j \le k_j$ for $j = 1, 2, \ldots, n$, let

$$A(m_1, m_2, \ldots, m_n) = \{f \in A: |f(x_j) - \xi_{jm_j}| < \tfrac{1}{4}\varepsilon \text{ for } j = 1, 2, \ldots, n\}. \tag{2}$$

If $A(m_1, m_2, \ldots, m_n)$ is non-empty choose a function $f_{m_1m_2\cdots m_n}$ in $A(m_1, m_2, \ldots, m_n)$; otherwise set $f_{m_1m_2\cdots m_n} = g$ where g is some function in A. We shall prove that the set

$$\{f_{m_1m_2\cdots m_n}: m_j = 1, 2, \ldots, k_j;\ j = 1, 2, \ldots, n\}$$

is an ε-net in A.

Let $f \in A$. Since $\{\xi_{j1}, \xi_{j2}, \ldots, \xi_{jk_j}\}$ is a $\frac{1}{4}\varepsilon$-net in $A[x_j]$ for each integer j with $1 \le j \le n$ there is an integer m_j, with $1 \le m_j \le k_j$, such that

$$|f(x_j) - \xi_{jm_j}| < \tfrac{1}{4}\varepsilon. \tag{3}$$

For brevity we shall write $f_0 = f_{m_1 m_2 \cdots m_n}$. Now let $x \in X$. Since $\{B(x_1,\delta_{x_1}), B(x_2,\delta_{x_2}),\ldots, B(x_n,\delta_{x_n})\}$ covers X there is an integer l with $1 \leq l \leq n$ such that $d(x,x_l) < \delta_{x_l}$. Inequalities (1), (2) and (3) now give

$$\begin{aligned} |f(x) - f_0(x)| &\leq |f(x) - f(x_l)| + |f(x_l) - \xi_{lm_l}| \\ &\quad + |\xi_{lm_l} - f_0(x_l)| + |f_0(x_l) - f_0(x)| \\ &< \varepsilon. \end{aligned}$$

Consequently $\|f - f_0\| = \max\{|f(x) - f_0(x)| : x \in X\} < \varepsilon$. This proves that $\{f_{m_1 m_2 \cdots m_n} : m_j = 1, 2,\ldots, k_j; j = 1, 2,\ldots, n\}$ is an ε-net in A and therefore A is totally bounded. □

As an illustration of the use of the Ascoli–Arzelà theorem in classical analysis, we shall now prove Peano's theorem on the existence of solutions of the differential equation

$$\frac{dy}{dx} = f(x,y). \tag{4}$$

Before proceeding with the proof of Peano's theorem let us compare it with Picard's theorem (2.11.9). It will be seen that the hypotheses of the two theorems differ only in that the Lipschitz condition (4) of Section 2.11 appears in Picard's theorem but not in Peano's theorem. Thus, viewed simply as an existence theorem for differential equations Peano's theorem contains Picard's theorem. However, there are two important differences between the theorems. First, Picard's theorem is not only an existence theorem but also a uniqueness theorem; there is no question of uniqueness in Peano's theorem (see Exercise (11)). Secondly, the proof of Picard's theorem is constructive in the sense explained in Section 2.11; the proof of Peano's theorem is not constructive in this sense.

4.6.3 THEOREM (Peano) *Let f be a continuous real-valued function on a non-empty open subset D of the Euclidean space $\mathbb{R}^2$, and let $(x_0,y_0) \in D$. Then there is a positive real number α such that the differential equation (4) has a solution ϕ on the interval $[x_0 - \alpha, x_0 + \alpha]$ which satisfies the boundary condition $\phi(x_0) = y_0$.*

PROOF By Lemma 2.11.5 it is sufficient to prove that there exist $\alpha > 0$ and a continuous real-valued function ϕ on the interval $[x_0 - \alpha, x_0 + \alpha]$, such that

$$\phi(x) = y_0 + \int_{x_0}^{x} f(t,\phi(t))\, dt \tag{5}$$

for $x_0 - \alpha \leq x \leq x_0 + \alpha$.

The proof begins exactly as that of Theorem 2.11.8. By the continuity of f we can find an open disc U, with centre (x_0,y_0), that is contained in D and such that

$$|f(x,y)| \leq K \tag{6}$$

for all $(x,y) \in U$, where K is some real number. Now choose $\alpha > 0$ so that the rectangle

$$R = \{(x,y): x_0 - \alpha \leq x \leq x_0 + \alpha, y_0 - \alpha K \leq y \leq y_0 + \alpha K\}$$

is contained in U. Let $I = [x_0 - \alpha, x_0 + \alpha]$ and let A be the set of all functions $\phi \in C_{\mathsf{R}}(I)$ that satisfy

$$|\phi(x) - y_0| \leq \alpha K \tag{7}$$

for all $x \in I$ and

$$|\phi(x_1) - \phi(x_2)| \leq K|x_1 - x_2| \tag{8}$$

for all $x_1,x_2 \in I$. It is obvious that A is non-empty. Inequality (7) shows that A is a bounded subset of $C_{\mathsf{R}}(I)$ and (8) shows that A is an equicontinuous subset of $C_{\mathsf{R}}(I)$. Therefore, by Theorem 4.6.2, the set A is relatively compact. An argument similar to that used in the proof of Lemma 2.11.6 shows that A is a closed subset of $C_{\mathsf{R}}(I)$ and therefore A is compact.

For each $\phi \in A$ the mapping $t \to f(t,\phi(t))$ is defined and continuous on I. Let

$$(T\phi)(x) = y_0 + \int_{x_0}^{x} f(t,\phi(t))\, dt \tag{9}$$

for all $x \in I$ and all $\phi \in A$. We shall prove first that $T\phi \in A$ for all $\phi \in A$. In fact, by (9) and (6), for all $x_1,x_2 \in I$

$$\begin{aligned} |(T\phi)(x_1) - (T\phi)(x_2)| &= \left|\int_{x_2}^{x_1} f(t,\phi(t))\, dt\right| \\ &\leq |x_1 - x_2| \sup\{|f(t,\phi(t))| : t \in I\} \\ &\leq K|x_1 - x_2|, \end{aligned}$$

and for all $x \in I$

$$\begin{aligned} |(T\phi)(x) - y_0| &= \left|\int_{x_0}^{x} f(t,\phi(t))\, dt\right| \\ &\leq K|x - x_0| \\ &\leq \alpha K. \end{aligned}$$

We shall prove next that T is a continuous mapping of A into itself. Let $\phi_0 \in A$ and $\varepsilon > 0$. By the Borel–Lebesgue theorem (4.2.11) the rectangle R is compact, so by Theorem 4.3.8 the function f is uniformly continuous on R and there exists $\delta > 0$ such that

$$|f(x_1,y_1) - f(x_2,y_2)| < \frac{\varepsilon}{\alpha} \tag{10}$$

for all $(x_1,y_1),(x_2,y_2) \in R$ with $((x_1 - x_2)^2 + (y_1 - y_2)^2)^{1/2} < \delta$.

Let $\phi \in A$ with $\| \phi - \phi_0 \| < \delta$. Then $| \phi(t) - \phi_0(t) | < \delta$ for all $t \in I$ and consequently by (10) and (9), for all $x \in I$,

$$\begin{aligned} | (T\phi)(x) - (T\phi_0)(x) | &= \left| \int_{x_0}^{x} (f(t,\phi(t)) - f(t,\phi_0(t)))\, dt \right| \\ &\leq | x - x_0 | \sup \{ | f(t,\phi(t)) - f(t,\phi_0(t)) | : t \in I\} \\ &\leq | x - x_0 | \frac{\varepsilon}{\alpha} \\ &\leq \varepsilon. \end{aligned}$$

Thus $\| T\phi - T\phi_0 \| = \max \{ | (T\phi)(x) - (T\phi_0)(x) | : x \in I\} \leq \varepsilon$. This proves that T is continuous at ϕ_0.

Let $F(\phi) = \| T\phi - \phi \|$ for all $\phi \in A$. Then for all $\phi_1,\phi_2 \in A$

$$\begin{aligned} | F(\phi_1) - F(\phi_2) | &= | \| T\phi_1 - \phi_1 \| - \| T\phi_2 - \phi_2 \| | \\ &\leq \| (T\phi_1 - \phi_1) - (T\phi_2 - \phi_2) \| \\ &\leq \| T\phi_1 - T\phi_2 \| + \| \phi_1 - \phi_2 \|, \end{aligned}$$

from which it follows that F is continuous on A. Since A is compact Corollary 4.3.2 shows that there is a function $\phi_0 \in A$ with

$$F(\phi_0) = \inf \{F(\phi) : \phi \in A\}.$$

Obviously $F(\phi_0) \geq 0$. If $F(\phi_0) = 0$, then $\| T\phi_0 - \phi_0 \| = 0$ and so $T\phi_0 = \phi_0$. But then, by definition of T, ϕ_0 satisfies (5). Thus to complete the proof of the theorem we have to show that $\inf \{F(\phi) : \phi \in A\} = 0$.

Let $\varepsilon > 0$. We shall construct a function $\phi \in A$, called a *Cauchy polygon,* such that $F(\phi) < \varepsilon$. Let $\delta > 0$ be chosen as in (10) and let x_j, $j = \pm 1, \pm 2, \ldots, \pm n$, be real numbers such that

$$\begin{aligned} x_0 - \alpha = x_{-n} < x_{-n+1} < \ldots < x_{-1} < x_0 \\ < x_1 < \ldots < x_{n-1} < x_n = x_0 + \alpha \end{aligned}$$

and

$$x_j - x_{j-1} < \frac{\delta}{(1 + K^2)^{1/2}} \tag{11}$$

for $j = -n + 1, -n + 2, \ldots, -1, 0, 1, \ldots, n$.

Before defining the Cauchy polygon ϕ explicitly, let us describe its construction. (See Fig. 4.1.) Draw the straight line through (x_0,y_0) with slope $f(x_0,y_0)$ and let it meet the line $x = x_1$ at the point (x_1,y_1). The line segment joining (x_0,y_0) and (x_1,y_1) lies inside the triangle bounded by the lines through (x_0,y_0) with slope $\pm K$ and the line $x = x_0 + \alpha$. Next draw the line through (x_1,y_1) with slope $f(x_1,y_1)$ and let it meet the line $x = x_2$ in (x_2,y_2), say. We continue drawing line segments in this way until we reach the point $x_n = x_0 + \alpha$ and then repeat the process to the left of x_0 until we reach $x_{-n} = x_0 - \alpha$. In this way we obtain a polygonal line lying inside the two triangular regions deter-

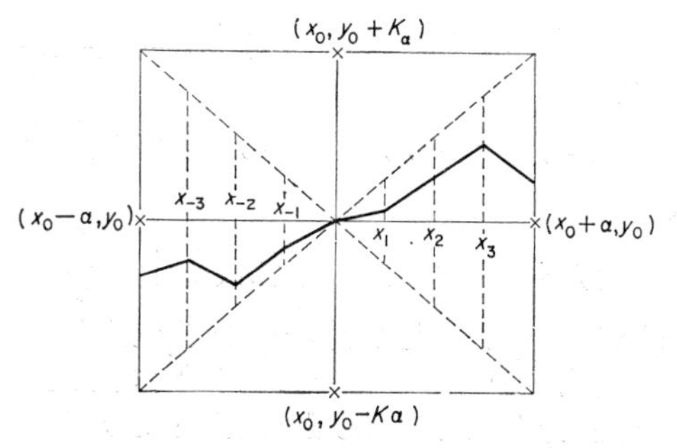

FIG. 4.1. The Cauchy polygon $y = \phi(x)$ with $n = 4$.

mined by the lines through (x_0,y_0) with slope $\pm K$ and the lines $x = x_0 \pm \alpha$. This polygonal line is the graph of the Cauchy polygon ϕ.

Explicitly ϕ is defined inductively as follows:

$$\phi(x_0) = y_0,$$
$$\phi(x) = \phi(x_{j-1}) + f(x_{j-1},\phi(x_{j-1}))(x - x_{j-1})$$

for $x_{j-1} < x \leq x_j$ and $j = 1, 2,..., n$, and

$$\phi(x) = \phi(x_{j+1}) + f(x_{j+1},\phi(x_{j+1}))(x - x_{j+1})$$

for $x_j \leq x < x_{j+1}$ and $j = -1, -2,..., -n$. Let σ be the step function on I defined by

$$\sigma(x) = \begin{cases} f(x_{j-1},\phi(x_{j-1})) \text{ for } x_{j-1} < x \leq x_j \text{ and } j = 1, 2,..., n, \\ f(x_{j+1},\phi(x_{j+1})) \text{ for } x_j \leq x < x_{j+1} \text{ and } j = -1, -2,..., -n, \\ f(x_0,y_0) \quad\quad \text{ for } x = x_0. \end{cases}$$

By (6) we have $|\sigma(x)| \leq K$ for all $x \in I$. It is easy to verify that, in the notation of Section 3.9, $\phi'_+(x) = \sigma(x,+)$ for all $x \in [x_{-n},x_n)$ and $\phi'_-(x) = \sigma(x,-)$ for all $x \in (x_{-n},x_n]$, and therefore Theorem 3.9.9 (or a direct calculation) shows that

$$\phi(x) = y_0 + \int_{x_0}^{x} \sigma(t)\,dt \tag{12}$$

for all $x \in I$. Using (12) we shall verify that $\phi \in A$. For all $x_1,x_2 \in I$,

$$\begin{aligned} |\phi(x_1) - \phi(x_2)| &= \left|\int_{x_2}^{x_1} \sigma(t)\,dt\right| \\ &\leq |x_1 - x_2| \sup \{|\sigma(t)| : t \in I\} \\ &\leq K|x_1 - x_2|, \end{aligned}$$

and for all $x \in I$,

$$\begin{aligned} |\phi(x) - y_0| &= \left|\int_{x_0}^{x} \sigma(t)\,dt\right| \\ &\leq |x - x_0| \sup \{|\sigma(t)| : t \in I\} \\ &\leq \alpha K. \end{aligned}$$

Finally we shall verify that $F(\phi) < \varepsilon$. By (8) and (11), for $j = 1, 2, \ldots, n$ and $x_{j-1} \leq t \leq x_j$ we have

$$((x_{j-1} - t)^2 + (\phi(x_{j-1}) - \phi(t))^2)^{1/2} \leq |x_{j-1} - t| (1 + K^2)^{1/2} < \delta$$

and hence by (10)

$$|f(x_{j-1},\phi(x_{j-1})) - f(t,\phi(t))| < \frac{\varepsilon}{\alpha}. \tag{13}$$

Let $x \in I$. We shall suppose that $x > x_0$. Only trivial modifications are needed in the case when $x < x_0$; these are left to the reader. Equations (9) and (12), inequality (13) and the definition of σ give

$$\begin{aligned}
&|\phi(x) - (T\phi)(x)| \\
&= \left| \int_{x_0}^{x} (\sigma(t) - f(t,\phi(t)))\, dt \right| \\
&\leq \int_{x_0}^{x} |\sigma(t) - f(t,\phi(t))|\, dt \\
&\leq \int_{x_0}^{x_0+\alpha} |\sigma(t) - f(t,\phi(t))|\, dt \\
&= \sum_{j=1}^{n} \int_{x_{j-1}}^{x_j} |f(x_{j-1},\phi(x_{j-1})) - f(t,\phi(t))|\, dt \\
&\leq \sum_{j=1}^{n} (x_j - x_{j-1}) \sup \{ |f(x_{j-1},\phi(x_{j-1})) - f(t,\phi(t))| : x_{j-1} \leq t \leq x_j \} \\
&\leq \frac{\varepsilon}{\alpha} \sum_{j=1}^{n} (x_j - x_{j-1}) \\
&= \varepsilon.
\end{aligned}$$

Thus

$$F(\phi) = \| T\phi - \phi \| = \max \{ |(T\phi)(x) - \phi(x)| : x \in I \} \leq \varepsilon.$$

This proves that $\inf \{F(\phi) : \phi \in A\} = 0$. □

The proofs of both Picard's theorem and Peano's theorem contain the same basic idea: we define a mapping T of a metric space into itself and show that T has a fixed point ϕ_0 which is then a solution of the given differential equation. In the proof of Picard's theorem the existence of a fixed point for T is a consequence of a general fixed point theorem—the contraction mapping theorem. In the proof of Peano's theorem the existence of a fixed point for T is established by a special argument which relies heavily on the form of T; however, there is a general fixed point theorem, due to J. Schauder, that can be used to prove Peano's theorem.

The most famous of all fixed point theorems is that of L. E. J. Brouwer which states that every continuous mapping of the closed unit ball in the Euclidean space R^n into itself has a fixed point. Schauder's theorem is a generalization of Brouwer's theorem to infinite dimensional normed linear spaces. A subset K of a real linear space is said to be *convex* if and only if $\alpha x + (1 - \alpha)y \in K$ for all $x,y \in K$ and all α with $0 \leq \alpha \leq 1$. Schauder's theorem states that every continuous mapping of a compact convex subset of a normed linear space into itself has a fixed point. It is readily verified that the set A of Theorem 4.6.3 is convex, and it therefore follows from Schauder's theorem that T has a fixed point.

The proofs of the theorems of Brouwer and Schauder are beyond the scope of this book. The reader who is interested will find elementary proofs of Brouwer's theorem in BERGE [5, pp. 168–76] and DUNFORD and SCHWARTZ [8, pp. 468–70], and a proof of Schauder's theorem in DUNFORD and SCHWARTZ [8, pp. 453–6].

EXERCISES

(1) A subset A of $C_{\mathsf{R}}(X)$ is said to be *uniformly equicontinuous on* X if for each $\varepsilon > 0$ there is a $\delta > 0$ such that $|f(x) - f(y)| < \varepsilon$ for all $x,y \in X$ with $d(x,y) < \delta$ and all $f \in A$. Prove that if X is compact and A is equicontinuous on X then A is uniformly equicontinuous on X.

(2) Give an example of an unbounded sequence (f_n) in $C_{\mathsf{R}}([0,1])$ which is pointwise convergent to a continuous function on $[0,1]$.

(3) Let X be a compact metric space and (f_n) a sequence in $C_{\mathsf{R}}(X)$. Prove that if the set $\{f_n: n = 1, 2,...\}$ is equicontinuous, and for each $x \in X$ the sequence $(f_n(x))$ is convergent, then (f_n) is convergent in $C_{\mathsf{R}}(X)$.

(4) Let $f_n(t) = \sin \sqrt{(t + 4n^2\pi^2)}$ for $t \geq 0$ and $n = 1, 2,....$ Prove that the set $\{f_n: n = 1, 2,...\}$ is a bounded and uniformly equicontinuous subset of $C_{\mathsf{R}}([0,\infty))$, but that it is not relatively compact. Prove also that the sequence (f_n) converges pointwise to 0 on $[0,\infty)$. (This shows that the Ascoli–Arzelà theorem and Exercise (3) may fail when X is not compact.)

(5) (a) Let A be a subset of $C_{\mathsf{R}}([a,b])$. Suppose that each $f \in A$ is differentiable on (a,b) and that there exists $M > 0$ such that $|f'(t)| \leq M$ for $a < t < b$ and all $f \in A$. Prove that A is equicontinuous.

(b) Prove that a subset A of the space $C^1_{\mathsf{R}}([a,b])$ of Example 2.1.13 is relatively compact if and only if A is bounded and $\{f' : f \in A\}$ is an equicontinuous subset of $C_{\mathsf{R}}([a,b])$.

(6) Prove that a subset A of the space c_0 of Exercise 3.4(8) is relatively compact if and only if A is bounded and given $\varepsilon > 0$ there is an integer N such that $|\xi_n| \leq \varepsilon$ for all $n \geq N$ and all $x = (\xi_n) \in A$.

(7) Prove that a subset A of the space ℓ^p is relatively compact if and only if A is bounded and given $\varepsilon > 0$ there is an integer N such that $\sum_{n=N}^{\infty} |\xi_n|^p < \varepsilon$ for all $x = (\xi_n) \in A$. (The compact subset $H = \{(\xi_n): |\xi_n| \leq 1/n \text{ for } n = 1, 2,\ldots\}$ of the space ℓ^2 is known as the *Hilbert cube*.)

(8) (a) Let X be a compact metric space and let $C_{\mathsf{R}^n}(X)$ denote the set of all continuous mappings of X into R^n. Prove that $C_{\mathsf{R}^n}(X)$ is a Banach space with linear space operations defined by

$$(\alpha f + \beta g)(x) = \alpha f(x) + \beta g(x)$$

for all $f,g \in C_{\mathsf{R}^n}(X)$, all $\alpha,\beta \in \mathsf{R}$, and all $x \in X$, and with norm defined by

$$\|f\| = \sup \{ \|f(x)\| : x \in X\}.$$

(b) A subset A of $C_{\mathsf{R}^n}(X)$ is said to be equicontinuous on X if and only if, given $x \in X$ and $\varepsilon > 0$ there exists $\delta > 0$ such that

$$\|f(y) - f(x)\| < \varepsilon$$

for all $y \in B(x,\delta)$ and all $f \in A$. Prove the following generalization of the Ascoli–Arzelà theorem: a non-empty subset of $C_{\mathsf{R}^n}(X)$ is relatively compact if and only if it is bounded and equicontinuous on X.

The generalized Ascoli–Arzelà theorem of Exercise (8) *can be used to extend Peano's theorem to systems of first order differential equations, and hence to higher order equations.*

(9) Let $f_1, f_2,\ldots, f_n$ be continuous real-valued functions on an open subset D of R^{n+1}. Given a point $(x_0, y_1,\ldots, y_n) \in D$, prove that for some $\alpha > 0$ there exist real-valued functions $\phi_1, \phi_2,\ldots, \phi_n$ on the interval $[x_0 - \alpha, x_0 + \alpha]$ such that

(a) $(x, \phi_1(x),\ldots, \phi_n(x)) \in D$ for $|x - x_0| \leq \alpha$,

(b) $\phi_j(x_0) = y_j$ for $j = 1, 2,\ldots, n$,

(c) $\phi_1, \phi_2,\ldots, \phi_n$ are differentiable on $[x_0 - \alpha, x_0 + \alpha]$ and

$$\phi_j'(x) = f_j(x, \phi_1(x),\ldots, \phi_n(x))$$

for $|x - x_0| \leq \alpha$ and $j = 1, 2,\ldots, n$. (Cf. Exercise 3.9(9).)

(10) Let f be a continuous real-valued function on an open subset D of R^{n+1}. Given a point $(x_0, y_1,\ldots, y_n) \in D$, prove that for some $\alpha > 0$ there exists a real-valued function ϕ on the interval $[x_0 - \alpha, x_0 + \alpha]$ such that

(a) ϕ is n-times differentiable on $[x_0 - \alpha, x_0 + \alpha]$,

(b) $(x, \phi(x), \phi'(x),\ldots, \phi^{(n-1)}(x)) \in D$ for $|x - x_0| \leq \alpha$,

(c) $\phi(x_0) = y_1, \phi'(x_0) = y_2,\ldots, \phi^{(n-1)}(x_0) = y_n$,

(d) $\phi^{(n)}(x) = f(x, \phi(x), \phi'(x),\ldots, \phi^{(n-1)}(x))$ for $|x - x_0| \leq \alpha$.

(Cf. Exercise 3.9(10).)

(11) The equation

$$\frac{dy}{dx} = 3xy^{1/3}$$

does not satisfy the conditions of Picard's theorem (2.11.9) in any neighbourhood of the point (0,0) in R^2, but does satisfy the conditions of Peano's theorem. Prove that in any interval $(-\alpha,\alpha)$ the equation has nine solutions which satisfy $y(0) = 0$.

CHAPTER 5

Bounded Linear Functionals

In Chapter 3 we introduced normed linear spaces, and derived some of their elementary properties. In the present chapter we shall begin the study of the deeper properties of normed linear spaces, and we shall continue it in Chapters 6, 8 and 9.

A bounded linear transformation of a normed linear space E over K into the normed linear space K is called a *bounded linear functional on E*. It is easy to give non-trivial examples of bounded linear functionals on concrete normed linear spaces, for example it is very easy to verify that the mapping $f \to \int_a^b f(s)\, ds$ is a bounded linear functional on $C_{\mathsf{R}}([a,b])$. However it is far from obvious that there are any non-zero bounded linear functionals on an arbitrary normed linear space.

The fact that there do exist non-zero bounded linear functionals on any normed linear space is a consequence of the Hahn–Banach theorem which is proved in Section 5.2. The Hahn–Banach theorem is one of the most important theorems in functional analysis.

In Sections 5.3, 5.4 and 5.5 we derive some consequences of the Hahn–Banach theorem that have many applications in the study of normed linear spaces and bounded linear transformations. We shall make extensive use of these results in Chapters 6 and 8. In Section 5.6 we use the Hahn–Banach theorem to prove the famous theorem of F. Riesz on the representation of the bounded linear functionals on $C_{\mathsf{R}}([a,b])$. As an illustration of the application of the Hahn–Banach theorem in classical analysis we use it and the Riesz representation theorem to derive a theorem of Herglotz on the representation of certain complex analytic functions by integrals.

5.1 Some Dual Spaces

The word *functional* is used to distinguish mappings of a linear space into the fields R and C.

5.1.1 DEFINITION A *linear functional* on a linear space E over K is a linear transformation of E into K.

If f is a linear functional on E then $f(\alpha x + \beta y) = \alpha f(x) + \beta f(y)$ for all $x,y \in E$ and all $\alpha,\beta \in \mathsf{K}$. The set of all linear functionals on E is a linear space over K, the algebraic operations being defined by $(f + g)(x) = f(x) + g(x)$ and $(\alpha f)(x) = \alpha f(x)$ for all $x \in E$. (See Definition 3.5.6.)

Suppose now that E is a normed linear space. By Theorem 3.5.4 a linear functional on E is continuous on E if and only if it is bounded. If f is a bounded linear functional on E then $|f(x)| \leq \|f\| \|x\|$ for all $x \in E$, where $\|f\| = \sup \{ |f(x)| : \|x\| \leq 1\}$ (Definition 3.5.5 and inequality (5) of Section 3.5).

The set $L(E, \mathsf{K})$ of all bounded linear functionals on E is a linear space and the mapping $f \to \|f\|$ is a norm on $L(E, \mathsf{K})$. Further, since K is complete $L(E, \mathsf{K})$ is a Banach space (Theorem 3.5.7).

5.1.2 DEFINITION The Banach space $L(E, \mathsf{K})$ of all bounded linear functionals on a normed linear space E over K is called the *dual space of* E. We shall write E^* in place of $L(E, \mathsf{K})$.

Typical elements of the space E^* will be denoted by x^*, y^*,.... This notation is now fairly common and has real advantages. The reader should notice, however, that the * is now being used in two different ways: the * in E^* indicates that E^* is the dual space of some normed linear space E; the * in x^* indicates only that x^* is an element of the space E^*—if $x \in E$ and $x^* \in E^*$ there is no implied relationship between x and x^*.

The space E^* is also called the *adjoint of* E or the *conjugate of* E. The dual space E^* always contains the zero linear functional, 0, given by $0(x) = 0$ for all $x \in E$. It is not obvious that there are non-zero bounded linear functionals on an arbitrary non-zero normed linear space. The existence of non-zero bounded linear functionals was established by Hahn in 1927. Hahn's theorem is a corollary of a more general theorem, now known as the Hahn–Banach theorem, which was proved by Banach in 1929.

For some particular normed linear spaces it is easy to establish the existence of non-zero bounded linear functionals without using the Hahn–Banach theorem. We shall obtain representations for the bounded linear functionals on certain normed linear spaces, and identify their dual spaces.

5.1.3 THEOREM *Let E be a finite dimensional normed linear space over K. Then every linear functional on E is bounded, the dual space E^* is finite dimensional and E and E^* have the same dimension.*

PROOF Suppose that E is n-dimensional and let $\{e_1, e_2, \ldots, e_n\}$ be a basis for E. Each $x \in E$ can be expressed uniquely in the form

$$x = \sum_{r=1}^{n} \xi_r e_r \tag{1}$$

where $\xi_r \in \mathsf{K}$ for $r = 1, 2, \ldots, n$. A simple calculation shows that the mapping $x \to \sum_{r=1}^{n} | \xi_r |$ is a norm on E. By Theorem 4.4.1 any two norms on E are equivalent; therefore there exists $M > 0$ such that

$$\sum_{r=1}^{n} | \xi_r | \leq M \| x \| \tag{2}$$

for all $x \in E$. (See p. 110.) For $r = 1, 2, \ldots, n$ let

$$e_r^*(x) = \xi_r$$

for all $x \in E$, where x and ξ_r are related by (1). It is easy to see that e_r^* is a linear functional on E. The inequality (2) gives

$$| e_r^*(x) | = | \xi_r | \leq M \| x \|$$

for all $x \in E$ and this shows that e_r^* is bounded. Thus $e_r^* \in E^*$.

The set $\{e_1^*, e_2^*, \ldots, e_n^*\}$ is linearly independent because if $\sum_{r=1}^{n} \alpha_r e_r^* = 0$ then, for $s = 1, 2, \ldots, n$, we have

$$0 = \left(\sum_{r=1}^{n} \alpha_r e_r^* \right)(e_s) = \sum_{r=1}^{n} \alpha_r e_r^*(e_s) = \alpha_s.$$

Finally let x^* be any linear functional on E. Then $x^* = \sum_{r=1}^{n} \beta_r e_r^*$, where $\beta_r = x^*(e_r)$. To see this let $x \in E$ and let $\xi_1, \xi_2, \ldots, \xi_n$ be given by (1), then

$$x^*(x) = x^*\left(\sum_{r=1}^{n} \xi_r e_r \right) = \sum_{r=1}^{n} \xi_r x^*(e_r) = \sum_{r=1}^{n} e_r^*(x) \beta_r = \left(\sum_{r=1}^{n} \beta_r e_r^* \right)(x).$$

This proves that x^* is a bounded linear functional and that $\{e_1^*, e_2^*, \ldots, e_n^*\}$ is a basis for E^*. □

If E is an n-dimensional linear space, bases $\{e_1, e_2, \ldots, e_n\}$ and $\{e_1^*, e_2^*, \ldots, e_n^*\}$ for E and E^*, respectively, for which $e_r^*(e_s) = \delta_{rs}$ for $r,s = 1, 2, \ldots, n$ are sometimes called *dual bases* for E and E^*.

5.1.4 EXAMPLE We shall obtain a representation for the bounded linear functionals on ℓ (Example 3.2.6) and use it to prove that ℓ^* is linearly isometric to m (Example 3.2.11).

Let $e_n = (\delta_{mn})_{m \geq 1}$. Clearly $e_n \in \ell$ and $\| e_n \| = 1$. Let $x = (\xi_n) \in \ell$ and set $s_n = \sum_{r=1}^{n} \xi_r e_r$. Then $\| x - s_n \| = \sum_{r=n+1}^{\infty} | \xi_r |$. Since $\sum_r | \xi_r |$ converges, given $\varepsilon > 0$ there exists an integer N such that $\| x - s_n \| < \varepsilon$ for all $n \geq N$. Now let $x^* \in \ell^*$ and let $\mu_n = x^*(e_n)$. For $n = 1, 2, \ldots,$

$$| \mu_n | = | x^*(e_n) | \leq \| x^* \| \; \| e_n \| = \| x^* \| \tag{3}$$

which shows that $(\mu_n) \in m$. Let $Tx^* = (\mu_n)$. This defines a mapping T of ℓ^* into m and it is easy to verify that T is linear. Inequality (3) gives

$$\| Tx^* \| = \sup \{ | \mu_n | : n = 1, 2, \ldots \} \leq \| x^* \| . \tag{4}$$

Also $| x^*(x - s_n) | \leq \| x^* \| \; \| x - s_n \| < \varepsilon \| x^* \|$ for all $n \geq N$, and

$$\begin{aligned} x^*(x - s_n) &= x^*(x) - x^*(s_n) \\ &= x^*(x) - \sum_{r=1}^{n} \xi_r x^*(e_r) \\ &= x^*(x) - \sum_{r=1}^{n} \xi_r \mu_r, \end{aligned}$$

and therefore $| x^*(x) - \sum_{r=1}^{n} \xi_r \mu_r | < \varepsilon \| x^* \|$ for all $n \geq N$. Thus we can represent x^* by the equation

$$x^*(x) = \sum_{r=1}^{\infty} \xi_r \mu_r .$$

It follows from this representation that $x^* = 0$ if $Tx^* = 0$. Consequently the mapping T is one-to-one.

Now suppose that $z = (\mu_n) \in m$. Since (μ_n) is a bounded sequence the series $\sum_n \xi_n \mu_n$ is absolutely convergent for each $x = (\xi_n) \in \ell$. We can define a mapping Sz of ℓ into K by $(Sz)(x) = \sum_{n=1}^{\infty} \xi_n \mu_n$ for all $x = (\xi_n) \in \ell$. It is easy to verify that Sz is a linear functional on ℓ. Also, for all $x = (\xi_n) \in \ell$,

$$\begin{aligned} | (Sz)(x) | &= \left| \sum_{n=1}^{\infty} \xi_n \mu_n \right| \\ &\leq \sum_{n=1}^{\infty} | \xi_n | \; | \mu_n | \\ &\leq \| z \| \sum_{n=1}^{\infty} | \xi_n | \\ &= \| z \| \; \| x \| . \end{aligned}$$

This proves that $Sz \in \ell^*$ and $\| Sz \| \leq \| z \|$. Obviously $T(Sz) = z$ so T maps ℓ^* onto m. Finally let $x^* \in \ell^*$ and let $z = Tx^*$. Since T is one-to-one, $x^* = Sz$ and

$$\| x^* \| = \| Sz \| \leq \| z \| = \| Tx^* \| \leq \| x^* \|,$$

where the final step follows from (4). This proves that T is a linear isometry of ℓ^* onto m.

5.1.5 EXAMPLE We shall obtain a representation for the bounded linear functionals on c_0 (Exercise 3.4(8)) and use it to prove that c_0^* is linearly isometric to ℓ.

Let $e_n = (\delta_{mn})_{m\geq 1}$. Clearly $e_n \in c_0$ and $\| e_n \| = 1$. Let $x = (\xi_n) \in c_0$ and set $s_n = \sum_{r=1}^{n} \xi_n e_n$. Since $\lim_{n\to\infty} \xi_n = 0$ there exists an integer N such that $| \xi_n | < \varepsilon$ for all $n \geq N$ and consequently

$$\| x - s_n \| = \sup \{ | \xi_r | : r = n+1, n+2, \ldots \} \leq \varepsilon$$

for $n \geq N$.

Now let $x^* \in c_0^*$ and let $\mu_n = x^*(e_n)$. As in Example 5.1.4 we obtain $| x^*(x) - \sum_{r=1}^{n} \xi_r \mu_r | \leq \varepsilon \| x^* \|$ for all $n \geq N$, and hence

$$x^*(x) = \sum_{r=1}^{\infty} \xi_r \mu_r. \tag{5}$$

We shall prove that $(\mu_n) \in \ell$. Let n be a positive integer, let

$$\alpha_r = \begin{cases} 0 & \text{if } \mu_r = 0, \\ \dfrac{| \mu_r |}{\mu_r} & \text{if } \mu_r \neq 0, \end{cases}$$

and let $x_n = \sum_{r=1}^{n} \alpha_r e_r$. Obviously $\| x_n \| \leq 1$ and (5) gives

$$x^*(x_n) = \sum_{r=1}^{n} | \mu_r |.$$

Therefore

$$\sum_{r=1}^{n} | \mu_r | = | x^*(x_n) | \leq \| x^* \| \; \| x_n \| \leq \| x^* \|.$$

This shows that $\sum_r | \mu_r |$ converges and that $\sum_{r=1}^{\infty} | \mu_r | \leq \| x^* \|$. Let $Tx^* = (\mu_n)$. We have just seen that T maps c_0^* into ℓ and $\| Tx^* \| \leq \| x^* \|$, and it is easy to verify that T is linear and one-to-one.

Now suppose that $(\mu_n) \in \ell$. It is obvious that $\sum_n \xi_n \mu_n$ converges absolutely for each $x = (\xi_n) \in c_0$. Let x^* be defined on c_0 by (5). It is easy to verify that $x^* \in c_0^*$ and $\| x^* \| \leq \sum_{n=1}^{\infty} | \mu_n |$. It now follows that T is a linear isometry of c_0^* onto ℓ.

5.1.6 EXAMPLE Let $p > 1$. We shall prove that $(\ell^p)^*$ is linearly isometric to ℓ^q where $q = p/(p-1)$. (The space ℓ^p is defined in Example 3.2.6.)

Let $e_n = (\delta_{mn})_{m\geq 1}$, let $x^* \in (\ell^p)^*$ and let $\mu_n = x^*(e_n)$. An argument similar to that used in Examples 5.1.4 and 5.1.5 shows that

$$x^*(x) = \sum_{n=1}^{\infty} \xi_n \mu_n \tag{6}$$

for all $x = (\xi_n) \in \ell^p$. We shall prove that $(\mu_n) \in \ell^q$. Let n be a positive integer, let

$$\alpha_r = \begin{cases} 0 & \text{if } \mu_r = 0, \\ \dfrac{|\mu_r|^q}{\mu_r} & \text{if } \mu_r \neq 0, \end{cases}$$

and let $x_n = \sum\limits_{r=1}^{n} \alpha_r e_r$. Then

$$\|x_n\|^p = \sum_{r=1}^{n} |\alpha_r|^p = \sum_{r=1}^{n} |\mu_r|^{p(q-1)} = \sum_{r=1}^{n} |\mu_r|^q,$$

and (6) gives

$$x^*(x_n) = \sum_{r=1}^{n} \alpha_r \mu_r = \sum_{r=1}^{n} |\mu_r|^q.$$

Therefore

$$\sum_{r=1}^{n} |\mu_r|^q = |x^*(x_n)| \leq \|x^*\| \; \|x_n\| = \|x^*\| \left(\sum_{r=1}^{n} |\mu_r|^q\right)^{1/p}.$$

It follows that

$$\left(\sum_{r=1}^{n} |\mu_r|^q\right)^{1/q} \leq \|x^*\|.$$

Consequently $\sum\limits_{r} |\mu_r|^q$ converges and

$$\left(\sum_{r=1}^{\infty} |\mu_r|^q\right)^{1/q} \leq \|x^*\|.$$

We leave the reader to verify, using Hölder's inequality (3.2.7), that the mapping $x^* \to (\mu_n)$ is a linear isometry of $(\ell^p)^*$ onto ℓ^q.

The case $p = 2$ is particularly interesting, for then $q = 2$ also, and the dual of ℓ^2 is linearly isometric to ℓ^2 itself. The space ℓ^2 is the most elementary example of a *Hilbert space*, and we shall see in Chapter 9 that the above result, for $p = 2$, is essentially a special case of a general theorem on Hilbert spaces (see Theorem 9.2.6 and the discussion that follows it).

EXERCISES

(1) A linear subspace M of a linear space E is said to be a *maximal linear subspace* if $M \neq E$ and, for any linear subspace N with $M \subseteq N \subseteq E$, either $N = M$ or $N = E$. Show that there is a natural correspondence between maximal linear subspaces and linear functionals:

(a) The *null space* $f^{-1}(\{0\})$ of a non-zero linear functional f is a maximal linear subspace.

(b) If M is a maximal linear subspace then $M = f^{-1}(\{0\})$ for some linear functional f.

(c) If f,g are linear functionals and $f^{-1}(\{0\}) = g^{-1}(\{0\})$ then $f = \alpha g$ for some $\alpha \in \mathsf{K}$.

(d) If E is a normed linear space, f is a linear functional on E and $M = f^{-1}(\{0\})$ then M is closed in E if and only if f is continuous.

(2) Let $x_1^*, x_2^*, \ldots, x_n^*, y^*$ be linear functionals on a linear space E over K with $y^*(x) = 0$ whenever $x_1^*(x) = x_2^*(x) = \ldots = x_n^*(x) = 0$. Prove that $y^* = \alpha_1 x_1^* + \ldots + \alpha_n x_n^*$ for some $\alpha_1, \alpha_2, \ldots, \alpha_n \in \mathsf{K}$.

(3) Let $\|\cdot\|_1$ and $\|\cdot\|_2$ be norms on a linear space E with $\|x\|_1 \leq M\|x\|_2$ for all $x \in E$. Prove that $(E, \|\cdot\|_1)^* \subseteq (E, \|\cdot\|_2)^*$.

(4) The real normed linear spaces c and c_0 are not linearly isometric (see Exercise 3.6(4)), but their dual spaces c^* and c_0^* are linearly isometric. Prove that there is a linear isometry of ℓ onto c^*, defined by

$$x^*(x) = \mu_1 \lim_{n\to\infty} \xi_n + \sum_{n=2}^{\infty} \mu_n \xi_{n-1}$$

for $x = (\xi_n) \in c$ and $(\mu_n) \in \ell$.

5.2 The Hahn-Banach Theorem

We turn now to the question of the existence of non-zero bounded linear functionals on an arbitrary non-zero normed linear space. Such functionals obviously exist when the space is one-dimensional, and every non-zero linear space has one-dimensional linear subspaces. Thus, if we can show that any bounded linear functional on a non-zero linear subspace of a normed linear space can be extended to a bounded linear functional defined on the whole space, then it will follow that non-zero bounded linear functionals exist in abundance (cf. 5.3.2 to 5.3.6).

In this section we derive a general extension theorem, called the Hahn–Banach theorem, for linear functionals on an arbitrary linear space. In the next section we apply the Hahn–Banach theorem to the problem of the existence of bounded linear functionals.

5.2.1 DEFINITION A *sublinear functional* on a linear space E over K is a mapping p of E into R that satisfies the following conditions:

(a) $p(x + y) \leq p(x) + p(y)$ for all $x,y \in E$, and
(b) $p(\alpha x) = \alpha p(x)$ for all $x \in E$ and $\alpha \in \mathsf{R}$ with $\alpha \geq 0$.

If $\|\cdot\|$ is a norm on a linear space E it is obvious that the mapping $x \to \| x \|$ is a sublinear functional on E. The following lemma provides the crucial step in the proof of the Hahn–Banach theorem.

5.2.2 Lemma *Let E_0 be a linear subspace of a linear space E over* R, *let $x_1 \in E \sim E_0$ and let E_1 be the linear hull of $E_0 \cup \{x_1\}$. Let p be a sublinear functional on E and f_0 a linear functional on E_0 such that*

$$f_0(x) \leq p(x)$$

for all $x \in E_0$. Then there exists a linear functional f_1 on E_1 such that

(a) $f_1(x) \leq p(x)$ *for all $x \in E_1$, and*
(b) $f_1(x) = f_0(x)$ *for all $x \in E_0$.*

Proof Since E_0 is a linear subspace, given $x,y \in E_0$ we have

$$\begin{aligned} f_0(x) + f_0(y) &= f_0(x + y) \\ &\leq p(x + y) \\ &= p((x + x_1) + (y - x_1)) \\ &\leq p(x + x_1) + p(y - x_1). \end{aligned}$$

Therefore

$$f_0(y) - p(y - x_1) \leq p(x + x_1) - f_0(x) \tag{1}$$

for all $x,y \in E_0$. Let

$$A = \{f_0(x) - p(x - x_1) : x \in E_0\}$$

and

$$B = \{p(x + x_1) - f_0(x) : x \in E_0\}.$$

Inequality (1) shows that $\alpha \leq \beta$ for all $\alpha \in A$ and $\beta \in B$. Thus A is bounded above, B is bounded below and $\sup A \leq \inf B$. Let γ be a real number with $\sup A \leq \gamma \leq \inf B$. Then

$$f_0(x) - \gamma \leq p(x - x_1) \text{ and } f_0(x) + \gamma \leq p(x + x_1) \tag{2}$$

for all $x \in E_0$.

Given $\lambda > 0$ and $x \in E_0$, we have $\lambda^{-1}x \in E_0$, and, by (2),

$$\begin{aligned} f_0(x) - \lambda\gamma &= \lambda(f_0(\lambda^{-1}x) - \gamma) \\ &\leq \lambda p(\lambda^{-1}x - x_1) \\ &= p(x - \lambda x_1), \end{aligned} \tag{3}$$

and

$$\begin{aligned} f_0(x) + \lambda\gamma &= \lambda(f_0(\lambda^{-1}x) + \gamma) \\ &\leq \lambda p(\lambda^{-1}x + x_1) \\ &= p(x + \lambda x_1). \end{aligned} \tag{4}$$

Inequalities (3) and (4) show that

$$f_0(x) + \lambda\gamma \leq p(x + \lambda x_1) \tag{5}$$

for all $x \in E_0$ and $\lambda \in \mathsf{R}$. Since $x_1 \notin E_0$ and E_1 is the linear hull of $E_0 \cup \{x_1\}$ each point of E_1 can be represented uniquely in the form $x + \lambda x_1$ with $x \in E_0$ and $\lambda \in \mathsf{R}$ (cf. Lemma 3.4.3). We can therefore define a functional f_1 on E_1 without ambiguity by

$$f_1(x + \lambda x_0) = f_0(x) + \lambda \gamma$$

for all $x \in E_0$ and $\lambda \in \mathsf{R}$. It is easy to verify that f_1 is a linear functional on E_1 that satisfies (b). The inequality (5) shows that f_1 also satisfies (a). □

The proof of the Hahn–Banach theorem follows from Lemma 5.2.2, but uses a sophisticated technique. The proof of Theorem 5.2.4, below, which is a special case of Theorem 5.3.2 and is obviously contained in the Hahn–Banach theorem, also uses Lemma 5.2.2, but is elementary and may help to explain and motivate the proof of the Hahn–Banach theorem itself. We need a simple lemma.

5.2.3 LEMMA *Let p be a sublinear functional on a normed linear space E. Then p is continuous on E if and only if there exists a real number M such that $| p(x) | \leq M \| x \|$ for all $x \in E$.*

PROOF We note first that $p(0) = 0$ because $p(0) = p(0.0) = 0\,p(0) = 0$. Suppose that p is continuous at 0. Then there exists $\delta > 0$ such that $| p(x) | < 1$ for $\| x \| < \delta$. If $x \in E$ and $x \neq 0$ then $\| (\frac{1}{2}\delta \| x \|^{-1})\, x \| < \delta$ and so $| p(\frac{1}{2}\delta \| x \|^{-1} x) | < 1$. Since $\frac{1}{2}\delta \| x \|^{-1} \geq 0$ we have

$$p(\tfrac{1}{2}\delta \| x \|^{-1} x) = \tfrac{1}{2}\delta \| x \|^{-1} p(x)$$

and hence $| p(x) | \leq 2\delta^{-1} \| x \|$. The last inequality is obviously true for $x = 0$ so $| p(x) | \leq 2\delta^{-1} \| x \|$ for all $x \in E$.

Suppose on the other hand that there is a number M with $| p(x) | \leq M \| x \|$ for all $x \in E$. For all $x, y \in E$ we have

$$p(x) = p(x - y + y) \leq p(x - y) + p(y)$$

and therefore

$$p(x) - p(y) \leq p(x - y) \leq M \| x - y \|.$$

By symmetry it follows that

$$| p(x) - p(y) | \leq M \| x - y \|.$$

This shows that p is continuous on E. □

5.2.4 THEOREM *Let E be a separable normed linear space over* R, *let p be a continuous sublinear functional on E, let E_0 be a linear subspace of E and let f_0 be a linear functional on E_0 such that*

$$f_0(x) \leq p(x)$$

for all $x \in E_0$. *Then there exists a linear functional* f *on* E *such that*

(a) $f(x) \leq p(x)$ *for all* $x \in E$, *and*
(b) $f(x) = f_0(x)$ *for all* $x \in E_0$.

PROOF Since E is separable there is a countable subset

$$A = \{x_n : n = 1, 2, \ldots\}$$

of E that is dense in E. We define a sequence (E_n) of linear subspaces of E inductively as follows: for $n \geq 1$ let $E_n = E_{n-1}$ if $A \subseteq E_{n-1}$, otherwise let E_n be the linear hull of $E_{n-1} \cup \{x_{m_n}\}$ where m_n is the least positive integer such that $x_{m_n} \notin E_{n-1}$. Let $F = \bigcup_{n=1}^{\infty} E_n$. Since $A \subseteq F$ the set F is dense in E and, since $E_n \subseteq E_{n+1}$ for $n = 0, 1, 2, \ldots$, it is easy to see that F is a linear subspace of E.

By Lemma 5.2.2 and induction we obtain, for each positive integer n, a linear functional f_n on E_n such that

$$f_n(x) \leq p(x) \tag{6}$$

for all $x \in E_n$ and

$$f_n(x) = f_{n-1}(x) \tag{7}$$

for all $x \in E_{n-1}$. Now $E_{n-1} \subseteq E_n$ and therefore by (7) we can define a functional g on F without ambiguity by $g(x) = f_n(x)$ for all $x \in E_n$ and $n = 0, 1, 2, \ldots$. Let $x, y \in F$ and $\alpha, \beta \in \mathsf{R}$. Then $x \in E_m$ and $y \in E_n$ for some non-negative integers m and n. We may suppose that $m \leq n$ so $E_m \subseteq E_n$ and therefore $\alpha x + \beta y \in E_n$. Thus, since f_n is linear,

$$\begin{aligned} g(\alpha x + \beta y) &= f_n(\alpha x + \beta y) \\ &= \alpha f_n(x) + \beta f_n(y) \\ &= \alpha g(x) + \beta g(y). \end{aligned}$$

This shows that g is a linear functional on F. Also, by (6),

$$g(x) \leq p(x) \tag{8}$$

for all $x \in F$. By Lemma 5.2.3 there is a real number M with $|p(x)| \leq M \| x \|$ for all $x \in E$ and so, by (8), we have $g(x) \leq M \| x \|$ for all $x \in F$. Then $-g(x) = g(-x) \leq M \| -x \| = M \| x \|$ and hence $|g(x)| \leq M \| x \|$ for all $x \in F$. This shows that g is a bounded linear functional on F. Since F is dense in E it now follows from Theorem 3.5.10 that there is a unique bounded linear functional f on E with $f(x) = g(x)$ for all $x \in F$. Further $p - f$ is continuous on E, $p(x) - f(x) \geq 0$ for all $x \in F$ and F is dense in E; therefore $p(x) - f(x) \geq 0$ for all $x \in E$. □

In the above proof, the hypothesis of separability has been used to construct the sequence (E_n)—with $\bigcup_{n=1}^{\infty} E_n$ dense in E—along which the functional can be extended step by step by using Lemma 5.2.2 and induction. If the space E is not separable, the inductive proof breaks down. However the general form of the Hahn–Banach theorem was obtained by an argument which is essentially an extension of the above proof. The technique by which this is achieved is known as 'transfinite induction'. The method of transfinite induction was commonly used until the 1940's. It is now customary, however, to use Zorn's lemma (1.5.2) where, previously, transfinite induction would have been used. The reason for doing so is that the application of Zorn's lemma is (usually) technically more efficient than the method of transfinite induction. The price which is paid for this efficiency is a greater sophistication in the tool used. Zorn's lemma is now an established and routine tool of mathematics, and the proof of the Hahn–Banach theorem which we now give provides a typical example of the way in which it is used.

5.2.5 THEOREM (The Hahn–Banach theorem) *Let E_0 be a linear subspace of a linear space E over* R, *let p be a sublinear functional on E and let f_0 be a linear functional on E_0 such that*

$$f_0(x) \leq p(x)$$

for all $x \in E_0$. Then there exists a linear functional f on E such that

(a) $f(x) \leq p(x)$ *for all* $x \in E$, *and*
(b) $f(x) = f_0(x)$ *for all* $x \in E_0$.

PROOF Let $\mathscr{G}$ be the family of all ordered pairs (G,g), in which G is a linear subspace of E, G contains E_0, and g is a linear functional on G such that $g(x) = f_0(x)$ for all $x \in E_0$ and $g(x) \leq p(x)$ for all $x \in G$. We define a relation $\leq$ on $\mathscr{G}$ by setting $(G_1,g_1) \leq (G_2,g_2)$ if and only if $G_1 \subseteq G_2$ and $g_1(x) = g_2(x)$ for all $x \in G_1$. It is easy to see that $\leq$ is a partial order relation on $\mathscr{G}$. We shall prove that every linearly ordered subset of $\mathscr{G}$ has an upper bound. (All these terms are defined in Section 1.3.)

Suppose that $\mathscr{F}$ is a non-empty subfamily of $\mathscr{G}$ that is linearly ordered by $\leq$. Let

$$G = \bigcup_{(G',g') \in \mathscr{F}} G'.$$

Suppose that (G_1,g_1) and (G_2,g_2) are pairs in $\mathscr{F}$ and that $x \in G_1 \cap G_2$. Since $\mathscr{F}$ is linearly ordered either $(G_1,g_1) \leq (G_2,g_2)$ or $(G_2,g_2) \leq (G_1,g_1)$. In either case, by definition of the relation $\leq$, we have $g_1(x) = g_2(x)$. This shows that we can define a functional g on G, without ambiguity, as follows: given $x \in G$, let $g(x) = g'(x)$ where (G',g') is any pair in $\mathscr{F}$ with $x \in G'$. We shall prove that G is a linear subspace of E and that g

is a linear functional on G. Let $x,y \in G$ and $\alpha,\beta \in \mathsf{R}$. Then $x \in G_1$ and $y \in G_2$ for some pairs (G_1,g_1) and (G_2,g_2) in $\mathscr{F}$. Since $\mathscr{F}$ is linearly ordered we may suppose that $(G_1,g_1) \leq (G_2,g_2)$. Then $G_1 \subseteq G_2$ and hence $x,y \in G_2$. Since G_2 is a linear subspace of E, we have $\alpha x + \beta y \in G_2 \subseteq G$. This shows that G is a linear subspace of E. Also, by definition of g,

$$g(\alpha x + \beta y) = g_2(\alpha x + \beta y) = \alpha g_2(x) + \beta g_2(y) = \alpha g(x) + \beta g(y).$$

This shows that g is a linear functional on G. It is obvious that $E_0 \subseteq G$, that $g(x) = f_0(x)$ for all $x \in E_0$ and that $g(x) \leq p(x)$ for all $x \in G$. Consequently $(G,g) \in \mathscr{G}$. Clearly $(G',g') \leq (G,g)$ for all $(G',g') \in \mathscr{F}$ so (G,g) is an upper bound for $\mathscr{F}$.

It follows now by Zorn's lemma (1.5.2) that $\mathscr{G}$ has at least one maximal element. Let (F,f) be a maximal element of $\mathscr{G}$. We shall prove that $F = E$. Suppose on the contrary that $F \neq E$ and choose $x_1 \in E \sim F$. Let F_1 be the linear hull of the set $F \cup \{x_1\}$. Since $(F,f) \in \mathscr{G}$ we have $f(x) \leq p(x)$ for all $x \in F$, and therefore, by Lemma 5.2.2, there is a linear functional f_1 on F_1 such that $f_1(x) \leq p(x)$ for all $x \in F_1$ and $f_1(x) = f(x)$ for all $x \in F$. It follows that $(F_1,f_1) \in \mathscr{G}$ and $(F,f) \leq (F_1,f_1)$. This contradicts the maximality of (F,f) in $\mathscr{G}$ because $(F,f) \neq (F_1,f_1)$, and we must therefore have $F = E$. It is now clear that the functional f satisfies the conditions of the theorem. □

Exercise

There are certain contexts (notably parts of the theory of Fourier series) in which it is possible, in a significant way, to associate a 'limit' with a sequence which is not convergent. Such generalized limits are the subject of summability theory (see Exercise 8.4(7)). It was shown by Banach that it is possible to associate a generalized limit with *every* bounded sequence.

Let K be the set of all elements of the form $(k_1, k_2,\ldots, k_r)$ with r and $k_1, k_2,\ldots, k_r$ positive integers. For each $x = (\xi_n) \in m$ let

$$p(x) = \inf\left\{ \limsup_{n\to\infty} \frac{1}{r}\sum_{i=1}^{r} \xi_{n+k_i} : (k_1, k_2,\ldots, k_r) \in K\right\}.$$

Prove that p is a sublinear functional on m and hence prove the existence of a linear functional l on m that has the properties:

(a) $l(x) \geq 0$ when $x = (\xi_n) \in m$ and $\xi_n \geq 0$ for $n = 1, 2,\ldots,$
(b) if $x = (\xi_n) \in m$ and $x' = (\eta_n)$ where $\eta_n = \xi_{n+1}$ then $l(x) = l(x')$,
(c) $\liminf \xi_n \leq l(x) \leq \limsup \xi_n$ for all $x = (\xi_n) \in m$,
(d) $l(x) = \lim_{n\to\infty} \xi_n$ for all $x = (\xi_n) \in c$.

5.3 The Existence of Bounded Linear Functionals

In this section we use the Hahn–Banach theorem to establish the existence of non-zero bounded linear functionals on an arbitrary normed linear space. The Hahn–Banach theorem is concerned with real linear spaces, so our first task is to extend it to complex linear spaces. This extension was obtained in 1938 by Bohnenblust and Sobczyk and, independently, by Soukhomlinof.

We have already remarked (p. 84) that associated with each complex linear space (E, C) there is a real linear space (E, R). A linear subspace of (E, C) will be called a *complex linear subspace of* E and a linear subspace of (E, R) will be called a *real linear subspace of* E. Thus F is a complex linear subspace of E if and only if $\alpha x + \beta y \in F$ for all $x,y \in F$ and all $\alpha,\beta \in \mathsf{C}$; and F is a real linear subspace of E if and only if $\alpha x + \beta y \in F$ for all $x,y \in F$ and all $\alpha,\beta \in \mathsf{R}$. Evidently each complex linear subspace of E is a real linear subspace of E; the converse is, in general, false. A linear transformation of (E, C) into C will be called a *complex linear functional on* E and a linear transformation of (E, R) into R will be called a *real linear functional on* E. Thus f is a complex linear functional on E if and only if $f(\alpha x + \beta y) = \alpha f(x) + \beta f(y)$ for all $x,y \in E$ and all $\alpha,\beta \in \mathsf{C}$; and f is a real linear functional on E if and only if $f(\alpha x + \beta y) = \alpha f(x) + \beta f(y)$ for all $x,y \in E$ and all $\alpha,\beta \in \mathsf{R}$.

5.3.1 THEOREM *Let E be a complex linear space, let E_0 be a complex linear subspace of E, let p be a sublinear functional on E and let f_0 be a complex linear functional on E_0 such that*

$$\operatorname{Re} f_0(x) \leq p(x)$$

for all $x \in E_0$. Then there is a complex linear functional f on E such that

(a) $\operatorname{Re} f(x) \leq p(x)$ *for all $x \in E$, and*
(b) $f(x) = f_0(x)$ *for all $x \in E_0$.*

PROOF Let $g_0(x) = \operatorname{Re} f_0(x)$ for all $x \in E_0$. Then g_0 is a real linear functional on E_0. In fact, if $x,y \in E_0$ and $\alpha,\beta \in \mathsf{R}$ then

$$\begin{aligned} g_0(\alpha x + \beta y) &= \operatorname{Re} f_0(\alpha x + \beta y) \\ &= \operatorname{Re} (\alpha f_0(x) + \beta f_0(y)) \\ &= \alpha \operatorname{Re} f_0(x) + \beta \operatorname{Re} f_0(y) \\ &= \alpha g_0(x) + \beta g_0(y). \end{aligned}$$

Since $g_0(x) = \operatorname{Re} f_0(x) \leq p(x)$ for all $x \in E_0$ and E_0 is a real linear subspace of the real linear space associated with the complex linear space E, it follows from the Hahn–Banach theorem (5.2.5) that there is a real linear functional g on E such that $g(x) \leq p(x)$ for all $x \in E$ and $g(x) = g_0(x)$ for all $x \in E_0$.

Let $x \in E_0$. Since E_0 is a complex linear subspace of E we have $ix \in E_0$ and consequently, since f_0 is a complex linear functional on E_0,

$$g_0(ix) = \operatorname{Re} f_0(ix) = \operatorname{Re}(if_0(x)) = -\operatorname{Im} f_0(x).$$

Hence

$$f_0(x) = \operatorname{Re} f_0(x) + i \operatorname{Im} f_0(x) = g_0(x) - ig_0(ix)$$

for all $x \in E_0$. Let $f(x) = g(x) - ig(ix)$ for all $x \in E$.

It is clear that f satisfies conditions (a) and (b) of the theorem. We shall prove that f is a complex linear functional on E. If $x,y \in E$ and $\alpha \in \mathsf{R}$ then, since g is a real linear functional on E,

$$\begin{aligned} f(x+y) &= g(x+y) - ig(ix+iy) \\ &= g(x) + g(y) - i(g(ix) + g(iy)) \\ &= f(x) + f(y), \end{aligned}$$

and

$$\begin{aligned} f(\alpha x) &= g(\alpha x) - ig(\alpha ix) \\ &= \alpha g(x) - i\alpha g(ix) \\ &= \alpha f(x). \end{aligned}$$

Next, if $x \in E$ then

$$\begin{aligned} f(ix) &= g(ix) - ig(-x) \\ &= g(ix) + ig(x) \\ &= i(g(x) - ig(ix)) \\ &= if(x). \end{aligned}$$

Finally, if $x \in E$ and $\alpha = \beta + i\gamma$ with $\beta,\gamma \in \mathsf{R}$ it follows from what we have just proved that

$$\begin{aligned} f(\alpha x) &= f(\beta x + i\gamma x) \\ &= f(\beta x) + f(i\gamma x) \\ &= \beta f(x) + \gamma f(ix) \\ &= \beta f(x) + \gamma i f(x) \\ &= \alpha f(x). \ \square \end{aligned}$$

We can now prove the fundamental extension theorem for bounded linear functionals. For real normed linear spaces it was proved by Hahn in 1927. (Some writers call Theorem 5.3.2 the 'Hahn–Banach theorem'.)

5.3.2 THEOREM *Let E be a normed linear space over* K, *let E_0 be a linear subspace of E, and let y^* be a bounded linear functional on E_0. Then there is a bounded linear functional x^* on E such that $x^*(x) = y^*(x)$ for all $x \in E_0$ and*† $\|x^*\| = \|y^*\|$.

† Notice that $\|x^*\| = \sup\{|x^*(x)| : x \in E \text{ and } \|x\| \leq 1\}$ and $\|y^*\| = \sup\{|y^*(x)| : x \in E_0 \text{ and } \|x\| \leq 1\}$.

PROOF We shall consider the cases $\mathsf{K} = \mathsf{R}$ and $\mathsf{K} = \mathsf{C}$ separately.

Suppose that $\mathsf{K} = \mathsf{R}$. Then, for all $x \in E_0$,

$$y^*(x) \leq |y^*(x)| \leq \|y^*\| \|x\|.$$

The mapping $x \to \|y^*\| \|x\|$ is obviously a sublinear functional on E. Therefore, by the Hahn–Banach theorem (5.2.5), there is a linear functional x^* on E such that $x^*(x) = y^*(x)$ for all $x \in E_0$ and $x^*(x) \leq \|y^*\| \|x\|$ for all $x \in E$. Then

$$-x^*(x) = x^*(-x) \leq \|y^*\| \|-x\| = \|y^*\| \|x\|$$

so $|x^*(x)| \leq \|y^*\| \|x\|$ for all $x \in E$. This shows that x^* is bounded and $\|x^*\| \leq \|y^*\|$. On the other hand

$$\begin{aligned}\|y^*\| &= \sup\{|y^*(x)| : x \in E_0 \text{ and } \|x\| \leq 1\}\\ &= \sup\{|x^*(x)| : x \in E_0 \text{ and } \|x\| \leq 1\}\\ &\leq \sup\{|x^*(x)| : x \in E \text{ and } \|x\| \leq 1\}\\ &= \|x^*\|\end{aligned}$$

so $\|x^*\| = \|y^*\|$. This completes the proof in the case $\mathsf{K} = \mathsf{R}$.

Now suppose that $\mathsf{K} = \mathsf{C}$. Then, for all $x \in E_0$,

$$\operatorname{Re} y^*(x) \leq |y^*(x)| \leq \|y^*\| \|x\|.$$

Again the mapping $x \to \|y^*\| \|x\|$ is a sublinear functional on E, and therefore, by Theorem 5.3.1, there exists on E a complex linear functional x^* such that $x^*(x) = y^*(x)$ for all $x \in E_0$ and $\operatorname{Re} x^*(x) \leq \|y^*\| \|x\|$ for all $x \in E$. Let $x \in E$ and let θ be a real number with

$$x^*(x) = e^{i\theta} |x^*(x)|.$$

Then

$$\begin{aligned}|x^*(x)| &= e^{-i\theta} x^*(x)\\ &= \operatorname{Re}(e^{-i\theta} x^*(x))\\ &= \operatorname{Re} x^*(e^{-i\theta} x)\\ &\leq \|y^*\| \|e^{-i\theta} x\|\\ &= \|y^*\| \|x\|.\end{aligned}$$

This shows that x^* is bounded, and $\|x^*\| \leq \|y^*\|$. Just as in the case $\mathsf{K} = \mathsf{R}$ we have $\|y^*\| \leq \|x^*\|$ so $\|x^*\| = \|y^*\|$. This completes the proof in the case $\mathsf{K} = \mathsf{C}$. □

We now derive some important consequences of Theorem 5.3.2.

5.3.3 THEOREM *Let E be a normed linear space over K, let F be a linear subspace of E and let u be a point of E with $\delta = d(u,F) > 0$. Then there is a bounded linear functional x^* on E such that $x^*(x) = 0$ for all $x \in F$, $x^*(u) = \delta$ and $\|x^*\| = 1$.*

PROOF Let E_0 be the linear hull of the set $F \cup \{u\}$. Since $\delta > 0$ we have $u \notin F$ and hence each point x of E_0 can be written uniquely in the form $x = y + \alpha u$ where $y \in F$ and $\alpha \in \mathsf{K}$. Let

$$y^*(y + \alpha u) = \alpha\delta$$

for all $y \in F$ and all $\alpha \in \mathsf{K}$. It is easy to verify that y^* is a linear functional on E_0. Also, if $\alpha \neq 0$,

$$\| y + \alpha u \| = | \alpha | \left\| \frac{1}{\alpha} y + u \right\| \geq | \alpha | \delta$$

and so

$$| y^*(y + \alpha u) | = | \alpha | \delta \leq \| y + \alpha u \|.$$

Since $| y^*(y) | = 0 \leq \| y \|$ it follows that $| y^*(x) | \leq \| x \|$ for all $x \in E_0$. This shows that y^* is a bounded linear functional on E_0 and $\| y^* \| \leq 1$. By definition of y^*

$$\delta = y^*(u - y) \leq \| y^* \| \; \| u - y \|$$

for all $y \in F$ and consequently

$$\delta \leq \inf \{ \| y^* \| \; \| u - y \| : y \in F\} = \delta \| y^* \|.$$

This shows that $\| y^* \| \geq 1$ and hence that $\| y^* \| = 1$. By Theorem 5.3.2 there is a bounded linear functional x^* on E such that $x^*(x) = y^*(x)$ for all $x \in E_0$ and $\| x^* \| = 1$. Therefore $x^*(y) = y^*(y) = 0$ for all $y \in F$ and $x^*(u) = y^*(u) = \delta$. □

5.3.4 Corollary *Let E be a normed linear space over K, let F be a closed linear subspace of E and let $u \in E \sim F$. Then there exists a bounded linear functional x^* on E such that $x^*(u) \neq 0$ and $x^*(x) = 0$ for all $x \in F$.*

Proof Let $\delta = d(u,F)$. Since $u \notin F$ and F is closed, we have $\delta > 0$ (Lemma 2.8.4). The existence of x^* now follows from Theorem 5.3.3. □

5.3.5 Corollary *Let E be a normed linear space over K and let x be a non-zero point of E. Then there is a bounded linear functional x^* on E such that $x^*(x) = \| x \|$ and $\| x^* \| = 1$.*

Proof Take $F = \{0\}$ and $u = x$ in Theorem 5.3.3. □

5.3.6 Corollary *Let E be a normed linear space over K and let x_1 and x_2 be distinct points of E. Then there is a bounded linear functional x^* on E with $x^*(x_1) \neq x^*(x_2)$.*

Proof Take $x = x_1 - x_2$ in Corollary 5.3.5. □

We have already remarked in Section 5.1 that the set E^* of all bounded linear functionals on a normed linear space E is a Banach space which is called the dual space of E. The norm on E^* is given by

$$\| x^* \| = \sup \{ | x^*(x) | : x \in E \text{ and } \| x \| \leq 1\}.$$

5.3.7 COROLLARY *Let E be a normed linear space over $\mathbb{K}$ and let $x \in E$. Then*

$$\| x \| = \sup \{ | x^*(x) | : x^* \in E^* \text{ and } \| x^* \| \leq 1\}.$$

PROOF This follows immediately from Corollary 5.3.5 and the definition of $\| x^* \|$. □

The preceding corollaries provide the tools for the study of the relations between the properties of a normed linear space and its dual space. As an illustration we consider the question of separability. It may happen that E is separable and E^* is not separable, for example the dual ℓ^* of ℓ is linearly isometric to m (Example 5.1.4), ℓ is separable (Example 3.4.9), but m, and hence ℓ^*, is not separable (Example 3.4.10). However the separability of E^* entails the separability of E.

5.3.8 THEOREM *Let E be a normed linear space and suppose that E^* is separable. Then E is separable.*

PROOF There is a countable subset, $\{x_n^*: n = 1, 2,...\}$ say, of E^* that is dense in E^*. For each positive integer n there is a point x_n in E with $\| x_n \| = 1$ and $| x_n^*(x_n) | \geq \frac{1}{2} \| x_n^* \|$. We shall prove that the closed linear hull F of the set $\{x_n: n = 1, 2,...\}$ is E. Suppose, on the contrary, that $F \neq E$. Then, by Corollary 5.3.4, there is a point $x^* \in E^*$ such that $x^*(x) = 0$ for all $x \in F$ and $x^* \neq 0$. Since $\{x_n^*: n = 1, 2,...\}$ is dense in E^* there is a positive integer N with $\| x^* - x_N^* \| < \frac{1}{4} \| x^* \|$. Then

$$\begin{aligned} \tfrac{1}{2} \| x_N^* \| &\leq | x_N^*(x_N) | \\ &= | x_N^*(x_N) - x^*(x_N) | \\ &\leq \| x_N^* - x^* \| \; \| x_N \| \\ &< \tfrac{1}{4} \| x^* \| \end{aligned}$$

and hence

$$\| x^* \| \leq \| x^* - x_N^* \| + \| x_N^* \| < \tfrac{3}{4} \| x^* \|$$

which is a contradiction. □

The dual E^* of a normed linear space E is a Banach space and therefore has a dual space $(E^*)^*$. We shall write E^{**} in place of $(E^*)^*$, and we shall call E^{**} the *second dual* of E.

5.3.9 THEOREM *Let E be a normed linear space. Given $x \in E$, let*

$$\hat{x}(x^*) = x^*(x)$$

for all $x^* \in E^*$. Then $\hat{x}$ is a bounded linear functional on E^* and the mapping $x \to \hat{x}$ is a linear isometry of E into its second dual E^{**}.

PROOF Let $x \in E$, $x_1^*, x_2^* \in E^*$ and $\alpha_1, \alpha_2 \in \mathsf{K}$. Then

$$\begin{aligned}\hat{x}(\alpha_1 x_1^* + \alpha_2 x_2^*) &= (\alpha_1 x_1^* + \alpha_2 x_2^*)(x)\\ &= \alpha_1 x_1^*(x) + \alpha_2 x_2^*(x)\\ &= \alpha_1 \hat{x}(x_1^*) + \alpha_2 \hat{x}(x_2^*).\end{aligned}$$

This shows that $\hat{x}$ is a linear functional on E^*. Further Corollary 5.3.7 gives

$$\begin{aligned}\| x \| &= \sup \{ | x^*(x) | : x^* \in E^* \text{ and } \| x^* \| \leq 1\}\\ &= \sup \{ | \hat{x}(x^*) | : x^* \in E^* \text{ and } \| x^* \| \leq 1\}.\end{aligned}$$

This shows that $\hat{x}$ is a bounded linear functional on E^* and that $\| \hat{x} \| = \| x \|$.

Finally let $x_1, x_2 \in E$, $\alpha_1, \alpha_2 \in \mathsf{K}$ and $x^* \in E^*$. Then

$$\begin{aligned}(\alpha_1 x_1 + \alpha_2 x_2)\hat{\ }(x^*) &= x^*(\alpha_1 x_1 + \alpha_2 x_2)\\ &= \alpha_1 x^*(x_1) + \alpha_2 x^*(x_2)\\ &= \alpha_1 \hat{x}_1(x^*) + \alpha_2 \hat{x}_2(x^*)\\ &= (\alpha_1 \hat{x}_1 + \alpha_2 \hat{x}_2)(x^*).\end{aligned}$$

Thus $(\alpha_1 x_1 + \alpha_2 x_2)\hat{\ } = \alpha_1 \hat{x}_1 + \alpha_2 \hat{x}_2$ and hence the mapping $x \to \hat{x}$ is linear. Since $\| \hat{x} \| = \| x \|$ the mapping $x \to \hat{x}$ is a linear isometry of E onto the linear subspace $\{\hat{x} : x \in E\}$ of E^{**}. □

5.3.10 DEFINITION Let E be a normed linear space. The linear isometry $x \to \hat{x}$ is called the *canonical mapping* of E into its second dual E^{**}.

The canonical mapping is a linear isometry of E onto the linear subspace $\hat{E} = \{\hat{x} : x \in E\}$ of the Banach space E^{**}. It is clear that if E is not a Banach space then we must have $\hat{E} \neq E^{**}$. The following example shows that we may have $\hat{E} \neq E^{**}$ even when E is a Banach space.

5.3.11 EXAMPLE We have seen in Example 5.1.5 that there is a linear isometry T of c_0^* onto ℓ. We shall write

$$\langle x,y \rangle = \sum_{n=1}^{\infty} \xi_n \eta_n$$

for all $x = (\xi_n) \in m$ and all $y = (\eta_n) \in \ell$. By definition of T we have

$$x^*(x) = \langle x, Tx^* \rangle$$

for all $x \in c_0$ and all $x^* \in c_0^*$. Let $x \in c_0$. Then for each $x^* \in c_0^*$

$$\hat{x}(x^*) = x^*(x) = \langle x, Tx^* \rangle. \tag{1}$$

Choose $y \in m \sim c_0$ and let

$$y^{**}(x^*) = \langle y, Tx^* \rangle \tag{2}$$

for all $x^* \in c_0^*$. Using Example 5.1.4, it is not difficult to verify that

$y^{**} \in c_0^{**}$. Suppose that $y^{**} = \hat{x}$ for some $x \in c_0$. Then (1) and (2) give

$$\langle x, Tx^* \rangle = \langle y, Tx^* \rangle$$

for all $x^* \in c_0^*$.

Since T maps c_0^* onto ℓ it follows that $\langle x, e_n \rangle = \langle y, e_n \rangle$ for $n = 1, 2, \ldots$, where $e_n = (\delta_{mn})_{m \geq 1}$, and hence that $x = y$ which is impossible. Thus $y^{**} \in c_0^{**}$ and $y^{**} \notin \hat{c}_0$.

5.3.12 Definition A normed linear space E is said to be *reflexive* if and only if the canonical mapping $x \to \hat{x}$ maps E onto E^{**}.

Obviously a necessary condition for E to be reflexive is that E be a Banach space. Example 5.3.11 shows that there are non-reflexive Banach spaces.

5.3.13 Example. We shall prove that the Banach space ℓ^p is reflexive if $p > 1$. Let $p > 1$ and $q = p/(p-1)$. We shall write

$$\langle x, y \rangle = \sum_{n=1}^{\infty} \xi_n \eta_n$$

for all $x = (\xi_n) \in \ell^p$ and all $y = (\eta_n) \in \ell^q$. By Example 5.1.6 there is a linear isometry T_p of ℓ^q onto $(\ell^p)^*$ such that

$$(T_p y)(x) = \langle x, y \rangle$$

for all $x \in \ell^p$ and all $y \in \ell^q$. Similarly there is a linear isometry T_q of ℓ^p onto $(\ell^q)^*$ such that

$$(T_q x)(y) = \langle x, y \rangle$$

for all $x \in \ell^p$ and all $y \in \ell^q$.

Now let $x^{**} \in (\ell^p)^{**}$ and let $y^* = x^{**} \circ T_p$. By Lemma 3.5.8 we have $y^* \in (\ell^q)^*$ so there exists a unique point $x \in \ell^p$ with $y^* = T_q x$. Let $x^* \in (\ell^p)^*$. Then $x^* = T_p y$ for some $y \in \ell^q$ and we have

$$\begin{aligned} x^{**}(x^*) &= x^{**}(T_p y) \\ &= (x^{**} \circ T_p)(y) \\ &= y^*(y) \\ &= (T_q x)(y) \\ &= \langle x, y \rangle \\ &= (T_p y)(x) \\ &= x^*(x) \\ &= \hat{x}(x^*). \end{aligned}$$

This shows that $x^{**} = \hat{x}$ and hence ℓ^p is reflexive.

We wish to emphasize that, for E to be reflexive, E and E^{**} must be linearly isometric under the canonical mapping $x \to \hat{x}$. In 1951 R. C. James exhibited a separable Banach space E that is linearly isometric to its second dual E^{**} but is not reflexive.

EXERCISES

(1) Let E be a normed linear space, $f \in E^*$, $M = f^{-1}(\{0\})$ and $x_0 \notin M$. Prove that there is a point $y \in M$ with $\| x_0 - y \| = d(x_0,M)$ if and only if there is a point $x \in E$ with $\| x \| = 1$ and $f(x) = \| f \|$. (Cf. Exercise 4.4(2).)

(2) Let E be a normed linear space, $x \in E$ and $\| x \| = 1$. Prove that there is a closed linear subspace M of E such that $E = M \oplus \{\lambda x : \lambda \in \mathsf{K}\}$ and $d(x,M) = 1$.

(3) Let E be an infinite-dimensional Banach space. Prove that there is a one-to-one continuous linear mapping of ℓ into E.

(4) (a) Let g be a mapping of the closed interval $[a,b]$ into a real normed linear space E and suppose that g is continuous on $[a,b]$ and differentiable on (a,b). Let $x^* \in E^*$. Prove that the real-valued function $t \to x^*(g(t))$ is continuous on $[a,b]$ and differentiable on (a,b), and calculate its derivative.

(b) Use the mean value theorem and (a) to give an alternative proof of Lemma 3.9.8.

(5) Prove that $\hat{E}$ is a closed subspace of E^{**} if and only if E is a Banach space.

(6) Prove that a normed linear space is finite-dimensional if and only if its dual is finite-dimensional.

(7) Prove that a finite-dimensional normed linear space is reflexive.

(8) Prove that if E is a Banach space then E is reflexive if and only if E^* is reflexive.

(9) Prove that the Banach spaces c and ℓ are not reflexive.

(10) (a) Let E be a linear space over K. A linearly independent subset A of E is said to be a *maximal* linearly independent set if it is a maximal member of the family of all linearly independent subsets of E partially ordered by inclusion. Show that a subset A of E is a basis for E if and only if A is a maximal linearly independent set. (The terminology is that of Section 3.1.)

(b) Use Zorn's lemma to prove that every linearly independent subset of E is a subset of some basis for E, and deduce that every non-zero linear space over K has a basis.

(c) Prove that if F is a linear subspace of E and $F \neq E$ then there is a linear subspace G of E such that $E = F \oplus G$.

(d) Prove that if A is a basis for E and f is any mapping of A into K then there is a unique linear functional F on E such that $F(x) = f(x)$ for all $x \in A$. Deduce that if E is an infinite-dimensional normed linear space, then there exists on E a linear functional which is not continuous.

(11) Let f be a continuous real-valued function on a compact subset X of a real normed linear space E.

(a) If $0 \notin X$ prove that for each $\varepsilon > 0$ there exist positive integers $N, n_1, \ldots, n_N$ and functionals $x^*_{jk} \in E^*$, where $1 \leq j \leq n_k$ and $1 \leq k \leq N$, with

$$\left| f(x) - \sum_{k=1}^{N} x^*_{1k}(x) x^*_{2k}(x) \ldots x^*_{n_k k}(x) \right| < \varepsilon$$

for all $x \in X$.

(b) How must (a) be modified if $0 \in X$?

(In (a) use Theorem 4.5.8 and in (b) use the result of Exercise 4.5(10).) For $E = C_{\mathsf{R}}([a,b])$ this approximation theorem was obtained by Fréchet in 1910. His proof depended on some results about Fourier series. Approximation theorems of the above type have been used in continuum mechanics, in the study of materials with memory.

5.4 Annihilators

Throughout this section E will denote a normed linear space over K.

5.4.1 Definition Let X and Y be non-empty subsets of E and E^*, respectively. We define

$$X^\perp = \{x^* \in E^* : x^*(x) = 0 \text{ for all } x \in X\}$$

and

$$Y_\perp = \{x \in E : x^*(x) = 0 \text{ for all } x^* \in Y\}.$$

The sets $X^\perp$ and $Y_\perp$ are called the *annihilators* of X and Y respectively.

The proof of the following lemma follows easily from the above definition. We leave the reader to supply the details.

5.4.2 Lemma *Let X and Y be non-empty subsets of E and E^*, respectively. Then $X^\perp$ and $Y_\perp$ are closed linear subspaces of E^* and E, respectively.*

5.4.3 Lemma *Let X be a non-empty subset of E. Then $(X^\perp)_\perp$ is the closed linear hull of X. In particular, $(X^\perp)_\perp = X$ if and only if X is a closed linear subspace of E.*

Proof It is obvious that $X \subseteq (X^\perp)_\perp$. Also, by Lemma 5.4.2, $(X^\perp)_\perp$ is a closed linear subspace of E. Consequently, if X_0 is the closed linear hull of X then $X_0 \subseteq (X^\perp)_\perp$. If $X_0 = E$ it follows that $X_0 = (X^\perp)_\perp$ and the proof is complete in this case. Suppose that $X_0 \neq E$, and let $x_0 \in E \sim X_0$. By Corollary 5.3.4 there exists $x^* \in E^*$ with $x^*(x_0) \neq 0$ and $x^*(x) = 0$ for all $x \in X_0$. Since $X \subseteq X_0$ we have $x^* \in X^\perp$. Since $x^*(x_0) \neq 0$ we now have $x_0 \notin (X^\perp)_\perp$. This proves that $X_0 = (X^\perp)_\perp$. □

The analogue of Lemma 5.4.3 for subsets Y of E^* is false. It is obvious that $Y \subseteq (Y_\perp)^\perp$ and so the closed linear hull of Y is contained in $(Y_\perp)^\perp$, but there may be closed linear subspaces Y of E^* such that $Y \neq (Y_\perp)^\perp$. (See Exercise (2).)

A linear subspace of a normed linear space is itself a normed linear space and so has a dual space. There is a close relationship between the dual spaces of linear subspaces, the annihilators introduced above, and the quotient spaces introduced in Definition 3.4.12.

5.4.4 Theorem *Let M be a linear subspace of E. Then there is a linear isometry J_1 of $E^*/M^\perp$ onto M^* which is given by*

$$(J_1(x^* + M^\perp))(x) = x^*(x)$$

for all $x^* \in E^*$ and all $x \in M$.

Proof Let $x^* \in E^*$. If $y^* \in E^*$ and $x^* + M^\perp = y^* + M^\perp$ then $x^* - y^* \in M^\perp$. Thus $(x^* - y^*)(x) = 0$ for all $x \in M$ and hence $x^*(x) = y^*(x)$ for all $x \in M$. This shows that we can define a mapping $J_1(x^* + M^\perp)$ of M into K without ambiguity by

$$(J_1(x^* + M^\perp))(x) = x^*(x)$$

for all $x \in M$. It is easy to verify that $J_1(x^* + M^\perp)$ is a linear functional on M. Further, for all $z^* \in M^\perp$, we have $x^* + M^\perp = (x^* + z^*) + M^\perp$ and therefore, for all $x \in M$,

$$|(J_1(x^* + M^\perp))(x)| = |(x^* + z^*)(x)| \leq \|x^* + z^*\| \, \|x\|.$$

This shows that $J_1(x^* + M^\perp) \in M^*$ and that

$$\|J_1(x^* + M^\perp)\| \leq \|x^* + z^*\|$$

for all $z^* \in M^\perp$. Consequently

$$\|J_1(x^* + M^\perp)\| \leq \inf\{\|x^* + z^*\| : z^* \in M^\perp\} = \|x^* + M^\perp\|. \quad (1)$$

It is easy to verify that J_1 is a linear mapping of $E^*/M^\perp$ into M^*. Let $x^* \in E^*$ be such that $J_1(x^* + M^\perp) = 0$. Then for all $x \in M$ we have $x^*(x) = 0$ and so $x^* \in M^\perp$. Hence $x^* + M^\perp = M^\perp$, which is the zero of $E^*/M^\perp$. This shows that J_1 is one-to-one.

Finally let $y^* \in M^*$. By Theorem 5.3.2 there exists $x^* \in E^*$ with $x^*(x) = y^*(x)$ for all $x \in M$ and $\|x^*\| = \|y^*\|$. Then, for all $x \in M$,

$$(J_1(x^* + M^\perp))(x) = x^*(x) = y^*(x)$$

which shows that $J_1(x^* + M^\perp) = y^*$. Thus J_1 maps $E^*/M^\perp$ onto M*. Also

$$\|x^* + M^\perp\| \leq \|x^*\| = \|y^*\| = \|J_1(x^* + M^\perp)\|. \quad (2)$$

It follows from (1) and (2) that J_1 is a linear isometry. □

5.4.5 THEOREM *Let M be a closed linear subspace of E. Then there is a linear isometry J_2 of $(E/M)^*$ onto $M^\perp$ which is given by*

$$(J_2z^*)(x) = z^*(x + M)$$

for all $z^* \in (E/M)^*$ and all $x \in E$.

PROOF Let $z^* \in (E/M)^*$ and let $(J_2z^*)(x) = z^*(x + M)$ for all $x \in E$. It is clear that J_2z^* is a linear functional on E. Further, for all $x \in E$,

$$|(J_2z^*)(x)| = |z^*(x + M)| \leq \|z^*\| \, \|x + M\| \leq \|z^*\| \, \|x\|.$$

This shows that $J_2z^* \in E^*$ and $\|J_2z^*\| \leq \|z^*\|$. If $x \in M$ then $x + M = M$ (which is the zero of E/M) and thus

$$(J_2z^*)(x) = z^*(x + M) = z^*(M) = 0.$$

This shows that $J_2z^* \in M^\perp$.

It is easy to verify that J_2 is a linear mapping of $(E/M)^*$ into $M^\perp$. The mapping J_2 is one-to-one because, if $z^* \in (E/M)^*$ and $J_2z^* = 0$, then $z^*(x + M) = 0$ for all $x \in E$ and hence $z^* = 0$.

Finally let $x^* \in M^\perp$. We shall show that there exists $z^* \in (E/M)^*$ with $x^* = J_2z^*$. If $x,y \in E$ and $x + M = y + M$ then $x - y \in M$. Thus $x^*(x - y) = 0$ and hence $x^*(x) = x^*(y)$. This shows that we can define a mapping z^* of E/M into K without ambiguity by

$$z^*(x + M) = x^*(x)$$

for all $x + M \in E/M$. It is easy to verify that z^* is a linear functional on E/M. Further, for all $x \in E$ and $y \in M$, we have $x + M = (x + y) + M$ and so

$$|z^*(x + M)| = |x^*(x + y)| \leq \|x^*\| \, \|x + y\|.$$

It follows that, for all $x \in E$,

$$|z^*(x + M)| \leq \|x^*\| \inf\{\|x + y\| : y \in M\} = \|x^*\| \, \|x + M\|.$$

This shows that $z^* \in (E/M)^*$ and $\|z^*\| \leq \|x^*\|$. Since

$$(J_2z^*)(x) = z^*(x + M) = x^*(x)$$

for all $x \in E$ we have $J_2z^* = x^*$. Thus J_2 maps $(E/M)^*$ onto $M^\perp$. Also $\|J_2z^*\| = \|x^*\| \geq \|z^*\|$. We have seen above that $\|J_2z^*\| \leq \|z^*\|$ for all $z^* \in (E/M)^*$. It now follows that J_2 is a linear isometry of $(E/M)^*$ onto $M^\perp$. □

5.4.6 THEOREM *Suppose that E is reflexive and let M be a closed linear subspace of E. Then M is reflexive.*

PROOF Let J_1 be the linear isometry of $E^*/M^\perp$ onto M^* constructed in Theorem 5.4.4 and let $y^{**} \in M^{**}$. We have to find $x \in M$ such that $y^{**}(y^*) = y^*(x)$ for all $y^* \in M^*$.

First let x^{**} be the functional on E^* defined by

$$x^{**}(x^*) = y^{**}(J_1(x^* + M^\perp))$$

for all $x^* \in E^*$. It is easy to verify that x^{**} is a linear functional. Also, for all $x^* \in E^*$,

$$\begin{aligned} | x^{**}(x^*) | &= | y^{**}(J_1(x^* + M^\perp)) | \\ &\leq \| y^{**} \| \; \| J_1(x^* + M^\perp) \| \\ &= \| y^{**} \| \; \| x^* + M^\perp \| \\ &\leq \| y^{**} \| \; \| x^* \|. \end{aligned}$$

This shows that $x^{**} \in E^{**}$. Since E is reflexive there exists $x \in E$ with $x^{**} = \hat{x}$.

For each $x^* \in M^\perp$ we have $x^* + M^\perp = M^\perp$ and so

$$x^*(x) = \hat{x}(x^*) = x^{**}(x^*) = y^{**}(J_1(x^* + M^\perp)) = 0.$$

This shows that $x \in (M^\perp)_\perp$ and hence, by Lemma 5.4.3, we have $x \in M$. Finally let $y^* \in M^*$. Then $y^* = J_1(x^* + M^\perp)$ for some $x^* \in E^*$ and it follows from the definitions of x^{**} and J_1 that

$$\begin{aligned} y^{**}(y^*) &= y^{**}(J_1(x^* + M^\perp)) \\ &= x^{**}(x^*) \\ &= \hat{x}(x^*) \\ &= x^*(x) \\ &= (J_1(x^* + M^\perp))(x) \\ &= y^*(x). \end{aligned}$$

This shows that M is reflexive. □

There is an alternative proof of the last theorem which uses the concept of adjoint operator that is introduced in Section 6.4. This proof yields a construction for the canonical mapping of M onto M^{**} in terms of the canonical mapping of E onto E^{**} and the linear isometries of $E^*/M^\perp$ onto M^* and $(E^*/M^\perp)^*$ onto $(M^\perp)^\perp$. (See Exercise 6.4(6).)

Exercises

(1) (a) Let E be a normed linear space and M a linear subspace of E. Prove that if $x_0^* \in E^*$ then $d(x_0^*, M^\perp) = \| x_0^* | M \|$ and there is a point $x^* \in M^\perp$ with $\| x_0^* - x^* \| = d(x_0^*, M^\perp)$. (Use Theorem 5.4.4 and Theorem 5.3.2 applied to the subspace M and the functional $x_0^* | M$.)

(*b*) Prove that if M is a closed linear subspace of a reflexive space E then for each $x_0 \in E$ there is a point $x \in M$ with $\| x_0 - x \| = d(x_0, M)$. (cf. Exercise 5.3(1).)

(2) (a) Let F be a non-reflexive Banach space and let $E = F^*$. Prove that $\hat{F}$ is a closed linear subspace of E^* with $(\hat{F}_\perp)^\perp \neq \hat{F}$.

(b) Prove that if E is a reflexive Banach space then $M = (M_{\perp})^{\perp}$ for every closed linear subspace M of E^*.

(c) Prove that if M is a finite dimensional linear subspace of a dual space E^* then $(M_{\perp})^{\perp} = M$. (Use Exercise 5.1(2).)

(3) (a) Prove that a subset X of a normed linear space E is fundamental in E if and only if $x^* = 0$ whenever $x^* \in E^*$ and $x^*(x) = 0$ for all $x \in X$.

(b) Let $\{x_n : n = 1, 2,...\}$ be a subset of ℓ with

$$\sup\{\| x_n - e_n \| : n = 1, 2,...\} < 1,$$

where $e_1, e_2,...$ are as in Example 5.1.4. Prove that the set

$$\{x_n : n = 1, 2,...\}$$

is fundamental in ℓ and is linearly independent.

(c) Let $\{x_n : n = 1, 2,...\}$ be a subset of c_0 such that the series $\sum_n \| x_n - e_n \|$ is convergent and $\sum_{n=1}^{\infty} \| x_n - e_n \| < 1$. Prove that $\{x_n : n = 1, 2,...\}$ is fundamental in c_0 and is linearly independent.

(4) Let $\Gamma = \{e^{i\theta} : 0 \leq \theta < 2\pi\}$ and let $g_n \in C_{\mathsf{C}}(\Gamma)$ be defined by $g_n(e^{i\theta}) = e^{in\theta}$. For each $f \in C_{\mathsf{C}}(\Gamma)$ and each integer n let $\phi_n(f)$ be the nth Fourier coefficient of f, that is

$$\phi_n(f) = \frac{1}{2\pi} \int_0^{2\pi} e^{-in\theta} f(e^{i\theta})\, d\theta.$$

For each $f \in C_{\mathsf{C}}(\Gamma)$ and each $\alpha \in [0,2\pi)$ let f_α be the function on Γ defined by $f_\alpha(e^{i\theta}) = f(e^{i(\theta+\alpha)})$ for $0 \leq \theta < 2\pi$.

(a) Prove that, for each integer n, ϕ_n is a bounded linear functional on $C_{\mathsf{C}}(\Gamma)$.

(b) Prove that $f_\alpha \in C_{\mathsf{C}}(\Gamma)$ for each $f \in C_{\mathsf{C}}(\Gamma)$ and $\alpha \in [0,2\pi)$.

(c) Let x^* be a bounded linear functional on $C_{\mathsf{C}}(\Gamma)$ and, for $f \in C_{\mathsf{C}}(\Gamma)$, let a function F on Γ be defined by $F(e^{i\alpha}) = x^*(f_\alpha)$ for $0 \leq \alpha < 2\pi$. Prove that $F \in C_{\mathsf{C}}(\Gamma)$ and that $\phi_n(F) = \phi_n(f)x^*(g_n)$ for each integer n.

(d) Let $f \in C_{\mathsf{C}}(\Gamma)$. Prove that the set $\{f_\alpha : 0 \leq \alpha < 2\pi\}$ is fundamental in $C_{\mathsf{C}}(\Gamma)$ if and only if $\phi_n(f) \neq 0$ for every integer n. (This result is the analogue for Fourier series of a celebrated theorem on Fourier transforms, due to N. Wiener.)

5.5 A Theorem on Convex Sets

The purpose of this section is to prove a theorem that will be needed in Chapter 8.

5.5.1 DEFINITION A subset K of a linear space E over K is said to be *convex* if and only if $\alpha x + \beta y \in K$ whenever $x,y \in K$ and α and β are non-negative real numbers with $\alpha + \beta = 1$. The simplest examples of convex sets are the open and closed balls in a normed linear space.

There is a close relationship between convex sets and linear functionals. The Minkowski functional, which we now define, is the key to the study of this relationship.

5.5.2 DEFINITION Let K be a convex neighbourhood of 0 in a normed linear space E over K and let

$$p_K(x) = \inf\left\{\xi : \xi > 0 \text{ and } \frac{1}{\xi}x \in K\right\}$$

for all $x \in E$. The functional p_K is called the *Minkowski functional* of K.

Since K is a neighbourhood of 0, there exists $\alpha > 0$ such that $\{x \in E: \| x \| \leq \alpha\} \subseteq K$. Therefore if $x \in E$ and $x \neq 0$, then the point $\alpha \| x \|^{-1} x \in K$. This shows that the set $\{\xi : \xi > 0 \text{ and } \xi^{-1}x \in K\}$ is non-empty for each $x \in E$. The closed unit ball of a normed linear space E is a convex neighbourhood of 0 and it is easy to see that its Minkowski functional is just the norm on E.

We summarize the most important properties of the Minkowski functional in the next lemma.

5.5.3 LEMMA *Let K be a convex neighbourhood of 0 in a normed linear space E over K. Then p_K is a continuous sublinear functional on E and*

(a) $p_K(x) \geq 1$ *for all* $x \in E \sim K$,
(b) $p_K(x) \leq 1$ *for all* $x \in K$,
(c) $p_K(x) < 1$ *if and only if* $x \in K^\circ$.

Further, if $\alpha x \in K$ for all $x \in K$ and all $\alpha \in \mathsf{K}$ with $| \alpha | = 1$, then

(d) $p_K(\alpha x) = | \alpha | p_K(x)$ *for all* $x \in E$ *and all* $\alpha \in \mathsf{K}$.

PROOF We begin by proving that if $x \in E$ and $\xi > p_K(x)$ then $\xi^{-1}x \in K$. In fact, by definition of p_K, if $\xi > p_K(x)$ there exists ξ_1 with $\xi > \xi_1 \geq p_K(x)$ and $\xi_1^{-1}x \in K$. Since $0 \in K$ and K is convex, we now see that

$$\frac{1}{\xi}x = \frac{\xi_1}{\xi}\left(\frac{1}{\xi_1}x\right) + \frac{\xi - \xi_1}{\xi}0 \in K.$$

The special case $\xi = 1 > p_K(x)$ proves (a).

Let us now prove that p_K is a continuous sublinear functional on E. It is obvious that $p_K(x) \geq 0$ for all $x \in E$ and that $p_K(0) = 0$. Let $x,y \in E$, $\xi > p_K(x)$ and $\eta > p_K(y)$. By the first part of the proof $\xi^{-1}x \in K$ and $\eta^{-1}y \in K$. By the convexity of K we have

$$\frac{1}{\xi + \eta}(x + y) = \frac{\xi}{\xi + \eta}\left(\frac{1}{\xi}x\right) + \frac{\eta}{\xi + \eta}\left(\frac{1}{\eta}y\right) \in K.$$

This shows that $p_K(x+y) \leq \xi + \eta$ and hence that

$$p_K(x+y) \leq p_K(x) + p_K(y).$$

Next let $x \in X$ and $\alpha > 0$. Then

$$\begin{aligned} p_K(\alpha x) &= \inf\left\{\xi : \xi > 0, \frac{1}{\xi}\alpha x \in K\right\} \\ &= \inf\left\{\alpha\eta : \eta > 0, \frac{1}{\alpha\eta}\alpha x \in K\right\} \\ &= \alpha \inf\left\{\eta : \eta > 0, \frac{1}{\eta}x \in K\right\} \\ &= \alpha\, p_K(x). \end{aligned}$$

This proves that p_K is sublinear. Since K is a neighbourhood of 0, there exists $\alpha > 0$ such that $\{x \in E : \| x \| \leq \alpha\} \subseteq K$. Thus if $x \in E$ and $x \neq 0$ we have $\alpha \| x \|^{-1} x \in K$ and hence $p_K(x) \leq \alpha^{-1} \| x \|$. It follows then from Lemma 5.2.3 that p_K is continuous on E.

We have already proved (a), and (b) is obvious. To prove (c) let $K_0 = \{x \in E : p_K(x) < 1\}$. By (a) we have $K_0 \subseteq K$. Since p_K is continuous and $K_0 = p_K^{-1}((-1,1))$ the set K_0 is open (Theorem 2.7.3). Consequently $K_0 \subseteq K^\circ$. On the other hand let $x \in K^\circ$ and $x \neq 0$. Since K° is open there exists $\delta > 0$ such that $y \in K$ whenever $\| y - x \| < \delta$. In particular $(1 + \frac{1}{2}\delta \| x \|^{-1})x \in K$ and hence $p_K(x) \leq (1 + \frac{1}{2}\delta \| x \|^{-1})^{-1} < 1$. This shows that $x \in K_0$. Since it is obvious that $0 \in K_0$ we have now proved that $K^\circ \subseteq K_0$. Thus $K^\circ = K_0$. This proves (c).

Finally suppose that $\lambda x \in K$ for all $x \in K$ and all $\lambda \in \mathsf{K}$ with $| \lambda | = 1$. Let $x \in E$ and $\alpha \in \mathsf{K}$ with $\alpha \neq 0$. Then

$$\begin{aligned} p_K(\alpha x) &= \inf\left\{\xi : \xi > 0, \frac{1}{\xi}\alpha x \in K\right\} \\ &= \inf\left\{ | \alpha | \eta : \eta > 0, \frac{\alpha}{| \alpha |}\frac{1}{\eta}x \in K\right\}. \end{aligned}$$

However $\| \alpha |^{-1}\alpha | = 1$ so $| \alpha |^{-1}\alpha y \in K$ if and only if $y \in K$. Therefore

$$\begin{aligned} p_K(\alpha x) &= \inf\left\{ | \alpha | \eta : \eta > 0, \frac{1}{\eta}x \in K\right\} \\ &= | \alpha |\, p_K(x). \end{aligned}$$

This proves (d). □

5.5.4 THEOREM *Let K be a non-empty convex subset of a normed linear space E over K and suppose that $\alpha x \in K$ for all $x \in K$ and all $\alpha \in \mathsf{K}$ with $| \alpha | = 1$. Let $x_0 \in E \sim \bar{K}$. Then there exists $x^* \in E^*$ such that $| x^*(x_0) | = 1$ and $| x^*(x) | < 1$ for all $x \in K$.*

PROOF We observe that $0 \in K$ because if $x \in K$ then $-x = (-1)x \in K$ and hence $0 = \frac{1}{2}x + \frac{1}{2}(-x) \in K$. Since $x_0 \notin \bar{K}$ there exists $r > 0$ such that $K \cap B(x_0,r) = \emptyset$. Let $K_1 = \{x + y : x \in K, y \in B(0,r)\}$. It is easy to verify that K_1 is convex and that $\alpha x \in K_1$ for all $x \in K_1$ and all $\alpha \in \mathsf{K}$ with $|\alpha| = 1$. Since $0 \in K$ we have $B(0,r) \subseteq K_1$ and so K_1 is a neighbourhood of 0. Also, if $k \in K$ and $\| x - k \| < r$ then $x - k \in B(0,r)$ and hence $x = k + (x - k) \in K_1$. This shows that $B(k,r) \subseteq K_1$ for each $k \in K$ and hence $K \subseteq K_1^\circ$. Finally it is clear that $x_0 \notin K_1$.

Let p_{K_1} be the Minkowski functional of K_1, let $E_0 = \{\lambda x_0 : \lambda \in \mathsf{K}\}$ and let

$$y^*(\lambda x_0) = \lambda p_{K_1}(x_0)$$

for all $\lambda \in \mathsf{K}$. It is obvious that y^* is a linear functional on E_0, and Lemma 5.5.3 (d) gives

$$\operatorname{Re} y^*(\lambda x_0) \leq | y^*(\lambda x_0) | = | \lambda | p_{K_1}(x_0) = p_{K_1}(\lambda x_0)$$

for all $\lambda \in \mathsf{K}$. Since p_{K_1} is a sublinear functional on E (Lemma 5.5.3) it follows from Theorem 5.2.5 in the case when $\mathsf{K} = \mathsf{R}$, and from Theorem 5.3.1 in the case when $\mathsf{K} = \mathsf{C}$, that there is a linear functional x_1^* on E such that

$$\operatorname{Re} x_1^*(x) \leq p_{K_1}(x) \tag{1}$$

for all $x \in E$ and

$$x_1^*(\lambda x_0) = y^*(\lambda x_0) \tag{2}$$

for all $\lambda \in \mathsf{K}$.

Let $x \in E$ and choose $\theta \in \mathsf{R}$ so that† $x_1^*(x) = e^{i\theta} | x_1^*(x) |$. Using (1) and Lemma 5.5.3 (d) we obtain

$$\begin{aligned} | x_1^*(x) | &= e^{-i\theta} x_1^*(x) \\ &= x_1^*(e^{-i\theta} x) \\ &= \operatorname{Re} x_1^*(e^{-i\theta} x) \\ &\leq p_{K_1}(e^{-i\theta} x) \\ &= p_{K_1}(x). \end{aligned}$$

Since p_{K_1} is continuous on E (Lemma 5.5.3) it follows now from Lemma 5.2.3 that $x_1^* \in E^*$. Further, since $K \subseteq K_1^\circ$ it follows from Lemma 5.5.3 (c) that $| x_1^*(x) | \leq p_{K_1}(x) < 1$ for all $x \in K$. Finally Lemma 5.5.3 (a) and (2) give

$$x_1^*(x_0) = y^*(x_0) = p_{K_1}(x_0) \geq 1.$$

It is now clear that the functional $x^* = (x_1^*(x_0))^{-1} x_1^*$ satisfies the conditions of the theorem. □

The systematic study of the relationship between convex sets and linear functionals is beyond the scope of this book; for further details

† Notice that $e^{i\theta} = \pm 1$ when $\mathsf{K} = \mathsf{R}$.

the reader may consult the treatise of DUNFORD and SCHWARTZ [8, Chapter V].

EXERCISES

(1) Prove that if E is a normed linear space and $B = \{x \in E : \| x \| \leq 1\}$ then $p_B(x) = \| x \|$ for all $x \in E$.

(2) Let E be a normed linear space over K and let K be a subset of E which is a convex neighbourhood of 0 such that $\alpha x \in K$ whenever $x \in K$ and $\alpha \in \mathsf{K}$ with $| \alpha | = 1$. Prove that if $x_0 \in E \sim K$ then there exists $x^* \in E^*$ with $x^*(x_0) = 1$ and $| x^*(x) | \leq 1$ for all $x \in K$.

(3) Let E be a normed linear space over K and let $\alpha_1, \alpha_2,..., \alpha_n \in \mathsf{K}$.

(a) Prove that if $x_1, x_2,..., x_n \in E$ then there is a point $x^* \in E^*$ with $\| x^* \| \leq M$ and $x^*(x_i) = \alpha_i$ for $i = 1, 2,..., n$ if and only if

$$\left| \sum_{i=1}^{n} \lambda_i \alpha_i \right| \leq M \left\| \sum_{i=1}^{n} \lambda_i x_i \right\|$$

for all $\lambda_1, \lambda_2,..., \lambda_n \in \mathsf{K}$.

(b) Prove that if $x_1^*, x_2^*,..., x_n^* \in E^*$ then for each $\varepsilon > 0$ there is a point $x \in E$ with $\| x \| \leq M + \varepsilon$ and $x_i^*(x) = \alpha_i$ for $i = 1, 2,..., n$, if and only if $| \sum_{i=1}^{n} \lambda_i \alpha_i | \leq M \| \sum_{i=1}^{n} \lambda_i x_i^* \|$ for all $\lambda_1, \lambda_2,..., \lambda_n \in \mathsf{K}$. (Apply Exercise (2) to the subset $\{(x_1^*(x), x_2^*(x),..., x_n^*(x)) : x \in E, \| x \| \leq M + \varepsilon\}$ of K^n.) This result is due to E. Helly (1921).

5.6 The Riesz Representation Theorem

Throughout this section a and b will denote real numbers with $a < b$. We shall derive a famous integral representation for the continuous linear functionals on $C_{\mathsf{R}}([a,b])$ that was obtained by F. Riesz in 1909.

5.6.1 DEFINITION A finite subset P of $[a,b]$ which contains a and b will be called a *partition* of $[a,b]$. When P is a partition of $[a,b]$ and we write $P = \{s_0, s_1,..., s_n\}$ it will be understood that

$$a = s_0 < s_1 < \ldots < s_n = b.$$

Let f and α be real-valued functions on $[a,b]$. We shall say that f is *α-integrable* if and only if there is a real number I with the following property: for each $\varepsilon > 0$ there exists $\delta > 0$ such that

$$\left| I - \sum_{k=1}^{n} f(t_k)(\alpha(s_k) - \alpha(s_{k-1})) \right| < \varepsilon \tag{1}$$

for all partitions $\{s_0, s_1, \ldots, s_n\}$ of $[a,b]$ with $s_k - s_{k-1} < \delta$ for $k = 1, 2, \ldots, n$, and all real numbers $t_1, t_2, \ldots, t_n$ with $s_{k-1} \leq t_k \leq s_k$ for $k = 1, 2, \ldots, n$.

It is clear that, if f is α-integrable, there is exactly one real number I which has the property (1); it will be denoted by $\int_a^b f\, d\alpha$ or $\int_a^b f(t)\, d\alpha(t)$.

The Riesz representation theorem states that to each bounded linear functional x^* on $C_{\mathrm{R}}([a,b])$ there corresponds a real-valued function α on $[a,b]$ with the property that each $f \in C_{\mathrm{R}}([a,b])$ is α-integrable and

$$x^*(f) = \int_a^b f\, d\alpha \tag{2}$$

for all $f \in C_{\mathrm{R}}([a,b])$. Of course, this representation is valueless until some properties of the 'integral', $\int_a^b f\, d\alpha$, are known.

5.6.2 Definition Let α be a real-valued function on $[a,b]$. Given a partition $P = \{s_0, s_1, \ldots, s_n\}$ of $[a,b]$, let

$$V(P,\alpha) = \sum_{k=1}^{n} |\, \alpha(s_k) - \alpha(s_{k-1})\, |.$$

The function α is said to be of *bounded variation on* $[a,b]$ if and only if the set $\{V(P,\alpha) : P \text{ a partition of } [a,b]\}$ is bounded. If α is of bounded variation the number

$$V(\alpha) = \sup \{V(P,\alpha) : P \text{ a partition of } [a,b]\}$$

is called the *total variation of* α *on* $[a,b]$.

It can be proved that if α is of bounded variation on $[a,b]$ then each $f \in C_{\mathrm{R}}([a,b])$ is α-integrable (see Rudin [24, Theorem 6.14 and the remark on pp. 121–2]); in this case the 'integral' $\int_a^b f\, d\alpha$ is called the *Riemann–Stieltjes integral of f with respect to* α.

We shall see that the function α that occurs in the Riesz representation (2) is of bounded variation on $[a,b]$; consequently (2) provides a representation of the bounded linear functionals on $C_{\mathrm{R}}([a,b])$ in terms of Riemann–Stieltjes integrals. The importance of the Riesz representation theorem derives from the existence of an extensive theory of the Riemann–Stieltjes integral analogous to that of the Riemann integral (see Rudin [24, Chapter 6]). However, the proof of the Riesz representation theorem depends only on Definition 5.6.1 and not on the theory of the Riemann–Stieltjes integral. We need one more definition before proceeding with the proof of the Riesz theorem.

5.6.3 DEFINITION A linear functional x^* on $C_{\mathsf{R}}([a,b])$ is said to be *positive* if and only if $x^*(f) \geq 0$ whenever $f \in C_{\mathsf{R}}([a,b])$ and $f \geq 0$.

5.6.4 LEMMA *Let x^* be a positive linear functional on $C_{\mathsf{R}}([a,b])$. Then x^* is bounded and $\| x^* \| = x^*(1)$.*

PROOF Obviously $\| f \| 1 \pm f \geq 0$. Consequently

$$\| f \| x^*(1) \pm x^*(f) = x^*(\| f \| 1 \pm f) \geq 0.$$

Since $x^*(1) \geq 0$ this gives $| x^*(f) | \leq \| f \| x^*(1)$. This proves that x^* is bounded and $\| x^* \| \leq x^*(1)$. On the other hand $x^*(1) \leq \| x^* \|$ because $\| 1 \| = 1$. Thus $\| x^* \| = x^*(1)$. □

5.6.5 THEOREM (The Riesz representation theorem) *Let x^* be a bounded linear functional on the Banach space $C_{\mathsf{R}}([a,b])$. Then there is a real-valued function α of bounded variation on $[a,b]$ such that*

$$x^*(f) = \int_a^b f \, d\alpha$$

for all $f \in C_{\mathsf{R}}([a,b])$ and

$$\| x^* \| = V(\alpha).$$

Further, if x^* is a positive linear functional, then α is increasing on $[a,b]$.

PROOF Since $C_{\mathsf{R}}([a,b])$ is a linear subspace of $B_{\mathsf{R}}([a,b])$, there is, by Theorem 5.3.2, a bounded linear functional y^* on $B_{\mathsf{R}}([a,b])$ such that $y^*(f) = x^*(f)$ for all $f \in C_{\mathsf{R}}([a,b])$ and $\| y^* \| = \| x^* \|$.

Let $\chi_a = 0$ and, for $a < s \leq b$, let

$$\chi_s(t) = \begin{cases} 1 \text{ if } a \leq t \leq s, \\ 0 \text{ if } s < t \leq b. \end{cases}$$

Obviously $\chi_s \in B_{\mathsf{R}}([a,b])$ for $a \leq s \leq b$. Let

$$\alpha(s) = y^*(\chi_s) \tag{3}$$

for $a \leq s \leq b$. We shall prove that α satisfies the conditions of the theorem.

Let $\{s_0, s_1, \ldots, s_n\}$ be a partition of $[a,b]$ and let $\lambda_1, \lambda_2, \ldots, \lambda_n$ be real numbers. Then

$$\sum_{k=1}^{n} \lambda_k(\chi_{s_k}(t) - \chi_{s_{k-1}}(t)) = \lambda_k \tag{4}$$

for $s_{k-1} < t \leq s_k$ and $k = 1, 2, \ldots, n$ and hence

$$\left\| \sum_{k=1}^{n} \lambda_k(\chi_{s_k} - \chi_{s_{k-1}}) \right\| = \max \{| \lambda_1 |, | \lambda_2 |, \ldots, | \lambda_n |\}. \tag{5}$$

In particular, if we set

$$\lambda_k = \begin{cases} 1 \text{ if } \alpha(s_k) - \alpha(s_{k-1}) \geq 0, \\ -1 \text{ if } \alpha(s_k) - \alpha(s_{k-1}) < 0, \end{cases}$$

then (3), (4) and (5) give

$$\begin{aligned} \sum_{k=1}^{n} | \alpha(s_k) - \alpha(s_{k-1}) | &= \sum_{k=1}^{n} \lambda_k(\alpha(s_k) - \alpha(s_{k-1})) \\ &= \sum_{k=1}^{n} \lambda_k(y^*(\chi_{s_k}) - y^*(\chi_{s_{k-1}})) \\ &= y^*\Big(\sum_{k=1}^{n} \lambda_k(\chi_{s_k} - \chi_{s_{k-1}})\Big) \\ &\leq \| y^* \| \Big\| \sum_{k=1}^{n} \lambda_k(\chi_{s_k} - \chi_{s_{k-1}}) \Big\| \\ &= \| x^* \|. \end{aligned}$$

This proves that α is of bounded variation on $[a,b]$ and that

$$V(\alpha) \leq \| x^* \|.$$

Now let $f \in C_{\mathsf{R}}([a,b])$ and $\varepsilon > 0$. By Theorem 4.3.8 the function f is uniformly continuous on $[a,b]$ so there exists $\delta > 0$ such that

$$| f(s) - f(s') | < \varepsilon \tag{6}$$

for all $s,s' \in [a,b]$ with $| s - s' | < \delta$. Let $\{s_0, s_1,..., s_n\}$ be a partition of $[a,b]$ with $s_k - s_{k-1} < \delta$ for $k = 1, 2,..., n$, and let $s_{k-1} \leq t_k \leq s_k$ for $k = 1, 2,..., n$. Let

$$g = \sum_{k=1}^{n} f(t_k)(\chi_{s_k} - \chi_{s_{k-1}}).$$

Clearly $g \in B_{\mathsf{R}}([a,b])$. Let $t \in (a,b]$. Then $s_{k-1} < t \leq s_k$ for some integer k with $1 \leq k \leq n$, and it follows from (4) and (6) that

$$| g(t) - f(t) | = | f(t_k) - f(t) | < \varepsilon.$$

Also $g(a) = f(t_1)$ so

$$| g(a) - f(a) | = | f(t_1) - f(a) | < \varepsilon.$$

This shows that $\| g - f \| \leq \varepsilon$. Consequently

$$| y^*(g) - y^*(f) | = | y^*(g - f) | \leq \| y^* \| \| g - f \| \leq \varepsilon \| x^* \|. \tag{7}$$

Also, by (3),

$$\begin{aligned} y^*(g) &= y^*\Big(\sum_{k=1}^{n} f(t_k)(\chi_{s_k} - \chi_{s_{k-1}})\Big) \\ &= \sum_{k=1}^{n} f(t_k)(y^*(\chi_{s_k}) - y^*(\chi_{s_{k-1}})) \end{aligned}$$

$$= \sum_{k=1}^{n} f(t_k)(\alpha(s_k) - \alpha(s_{k-1})). \tag{8}$$

Since $y^*(f) = x^*(f)$ inequalities (7) and (8) give

$$\left| \sum_{k=1}^{n} f(t_k)(\alpha(s_k) - \alpha(s_{k-1})) - x^*(f) \right| \leq \varepsilon \| x^* \|.$$

This proves that

$$x^*(f) = \int_a^b f \, d\alpha.$$

It remains to prove that $V(\alpha) = \| x^* \|$ and to verify the assertion about positive linear functionals. Let $f \in C_{\mathsf{R}}([a,b])$. We have seen that $x^*(f) = \int_a^b f \, d\alpha$ and consequently, given $\varepsilon > 0$, there is a partition $\{s_0, s_1, \ldots, s_n\}$ of $[a,b]$ such that $| x^*(f) - \sum_{k=1}^{n} f(s_k)(\alpha(s_k) - \alpha(s_{k-1})) | < \varepsilon$. Therefore

$$\begin{aligned} | x^*(f) | &< \varepsilon + \left| \sum_{k=1}^{n} f(s_k)(\alpha(s_k) - \alpha(s_{k-1})) \right| \\ &\leq \varepsilon + \sum_{k=1}^{n} | f(s_k) | \; | \alpha(s_k) - \alpha(s_{k-1}) | \\ &\leq \varepsilon + \| f \| \sum_{k=1}^{n} | \alpha(s_k) - \alpha(s_{k-1}) | \\ &\leq \varepsilon + \| f \| \, V(\alpha). \end{aligned}$$

It follows that $| x^*(f) | \leq \| f \| \, V(\alpha)$ and hence that $\| x^* \| \leq V(\alpha)$. We have already seen that $\| x^* \| \geq V(\alpha)$ so we have $\| x^* \| = V(\alpha)$.

Suppose finally that x^* is a positive linear functional. We shall show that $y^*(f) \geq 0$ for all $f \in B_{\mathsf{R}}([a,b])$ with $f \geq 0$. Suppose, on the contrary, that there exists $f_0 \in B_{\mathsf{R}}([a,b])$ with $f_0 \geq 0$ and $y^*(f_0) < 0$. Replacing f_0 by $\| f_0 \|^{-1} f_0$ if necessary, we may suppose that $0 \leq f_0 \leq 1$. Thus $\| 1 - f_0 \| \leq 1$ and hence

$$y^*(1) < y^*(1) - y^*(f_0) = y^*(1 - f_0) \leq \| y^* \|.$$

On the other hand we have $\| x^* \| = x^*(1)$ by Lemma 5.6.4, so $\| y^* \| = \| x^* \| = x^*(1) = y^*(1)$. This contradiction proves that $y^*(f) \geq 0$ for all $f \in B_{\mathsf{R}}([a,b])$ with $f \geq 0$. Now let $a \leq s_1 \leq s_2 \leq b$. Then

$$\alpha(s_2) - \alpha(s_1) = y^*(\chi_{s_2} - \chi_{s_1}) \geq 0$$

because $\chi_{s_2} \geq \chi_{s_1}$. This shows that α is increasing on $[a,b]$. □

In Exercise (1) we sketch an alternative proof of Theorem 5.6.5 that does not depend on the Hahn–Banach theorem.

It follows from the theory of the Riemann–Stieltjes integral that if α is a real-valued function of bounded variation on $[a,b]$ and $x^*(f) = \int_a^b f\,d\alpha$ for all $f \in C_{\mathsf{R}}([a,b])$ then x^* is a bounded linear functional on $C_{\mathsf{R}}([a,b])$ and $\| x^* \| \leq V(\alpha)$. (See RUDIN [24, Corollary to Theorem 6.29].) However the correspondence between bounded linear functionals on $C_{\mathsf{R}}([a,b])$ and functions of bounded variation on $[a,b]$ is not one-to-one. Indeed, if $a < c < b$ and

$$\alpha(t) = \begin{cases} 0 \text{ if } t \neq c, \\ 1 \text{ if } t = c \end{cases}$$

then $\int_a^b f\,d\alpha = 0$ for all $f \in C_{\mathsf{R}}([a,b])$ and $V(\alpha) = 2$. However it can be shown that there is a one-to-one correspondence between the bounded linear functionals on $C_{\mathsf{R}}([a,b])$ and a suitable subclass of the class of functions of bounded variation on $[a,b]$. (See TAYLOR [26, pp. 197–200].)

The Riesz representation theorem provided the inspiration for one of the most important modern approaches to the theory of integration. A comprehensive account of this approach to integration can be found in the treatise of BOURBAKI [6]. The first chapter of the book of NACHBIN [20] provides an excellent introduction to the Bourbaki theory of integration, and contains a discussion of its relationship to the Riesz representation theorem.

EXERCISES

(1) The following exercises present an alternative proof of the Riesz representation theorem which is independent of the Hahn–Banach theorem. In these exercises C will denote the Banach space $C_{\mathsf{R}}([a,b])$.

(a) Let $x^* \in C^*$. Prove that there are positive linear functionals x_1^* and x_2^* on C with $x^* = x_1^* - x_2^*$. (Hint: for $f \in C$ with $f \geq 0$, set $x_1^*(f) = \sup\{x^*(g) : g \in C \text{ and } 0 \leq g \leq f\}$.)

(b) Let $\mathscr{L}_1$ be the subset of $B_{\mathsf{R}}([a,b])$ consisting of those functions $f \in B_{\mathsf{R}}([a,b])$ with the property that there is a sequence (f_n) in C with $0 \leq f_n \leq f_{n+1}$ for $n = 1, 2,\ldots$ and $\lim_{n\to\infty} f_n(t) = f(t)$ for $a \leq t \leq b$. Let $\mathscr{L} = \{f - g : f,g \in \mathscr{L}_1\}$. Prove that each positive linear functional on C has a unique extension to a positive linear functional on $\mathscr{L}$. (Hint: for $f \in \mathscr{L}_1$ let $x^*(f) = \lim_{n\to\infty} x^*(f_n)$ where (f_n) is a sequence in C with $0 \leq f_n \leq f_{n+1}$ and $\lim_{n\to\infty} f_n(t) = f(t)$.)

(c) Use (a) and (b) to give a proof of the Riesz representation theorem which is independent of the Hahn–Banach theorem.

(2) Let p be a positive integer and let $C^p([a,b])$ be the set of all real-valued functions f on $[a,b]$ that are p-times differentiable on $[a,b]$ and whose pth derivative, $f^{(p)}$, is continuous on $[a,b]$ (cf. Example 2.1.13). Verify that $C^p([a,b])$ is a Banach space with respect to the norm

$$\| f \| = \sum_{k=0}^{p} \sup \{ | f^{(k)}(t) | : a \leq t \leq b\}$$

where $f^{(0)} = f$. Let x^* be a bounded linear functional on $C^p([a,b])$. Prove that there are real numbers $\beta_0, \beta_1, \ldots, \beta_{p-1}$ and a real-valued function α of bounded variation on $[a,b]$ such that

$$x^*(f) = \sum_{k=0}^{p-1} \beta_k f^{(k)}(a) + \int_a^b f^{(p)}(t)\, d\alpha(t)$$

for all $f \in C^p([a,b])$.

(3) The result of Exercise 4.5(2) has a simple and elegant generalization due to P. P. Korovkin (1953).

A linear operator T on $C_{\mathrm{R}}([a,b])$ is said to be *positive* if $Tf \geq 0$ whenever $f \geq 0$. Let f_0 and f_0^2 be the functions given by $f_0(t) = t$ and $f_0^2(t) = t^2$ for all $t \in [a,b]$. Prove that if (T_n) is a sequence of positive linear operators on $C_{\mathrm{R}}([a,b])$ such that $1 = \lim_{n\to\infty} T_n 1$, $f_0 = \lim_{n\to\infty} T_n f_0$ and $f_0^2 = \lim_{n\to\infty} T_n f_0^2$ then $f = \lim_{n\to\infty} T_n f$ for every $f \in C_{\mathrm{R}}([a,b])$. (The proof is achieved by applying the Riesz representation theorem (5.6.5) to the functionals $f \to (T_n f)(t)$ and then mimicking the final steps of Exercise 4.5(2).)

(4) Let f_0 be the function (on $[0,1]$) of Exercise (3). The classical *moment problem* is that of determining which sequences $(\mu_n)_{n\geq 0}$ have, for some increasing function α,

$$\mu_n = \int_0^1 t^n\, d\alpha(t)$$

for $n = 0, 1, 2, \ldots$. This problem is equivalent to that of determining which sequences have, for some positive linear functional x^* on $C_{\mathrm{R}}([0,1])$, $\mu_n = x^*(f_0^n)$ for $n = 0, 1, 2, \ldots$. If (μ_n) is a sequence and x^* a positive linear functional which satisfy these equations, then

$$\sum_{r=0}^{n} (-1)^r \binom{n}{r} \mu_{k+r} \geq 0 \tag{*}$$

for $n,k = 0, 1, 2, \ldots$ because $\sum_{r=0}^{n} (-1)^r \binom{n}{r} \mu_{k+r} = x^*(f_0^k(1 - f_0)^n)$. Thus the inequalities (*) are a necessary condition on (μ_n) for the existence of x^*. The condition is also sufficient:

(a) Prove that if f is a polynomial of degree m then each of the Bernstein polynomials $B_n(f, \cdot)$ of Exercise 4.5(2) is of degree $\leq m$. (Consider the $(m+1)$st derivative of $B_n(f_0^m, \cdot)$.)

(b) Use the result of Exercise 4.5.2, Theorem 5.1.3 and (a) to prove that if a linear functional x^* is defined on the set of all polynomials on [0,1] by $x^*\left(\sum_{r=0}^{m} \lambda_r f_0^r\right) = \sum_{r=0}^{m} \lambda_r \mu_r$, where (μ_n) satisfies (*), then x^* is a positive linear functional. Hence prove that x^* is bounded and has a positive extension to all of $C_{\mathsf{R}}([0,1])$.

(5) For a sequence $(\mu_n)_{n \geq 0}$ let $\Delta_k^n = \sum_{r=0}^{n} (-1)^r \binom{n}{r} \mu_{k+r}$ Prove that there is a bounded linear functional x^* on $C_{\mathsf{R}}([0,1])$ with $\mu_n = x^*(f_0^n)$ for $n = 0, 1, 2, \ldots$ if and only if, for some M, $\sum_{k=0}^{n} \binom{n}{k} \mid \Delta_k^{n-k} \mid \leq M$ for $n = 0, 1, 2, \ldots$. (As in Exercise (4) use the Bernstein polynomials.)

(6) Let X be a metric space and let $f_0 \in C_{\mathsf{R}}(X)$ and $\phi \in C_{\mathsf{R}}(X)^*$ satisfy $\| f_0 \| = 1$ and $\phi(f_0) = \| \phi \|$. Prove that if Y is a subset of X with $\sup \{ \mid f_0(x) \mid : x \in Y\} < 1$ then $\phi(f) = 0$ whenever $f \in C_{\mathsf{R}}(X)$ and $f(x) = 0$ for all $x \in X \sim Y$. Prove further that if the space X is compact then $\phi(f) = 0$ whenever $f \in C_{\mathsf{R}}(X)$ and $f(x) = 0$ for all those $x \in X$ for which $\mid f_0(x) \mid = 1$.

(7) Prove that if ϕ is a real-valued continuous function on $[a,b]$ then Φ, defined by $\Phi(f) = \int_a^b f(t)\, \phi(t)\, dt$ for $f \in C_{\mathsf{R}}([a,b])$, is a bounded linear functional on $C_{\mathsf{R}}([a,b])$ with $\| \Phi \| = \int_a^b \mid \phi(t) \mid dt$.

5.7 Herglotz's Theorem

In this section we shall use the Hahn–Banach theorem and the Riesz representation theorem to prove a theorem of G. Herglotz (1911) on analytic functions with non-negative real part on the unit disc. Herglotz's theorem has been used in the theory of linear operators.

5.7.1 LEMMA *Let x^* be a bounded linear functional on $C_{\mathsf{R}}([0,2\pi])$ and let*

$$x_1^*(f) = x^*(\mathrm{Re}\, f) + ix^*(\mathrm{Im}\, f) \tag{1}$$

for all $f \in C_{\mathsf{C}}([0,2\pi])$. Then x_1^* is a bounded linear functional on $C_{\mathsf{C}}([0,2\pi])$.

PROOF A routine calculation shows that x_1^* is linear. Also, if $f \in C_{\mathsf{C}}([0,2\pi])$ then

$$\begin{aligned}|x_1^*(f)| &\leq |x^*(\mathrm{Re}\, f)| + |x^*(\mathrm{Im}\, f)| \\ &\leq \|x^*\| (\|\mathrm{Re}\, f\| + \|\mathrm{Im}\, f\|) \\ &\leq 2\|x^*\| \, \|f\|.\end{aligned}$$

This shows that x_1^* is a bounded linear functional. □

Let x^* be a bounded linear functional on $C_{\mathsf{R}}([0,2\pi])$. By Theorem 5.6.5 there is a real-valued function α of bounded variation on $[0,2\pi]$ such that

$$x^*(f) = \int_0^{2\pi} f \, d\alpha$$

for all $f \in C_{\mathsf{R}}([0,2\pi])$. Given $f \in C_{\mathsf{C}}([0,2\pi])$ we shall write

$$\int_0^{2\pi} f \, d\alpha = \int_0^{2\pi} \mathrm{Re}\, f \, d\alpha + i \int_0^{2\pi} \mathrm{Im}\, f \, d\alpha. \tag{2}$$

Obviously if x_1^* is the functional defined by (1) then

$$x_1^*(f) = \int_0^{2\pi} f \, d\alpha \tag{3}$$

for all $f \in C_{\mathsf{C}}([0,2\pi])$.

Recall that a real trigonometric polynomial on $[0,2\pi]$ is a real-valued function g of the form

$$g(\theta) = \tfrac{1}{2}\lambda_0 + \sum_{n=1}^{k} (\lambda_n \cos n\theta + \mu_n \sin n\theta) \tag{4}$$

for $0 \leq \theta \leq 2\pi$, where $\lambda_0, \lambda_1, \ldots, \lambda_k, \mu_1, \ldots, \mu_k$ are real numbers and k is a non-negative integer. It is clear that the set E_0 of all real trigonometric polynomials on $[0,2\pi]$ is a linear subspace of $C_{\mathsf{R}}([0,2\pi])$. If g is given by (4) an elementary calculation gives

$$\lambda_n = \frac{1}{\pi} \int_0^{2\pi} g(\theta) \cos n\theta \, d\theta, \; n = 0, 1, \ldots, k,$$

and
$$\mu_n = \frac{1}{\pi} \int_0^{2\pi} g(\theta) \sin n\theta \, d\theta, \; n = 1, 2, \ldots, k.$$

Consequently each function in E_0 has a unique representation in the form (4). We can write (4) as

$$g = \tfrac{1}{2}\lambda_0 1 + \sum_{n=1}^{k} (\lambda_n u_n + \mu_n v_n), \tag{5}$$

where $u_n(\theta) = \cos n\theta$ and $v_n(\theta) = \sin n\theta$ for $0 \leq \theta \leq 2\pi$ and $n = 1, 2, \ldots$.

5.7.2 THEOREM (Herglotz) *Let f be an analytic function on the open disc $\Delta = \{z \in \mathsf{C} : |z| < 1\}$ such that $\mathrm{Re}\, f(z) \geq 0$ for all $z \in \Delta$. Then there is an increasing real-valued function α on $[0,2\pi]$ such that*

$$f(z) = i \operatorname{Im} f(0) + \int_0^{2\pi} \frac{e^{it} + z}{e^{it} - z} \, d\alpha(t)$$

for all $z \in \Delta$.

PROOF It is well-known† that the Taylor series, $\sum_n \gamma_n z^n$ say, of f is absolutely convergent on Δ and that $f(z) = \sum_{n=0}^{\infty} \gamma_n z^n$ for all $z \in \Delta$. Let $\gamma_n = \alpha_n + i\beta_n$, where α_n and β_n are real. Then for all $z = re^{i\theta} \in \Delta$ we have

$$\begin{aligned} \operatorname{Re} f(z) &= \operatorname{Re}\left(\sum_{n=0}^{\infty} r^n(\alpha_n + i\beta_n)(\cos n\theta + i \sin n\theta)\right) \\ &= \sum_{n=0}^{\infty} r^n(\alpha_n \cos n\theta - \beta_n \sin n\theta) \\ &= \alpha_0 + \sum_{n=1}^{\infty} r^n(\alpha_n \cos n\theta - \beta_n \sin n\theta). \end{aligned} \tag{6}$$

Let x_0^* be the functional on E_0 defined by

$$x_0^*(g) = \tfrac{1}{2}\alpha_0\lambda_0 + \tfrac{1}{2} \sum_{n=1}^{k} (\alpha_n\lambda_n - \beta_n\mu_n), \tag{7}$$

where g is given by (5). It is obvious that x_0^* is a linear functional on E_0 and

$$x_0^*(1) = \alpha_0, \; x_0^*(u_n) = \tfrac{1}{2}\alpha_n, \text{ and } x_0^*(v_n) = -\tfrac{1}{2}\beta_n \tag{8}$$

for $n = 1, 2, \ldots$.

Now let $g \in E_0$ be given by (5) and let $0 \leq r < 1$. Since

$$\| r^n(\alpha_n u_n - \beta_n v_n) \| \leq (\, | \alpha_n | + | \beta_n | \,) r^n \leq 2 \, | \gamma_n | \, r^n$$

and the series $\sum_n | \gamma_n | \, r^n$ converges, the series $\sum_n r^n(\alpha_n u_n - \beta_n v_n)$ converges uniformly on $[0,2\pi]$ (Lemma 3.3.8). It now follows from a routine calculation using (4) and (6) that

$$\int_0^{2\pi} g(\theta) \operatorname{Re} f(re^{i\theta}) \, d\theta = \pi\alpha_0\lambda_0 + \pi \sum_{n=1}^{k} r^n(\lambda_n\alpha_n - \mu_n\beta_n). \tag{9}$$

Equations (7) and (9) give

$$x_0^*(g) = \frac{1}{2\pi} \lim_{r \to 1} \int_0^{2\pi} g(\theta) \operatorname{Re} f(re^{i\theta}) \, d\theta.$$

† The reader will find this result in any textbook on complex function theory (e.g. AHLFORS [2, p. 177] and HILLE [14, p. 196]).

By hypothesis $\operatorname{Re} f(re^{i\theta}) \geq 0$ for $0 \leq r < 1$ and $0 \leq \theta \leq 2\pi$, so it follows that

$$x_0^*(g) \geq 0 \tag{10}$$

for all $g \in E_0$ with $g \geq 0$. Let p be the functional on $C_{\mathsf{R}}([0,2\pi])$ given by

$$p(g) = \sup \{g(\theta) : 0 \leq \theta \leq 2\pi\}$$

for all $g \in C_{\mathsf{R}}([0,2\pi])$. Obviously, $p(g)1 - g \geq 0$ for all $g \in E_0$, so we have $x_0^*(p(g)1 - g) \geq 0$ for all $g \in E_0$. Thus, by (8),

$$x_0^*(g) \leq x_0^*(1)p(g) = \alpha_0 p(g)$$

for all $g \in E_0$. Since $\alpha_0 = \operatorname{Re} f(0) \geq 0$ it is easy to verify that the map $g \rightarrow \alpha_0 p(g)$ is sublinear on $C_{\mathsf{R}}([0,2\pi])$. By the Hahn–Banach theorem (5.2.5) there is a linear functional x^* on $C_{\mathsf{R}}([0,2\pi])$ such that

$$x^*(g) \leq \alpha_0 p(g) \tag{11}$$

for all $g \in C_{\mathsf{R}}([0,2\pi])$ and

$$x^*(g) = x_0^*(g) \tag{12}$$

for all $g \in E_0$.

Let $g \in C_{\mathsf{R}}([0,2\pi])$ with $g \geq 0$. Then $p(-g) \leq 0$ and by (11) we have $x^*(-g) \leq 0$. Hence $x^*(g) \geq 0$. This shows that x^* is a positive linear functional on $C_{\mathsf{R}}([0,2\pi])$. By Lemma 5.6.4 the functional x^* is bounded. Let x_1^* be the bounded linear functional on $C_{\mathsf{C}}([0,2\pi])$ defined by (1).

For $z \in \Delta$ and $0 \leq t \leq 2\pi$, let

$$H_z(t) = \frac{e^{it} + z}{e^{it} - z}.$$

It is clear that $H_z \in C_{\mathsf{C}}([0,2\pi])$ for each $z \in \Delta$. Also, if $0 \leq t \leq 2\pi$ and $z \in \Delta$

$$\begin{aligned} H_z(t) &= 1 + \frac{2ze^{-it}}{1 - ze^{-it}} \\ &= 1 + 2\sum_{n=1}^{\infty} e^{-int} z^n \\ &= 1 + 2\sum_{n=1}^{\infty} (\cos nt - i \sin nt) z^n. \end{aligned} \tag{13}$$

Let $z \in \Delta$. Since $\| z^n(u_n - iv_n) \| \leq 2 \mid z \mid^n$ the series $\sum_n z^n(u_n - iv_n)$ converges uniformly on $[0,2\pi]$ (Lemma 3.3.8). Equation (13) shows that

$$H_z = 1 + 2\sum_{n=1}^{\infty} z^n(u_n - iv_n). \tag{14}$$

Since x_1^* is a bounded linear functional on $C_{\mathsf{C}}([0,2\pi])$, by using (14), (1), (12) and (8), we obtain

$$
\begin{aligned}
x_1^*(H_z) &= x_1^*(1) + 2\sum_{n=1}^{\infty} z^n x_1^*(u_n - iv_n) \\
&= x_0^*(1) + 2\sum_{n=1}^{\infty} z^n(x_0^*(u_n) - ix_0^*(v_n)) \\
&= \alpha_0 + 2\sum_{n=1}^{\infty} \tfrac{1}{2}(\alpha_n + i\beta_n)z^n \\
&= \sum_{n=0}^{\infty} (\alpha_n + i\beta_n)z^n - i\beta_0 \\
&= f(z) - i\,\mathrm{Im}\, f(0). \qquad (15)
\end{aligned}
$$

Since x^* is a positive linear functional on $C_{\mathsf{R}}([0,2\pi])$, Theorem 5.6.5 shows that there is an increasing real-valued function α on $[0,2\pi]$ such that $x^*(g) = \int_0^{2\pi} g\, d\alpha$ for all $g \in C_{\mathsf{R}}([0,2\pi])$. It now follows from (15) and (3) that, for all $z \in \Delta$,

$$
\begin{aligned}
f(z) &= i\,\mathrm{Im}\, f(0) + x_1^*(H_z) \\
&= i\,\mathrm{Im}\, f(0) + \int_0^{2\pi} H_z\, d\alpha. \; \square
\end{aligned}
$$

The reader will notice that if $\mathrm{Re}\, f(0) = 0$ then $x^* = 0$, and consequently $f(z) = i\,\mathrm{Im}\, f(0)$ for all $z \in \Delta$. Thus any analytic function f on Δ such that $\mathrm{Re}\, f(z) \geq 0$ for all $z \in \Delta$ and $\mathrm{Re}\, f(0) = 0$ is constant on Δ.

CHAPTER 6

Bounded Linear Operators

The reader will be familiar with a number of examples of linear equations and systems of linear equations. In applications, the equations are usually differential equations but the most elementary example is a system of m simultaneous linear equations in n unknowns of the form

$$\begin{aligned} \alpha_{11}\xi_1 + \alpha_{12}\xi_2 + \ldots + \alpha_{1n}\xi_n &= \eta_1 \\ \alpha_{21}\xi_1 + \alpha_{22}\xi_2 + \ldots + \alpha_{2n}\xi_n &= \eta_2 \\ \cdot\ \ \cdot\ \ \cdot\ \ \cdot\ \ \cdot\ \ \cdot\ \ \cdot\ \ \cdot\ \ \cdot\ \ \cdot\ \ \cdot\ \ \cdot\ \ \cdot \\ \alpha_{m1}\xi_1 + \alpha_{m2}\xi_2 + \ldots + \alpha_{mn}\xi_n &= \eta_m. \end{aligned} \tag{*}$$

Such a system defines a mapping T of R^n into R^m as follows: given $x = (\xi_1,\ldots, \xi_n)$ let $Tx = y$, where $y = (\eta_1,\ldots, \eta_m)$ and $\eta_1,\ldots, \eta_m$ are determined by (*). It is easy to see that T is a linear transformation of R^n into R^m. Questions about the system (*) can be formulated as questions about T. For example to ask for which $\eta_1,\ldots, \eta_m$ the system (*) has a solution is to ask for a description of the range of T.

More generally a linear equation is an equation of the form $Tx = y$ where T is a linear transformation of one linear space into another. Many of the equations which arise in applied mathematics and physics are linear equations, and the study of linear equations is a prerequisite for the study of non-linear equations. (Cf. Section 7.5.)

Just as the most complete results about systems of the form (*) can be obtained in the case when $n = m$, in which case T is a mapping of R^n into itself, so, in general, more can be said about linear equations $Tx = y$ when T is a linear transformation of a linear space into itself; such linear transformations are usually called *linear operators*.

We shall begin with the description of some important examples, and then derive some of the general properties of linear operators. As an illustration of the general theory we shall prove the famous results of F. Riesz on compact linear operators and then use them to obtain some classical results of Fredholm on linear integral equations.

6.1 Examples

First we consider linear transformations on finite dimensional spaces.

6.1.1 THEOREM *Let E be an n-dimensional linear space over* K *and let F be an m-dimensional linear space over* K. *Associated with each choice of bases $\{e_1, e_2, \ldots, e_n\}$ for E and $\{f_1, f_2, \ldots, f_m\}$ for F there is a one-to-one correspondence between the set of all linear transformations T of E into F and the set of all $m \times n$ matrices (τ_{jk}) of elements of* K. *This correspondence is given by*

$$Te_k = \sum_{j=1}^{m} \tau_{jk} f_j$$

for $k = 1, 2, \ldots, n$.

PROOF Let $\{e_1, e_2, \ldots, e_n\}$ and $\{f_1, f_2, \ldots, f_m\}$ be bases for E and F, respectively, and let T be a linear transformation of E into F. Then for each integer k with $1 \leq k \leq n$, there are unique elements $\tau_{1k}, \tau_{2k}, \ldots, \tau_{mk}$ of K such that

$$Te_k = \sum_{j=1}^{m} \tau_{jk} f_j. \tag{1}$$

Each point $x \in E$ has a unique representation in the form $x = \sum_{k=1}^{n} \xi_k e_k$, where $\xi_1, \xi_2, \ldots, \xi_n$ are in K. Hence

$$\begin{aligned} Tx &= \sum_{k=1}^{n} \xi_k Te_k \\ &= \sum_{k=1}^{n} \xi_k \left(\sum_{j=1}^{m} \tau_{jk} f_j \right) \\ &= \sum_{j=1}^{m} \left(\sum_{k=1}^{n} \tau_{jk} \xi_k \right) f_j. \end{aligned} \tag{2}$$

Conversely let (τ_{jk}) be an $m \times n$ matrix of elements of K. We can define a mapping T of E into F by equation (2). It is easy to check that T is linear and satisfies (1). □

It is left to the reader to verify that the correspondence established in the above theorem is a one-to-one linear mapping of the linear space of all linear transformations of E into F onto the linear space of all $m \times n$ matrices. It follows that all the classical results on canonical forms of matrices have analogues for linear operators on finite dimensional spaces. (See HALMOS [12, pp. 102–15].) These classical results serve as models for the theory of linear operators on infinite dimensional spaces.

The following theorem shows that in the study of linear operators on finite-dimensional spaces continuity is, in a sense, irrelevant.

6.1.2 THEOREM *Let E and F be normed linear spaces over K and suppose that E is finite-dimensional. Then every linear transformation of E into F is bounded.*

PROOF Let $\{e_1, e_2, \ldots, e_n\}$ be a basis for E. Each point $x \in E$ can be expressed uniquely in the form

$$x = \sum_{k=1}^{n} \xi_k e_k, \tag{3}$$

where $\xi_1, \xi_2, \ldots, \xi_n$ are in K. A simple calculation shows that the mapping $x \to \sum_{k=1}^{n} | \xi_k |$ is a norm on E. By Theorem 4.4.1 any two norms on E are equivalent so there exists $M > 0$ such that

$$\sum_{k=1}^{n} | \xi_k | \leq M \| x \| \tag{4}$$

for all $x \in E$. (See Definition 3.6.2.)

Let T be a linear transformation of E into F. Then, for all $x \in E$, (3) and (4) give

$$\begin{aligned} \| Tx \| &= \left\| \sum_{k=1}^{n} \xi_k T e_k \right\| \\ &\leq \sum_{k=1}^{n} | \xi_k | \, \| T e_k \| \\ &\leq M \| x \| \max \{ \| Te_1 \|, \| Te_2 \|, \ldots, \| Te_n \| \}. \end{aligned}$$

This shows that T is bounded. □

It follows from the above theorem that all results concerning linear transformations on finite-dimensional normed linear spaces must be independent of any considerations of continuity, and so they must be essentially algebraic. It follows also that all results concerning bounded linear transformations on normed linear spaces can be applied to linear transformations on finite-dimensional spaces, even though the finite-dimensional space has no specified norm. In fact we saw in the proof of Theorem 6.1.2 that the mapping $x \to \sum_{k=1}^{n} | \xi_k |$, where $\xi_1, \xi_2, \ldots, \xi_n$ are defined by (3), is a norm on any n-dimensional linear space E, and that all linear transformations on E are bounded with respect to this norm.

Just as we can associate a linear transformation with a system of algebraic equations, so we can associate a linear transformation with a linear differential equation. However linear differential equations give rise to linear transformations on infinite-dimensional linear spaces, and these linear transformations may not be bounded with respect to the natural norm on the space.

6.1.3 Example Let E be the linear subspace of $C_{\mathsf{R}}([0,2\pi])$ consisting of those functions on $[0,2\pi]$ that have continuous first derivatives on $[0,2\pi]$ (see Example 2.1.13) and let $\|\cdot\|$ be the uniform norm on $C_{\mathsf{R}}([0,2\pi])$. The study of linear differential equations of the form

$$\frac{dy}{dx} + y = F(x),$$

where $F \in C_{\mathsf{R}}([0,2\pi])$, leads naturally to the study of the linear transformation T of E into $C_{\mathsf{R}}([0,2\pi])$ defined by

$$(Tf)(t) = f'(t) + f(t)$$

for $f \in E$ and $0 \leq t \leq 2\pi$. However, T is not a bounded linear mapping of $(E, \|\cdot\|)$ into $C_{\mathsf{R}}([0,2\pi])$ because, if $f_n(t) = \sin nt$, then $\| f_n \| = 1$ and $\| Tf_n \| \geq n$.

However, if we take on E the norm $\| \cdot \|_1$ defined by

$$\| f \|_1 = \sup \{ | f(t) | : 0 \leq t \leq 2\pi\} + \sup \{ | f'(t) | : 0 \leq t \leq 2\pi\},$$

(cf. Example 2.1.13), then, for all $f \in E$,

$$\| Tf \| = \sup \{ | f'(t) + f(t) | : 0 \leq t \leq 2\pi\} \leq \| f \|_1.$$

Thus T is a bounded linear mapping of $(E, \|\cdot\|)$ into $C_{\mathsf{R}}([0,2\pi])$.

It is often possible to replace a differential equation by an integral equation. (See Lemma 2.11.5.) Much of the early work on linear operators was motivated by the theory of linear integral equations. We shall now define two linear operators that arise from the study of linear integral equations.

6.1.4 Theorem *Let k be a complex-valued continuous function on the square $S = \{ (s,t) : a \leq s \leq b, a \leq t \leq b\}$ and for each $f \in C_{\mathsf{C}}([a,b])$ let*

$$(Kf)(s) = \int_a^b k(s,t)f(t)\, dt \tag{5}$$

for $a \leq s \leq b$. This defines a mapping K which is a bounded linear operator on $C_{\mathsf{C}}([a,b])$ with

$$\| K \| \leq (b - a) \sup \{ | k(s,t) | : (s,t) \in S\}.$$

Proof Let $f \in C_{\mathsf{C}}([a,b])$. The function Kf is defined because, for each $s \in [a,b]$, the function $t \to k(s,t)f(t)$ is continuous on $[a,b]$. We shall prove that $Kf \in C_{\mathbf{C}}([a,b])$. Since $K0 = 0$, we may suppose that $f \neq 0$. By the Borel–Lebesgue theorem (4.2.11) S is compact, and therefore, by Theorem 4.3.8, k is uniformly continuous on S. Thus, given $\varepsilon > 0$, there exists $\delta > 0$ such that

$$| k(s_1,t_1) - k(s_2,t_2) | < \frac{\varepsilon}{\| f \| (b-a)}$$

for all (s_1,t_1) and (s_2,t_2) in S with $((s_1 - s_2)^2 + (t_1 - t_2)^2)^{1/2} < \delta$. Consequently, if $s_1,s_2 \in [a,b]$ and $| s_1 - s_2 | < \delta$ then

$$\begin{aligned}
| (Kf)(s_1) - (Kf)(s_2) | & \\
&= \left| \int_a^b (k(s_1,t) - k(s_2,t))f(t)\, dt \right| \\
&\leq (b-a) \sup \{ | k(s_1,t) - k(s_2,t) | \, | f(t) | : a \leq t \leq b \} \\
&\leq (b-a) \frac{\varepsilon}{\| f \| (b-a)} \| f \| \\
&= \varepsilon.
\end{aligned}$$

This shows that $Kf \in C_{\mathbf{C}}([a,b])$. It is now easy to verify that K is a bounded linear operator, and to obtain the estimate for the norm of K. □

6.1.5 THEOREM *Let k be a complex-valued continuous function on the triangle $\{ (s,t) : a \leq t \leq s, a \leq s \leq b\}$ and for each $f \in C_{\mathbf{C}}([a,b])$ let*

$$(Kf)(s) = \int_a^s k(s,t)f(t)\, dt \tag{6}$$

for $a \leq s \leq b$. This defines a mapping K which is a bounded linear operator on $C_{\mathbf{C}}([a,b])$ with

$$\| K \| \leq (b-a) \sup \{ | k(s,t) | : a \leq t \leq s, a \leq s \leq b\}.$$

PROOF The proof is similar to, but slightly more complicated than, that of Theorem 6.1.4. It is left as an exercise for the reader. □

The linear operators K defined by (5) and (6) are called *Fredholm and Volterra integral operators*, respectively, after I. Fredholm (1866–1927) and V. Volterra (1860–1940) who were the first to study the associated linear integral equations systematically. The function k is called the *kernel* of the operator K.

Finally we shall give an example of a linear operator defined by a double sequence (or infinite matrix). Such operators are historically important; they arise in the study of quadratic forms in infinitely many variables, which was initiated by D. Hilbert (1862–1943) at the beginning of this century and which marks the beginning of the study of functional analysis. Operators defined by double sequences are an important source of illustrative and theoretical examples. (See the Exercises in this chapter.) For the sake of completeness we recall some facts about double sequences and series.

6.1.6 Definition Let N denote the set of all positive integers. A mapping $(m,n) \to \alpha_{mn}$ of $\mathsf{N} \times \mathsf{N}$ into K is called a *double sequence* and will be denoted by (α_{mn}). (Some writers call a double sequence an *infinite matrix*.) The double sequence (α_{mn}) is said to *converge to the limit* α if and only if for each $\varepsilon > 0$ there exists an integer N such that $|\alpha_{mn} - \alpha| < \varepsilon$ whenever $m,n \geq N$. It is clear that a double sequence can have at most one limit.

A pair of double sequences (α_{mn}) and (σ_{mn}), where

$$\sigma_{mn} = \sum_{j=1}^{m} \sum_{k=1}^{n} \alpha_{jk}$$

for $m,n = 1, 2, \ldots$, is called a *double series* and will be denoted by $\sum_{mn} \alpha_{mn}$. The double series $\sum_{mn} \alpha_{mn}$ is said to *converge* if and only if the double sequence (σ_{mn}) converges. When the double series $\sum_{mn} \alpha_{mn}$ converges the limit, σ say, of the double sequence (σ_{mn}) is called the *sum* of the series and we shall write $\sigma = \sum_{m,n=1}^{\infty} \alpha_{mn}$.

The following lemma contains all the information we shall need about the convergence of double series.

6.1.7 Lemma *Let $\sum_{mn} \alpha_{mn}$ be a double series of non-negative real numbers. If $\sum_{mn} \alpha_{mn}$ converges then the series $\sum_{n} \alpha_{mn}$ converges for each positive integer m, the series $\sum_{m}(\sum_{n=1}^{\infty} \alpha_{mn})$ converges and*

$$\sum_{m,n=1}^{\infty} \alpha_{mn} = \sum_{m=1}^{\infty} \sum_{n=1}^{\infty} \alpha_{mn}.$$

Proof Suppose that $\sum_{mn} \alpha_{mn}$ converges and has sum σ. Let $\varepsilon > 0$. Then there exists an integer N_0 with

$$\left| \sigma - \sum_{m=1}^{M} \sum_{n=1}^{N} \alpha_{mn} \right| < \varepsilon \tag{7}$$

whenever $M,N \geq N_0$. Let M be a positive integer. It follows from (7) that

$$\sum_{n=1}^{N} \alpha_{Mn} \leq \sum_{m=1}^{M_0} \sum_{n=1}^{N} \alpha_{mn} < \sigma + \varepsilon$$

for $N \geq N_0$, where $M_0 = \max\{M, N_0\}$. Since $\alpha_{Mn} \geq 0$ this proves that $\sum_n \alpha_{Mn}$ converges. Further, if $M \geq N_0$ inequality (7) gives

$$\left| \sigma - \sum_{m=1}^{M} \sum_{n=1}^{\infty} \alpha_{mn} \right| = \lim_{N \to \infty} \left| \sigma - \sum_{m=1}^{M} \sum_{n=1}^{N} \alpha_{mn} \right| \leq \varepsilon.$$

This proves that $\sum_m \left(\sum_{n=1}^{\infty} \alpha_{mn} \right)$ converges and that $\sigma = \sum_{m=1}^{\infty} \left(\sum_{n=1}^{\infty} \alpha_{mn} \right)$. □

6.1.8 EXAMPLE Let (α_{mn}) be a double sequence in K and suppose that the double series $\sum_{mn} |\alpha_{mn}|^2$ converges. By Lemma 6.1.7 the series $\sum_n |\alpha_{mn}|^2$ converges for each positive integer m. Consequently, by Hölder's inequality (3.2.7), the series $\sum_n \alpha_{mn}\xi_n$ converges absolutely for each $x = (\xi_n) \in \ell^2$ and, for each positive integer m,

$$\left| \sum_{n=1}^{\infty} \alpha_{mn}\xi_n \right|^2 \leq \left(\sum_{n=1}^{\infty} |\alpha_{mn}|^2 \right) \left(\sum_{n=1}^{\infty} |\xi_n|^2 \right)$$

$$= \|x\|^2 \sum_{n=1}^{\infty} |\alpha_{mn}|^2. \tag{8}$$

By Lemma 6.1.7 the series $\sum_m \left(\sum_{n=1}^{\infty} |\alpha_{mn}|^2 \right)$ converges, and therefore it follows from (8) that the sequence $Tx = \left(\sum_{n=1}^{\infty} \alpha_{mn}\xi_n \right)_{m \geq 1}$ is in ℓ^2 and

$$\|Tx\| = \left(\sum_{m=1}^{\infty} \left| \sum_{n=1}^{\infty} \alpha_{mn}\xi_n \right|^2 \right)^{1/2}$$

$$\leq \|x\| \left(\sum_{m=1}^{\infty} \left(\sum_{n=1}^{\infty} |\alpha_{mn}|^2 \right) \right)^{1/2}$$

$$= \|x\| \left(\sum_{m,n=1}^{\infty} |\alpha_{mn}|^2 \right)^{1/2}.$$

It is easy to verify that T is a linear mapping of ℓ^2 into itself. Thus T is a bounded linear operator on ℓ^2 and

$$\|T\| \leq \left(\sum_{m,n=1}^{\infty} |\alpha_{mn}|^2 \right)^{1/2}.$$

EXERCISES

(1) Prove that the mapping $T \to (\tau_{jk})$ constructed in Theorem 6.1.1 is a one-to-one linear mapping of the set of all linear transformations of E into F onto the set of all $m \times n$ matrices with entries from K.

(2) Let (α_n) be a bounded sequence in K and for each $x = (\xi_n) \in \ell$ let $Tx = (\alpha_n \xi_n)$.

(a) Prove that $T \in L(\ell)$ and $\| T \| = \sup \{ | \alpha_n | : n = 1, 2, \ldots \}$.

(b) Prove that $\mathscr{N}(T) = \{0\}$ if and only if $\alpha_n \neq 0$ for $n = 1, 2, \ldots$.

(c) Prove that, if $\mathscr{N}(T) = \{0\}$, then $\mathscr{R}(T)$ is dense in ℓ.

(d) Prove that $\mathscr{R}(T) = \ell$ and T is a linear homeomorphism if and only if $\inf \{ | \alpha_n | : n = 1, 2, \ldots \} > 0$.

(3) (a) Let T be a bounded linear operator on the space c of convergent sequences. Use the result of Exercise 5.1(4) to prove that there is a double sequence $(\tau_{mn})_{m \geq 1, n \geq 0}$ such that (i) if $x = (\xi_n) \in c$ and $Tx = (\eta_n)$ then

$$\eta_m = \tau_{m0} \lim_{n \to \infty} \xi_n + \sum_{n=1}^{\infty} \tau_{mn} \xi_n$$

for $m = 1, 2, \ldots$, (ii) the series $\sum_n \tau_{mn}$ is absolutely convergent for each positive integer m and the set $\left\{ \sum_{n=0}^{\infty} | \tau_{mn} | : m = 1, 2, \ldots \right\}$ is bounded, (iii) the sequence $(\tau_{mn})_{m \geq 1}$ is convergent for each positive integer n and (iv) the sequence $\left(\sum_{n=0}^{\infty} \tau_{mn} \right)_{m \geq 1}$ is convergent.

Calculate $\lim_{m \to \infty} \eta_m$. (Hint: $(\xi_n) \to \lim_{n \to \infty} \xi_n$ is a bounded linear functional on c.)

(b) Let $(\tau_{mn})_{m \geq 1, n \geq 0}$ be a double sequence which satisfies conditions (ii), (iii) and (iv). Prove that the equations $x = (\xi_n)$,

$$\eta_m = \tau_{m0} \lim_{n \to \infty} \xi_n + \sum_{n=1}^{\infty} \tau_{mn} \xi_n$$

for $m = 1, 2, \ldots$ and $Tx = (\eta_n)$ define a bounded linear operator T on c.

The reader is referred to TAYLOR [26] for further information about the relations between double sequences and bounded linear operators on sequence spaces.

6.2 The Algebra of Bounded Linear Operators

We have seen (Theorem 3.5.7) that the set $L(E,F)$ of all bounded linear transformations of a normed linear space E into a normed linear space F is itself a normed linear space. Further, if F is a Banach space then so is $L(E,F)$.

In this and the following section we shall be concerned with bounded linear transformations of a normed linear space into itself. Throughout this section E will denote a normed linear space over K. We shall write

$L(E)$ for $L(E,E)$ and we shall call the elements of $L(E)$ *bounded linear operators on* E.

The set $L(E)$ is, of course, a normed linear space; it is also a linear associative algebra (Definition 4.5.2). The product ST of two bounded linear operators on E is defined by

$$ST = S \circ T.$$

Lemma 3.5.8 shows that $ST \in L(E)$ and also that

$$\| ST \| \leq \| S \| \| T \|. \tag{1}$$

It is easy to verify that conditions (a), (b) and (c) of Definition 4.5.2 are satisfied. Suppose now that $E \neq \{0\}$. Then the mapping I defined by $Ix = x$ for all $x \in E$ is obviously in $L(E)$ and $\| I \| = 1$. Since $IT = TI = T$ for all $T \in L(E)$ the operator I is the identity of the algebra $L(E)$. It is an important fact that if E is not one-dimensional then $L(E)$ is not commutative. (See Exercise (3).)

We shall suppose throughout the rest of this section that $E \neq \{0\}$.

6.2.1 DEFINITION An operator $T \in L(E)$ is said to be *regular* if and only if there exists $S \in L(E)$ such that

$$TS = ST = I.$$

Such an operator S is called an *inverse of* T.

An operator T can have at most one inverse. In fact, if $S_1, S_2 \in L(E)$ and $TS_1 = S_1T = I$ and $TS_2 = S_2T = I$, then

$$S_1 = IS_1 = (S_2T)S_1 = S_2(TS_1) = S_2I = S_2.$$

A bounded linear operator T on E is a mapping of E into itself, and consequently if T is a one-to-one mapping of E onto itself then it has an inverse mapping T^{-1}, which also maps E onto itself. The following two lemmas will clarify the relationship between the concept of the inverse mapping and that of the inverse of a regular element of $L(E)$.

6.2.2 LEMMA *Let T be a linear transformation of a linear space F into a linear space G.*

(a) *If there is a mapping S of G into F such that $T \circ S = I_G$, where I_G is the identity mapping on G, then the range of T is G.*

(b) *If there is a linear transformation U of G into F such that $U \circ T = I_F$, where I_F is the identity mapping on F, then T is one-to-one.*

(c) *If T is a one-to-one mapping of F onto G then the inverse mapping T^{-1} is a linear mapping of G onto F and $T^{-1} \circ T = I_F$ and $T \circ T^{-1} = I_G$.*

PROOF (a) Given $y \in G$ we have $y = I_G y = (T \circ S)y = T(Sy)$.

(b) Given $x \in F$ we have $x = I_F x = (U \circ T)x = U(Tx)$. Since U is linear, $x = 0$ whenever $Tx = 0$ and therefore T is one-to-one, because it is linear.

(c) We leave the simple verification to the reader. □

6.2.3 LEMMA *Let $T \in L(E)$. The operator T is regular if and only if it is a linear homeomorphism of E onto itself. If T is regular then the inverse mapping T^{-1} is in $L(E)$ and is the inverse of T in the sense of Definition* 6.2.1.

PROOF Suppose that T is regular and let $S \in L(E)$ be the inverse of T. By definition, $ST = TS = I$ and it follows from (a) and (b) of Lemma 6.2.2 that T is a one-to-one mapping of E onto itself. It is clear that $T^{-1} = S$. This proves that T is a linear homeomorphism of E onto itself.

Suppose conversely that T is a linear homeomorphism of E onto itself. By definition of a homeomorphism T^{-1} is a continuous mapping of E onto itself and, by Lemma 6.2.2 (c), T^{-1} is linear. Thus $T^{-1} \in L(E)$ (Theorem 3.5.4). Lemma 6.2.2 (c) now shows that T is regular and that T^{-1} is the inverse of T. □

In future if T is regular in $L(E)$ we shall denote the inverse of T by T^{-1}. It follows from Lemma 6.2.3 that if T is not regular then at least one of the following conditions must hold:

(a) $T(E) \neq E$,

(b) T is not one-to-one,

(c) $T(E) = E$ and T is one-to-one, but the inverse mapping T^{-1} is not bounded.

It is an important property of Banach spaces that condition (c) cannot occur when E is a Banach space. This will be proved in Chapter 8. (See Theorem 8.5.4.)

The following lemma is quite elementary. We leave the reader to supply the proof.

6.2.4 LEMMA *Let $S, T \in L(E)$ and $\alpha \in \mathsf{K}$.*

(a) *I is regular and $I^{-1} = I$.*

(b) *If T is regular then T^{-1} is regular and $(T^{-1})^{-1} = T$.*

(c) *If $\alpha \neq 0$ then αT is regular if and only if T is regular. If T is regular and $\alpha \neq 0$, then $(\alpha T)^{-1} = \alpha^{-1}T^{-1}$.*

(d) *If S and T are regular then ST is regular and $(ST)^{-1} = T^{-1}S^{-1}$.*

(e) *If S and T are regular and $ST = TS$ then $S^{-1}T^{-1} = T^{-1}S^{-1}$.*

(f) *If ST is regular and $ST = TS$ then T and S are regular, $S^{-1} = T(ST)^{-1}$ and $T^{-1} = S(ST)^{-1}$.*

6.2.5 LEMMA *Let (S_n) and (T_n) be convergent sequences in $L(E)$ with $\lim_{n\to\infty} S_n = S$ and $\lim_{n\to\infty} T_n = T$. Then $ST = \lim_{n\to\infty} S_nT_n$.*

PROOF The proof, which uses the inequality (1), is very similar to the argument used in the proof of Lemma 4.5.3. We leave the reader to supply the details. □

We turn now to the question of the existence of regular elements in $L(E)$. The identity operator I is always regular; we shall see that if E is a Banach space, every operator 'sufficiently near to I' is also regular. The proof is quite elementary, and is based on the well-known identity

$$(1 - x)^{-1} = \sum_{n=0}^{\infty} x^n$$

when $|x| < 1$.

We recall that if $T \in L(E)$ the positive integral powers of T are defined inductively by $T^1 = T$ and $T^n = T(T^{n-1})$ for $n = 2, 3,\ldots$. If $T \neq 0$ we define $T^0 = I$. It is easy to check that

$$T^mT^n = T^nT^m = T^{m+n}$$

for all positive integers m and n. We shall have to consider the convergence of the series $\sum_n T^n$ in $L(E)$. It follows, by induction, from the inequality (1) that

$$\| T^n \| \leq \| T \|^n \tag{2}$$

for $n = 1, 2,\ldots$. The inequalities (2) and Lemma 3.3.8 show that, if E is a Banach space and $\| T \| < 1$, then the series $\sum_n T^n$ converges in $L(E)$. A more careful argument will allow us to improve on the condition $\| T \| < 1$. We shall need the following lemma.

6.2.6 LEMMA *Let $T \in L(E)$. Then the sequence $(\| T^n \|^{1/n})$ converges and*

$$\lim_{n\to\infty} \| T^n \|^{1/n} = \inf \{ \| T^n \|^{1/n} : n = 1, 2,\ldots\}.$$

PROOF Let $\varepsilon > 0$ and let $\nu = \inf \{ \| T^n \|^{1/n} : n = 1, 2,\ldots\}$. There is a positive integer m such that $\| T^m \| < (\nu + \varepsilon)^m$. For each positive integer n there are integers p_n and q_n such that $n = p_nm + q_n$, $p_n \geq 0$ and $1 \leq q_n \leq m$. Then using (2) we obtain

$$\| T^n \| = \| (T^m)^{p_n}T^{q_n} \| \leq \| T^m \|^{p_n} \| T \|^{q_n} < (\nu + \varepsilon)^{mp_n} \| T \|^{q_n}.$$

Hence

$$\nu \leq \| T^n \|^{1/n} < (\nu + \varepsilon)^{mp_n/n} \| T \|^{q_n/n} = (\nu + \varepsilon) \left(\frac{\| T \|}{\nu + \varepsilon}\right)^{q_n/n}.$$

But $\lim_{n\to\infty} q_n/n = 0$ so there is an integer N such that

$$\left(\frac{\| T \|}{\nu + \varepsilon}\right)^{q_n/n} < \frac{\nu + 2\varepsilon}{\nu + \varepsilon}$$

for all $n \geq N$. Consequently $\nu \leq \| T^n \|^{1/n} < \nu + 2\varepsilon$ for all $n \geq N$. This proves that $\lim_{n\to\infty} \| T^n \|^{1/n} = \nu$. □

6.2.7 THEOREM *Suppose that E is a Banach space and let $T \in L(E)$ be such that $\lim_{n\to\infty} \| T^n \|^{1/n} < 1$. Then $I - T$ is regular and*

$$(I - T)^{-1} = I + \sum_{n=1}^{\infty} T^n,$$

where the series $\sum_n T^n$ converges in $L(E)$.

PROOF Choose α with $\lim_{n\to\infty} \| T^n \|^{1/n} < \alpha < 1$. There is an integer N such that $\| T^n \| < \alpha^n$ for $n \geq N$. Consequently the series $\sum_n \| T^n \|$ converges. By Theorem 3.5.7 $L(E)$ is a Banach space so Lemma 3.3.8 shows that the series $\sum_n T^n$ converges in $L(E)$. Let $S = I + \sum_{n=1}^{\infty} T^n$. By definition $S = \lim_{n\to\infty} S_n$, where $S_n = I + \sum_{m=1}^{n} T^m$. An elementary calculation shows that

$$(I - T)S_n = S_n(I - T) = I - T^{n+1}.$$

Since $\| T^n \| < \alpha^n$ for $n \geq N$ and $\alpha < 1$ we have $\lim_{n\to\infty} \| T^n \| = 0$, and it now follows from Lemma 6.2.5 that

$$(I - T)S = \lim_{n\to\infty} (I - T)S_n = \lim_{n\to\infty} (I - T^{n+1}) = I.$$

Similarly $S(I - T) = I$. This proves that $I - T$ is regular and $(I - T)^{-1} = S$. □

In practice it is often difficult to calculate $\lim_{n\to\infty} \| T^n \|^{1/n}$ (for an example in which the calculation is possible see Theorem 6.2.9), and the following, weaker, version of Theorem 6.2.7 is often more useful.

6.2.8 COROLLARY *Suppose that E is a Banach space and let $T \in L(E)$ be such that $\| T \| < 1$. Then $I - T$ is regular and*

$$(I - T)^{-1} = I + \sum_{n=1}^{\infty} T^n,$$

where the series $\sum_n T^n$ converges in $L(E)$.

PROOF By inequality (2) we have $\| T^n \|^{1/n} \leq \| T \|$ for $n = 1, 2, \ldots$, so $\lim_{n\to\infty} \| T^n \|^{1/n} \leq \| T \| < 1$. □

We remark that, if E is a Banach space, the open ball

$$\{T \in L(E) : \| I - T \| < 1\}$$

consists entirely of regular elements because $T = I - (I - T)$.

Let $T \in L(E)$ and let $\lambda \in \mathsf{K}$ be such that $| \lambda |^{-1} > \lim_{n\to\infty} \| T^n \|^{1/n}$. Then $\lim_{n\to\infty} \| (\lambda T)^n \|^{1/n} < 1$ and hence, if E is a Banach space, Theorem 6.2.7 shows that $I - \lambda T$ is regular and

$$(I - \lambda T)^{-1} = I + \sum_{n=1}^{\infty} \lambda^n T^n. \tag{3}$$

The series $\sum_n \lambda^n T^n$ is called the *Neumann series of* T after C. Neumann who used it systematically in potential theory towards the end of the last century (cf. KELLOGG [19, pp. 281–2]). In the following theorem we shall use the Neumann series to obtain a fundamental property of Volterra integral equations.

6.2.9 THEOREM *Let k be a complex-valued continuous function on the triangle $\{(s,t) : a \leq t \leq s,\ a \leq s \leq b\}$. Then the Volterra integral equation*

$$f(s) = g(s) + \lambda \int_a^s k(s,t) f(t)\, dt \tag{4}$$

for $a \leq s \leq b$, has a unique solution $f \in C_{\mathsf{C}}([a,b])$ for each complex number λ and each $g \in C_{\mathsf{C}}([a,b])$.

PROOF Equation (4) can be written as

$$(I - \lambda K) f = g \tag{5}$$

where K is the Volterra integral operator with kernel k. (See p. 218.) We saw in Theorem 6.1.5 that K is a bounded linear operator on $C_{\mathsf{C}}([a,b])$. Also, by Theorem 2.8.3, the space $C_{\mathsf{C}}([a,b])$ is complete. Consequently, if we can prove that $\lim_{n\to\infty} \| K^n \|^{1/n} = 0$ the conclusion of the theorem will follow directly from (5), the remarks immediately preceding the statement of the theorem and Lemma 6.2.3.

Let $M = \sup \{ | k(s,t) | : a \leq t \leq s,\ a \leq s \leq b\}$ and let $f \in C_{\mathsf{C}}([a,b])$. Then, for $a \leq s \leq b$,

$$\begin{aligned} | (Kf)(s) | &= \left| \int_a^s k(s,t) f(t)\, dt \right| \\ &\leq (s - a) \sup \{ | k(s,t) | \, | f(t) | : a \leq t \leq s\} \\ &\leq M \| f \| (s - a). \end{aligned}$$

We shall prove by induction that

$$|(K^n f)(s)| \leq M^n \| f \| \frac{(s-a)^n}{n!} \tag{6}$$

for $a \leq s \leq b$ and $n = 1, 2, \ldots$. We have just seen that (6) is true for $n = 1$. Suppose that (6) is true for some positive integer n. Then since $K^{n+1} = K(K^n)$ we have

$$\begin{aligned} |(K^{n+1}f)(s)| &= |(K(K^n f))(s)| \\ &= \left| \int_a^s k(s,t)(K^n f)(t)\, dt \right| \\ &\leq \frac{M^{n+1} \| f \|}{n!} \int_a^s (t-a)^n\, dt \\ &= M^{n+1} \| f \| \frac{(s-a)^{n+1}}{(n+1)!}. \end{aligned}$$

It follows, by induction, that (6) is true for all positive integers n. Using (6) we obtain, for $n = 1, 2, \ldots$,

$$\| K^n f \| = \max \{ |(K^n f)(s)| : a \leq s \leq b \} \leq M^n \| f \| \frac{(b-a)^n}{n!}. \tag{7}$$

Since (7) holds for all $f \in C_{\mathbf{C}}([a,b])$ we have, for $n = 1, 2, \ldots$,

$$\| K^n \| = \sup \{ \| K^n f \| : f \in C_{\mathbf{C}}([a,b]), \| f \| \leq 1 \} \leq M^n \frac{(b-a)^n}{n!}.$$

Therefore $\lim_{n\to\infty} \| K^n \|^{1/n} = 0$ because $\lim_{n\to\infty} (n!)^{-1/n} = 0$. □

Using the Neumann series (3) (for K) we can obtain the solution of the Volterra integral equation (4) as the sum of an infinite series. (See Exercise (10).)

The following properties of the set of regular elements of $L(E)$ will be needed in the next section. The first is a consequence of Corollary 6.2.8.

6.2.10 THEOREM *Suppose that E is a Banach space. Then the set $\mathscr{G}$ of regular elements of $L(E)$ is an open subset of $L(E)$.*

PROOF Let $S \in \mathscr{G}$ and $T \in L(E)$. Then

$$\| I - S^{-1}T \| = \| S^{-1}(S - T) \| \leq \| S^{-1} \| \| S - T \|.$$

It follows now from Corollary 6.2.8 that $S^{-1}T \in \mathscr{G}$ whenever $\| S - T \| < \| S^{-1} \|^{-1}$. But $T = S(S^{-1}T)$ so Lemma 6.2.4(d) shows that $T \in \mathscr{G}$ whenever $S^{-1}T \in \mathscr{G}$. Consequently we have

$$\{T \in L(E) : \| S - T \| < \| S^{-1} \|^{-1}\} \subseteq \mathscr{G}.$$

This proves that $\mathscr{G}$ is open. □

6.2.11 THEOREM *The mapping $T \to T^{-1}$ is a homeomorphism of the set $\mathscr{G}$ of regular elements of $L(E)$ onto itself.*

PROOF Lemma 6.2.4(b) shows that the mapping $T \to T^{-1}$ is its own inverse mapping, so we need only prove that it is continuous on $\mathscr{G}$. Let $S,T \in \mathscr{G}$. Then

$$\| S^{-1} - T^{-1} \| = \| S^{-1}(T - S)T^{-1} \| \leq \| S^{-1} \| \| T^{-1} \| \| T - S \|, \quad (8)$$

and hence

$$\begin{aligned} \| T^{-1} \| &\leq \| S^{-1} - T^{-1} \| + \| S^{-1} \| \\ &\leq \| S^{-1} \| \ \| T^{-1} \| \ \| T - S \| + \| S^{-1} \|. \end{aligned} \quad (9)$$

Suppose that $\| T - S \| \leq \frac{1}{2} \| S^{-1} \|^{-1}$. Then (9) gives $\| T^{-1} \| \leq 2 \| S^{-1} \|$ and hence (8) gives

$$\| S^{-1} - T^{-1} \| \leq 2 \| S^{-1} \|^2 \| T - S \|.$$

It is now obvious that the mapping $T \to T^{-1}$ is continuous at S. □

EXERCISES

(1) Let E be an n-dimensional linear space. Prove that $L(E)$ is finite-dimensional with dimension n^2.

(2) Let E be a normed linear space. Given $x \in E$ and $x^* \in E^*$ let $x \otimes x^*$ be the mapping of E into itself defined by $(x \otimes x^*)y = x^*(y)x$ for all $y \in E$. Prove that $x \otimes x^* \in L(E)$ and $\| x \otimes x^* \| = \| x \| \ \| x^* \|$.

(3) Let E be a normed linear space. Prove that, if E is not one-dimensional, then $L(E)$ is not commutative. (Hint: Choose points $x,y \in E$ and a functional $x^* \in E^*$ with $x^*(x) = 1$ and $x^*(y) = 0$. Then consider the operators $x \otimes x^*$ and $y \otimes x^*$.)

(4) Let E be a normed linear space and let $T \in L(E)$. Prove that if $\mathscr{R}(T)$ is finite dimensional then

$$T = x_1 \otimes x_1^* + x_2 \otimes x_2^* + \ldots + x_n \otimes x_n^*$$

where $x_1, x_2, \ldots, x_n \in E$, $x_1^*, x_2^*, \ldots, x_n^* \in E^*$ and $n = \dim \mathscr{R}(T)$. (Hint: to prove that $x_1^*, x_2^*, \ldots, x_n^*$ are bounded use Theorem 4.4.1.)

(5) Let E be an infinite-dimensional normed linear space. Prove that there is a linear operator on E that is not bounded. (Hint: see Exercise 5.3(10).)

(6) Let E be the linear hull of the subset $\{e_n : n = 1, 2, \ldots\}$ of the normed linear space ℓ. (See Example 3.4.9.) Prove that there is a unique operator $T \in L(E)$ with $Te_n = 2^{-n}e_n$ for $n = 1, 2, \ldots$. Prove that T is a one-to-one mapping of E onto itself but that T is not a linear homeomorphism of E onto itself. (This example shows that condition (c) on p. 223 can occur when E is not a Banach space.)

(7) Let T be a bounded linear operator on a non-zero normed linear space E and let $\lambda,\mu \in \mathsf{K}$ be such that $\lambda I - T$ and $\mu I - T$ are regular. Prove that

$$(\lambda I - T)^{-1} - (\mu I - T)^{-1} = (\mu - \lambda)(\lambda I - T)^{-1}(\mu I - T)^{-1}$$
$$= (\mu - \lambda)(\mu I - T)^{-1}(\lambda I - T)^{-1}.$$

(8) Let T be a bounded linear operator on a non-zero Banach space E and let $\lambda \in \mathsf{K}$ be such that $\lambda I - T$ is regular. Prove that, if $\mu \in \mathsf{K}$ and $|\lambda - \mu| < \|(\lambda I - T)^{-1}\|^{-1}$, then $\mu I - T$ is regular and

$$(\mu I - T)^{-1} = \sum_{n=0}^{\infty} (\lambda - \mu)^n R_\lambda^{n+1},$$

where $R_\lambda = (\lambda I - T)^{-1}$.

(9) Let E be a non-zero normed linear space and let $T \in L(E)$ with $\lim_{n\to\infty} \|T^n\|^{1/n} = 0$. Prove that T is not regular.

(10) Let K be a Volterra integral operator on $C_{\mathsf{C}}([a,b])$ with kernel k.

(a) Prove that for each positive integer n the operator K^n is a Volterra integral operator with kernel k_n where $k_1 = k$ and, for $n = 2, 3,\ldots,$

$$k_n(s,t) = \int_t^s k(s,u)k_{n-1}(u,t)\,du$$

for $a \leq t \leq s \leq b$.

(b) Prove that

$$|k_n(s,t)| \leq M^n \frac{(s - t)^{n-1}}{(n - 1)!}$$

for $n = 1, 2,\ldots,$ and $a \leq t \leq s \leq b$, where

$$M = \sup\{|k(s,t)| : a \leq t \leq s \leq b\},$$

and deduce that for each complex number λ the series $\sum_n \lambda^n k_n$ is uniformly convergent on the triangle $\{(s,t) : a \leq t \leq s \leq b\}$.

(c) Prove that the unique solution of the integral equation

$$f(s) = g(s) + \lambda \int_a^s k(s,t)f(t)\,dt \quad \text{for } a \leq s \leq b$$

is

$$f(s) = g(s) + \int_a^s h_\lambda(s,t)g(t)\,dt \quad \text{for } a \leq s \leq b,$$

where $h_\lambda = \sum_{n=1}^{\infty} \lambda^n k_n$.

(d) Solve the integral equations

$$f(s) = g(s) + \lambda \int_0^s e^{s-t}f(t)\,dt \quad \text{for } 0 \leq s \leq 1,$$

and

$$f(s) = g(s) + \lambda \int_0^s (s - t)f(t)\,dt \quad \text{for } 0 \leq s \leq 1.$$

6.3 The Spectrum of a Bounded Linear Operator

Much information about the behaviour of a linear operator T can be obtained by studying the family of linear operators $\lambda I - T$, where λ is a real or complex parameter. At this point in the theory a fundamental difference appears between real linear spaces and complex linear spaces. We shall see that the difference arises from the fact that the complex field is algebraically closed while the real field is not; that is, every non-constant polynomial with complex coefficients has a complex zero, but there are non-constant polynomials with real coefficients that have no real zeros.

6.3.1 Definition Let T be a bounded linear operator on a non-zero complex normed linear space E. A complex number λ is said to be a *regular point of T* if and only if the operator $\lambda I - T$ is regular. The set of all regular points of T is called the *resolvent set of T*. The mapping $\lambda \to (\lambda I - T)^{-1}$ of the resolvent set of T into $L(E)$ is called the *resolvent of T*. The complement in the complex plane C of the resolvent set of T is called the *spectrum of T* and is denoted by $\mathrm{sp}(T)$.

According to Definition 6.3.1 a complex number λ is in the spectrum of T if and only if the operator $\lambda I - T$ is not regular. The reader is warned that there are other definitions of the spectrum which differ from 6.3.1 when E is not a Banach space (see Taylor [26, p. 253] and Hille and Phillips [15, p. 54]; see also the remarks on p. 223 and Theorem 8.5.4).

The definition of $\mathrm{sp}(T)$ can also be formulated for a bounded linear operator T on a real normed linear space, but the concept is not useful in this case because $\mathrm{sp}(T)$ may be empty even when E is finite-dimensional (Exercise (1)). We shall see that this cannot happen when E is a complex normed linear space.

Our aim in this section is to obtain as complete a description as possible of the spectrum of an arbitrary bounded linear operator on a complex Banach space. That part of the theory of bounded linear operators which is centred on the concept of the spectrum is called *spectral theory*.

6.3.2 Definition Let T be a linear operator on a complex linear space E. A complex number λ is said to be an *eigenvalue of T* if and only if there is a non-zero point $x \in E$ such that $Tx = \lambda x$. If λ is an eigenvalue of T the non-zero points $x \in E$ that satisfy $Tx = \lambda x$ are called the *eigenvectors of T corresponding to the eigenvalue λ*.

It is clear that the complex number λ is an eigenvalue of T if and only if $\lambda I - T$ is not a one-to-one mapping. Consequently the eigenvalues, if any, of a bounded linear operator are in its spectrum. Eigenvalues are important both historically and in applications.

6.3.3 LEMMA *Let $\lambda_1, \lambda_2, \ldots, \lambda_n$ be distinct eigenvalues of a linear operator T on a complex linear space E. If $x_1, x_2, \ldots, x_n$ are eigenvectors of T corresponding to the eigenvalues $\lambda_1, \lambda_2, \ldots, \lambda_n$, respectively, then the set $\{x_1, x_2, \ldots, x_n\}$ is linearly independent.*

PROOF Suppose that $\{x_1, x_2, \ldots, x_n\}$ is not linearly independent. Then there is a positive integer m such that $\{x_1, x_2, \ldots, x_m\}$ is linearly independent, while $\{x_1, x_2, \ldots, x_m, x_{m+1}\}$ is not linearly independent. There are complex numbers $\alpha_1, \alpha_2, \ldots, \alpha_{m+1}$, not all zero, such that $\sum_{j=1}^{m+1} \alpha_j x_j = 0$. Consequently

$$0 = T\left(\sum_{j=1}^{m+1} \alpha_j x_j\right) = \sum_{j=1}^{m+1} \alpha_j T x_j = \sum_{j=1}^{m+1} \alpha_j \lambda_j x_j$$

and hence

$$0 = \sum_{j=1}^{m+1} \alpha_j(\lambda_j - \lambda_{m+1})x_j = \sum_{j=1}^{m} \alpha_j(\lambda_j - \lambda_{m+1})x_j.$$

Since $\lambda_j \neq \lambda_{m+1}$ for $1 \leq j \leq m$, and $\{x_1, x_2, \ldots, x_m\}$ is linearly independent, we must have $\alpha_1 = \alpha_2 = \ldots = \alpha_m = 0$. But then $\alpha_{m+1}x_{m+1} = 0$, and hence $\alpha_{m+1} = 0$, because $x_{m+1} \neq 0$. We have now arrived at a contradiction. □

Suppose now that E is a finite-dimensional linear space. By Theorem 6.1.2 all linear operators on E are bounded with respect to any norm on E, so a linear operator T on E is regular if and only if T is a one-to-one mapping of E onto itself. We shall now prove that if T is one-to-one then, necessarily, it maps E onto itself.

6.3.4 LEMMA *Let T be a linear operator on a finite-dimensional linear space E and suppose that T is one-to-one. Then T is regular.*

PROOF Let $\{e_1, e_2, \ldots, e_n\}$ be a basis for E. If $Te_1, Te_2, \ldots, Te_n$ are linearly dependent there are complex numbers $\alpha_1, \alpha_2, \ldots, \alpha_n$, not all zero, such that $\sum_{j=1}^{n} \alpha_j Te_j = 0$. Then we have

$$T\left(\sum_{j=1}^{n} \alpha_j e_j\right) = \sum_{j=1}^{n} \alpha_j Te_j = 0$$

and hence $\sum_{j=1}^{n} \alpha_j e_j = 0$, because T is one-to-one. Since $\{e_1, e_2, \ldots, e_n\}$ is linearly independent we must have $\alpha_1 = \alpha_2 = \ldots = \alpha_n = 0$. This contradiction proves that $\{Te_1, Te_2, \ldots, Te_n\}$ is linearly independent and hence is a basis for E.

Let $x \in E$. Then there are complex numbers $\beta_1, \beta_2, \ldots, \beta_n$ with $x = \sum_{j=1}^{n} \beta_j Te_j = T\left(\sum_{j=1}^{n} \beta_j e_j\right)$. This proves that T maps E onto itself. □

In order to be able to state certain theorems in Chapter 7 in their classical form, we shall need a classical criterion for a linear operator T on a finite-dimensional linear space E to be regular. Let $\{e_1, e_2, \ldots, e_n\}$ be any basis for E and let (τ_{jk}) be the matrix which represents T with respect to this basis—so that $Te_k = \sum_{j=1}^{n} \tau_{jk} e_j$ for $k = 1, 2, \ldots, n$. (See Theorem 6.1.1.) Then T is regular if and only if the matrix (τ_{jk}) has non-zero determinant (see HALMOS [12, p. 99]). A matrix with non-zero determinant is said to be *non-singular*.

Suppose now that E is a finite-dimensional complex linear space. It follows from Lemma 6.3.4 that the spectrum of a linear operator T on E consists of those complex numbers λ such that $\lambda I - T$ is not one-to-one, that is, it consists of the eigenvalues of T. For the purposes of reference and comparison with the more complicated situation in infinite-dimensional spaces we shall now obtain a complete description of the spectrum of a linear operator on a finite-dimensional space.

6.3.5 THEOREM *Let T be a linear operator on an n-dimensional complex linear space E. Then* sp(T) *is a non-empty finite set consisting of not more than n points, each of which is an eigenvalue of T.*

PROOF We have remarked above that sp(T) consists entirely of eigenvalues of T, and Lemma 6.3.3 shows that T cannot have more than n eigenvalues. It remains only to prove that T has at least one eigenvalue. This follows from Theorem 6.3.9. □

An alternative, and elementary, proof that T has at least one eigenvalue is outlined in Exercise (4). In general, nothing further can be said about the spectrum of a linear operator on a finite-dimensional space. (See Exercise (5).) The following Examples show that in infinite-dimensional spaces the spectrum may not be a finite set, and may contain points that are not eigenvalues.

6.3.6 EXAMPLE Let T be the mapping of ℓ^2 into itself defined by $Tx = y$, where $x = (\xi_n)$ and $y = (\eta_n)$, with $\eta_1 = 0$ and $\eta_n = \xi_{n-1}$ for $n = 2, 3, \ldots$. It is easy to see that T is a bounded linear operator on ℓ^2. Also $\| Tx \| = \| x \|$ for all $x \in \ell^2$ so T is one-to-one. Since the point

$e_1 = (\delta_{1n})_{n\geq 1}$ is in ℓ^2 but not in the range of T the operator T is not regular. This shows that 0 is not an eigenvalue of T but that $0 \in \mathrm{sp}(T)$.

6.3.7 Example Let T be the mapping of ℓ^2 into itself defined by $Tx = y$, where $x = (\xi_n)$ and $y = (\eta_n)$, with $\eta_n = \xi_{n+1}$ for $n = 1, 2,\ldots$. Again T is a bounded linear operator on ℓ^2. Let λ be a complex number with $|\lambda| < 1$. Then the sequence $x_\lambda = (\xi_n)$, where $\xi_1 = 1$ and $\xi_n = \lambda^{n-1}$ for $n = 2, 3,\ldots$ is in ℓ^2, $x_\lambda \neq 0$, and a simple calculation shows that $Tx_\lambda = \lambda x_\lambda$. Thus each point of the open disc $\{\lambda \in \mathsf{C} : |\lambda| < 1\}$ is an eigenvalue of T. The operator T is called the *unilateral shift*.

Let T be a bounded linear operator on a non-zero complex normed linear space E. We shall prove that $\mathrm{sp}(T)$ is non-empty. The proof we give is due to C. E. Rickart. (See [22, pp. 28–9].)

Given complex numbers $\alpha_0, \alpha_1,\ldots, \alpha_n$ the complex-valued function p defined by

$$p(t) = \alpha_0 + \alpha_1 t + \ldots + \alpha_n t^n, \tag{1}$$

for all $t \in \mathsf{C}$, is called a polynomial function of degree n. The set of all polynomial functions (of arbitrary degree) is a commutative complex linear associative algebra with algebraic operations defined by

$$(p + q)(t) = p(t) + q(t),$$
$$(\alpha p)(t) = \alpha p(t),$$

and

$$(pq)(t) = p(t)q(t)$$

for all $t \in \mathsf{C}$. Let p be the polynomial function defined by (1) and let

$$p(T) = \alpha_0 I + \alpha_1 T + \ldots + \alpha_n T^n.$$

It is clear that $p(T) \in L(E)$; we shall say that $p(T)$ is a *polynomial in* T. It is easy to verify that the mapping $p \to p(T)$ is a 'homomorphism' of the algebra of all polynomial functions into the algebra $L(E)$, that is

$$(p + q)(T) = p(T) + q(T),$$
$$(\alpha p)(T) = \alpha p(T),$$

and

$$(pq)(T) = p(T)q(T)$$

for all polynomial functions p and q and all $\alpha \in \mathsf{C}$. Since $pq = qp$ we also have

$$p(T)q(T) = q(T)p(T).$$

6.3.8 Lemma *Let T be a bounded linear operator on a non-zero complex normed linear space E and suppose that there is a positive real number ν such that $I - \lambda^{-1}T$ is regular for all complex numbers λ with $|\lambda| = \nu$. Then, for each positive integer n and for each complex number λ with $|\lambda| = \nu$, the operator $I - \lambda^{-n}T^n$ is regular and*

$$\left(I - \frac{1}{\lambda^n}T^n\right)^{-1} = \frac{1}{n}\sum_{k=0}^{n-1}\left(I - \frac{1}{\varepsilon_k\lambda}T\right)^{-1}, \tag{2}$$

where $\varepsilon_k = \exp(2\pi ik/n)$.

PROOF The partial fraction expansion of $(1 - t^n)^{-1}$ is

$$\frac{1}{1 - t^n} = \frac{1}{n}\sum_{k=0}^{n-1}\frac{1}{1 - \varepsilon_k^{-1}t} \tag{3}$$

for all $t \in \mathrm{C}$ with $t \neq \varepsilon_k$ for $k = 0, 1, 2, \ldots, n - 1$. We have also

$$1 - t^n = \left(1 - \frac{1}{\varepsilon_0}t\right)\left(1 - \frac{1}{\varepsilon_1}t\right)\ldots\left(1 - \frac{1}{\varepsilon_{n-1}}t\right) \tag{4}$$

for all $t \in \mathrm{C}$. From (3) and (4) we obtain, for all $t \in \mathrm{C}$,

$$1 = \frac{1}{n}\sum_{k=0}^{n-1}\left(1 - \frac{1}{\varepsilon_0}t\right)\ldots\left(1 - \frac{1}{\varepsilon_{k-1}}t\right)\left(1 - \frac{1}{\varepsilon_{k+1}}t\right)\ldots\left(1 - \frac{1}{\varepsilon_{n-1}}t\right). \tag{5}$$

Let $\lambda \in \mathrm{C}$ with $|\lambda| = \nu$ and let n be a positive integer. Applying the remarks immediately preceding the statement of the lemma to equations (4) and (5) we obtain

$$I - \frac{1}{\lambda^n}T^n = \left(I - \frac{1}{\varepsilon_0\lambda}T\right)\left(I - \frac{1}{\varepsilon_1\lambda}T\right)\ldots\left(I - \frac{1}{\varepsilon_{n-1}\lambda}T\right) \tag{6}$$

and

$$I = \frac{1}{n}\sum_{k=0}^{n-1}\left[\left(I - \frac{1}{\varepsilon_0\lambda}T\right)\ldots\left(I - \frac{1}{\varepsilon_{k-1}\lambda}T\right)\right.$$
$$\left.\times\left(I - \frac{1}{\varepsilon_{k+1}\lambda}T\right)\ldots\left(I - \frac{1}{\varepsilon_{n-1}\lambda}T\right)\right]. \tag{7}$$

By hypothesis $I - (\varepsilon_k\lambda)^{-1}T$ is regular for $k = 0, 1, 2, \ldots, n-1$, so it follows from (6), (7), and the observation that the terms in the product (6) can be written in any order, that

$$I = \left(I - \frac{1}{\lambda^n}T^n\right)\left(\frac{1}{n}\sum_{k=0}^{n-1}\left(I - \frac{1}{\varepsilon_k\lambda}T\right)^{-1}\right)$$
$$= \left(\frac{1}{n}\sum_{k=0}^{n-1}\left(I - \frac{1}{\varepsilon_k\lambda}T\right)^{-1}\right)\left(I - \frac{1}{\lambda^n}T^n\right). \square$$

We remark that it is essential in the above lemma that E should be a complex normed linear space.

6.3.9 THEOREM *Let T be a bounded linear operator on a non-zero* *complex normed linear space E. Then there is a point* $\lambda \in \mathrm{sp}(T)$ *such that* $|\lambda| \geq \lim_{n\to\infty} \|T^n\|^{1/n}$.

PROOF Suppose first that $\lim_{n\to\infty} \|T^n\|^{1/n} = 0$. We shall prove that $0 \in \mathrm{sp}(T)$. Suppose, on the contrary, that 0 is in the resolvent set of T. Then T is regular. Thus $I = I^n = (TT^{-1})^n = T^n(T^{-1})^n$ and hence

$$1 = \|I\| = \|T^n(T^{-1})^n\| \leq \|T^n\| \; \|(T^{-1})^n\| \leq \|T^n\| \; \|T^{-1}\|^n$$

for $n = 1, 2, \ldots$. Consequently $1 \leq \|T^n\|^{1/n} \|T^{-1}\|$ for $n = 1, 2, \ldots$, which is impossible because $\lim_{n\to\infty} \|T^n\|^{1/n} = 0$. This completes the proof in the case when $\lim_{n\to\infty} \|T^n\|^{1/n} = 0$.

Suppose now that $\lim_{n\to\infty} \|T^n\|^{1/n} = \nu > 0$ and that the set

$$\{\lambda \in \mathsf{C} : |\lambda| \geq \nu\}$$

is contained in the resolvent set of T. We shall again derive a contradiction.

The mapping $\lambda \to I - \lambda^{-1}T$ is obviously continuous on $\mathsf{C} \sim \{0\}$ and, by Theorem 6.2.11, the mapping $S \to S^{-1}$ is continuous on the set of regular elements of $L(E)$. Consequently, by Theorem 2.7.4, the mapping $\lambda \to (I - \lambda^{-1}T)^{-1}$, which is the composition of two continuous mappings, is itself continuous on the set $\{\lambda \in \mathsf{C} : |\lambda| \geq \nu\}$. By the Borel–Lebesgue theorem (4.2.11) the annulus

$$A = \{\lambda \in \mathsf{C} : \nu \leq |\lambda| \leq 2\nu\}$$

is compact. Therefore by Theorem 4.3.8 the mapping $\lambda \to (I - \lambda^{-1}T)^{-1}$ is uniformly continuous on A. Let $\varepsilon > 0$. There exists $\delta > 0$ such that

$$\left\| \left(I - \frac{1}{\lambda}T\right)^{-1} - \left(I - \frac{1}{\mu}T\right)^{-1} \right\| < \varepsilon \tag{8}$$

for all $\lambda, \mu \in A$ with $|\lambda - \mu| < \delta$.

Choose a complex number λ with $|\nu - \lambda| < \delta$ and $\nu < |\lambda| \leq 2\nu$. For each positive integer n Lemma 6.3.8 and inequality (8) give

$$\begin{aligned}
&\left\| \left(I - \frac{1}{\nu^n}T^n\right)^{-1} - \left(I - \frac{1}{\lambda^n}T^n\right)^{-1} \right\| \\
&\qquad = \frac{1}{n} \left\| \sum_{k=0}^{n-1} \left(\left(I - \frac{1}{\varepsilon_k \nu}T\right)^{-1} - \left(I - \frac{1}{\varepsilon_k \lambda}T\right)^{-1} \right) \right\| \\
&\qquad \leq \frac{1}{n} \sum_{k=0}^{n-1} \left\| \left(I - \frac{1}{\varepsilon_k \nu}T\right)^{-1} - \left(I - \frac{1}{\varepsilon_k \lambda}T\right)^{-1} \right\| \\
&\qquad < \varepsilon,
\end{aligned} \tag{9}$$

because $\varepsilon_k \nu \in A$, $\varepsilon_k \lambda \in A$ and $|\varepsilon_k \nu - \varepsilon_k \lambda| < \delta$ for $k = 0, 1, 2, \ldots, n-1$.

Since $|\lambda| > \nu$ we have $\lim_{n\to\infty} \| \lambda^{-n} T^n \|^{1/n} < 1$ and hence

$$\lim_{n\to\infty} \| \lambda^{-n} T^n \| = 0.$$

It follows now from Theorem 6.2.11 that $\lim_{n\to\infty} (I - \lambda^{-n}T^n)^{-1} = I$ so there is an integer N such that

$$\left\| I - \left(I - \frac{1}{\lambda^n} T^n \right)^{-1} \right\| < \varepsilon \tag{10}$$

for $n \geq N$. From inequalities (9) and (10) we obtain, for $n \geq N$,

$$\begin{aligned} \left\| I - \left(I - \frac{1}{\nu^n} T^n \right)^{-1} \right\| &\leq \left\| I - \left(I - \frac{1}{\lambda^n} T^n \right)^{-1} \right\| \\ &\quad + \left\| \left(I - \frac{1}{\lambda^n} T^n \right)^{-1} - \left(I - \frac{1}{\nu^n} T^n \right)^{-1} \right\| \\ &< 2\varepsilon. \end{aligned}$$

The last inequality shows that $\lim_{n\to\infty} (I - \nu^{-n}T^n)^{-1} = I$. Another application of Theorem 6.2.11 now gives $\lim_{n\to\infty} (I - \nu^{-n}T^n) = I$ and consequently $\lim_{n\to\infty} \nu^{-n}T^n = 0$.

On the other hand it follows from Lemma 6.2.6 that $\nu^{-n} \| T^n \| \geq 1$ for $n = 1, 2, \ldots$ so we have arrived at a contradiction. This proves that there is a complex number λ with $\lambda \in \mathrm{sp}(T)$ and $|\lambda| \geq \lim_{n\to\infty} \| T^n \|^{1/n}$. □

When E is a Banach space we can improve the preceding theorem.

6.3.10 THEOREM *Let T be a bounded linear operator on a non-zero complex Banach space E. Then* $\mathrm{sp}(T)$ *is a non-empty compact subset of the closed disc* $\{\lambda \in \mathbb{C} : |\lambda| \leq \lim_{n\to\infty} \| T^n \|^{1/n}\}$ *and there is a point λ_0 in* $\mathrm{sp}(T)$ *such that* $|\lambda_0| = \lim_{n\to\infty} \| T^n \|^{1/n}$.

PROOF Let $\lambda \in \mathbb{C}$ with $|\lambda| > \lim_{n\to\infty} \| T^n \|^{1/n}$. Theorem 6.2.7 shows that $I - \lambda^{-1}T$ is regular and hence, by Lemma 6.2.4(c), $\lambda I - T$ is regular. This proves that

$$\mathrm{sp}(T) \subseteq \{\lambda \in \mathbb{C} : |\lambda| \leq \lim_{n\to\infty} \| T^n \|^{1/n}\}.$$

The existence of a point λ_0 in $\mathrm{sp}(T)$ with $|\lambda_0| = \lim_{n\to\infty} \| T^n \|^{1/n}$ now follows from Theorem 6.3.9.

By Theorem 6.2.10 the set $\mathcal{G}$ of regular elements of $L(E)$ is open. Consequently the set $\{\lambda \in \mathbb{C} : \lambda I - T \text{ is regular}\}$ is open, because it is the inverse image of $\mathcal{G}$ under the continuous mapping $\lambda \to \lambda I - T$. Since $\mathrm{sp}(T) = \mathbb{C} \sim \{\lambda \in \mathbb{C} : \lambda I - T \text{ is regular}\}$ $\mathrm{sp}(T)$ is closed. The Borel–Lebesgue theorem (4.2.11) now shows that $\mathrm{sp}(T)$ is compact. □

The number $\nu(T) = \sup \{ | \lambda | : \lambda \in \text{sp}(T) \}$ is called the *spectral radius of* T. By the above theorem we have also $\nu(T) = \lim_{n \to \infty} \| T^n \|^{1/n}$, when E is a Banach space.

The reader may have noticed that the proofs of many of the results of this and the preceding section use only the facts that $L(E)$ is a linear associative algebra and a normed linear space (or a Banach space), and the norm inequality (1) of Section 6.2. In particular the points of the space E do not appear explicitly in these proofs.

A Banach space that is also a linear associative algebra, and in which the norm inequality (1) is satisfied, is called a *Banach algebra*. The results of this and the preceding section concerning regular elements and the spectrum carry over, without change of proof, to a general Banach algebra and provide the basic tools for the study of Banach algebras.

Banach algebras were introduced in 1940 by the Russian mathematician I. M. Gel'fand. Their study is now flourishing and has exercised a unifying influence on modern analysis. There is an introduction to the theory of commutative Banach algebras in SIMMONS [25]. The standard references for the theory of Banach algebras are NAIMARK [21] and RICKART [22].

EXERCISES

(1) Give an example of a bounded linear operator T on some real normed linear space E such that $\lambda I - T$ is regular for all $\lambda \in \mathsf{R}$.

(2) Let E be a complex normed linear space, let $T \in L(E)$ and let λ be an eigenvalue of T. Let F be the closed linear hull of the set of eigenvectors of T corresponding to λ. Prove that $T(F) \subseteq F$.

(3) Let F be a one-dimensional linear subspace of a complex linear space E and let T be a linear operator on E with $T(F) \subseteq F$. Prove that there is an eigenvalue λ of T such that each non-zero vector in F is an eigenvector of T corresponding to λ.

(4) Let T be a non-zero linear operator on an n-dimensional complex linear space E.

(a) Prove that there is an integer $m \geq 1$ and a unique polynomial p of the form

$$p(t) = t^m + \alpha_1 t^{m-1} + \ldots + \alpha_{m-1} t + \alpha_m$$

such that $p(T) = 0$ and $q(T) \neq 0$ for all polynomials q with degree $\leq m - 1$. The polynomial p is called the *minimal polynomial* of T. (Hint: use Exercise 6.2(1).)

(b) Prove that each zero of the minimal polynomial of T is an eigenvalue of T. Deduce that sp(T) is non-empty.

(5) Let E be an n-dimensional linear space, let $m \leq n$ and let

$\lambda_1, \lambda_2, \ldots, \lambda_m$ be distinct complex numbers. Prove that there is a linear operator T on E with $\text{sp}(T) = \{\lambda_1, \lambda_2, \ldots, \lambda_m\}$.

(6) Let T be a bounded linear operator on a finite-dimensional normed linear space E. Prove that if $\lim_{n\to\infty} \| T^n \|^{1/n} = 0$ then $T^k = 0$ for some positive integer k. (One possible proof is to show that if M is a non-zero linear subspace of E and $T(M) \subseteq M$ then, by Exercise 6.2(9), $T|M$ is not regular; then to deduce, using Lemma 6.3.4, that, for some positive integer k, $\mathscr{R}(T^k) = \{0\}$.)

(7) Let (α_n) be a bounded sequence of complex numbers and let $T \in L(\ell)$ be defined by $T(\xi_n) = (\alpha_n \xi_n)$ for all $(\xi_n) \in \ell$. (See Exercise 6.1(2).) Determine $\text{sp}(T)$. What are the eigenvalues of T?

(8) Let A be a compact subset of the complex plane and let T be the mapping of $C_{\mathsf{C}}(A)$ into itself defined by $(Tf)(z) = zf(z)$ for all $z \in A$ and all $f \in C_{\mathsf{C}}(A)$. Prove that T is a bounded linear operator on $C_{\mathsf{C}}(A)$ and determine its spectrum and eigenvalues.

(9) Determine the spectra of the linear operators defined in Examples 6.3.6 and 6.3.7.

(10) Let E be a complex normed linear space and let $S,T \in L(E)$. Prove that, if $I - ST$ is regular and $U = (I - ST)^{-1}$, then $I - TS$ is also regular and $(I - TS)^{-1} = I + TUS$. Deduce that

$$\text{sp}(ST) \sim \{0\} = \text{sp}(TS) \sim \{0\}.$$

(11) Let T be a bounded linear operator on a complex normed linear space E and let p be a polynomial. Prove that

$$\text{sp}(p(T)) = \{p(\lambda) : \lambda \in \text{sp}(T)\}.$$

(12) (a) Prove that if $T \in L(E)$, where E is a complex Banach space, and $\lambda \in \text{Fr}\,(\text{sp}(T))$ then for each $\varepsilon > 0$ there is a point $x \in E$ with $\| (\lambda I - T)x \| < \varepsilon \| x \|$.

(b) Let E be a complex Banach space, $T \in L(E)$ and M a closed linear subspace of E with $T(M) \subseteq M$. Prove that $\text{Fr}\,(\text{sp}(T|M)) \subseteq \text{sp}(T)$.

(13) Let E be the set of complex-valued functions f on the integers for which the series $\sum_n | f(n) |^2$ is convergent. Prove that E is a Banach space with respect to the norm $\| f \| = \left(\sum_{n=-\infty}^{\infty} | f(n) |^2 \right)^{1/2}$. (Note that E is linearly isometric to the space ℓ^2.)

Let M be the linear subspace $\{f \in E : f(n) = 0 \text{ for } n \leq 0\}$ of E and let T be the bounded linear operator on E defined by $(Tf)(n) = f(n-1)$. Show that $T(M) \subseteq M$ and determine $\text{sp}(T)$ and $\text{sp}(T|M)$. (Hint: the operator $T|M$ can be identified with the operator of Example 6.3.6 and Exercise (9).) The operator T is called the *bilateral shift*.

6.4 The Adjoint of a Bounded Linear Transformation

Throughout this section E and F will denote normed linear spaces over K.

6.4.1 Definition Let $T \in L(E,F)$. Lemma 3.5.8 shows that the functional $y^* \circ T$ is in E^* for each $y^* \in F^*$. Let T^* be the mapping of F^* into E^* defined by $T^*y^* = y^* \circ T$ for all $y^* \in F^*$. The mapping T^* is called the *adjoint* of T.

For all $y^* \in F^*$ and all $x \in E$, we have

$$(T^*y^*)(x) = (y^* \circ T)(x) = y^*(Tx). \tag{1}$$

All the properties of T^* are derived from equation (1) which, henceforth, we shall use without comment. There is a far-reaching duality between the properties of T and T^* which provides an important tool in the study of bounded linear operators.

6.4.2 Theorem *Let $T \in L(E,F)$. Then $T^* \in L(F^*,E^*)$ and*

$$\| T^* \| = \| T \|.$$

Proof It follows easily from (1) that T^* is linear, and Lemma 3.5.8 gives

$$\| T^*y^* \| = \| y^* \circ T \| \leq \| y^* \| \; \| T \|$$

for all $y^* \in F^*$. This shows that $T^* \in L(F^*,E^*)$ and $\| T^* \| \leq \| T \|$.

For each $x \in E$ there exists, by Corollary 5.3.5, a functional $y^* \in F^*$ with $\| y^* \| = 1$ and $y^*(Tx) = \| Tx \|$. Then

$$\begin{aligned} \| Tx \| = y^*(Tx) &= (T^*y^*)(x) \leq \| T^*y^* \| \; \| x \| \\ &\leq \| T^* \| \; \| y^* \| \; \| x \| = \| T^* \| \; \| x \|. \end{aligned}$$

This shows that $\| T \| \leq \| T^* \|$. □

The proof of the following lemma is quite elementary, and is left to the reader.

6.4.3 Lemma *Let $S,T \in L(E,F)$ and $\alpha \in \mathsf{K}$. Then $(\alpha T)^* = \alpha T^*$ and $(S + T)^* = S^* + T^*$.*

It is also easy to verify that if I is the identity operator on E then I^* is the identity operator on E^*.

6.4.4 Lemma *Let $T \in L(E,F)$ and $S \in L(F,G)$, where G is a normed linear space over K. Then $(S \circ T)^* = T^* \circ S^*$.*

PROOF For all $z^* \in G^*$ and all $x \in E$ we have

$$\begin{aligned}((S \circ T)^* z^*)(x) &= z^*((S \circ T)x) \\ &= z^*(S(Tx)) \\ &= (S^* z^*)(Tx) \\ &= (T^*(S^* z^*))(x) \\ &= ((T^* \circ S^*) z^*)(x). \quad \square\end{aligned}$$

6.4.5 EXAMPLE Suppose that E and F are finite-dimensional and let T be a linear transformation of E into F. Let $\{e_1, e_2, \ldots, e_n\}$ and $\{e_1^*, e_2^*, \ldots, e_n^*\}$ be dual bases for E and E^*, and let $\{f_1, f_2, \ldots, f_m\}$ and $\{f_1^*, f_2^*, \ldots, f_m^*\}$ be dual bases for F and F^*. (See p. 176.) Let (τ_{jk}) and (τ_{jk}^*) be the matrices that represent T and T^* relative to the above bases. (See Theorem 6.1.1.) Then $Te_k = \sum_{j=1}^{m} \tau_{jk} f_j$ for $k = 1, 2, \ldots, n$ and $T^* f_k^* = \sum_{j=1}^{n} \tau_{jk}^* e_j^*$ for $k = 1, 2, \ldots, m$. Consequently, for $j = 1, 2, \ldots, n$ and $k = 1, 2, \ldots, m$,

$$(T^* f_k^*)(e_j) = \left(\sum_{r=1}^{n} \tau_{rk}^* e_r^*\right)(e_j) = \sum_{r=1}^{n} \tau_{rk}^* e_r^*(e_j) = \tau_{jk}^*.$$

On the other hand,

$$(T^* f_k^*)(e_j) = f_k^*(Te_j) = f_k^*\left(\sum_{r=1}^{m} \tau_{rj} f_r\right) = \sum_{r=1}^{m} \tau_{rj} f_k^*(f_r) = \tau_{kj}.$$

Thus $\tau_{jk}^* = \tau_{kj}$ which shows that the matrix that represents T^* is the transpose of the matrix that represents T.

Let $T \in L(E,F)$. Then $T^* \in L(F^*,E^*)$ and consequently T^* has an adjoint $(T^*)^*$ which is in $L(E^{**},F^{**})$. We shall write T^{**} for $(T^*)^*$. The transformations T and T^{**} are closely related. We recall that there is a canonical mapping $x \to \hat{x}$ of E onto a linear subspace $\hat{E}$ of E^{**} (Theorem 5.3.9). We shall also denote the canonical mapping of F into F^{**} by $y \to \hat{y}$; this should cause no confusion. For each $x \in E$, let $\hat{T}\hat{x} = (Tx)\hat{}$. It is obvious that $\hat{T} \in L(\hat{E},\hat{F})$ and $\| \hat{T} \| = \| T \|$. The following lemma shows that T^{**} is an extension of $\hat{T}$.

6.4.6 LEMMA *Let $T \in L(E,F)$. Then $T^{**}\hat{x} = (Tx)\hat{}$ for all $x \in E$.*

PROOF For all $x \in E$ and all $y^* \in F^*$ we have

$$(T^{**}\hat{x})(y^*) = \hat{x}(T^* y^*) = (T^* y^*)(x) = y^*(Tx) = (Tx)\hat{}(y^*)$$

and therefore $T^{**}\hat{x} = (Tx)\hat{}$. $\square$

It follows from Theorem 6.4.2 that $\| T^{**} \| = \| T^* \| = \| T \|$.

We shall now begin the study of the duality between a linear transformation and its adjoint. This study will be carried further in Chapter 8.

6.4.7 DEFINITION Let $T \in L(E,F)$. The set $\mathcal{N}(T) = \{x \in E : Tx = 0\}$ is called the *null-space of T*.

The range of T will be denoted by $\mathcal{R}(T)$. Obviously $\mathcal{N}(T)$ and $\mathcal{R}(T)$ are linear subspaces of E and F, respectively. Since T is continuous and $\mathcal{N}(T) = T^{-1}(\{0\})$ Theorem 2.7.5 shows that $\mathcal{N}(T)$ is closed; simple examples show that $\mathcal{R}(T)$ need not be closed.

6.4.8 LEMMA *Let $T \in L(E,F)$. Then*

(a) $\mathcal{N}(T) = \mathcal{R}(T^*)_\perp$,

(b) $\mathcal{N}(T^*) = \mathcal{R}(T)^\perp$,

(c) $\mathcal{R}(T)^- = \mathcal{N}(T^*)_\perp$.

PROOF Let $x \in \mathcal{N}(T)$. Then for all $y^* \in F^*$

$$0 = y^*(0) = y^*(Tx) = (T^*y^*)(x)$$

which shows that $x \in \mathcal{R}(T^*)_\perp$. Now let $x \in \mathcal{R}(T^*)_\perp$. If $y^* \in F^*$ then $y^*(Tx) = (T^*y^*)(x) = 0$, and it follows from Corollary 5.3.5 that $Tx = 0$. This proves (a). The proof of (b) is similar, but even simpler because it does not use Corollary 5.3.5. Finally, by Lemma 5.4.3 and (b) we have $\mathcal{R}(T)^- = (\mathcal{R}(T)^\perp)_\perp = \mathcal{N}(T^*)_\perp$. □

The remark on p. 195, and (a), gives $\mathcal{R}(T^*)^- \subseteq (\mathcal{R}(T^*)_\perp)^\perp = \mathcal{N}(T)^\perp$. It may happen that $\mathcal{R}(T^*)^- \neq \mathcal{N}(T)^\perp$ (Exercise (7)).

Lemma 6.4.8 does not lie very deep, but it gives some important information about the relation between $\mathrm{sp}(T)$ and $\mathrm{sp}(T^*)$.

6.4.9 THEOREM *Let $T \in L(E,F)$.*

(a) *If T is a linear homeomorphism of E onto F then T^* is a linear homeomorphism of F^* onto E^* and $(T^*)^{-1} = (T^{-1})^*$.*

(b) *If E is a Banach space and T^* is a linear homeomorphism of F^* onto E^* then T is a linear homeomorphism of E onto F.*

PROOF Let T be a linear homeomorphism of E onto F. By Lemma 6.2.2 we have $T \circ T^{-1} = I_F$ and $T^{-1} \circ T = I_E$, where I_E and I_F are the identity operators on E and F respectively, and consequently by Lemma 6.4.4 and the remark immediately preceding it

$$I_{F^*} = I_F^* = (T \circ T^{-1})^* = (T^{-1})^* \circ T^*$$

and

$$I_{E^*} = I_E^* = (T^{-1} \circ T)^* = T^* \circ (T^{-1})^*.$$

It now follows from Lemma 6.2.2 that T^* is a one-to-one mapping of F^* onto E^* and $(T^*)^{-1} = (T^{-1})^*$. Since T^{-1} is bounded $(T^{-1})^*$ is also bounded and therefore T^* is a linear homeomorphism of F^* onto E^*. This proves (a).

Now suppose that E is a Banach space and that T^* is a linear homeomorphism of F^* onto E^*. By (a) T^{**} is a linear homeomorphism of E^{**} onto F^{**} so there exist $M > 0$ and $m > 0$ with

$$m \| x^{**} \| \leq \| T^{**}x^{**} \| \leq M \| x^{**} \|$$

for all $x^{**} \in E^{**}$. (See p. 110.) Lemma 6.4.6 and Theorem 5.3.9 now give, for all $x \in E$,

$$m \| x \| = m \| \hat{x} \| \leq \| (Tx)\hat{x} \| = \| Tx \| \leq M \| \hat{x} \| = M \| x \|,$$

which shows that T is a linear homeomorphism of E onto $\mathscr{R}(T)$. Since E is complete, it follows from Lemma 2.9.4 that $\mathscr{R}(T)$ is complete, and therefore, by Theorem 2.6.8, $\mathscr{R}(T)$ is closed. But then Lemma 6.4.8 shows that $\mathscr{R}(T) = \mathscr{N}(T^*)_\perp$. Since $\mathscr{N}(T^*) = \{0\}$ it follows that $\mathscr{R}(T) = F$. □

6.4.10 COROLLARY *Let E be a non-zero complex normed linear space and let $T \in L(E)$. Then* $\operatorname{sp}(T^*) \subseteq \operatorname{sp}(T)$. *Further, if E is a Banach space then* $\operatorname{sp}(T^*) = \operatorname{sp}(T)$.

PROOF Apply Theorem 6.4.9 to the operator $\lambda I - T$. □

EXERCISES

(1) Let M be a linear subspace of a normed linear space E and let $J \in L(M,E)$ be defined by $Jx = x$ for all $x \in M$. Verify that $J^* = J_1 \circ \pi_1$, where J_1 is the linear isometry of $E^*/M^\perp$ onto M^* constructed in Theorem 5.4.4 and π_1 is the canonical mapping of E^* onto $E^*/M^\perp$ given by $\pi_1 x^* = x^* + M^\perp$.

(2) Let M be a closed linear subspace of a normed linear space E and let π be the canonical mapping of E onto E/M given by $\pi x = x + M$. Verify that $\pi^* = J \circ J_2$, where $J \in L(M^\perp, E^*)$ is defined by $Jx^* = x^*$ for all $x^* \in M^\perp$ and J_2 is the linear isometry of $(E/M)^*$ onto $M^\perp$ constructed in Theorem 5.4.5.

(3) For $p > 1$ and $q = p/(p-1)$ let T_p be the linear isometry of ℓ^q onto $(\ell^p)^*$ defined in Example 5.1.6 and let J be the canonical mapping of ℓ^p into $(\ell^p)^{**}$. Verify that $J = (T_p^*)^{-1} \circ T_q$ and deduce that ℓ^p is reflexive. (Cf. Example 5.3.13.)

(4) Let J be the canonical mapping of c_0 into c_0^{**}, let S be the linear isometry of m onto ℓ^* defined in Example 5.1.4 and let T be the linear isometry of ℓ onto c_0^* defined in Example 5.1.5. Verify that

$$J = (T^*)^{-1} \circ (S|c_0)$$

and deduce that c_0 is not reflexive. (Cf. Example 5.3.11.)

(5) Let E and F be normed linear spaces that are linearly homeomorphic. Prove that if E is reflexive then so also is F.

(6) Let M be a closed linear subspace of a normed linear space E. The following exercises describe the relations between the second dual spaces of E, M and E/M and provide an alternative proof of Theorem 5.4.6.

(a) Let $M^{\perp\perp} = \{x^{**} \in E^{**}: x^{**}(x^*) = 0 \text{ for all } x^* \in M^{\perp}\}$ and let T be the canonical mapping of E into E^{**}. Prove that $T(M) \subseteq M^{\perp\perp}$ and that, if E is reflexive, then $T(M) = M^{\perp\perp}$.

(b) Let S be the canonical mapping of M into M^{**}. Prove that $S = (J_1^*)^{-1} \circ J_2^{-1} \circ (T|M)$, where J_1 is the linear isometry of $E^*/M^{\perp}$ onto M^* given by Theorem 5.4.4 and J_2 is the linear isometry of $(E^*/M^{\perp})^*$ onto $M^{\perp\perp}$ given by Theorem 5.4.5. Deduce that, if E is reflexive, then so is M.

(c) Prove that the equation $V(x + M) = Tx + M^{\perp\perp}$ defines a bounded linear mapping V of E/M into $E^{**}/M^{\perp\perp}$.

(d) Let U be the canonical mapping of E/M into $(E/M)^{**}$. Prove that $U = J_3^* \circ J_4 \circ V$, where J_3 is the linear isometry of $(E/M)^*$ onto $M^{\perp}$ given by Theorem 5.4.5, and J_4 is the linear isometry of $E^{**}/M^{\perp\perp}$ onto $(M^{\perp})^*$ given by Theorem 5.4.4. Deduce that, if E is reflexive, then so is E/M.

(e) Prove that, if M and E/M are both reflexive, then E is also reflexive.

(7) (a) A mapping T of c into itself is defined by $Tx = (\xi_n/n)$ for all $x = (\xi_n) \in c$. Prove that $T \in L(c)$ and that $\mathscr{R}(T)$ is not closed.

(b) A mapping T' of ℓ into itself is defined by $Tx = (\eta_n)$, where $\eta_1 = 0$ and $\eta_n = \xi_n/n$ for $n = 2, 3, \ldots$, and $x = (\xi_n) \in \ell$. Prove that $T' \in L(\ell)$ and that $T' = J^{-1} \circ T^* \circ J$, where J is the linear isometry of ℓ onto c^* constructed in Exercise 5.1(4). Deduce that $\mathscr{R}(T^*)^- \neq \mathscr{N}(T)^{\perp}$. (Compare this example with Lemma 6.4.8 and Theorem 8.6.2.)

(8) Let A be a compact subset of the complex plane and let T be the bounded linear operator on $C_{\mathbf{C}}(A)$ defined in Exercise 6.3(8). Prove that $\mathrm{sp}(T^*) = A$ and that each point of $\mathrm{sp}(T^*)$ is an eigenvalue of T^*.

(9) Let E be a Banach space, let F be a linear subspace of E that is dense in E and let $T \in L(F,E)$ be given by $Tx = x$ for all $x \in F$. Prove that T^* is a linear homeomorphism of E^* onto F^* but that T is not a linear homeomorphism of F onto E unless $F = E$. (Cf. Theorem 6.4.9.)

6.5 Compact Linear Operators

One of the earliest successes of functional analysis dates from 1916 when F. Riesz used linear space methods to prove some of Fredholm's results on linear integral equations. The concept of a normed linear space had not been formulated in 1916, and Riesz worked with integral equations, but his techniques generalize directly and can be applied to a special class of linear operators, now called compact operators. In this section we shall obtain some of the elementary properties of compact operators. Throughout the section E will denote a normed linear space over K; we shall not assume that E is complete.

6.5.1 DEFINITION A linear mapping T of E into itself is said to be *compact* if and only if, for each bounded sequence (x_n) in E, the sequence (Tx_n) has a convergent subsequence.

The following lemma, whose proof is almost trivial, explains the use of the term 'compact linear mapping'.

6.5.2 LEMMA *Let T be a linear mapping of E into itself. Then T is compact if and only if, for each bounded subset A of E, the set $T(A)$ is relatively sequentially compact.*

PROOF Suppose that T is compact and let A be a bounded subset of E. Let (y_n) be a sequence in $T(A)^-$. For $n = 1, 2,...$ choose $x_n \in A$ so that $\| y_n - Tx_n \| < 1/n$. The sequence (Tx_n) has a convergent subsequence, (Tx_{n_k}) say, with limit y. Obviously $\lim_{k\to\infty} y_{n_k} = y$. This shows that $T(A)$ is relatively sequentially compact.

Suppose conversely that $T(A)$ is relatively sequentially compact for each bounded subset A of E and let (x_n) be a bounded sequence in E. By hypothesis the set $\{Tx_n : n = 1, 2,...\}$ is relatively sequentially compact, and it follows, in particular, that the sequence (Tx_n) has a convergent subsequence. □

Some writers call a linear mapping T compact if and only if, for each bounded subset A of E, the set $T(A)$ is relatively compact. By Lemma 6.5.2 and Corollary 4.2.10 this is equivalent to Definition 6.5.1. Most of the results of this and the following two sections can be derived from Definition 6.5.1 without appealing to the deep properties of compactness obtained in Section 4.2.

Before beginning the study of compact linear mappings we prove one more very elementary lemma which will be useful.

6.5.3 LEMMA *Let T be a linear mapping of E into itself and suppose that, for each sequence (x_n) in the closed unit ball of E, the sequence (Tx_n) has a convergent subsequence. Then T is compact.*

PROOF Let (x_n) be a bounded sequence in E and let $r > 0$ be such that $\| x_n \| \leq r$ for $n = 1, 2, \ldots$. Then the sequence $(T(r^{-1}x_n))$ has a convergent subsequence, $(T(r^{-1}x_{n_k}))$ say. Since $Tx_{n_k} = rT(r^{-1}x_{n_k})$ the sequence (Tx_{n_k}) also converges. □

Lemma 6.5.3 can be reformulated in terms of sequential compactness: if T is a linear mapping of E into itself such that $T(B)$ is relatively sequentially compact, where B is the closed unit ball of E, then T is compact.

6.5.4 LEMMA *A compact linear mapping of E into itself is bounded.*

PROOF If T is a linear mapping of E into itself that is not bounded then, for each positive integer n, there is a point $x_n \in E$ with $\| x_n \| = 1$ and $\| Tx_n \| \geq n$. Obviously the sequence (Tx_n) cannot have a convergent subsequence, so T is not compact. □

6.5.5 LEMMA *The set of all compact linear operators on E is a linear subspace of $L(E)$.*

PROOF Let S and T be compact linear operators on E and let $\alpha, \beta \in \mathsf{K}$. Let (x_n) be a bounded sequence in E. Then (Tx_n) has a convergent subsequence, (Tx_{n_k}) say, and (Sx_{n_k}) has a convergent subsequence, $(Sx_{n_{k(j)}})$ say. The sequence $(Tx_{n_{k(j)}})$ converges because it is a subsequence of a convergent sequence. It is now clear that the sequence $(\alpha Tx_{n_{k(j)}} + \beta Sx_{n_{k(j)}})$ converges. □

6.5.6 LEMMA *Let T be a compact linear operator on E and let $S \in L(E)$. Then ST and TS are compact.*

PROOF Let (x_n) be a bounded sequence in E. The sequence (Sx_n) is also bounded, because $\| Sx_n \| \leq \| S \| \| x_n \|$, and hence $(T(Sx_n))$ has a convergent subsequence. Since $(TS)x_n = T(S)x_n$this proves that TS is compact. On the other hand, (Tx_n) has a convergent subsequence, (Tx_{n_k}) say, and, consequently, since S is continuous $(S(Tx_{n_k}))$ also converges (Theorem 2.7.7). This proves that ST is compact. □

6.5.7 THEOREM *Suppose that E is a Banach space. Then the set of all compact linear operators on E is a closed subset of $L(E)$.*

PROOF Let (T_n) be a convergent sequence of compact linear operators on E with $\lim_{n\to\infty} T_n = T$ and let B be the closed unit ball in E. For each $\varepsilon > 0$ there is an integer N with $\| T - T_n \| < \frac{1}{3}\varepsilon$ for all $n \geq N$ and therefore

$$\| Tx - T_n x \| < \tfrac{1}{3}\varepsilon \tag{1}$$

for all $n \geq N$ and all $x \in B$. By Lemma 6.5.2 the set $T_N(B)$ is relatively sequentially compact and so, by Lemma 4.2.6, it is totally bounded. Thus there is a finite subset, $\{x_1, x_2, \ldots, x_m\}$ say, of B such that $\{T_N x_1, T_N x_2, \ldots, T_N x_m\}$ is a $\frac{1}{3}\varepsilon$-net in $T_N(B)$. Given $x \in B$ there is an integer k, with $1 \leq k \leq m$, such that $\| T_N x - T_N x_k \| < \frac{1}{3}\varepsilon$ and therefore by (1)

$$\| Tx - Tx_k \| \leq \| Tx - T_N x \| + \| T_N x - T_N x_k \| + \| T_N x_k - Tx_k \| < \varepsilon.$$

This proves that $T(B)$ is totally bounded. Since E is complete Corollary 4.2.10 shows that $T(B)$ is relatively sequentially compact. Lemma 6.5.3 now shows that T is compact. □

We shall now consider some examples of compact operators.

6.5.8 THEOREM *Let $T \in L(E)$ and suppose that the range of T is a finite-dimensional linear subspace of E. Then T is compact.*

PROOF Let A be a non-empty bounded subset of E. The set $T(A)$ is bounded because T is bounded. By Corollary 4.4.3 the range $\mathscr{R}(T)$ of T is closed so we have $T(A)^- \subseteq \mathscr{R}(T)$. Thus $T(A)^-$ is a bounded closed subset of the finite-dimensional normed linear space $\mathscr{R}(T)$. It follows from Corollary 4.4.4 that $T(A)$ is relatively compact and hence relatively sequentially compact. □

A complete description of the operators in $L(E)$ with finite-dimensional range is given in Exercise 6.2(4). Theorems 6.1.2 and 6.5.8 show that all linear operators on a finite-dimensional normed linear space are compact.

6.5.9 THEOREM *A Fredholm integral operator on $C_{\mathbf{C}}([a,b])$ is compact.*

PROOF Let K be a Fredholm integral operator with kernel k (see p. 218) and let B be the closed unit ball in $C_{\mathbf{C}}([a,b])$. We have to prove that $K(B)$ is relatively sequentially compact (Lemma 6.5.3). By the Ascoli–Arzelà theorem (4.6.2) and Corollary 4.2.10 this will follow if we can prove that $K(B)$ is bounded and equicontinuous. Since K is a bounded linear operator $K(B)$ is bounded. Let $\varepsilon > 0$ and $f \in B$. Theorem 4.3.8 shows that k is uniformly continuous on the square $\{(s,t)\colon a \leq s \leq b,\ a \leq t \leq b\}$ so there exists $\delta > 0$ such that

$$| k(s_1,t) - k(s_2,t) | < \varepsilon(b - a)^{-1}$$

for all $t \in [a,b]$ and all $s_1,s_2 \in [a,b]$ with $|s_1 - s_2| < \delta$. Therefore, for all $s_1,s_2 \in [a,b]$ with $|s_1 - s_2| < \delta$,

$$\begin{aligned}
&|(Kf)(s_1) - (Kf)(s_2)| \\
&\qquad = \left| \int_a^b (k(s_1,t) - k(s_2,t))f(t)\,dt \right| \\
&\qquad \le (b-a) \sup \{ |k(s_1,t) - k(s_2,t)|\ |f(t)| : a \le t \le b\} \\
&\qquad \le \varepsilon \sup \{ |f(t)| : a \le t \le b\} \\
&\qquad \le \varepsilon.
\end{aligned}$$

This proves that $K(B)$ is equicontinuous. □

The above theorem provides the connection between the study of Fredholm integral equations and compact operators. (See Section 6.8.) An alternative proof of Theorem 6.5.9 is outlined in Exercise 6.8(2).

6.5.10 THEOREM (Schauder) *Let T be a compact linear operator on E. Then T^* is also compact.*

PROOF Let (x_n^*) be a sequence in the closed unit ball of E^*, let B be the closed unit ball of E and let $X = T(B)^-$. By Lemma 6.5.2 and Theorem 4.2.9 the subspace X of E is a compact metric space. We shall prove that the set $A = \{x_n^*|X: n = 1, 2,...\}$ is relatively compact in $C_K(X)$. By Lemma 4.1.4 the set X is bounded. Let

$$M = \sup \{ \|y\| : y \in X\}.$$

Then for all $y \in X$ and $n = 1, 2,...$ we have

$$|x_n^*(y)| \le \|x_n^*\| \|y\| \le M.$$

This shows that A is bounded. Also, for all $y_1,y_2 \in X$ and $n = 1, 2,...$

$$|x_n^*(y_1) - x_n^*(y_2)| = |x_n^*(y_1 - y_2)| \le \|x_n^*\|\ \|y_1 - y_2\| \le \|y_1 - y_2\|.$$

This shows that A is equicontinuous. It follows now from the Ascoli–Arzelà theorem (4.6.2) that A is relatively compact and hence relatively sequentially compact (Corollary 4.2.10). Therefore there is a subsequence $(x_{n_k}^*)$ of (x_n^*) such that the sequence $(x_{n_k}^*|X)$ converges in $C_K(X)$. We shall prove that $(T^*x_{n_k}^*)$ converges.

Let $\varepsilon > 0$. The sequence $(x_{n_k}^*|X)$ is a Cauchy sequence in $C_K(X)$, and so there is an integer K such that $|x_{n_j}^*(y) - x_{n_k}^*(y)| < \varepsilon$ for all $j,k \ge K$ and all $y \in X$. If $x \in B$ then $Tx \in X$ and therefore, if $j,k \ge K$,

$$|(T^*x_{n_j}^*)(x) - (T^*x_{n_k}^*)(x)| = |x_{n_j}^*(Tx) - x_{n_k}^*(Tx)| < \varepsilon$$

for all $x \in B$. It follows that $\|T^*x_{n_j}^* - T^*x_{n_k}^*\| \le \varepsilon$ for all $j,k \ge K$. This proves that $(T^*x_{n_k}^*)$ is a Cauchy sequence in E^*. Since E^* is complete the sequence $(T^*x_{n_k}^*)$ converges and hence T^* is compact. □

EXERCISES

(1) Let K be a Volterra integral operator on $C_{\mathsf{C}}([a,b])$. Prove that K is compact and $\mathrm{sp}(K) = \{0\}$.

(2) Prove that the linear operator T defined in Example 6.1.8 is compact. (Use Theorems 6.5.8 and 6.5.7.)

(3) Let E be a Banach space and let $T \in L(E)$ be such that T^* is compact. Prove that T is also compact. (Use Theorem 6.5.10 and Lemma 6.4.6.)

(4) Let E be a normed linear space, $x \in E$ and $x^* \in E^*$. Without using Theorem 6.5.8 prove that the operator $x \otimes x^*$ of Exercise 6.2(2) is compact. Hence give another proof of Theorem 6.5.8. (Hint: use Exercise 6.2(4).)

6.6 The Riesz–Schauder Theory of Compact Linear Operators

The results of this section form the basis for the discussions, in the two subsequent sections, of the spectral theory of compact linear operators and of Fredholm integral equations. The methods of proof are due to F. Riesz, and the results concerning the adjoint operator are due to J. Schauder.

Throughout this section E will denote a non-zero normed linear space over K and T will denote a compact linear operator on E. To avoid having to consider exceptional cases in some of the proofs, we shall suppose that E is infinite-dimensional.

6.6.1 LEMMA *The null-space $\mathcal{N}(I - T)$ is a closed and finite dimensional linear subspace of E.*

PROOF Since T is continuous (Lemma 6.5.4) and

$$\mathcal{N}(I - T) = (I - T)^{-1}(\{0\}),$$

the set $\mathcal{N}(I - T)$ is closed. Let $B = \{x \in \mathcal{N}(I - T) : \| x \| \leq 1\}$. Since $\mathcal{N}(I - T) = \{x \in E : Tx = x\}$ we have $T(B) = B$ and hence, since T is compact and B is bounded, the set $B = T(B)$ is relatively sequentially compact. By Lemma 4.2.6, B is totally bounded and it now follows from the proof of Theorem 4.4.6 that $\mathcal{N}(I - T)$ is finite-dimensional. □

6.6.2 LEMMA *There is a number $M > 0$ with the following property: for each $y \in \mathcal{R}(I - T)$ there exists $x \in E$ such that $y = (I - T)x$ and $\| x \| \leq M \| y \|$.*

PROOF Suppose that the lemma is false. Then for each positive integer n there is a point $y_n \in \mathscr{R}(I - T)$ such that, if $x_n \in E$ and $(I - T)x_n = y_n$, then $\| x_n \| > n \| y_n \|$. Obviously $y_n \neq 0$ (because $0 = (I - T)0$). Choose $w_n \in E$ such that $(I - T)w_n = y_n$. Then $w_n \notin \mathscr{N}(I - T)$. Let $d_n = d(w_n, \mathscr{N}(I - T))$. By Lemma 6.6.1 the subspace $\mathscr{N}(I - T)$ is closed so, by Lemma 2.8.4, $d_n > 0$. Thus there exists $v_n \in \mathscr{N}(I - T)$ such that $d_n \leq \| w_n - v_n \| < 2d_n$.

Let $z_n = \| w_n - v_n \|^{-1}(w_n - v_n)$. Then $\| z_n \| = 1$ and hence the sequence (Tz_n) has a convergent subsequence, (Tz_{n_k}) say. From the equation

$$(I - T)(\| w_n - v_n \| z_n) = (I - T)(w_n - v_n) = (I - T)w_n = y_n$$

we conclude that $\| w_n - v_n \| \| z_n \| > n \| y_n \|$. Therefore

$$\| (I - T)z_n \| = \| y_n \| \| w_n - v_n \|^{-1} < 1/n$$

for $n = 1, 2, \ldots$ and consequently $\lim_{n\to\infty} (I - T)z_n = 0$. Since we have $z_{n_k} = (I - T)z_{n_k} + Tz_k$ it follows that (z_{nk}) converges. Let $z = \lim_{k\to\infty} z_{n_k}$. Then

$$(I - T)z = \lim_{k\to\infty} (I - T)z_{n_k} = 0$$

and so $z \in \mathscr{N}(I - T)$. But now for $n = 1, 2, \ldots$ we have

$$\begin{aligned} \| z_n - z \| &= \| \| w_n - v_n \|^{-1}(w_n - v_n) - z \| \\ &= \| w_n - (v_n + \| w_n - v_n \| z) \| \; \| w_n - v_n \|^{-1} \\ &\geq d_n \| w_n - v_n \|^{-1} \\ &\geq \tfrac{1}{2}, \end{aligned}$$

because $v_n + \| w_n - v_n \| z \in \mathscr{N}(I - T)$. This contradicts the fact that $\lim_{k\to\infty} z_{n_k} = z$. □

6.6.3 LEMMA *The linear subspace $\mathscr{R}(I - T)$ is closed.*

PROOF Let (y_n) be a sequence in $\mathscr{R}(I - T)$ with $\lim_{n\to\infty} y_n = y$, and let M be as in Lemma 6.6.2. Choose $x_n \in E$ with $(I - T)x_n = y_n$ and $\| x_n \| \leq M \| y_n \|$ for $n = 1, 2, \ldots$. Since the sequence (y_n) converges it is bounded, and hence (x_n) is also bounded. Thus (Tx_n) has a convergent subsequence, (Tx_{n_k}) say. Since $x_n = y_n + Tx_n$, the sequence (x_{n_k}) also converges. Let $x = \lim_{k\to\infty} x_{n_k}$. Then $(I - T)x = \lim_{k\to\infty} (I - T)x_{n_k} = y$ which shows that $y \in \mathscr{R}(I - T)$. □

6.6.4 LEMMA *For each positive integer n we have $(I - T)^n = I - S_n$, where S_n is a compact linear operator on E.*

PROOF By the binomial theorem and the remarks on p. 233, we have

$$(I - T)^n = \sum_{r=0}^{n} (-1)^r \binom{n}{r} T^r = I - S_n,$$

where $S_n = T\left(\sum_{r=1}^{n} (-1)^{r-1} \binom{n}{r} T^{r-1}\right)$ is compact by Lemmas 6.5.4 and 6.5.6. □

In what follows we shall write

$$\mathcal{N}_n = \mathcal{N}((I - T)^n) \text{ and } \mathcal{R}_n = \mathcal{R}((I - T)^n).$$

The sets $\mathcal{N}_n$ and $\mathcal{R}_n$ are linear subspaces of E and it is clear that

$$\{0\} = \mathcal{N}_0 \subseteq \mathcal{N}_1 \subseteq \ldots \subseteq \mathcal{N}_n \subseteq \mathcal{N}_{n+1} \subseteq \ldots \tag{1}$$

and

$$E = \mathcal{R}_0 \supseteq \mathcal{R}_1 \supseteq \ldots \supseteq \mathcal{R}_n \supseteq \mathcal{R}_{n+1} \supseteq \ldots . \tag{2}$$

6.6.5 THEOREM *The null-spaces $\mathcal{N}_n$, $n = 0, 1, 2, \ldots$, are finite-dimensional linear subspaces of E and there is a non-negative integer ν such that*

$$\mathcal{N}_n \neq \mathcal{N}_{n+1} \textit{ for } n = 0, 1, 2, \ldots, \nu - 1$$

and

$$\mathcal{N}_n = \mathcal{N}_{n+1} \textit{ for } n = \nu, \nu + 1, \ldots .$$

PROOF It follows from Lemmas 6.6.1 and 6.6.4 that $\mathcal{N}_n$ is finite-dimensional for $n \geq 1$. Suppose that $\mathcal{N}_n \neq \mathcal{N}_{n+1}$ for all non-negative integers n. Since $\mathcal{N}_n$ is a closed linear subspace of $\mathcal{N}_{n+1}$ it follows from Lemma 4.4.5 that there exists $x_n \in \mathcal{N}_{n+1}$ with $\| x_n \| = 1$ and $d(x_n, \mathcal{N}_n) \geq \frac{1}{2}$. If $n > m$ then

$$\| Tx_n - Tx_m \| = \| x_n - ((I - T)x_n + x_m - (I - T)x_m) \| \geq \tfrac{1}{2}$$

because, by (1), we have $(I - T)x_n + x_m - (I - T)x_m \in \mathcal{N}_n$. Since (Tx_n) must have a convergent subsequence we have derived a contradiction, and, therefore, $\mathcal{N}_m = \mathcal{N}_{m+1}$ for some non-negative integer m. Let ν be the least such integer. Consider any integer $n > \nu$. If $x \in \mathcal{N}_n$ then

$$(I - T)^{\nu+1}((I - T)^{n-\nu-1}x) = (I - T)^n x = 0$$

and so $(I - T)^{n-\nu-1}x \in \mathcal{N}_{\nu+1} = \mathcal{N}_\nu$. Therefore

$$(I - T)^{n-1}x = (I - T)^\nu((I - T)^{n-\nu-1}x) = 0$$

and hence $x \in \mathcal{N}_{n-1}$. This proves that $\mathcal{N}_n = \mathcal{N}_{n-1}$. □

6.6.6 THEOREM *The ranges $\mathcal{R}_n$, $n = 0, 1, 2, \ldots$, are closed linear subspaces of E and there is a non-negative integer μ such that*

$$\mathcal{R}_n \neq \mathcal{R}_{n+1} \textit{ for } n = 0, 1, 2, \ldots, \mu - 1$$

and

$$\mathcal{R}_n = \mathcal{R}_{n+1} \textit{ for } n = \mu, \mu + 1, \ldots .$$

PROOF It follows from Lemmas 6.6.3 and 6.6.4 that $\mathscr{R}_n$ is closed for $n \geq 1$. Suppose that $\mathscr{R}_n \neq \mathscr{R}_{n+1}$ for all non-negative integers n. Since $\mathscr{R}_{n+1}$ is a closed linear subspace of $\mathscr{R}_n$ it follows from Lemma 4.4.5 that there exists $x_n \in \mathscr{R}_n$ with $\| x_n \| = 1$ and $d(x_n, \mathscr{R}_{n+1}) \geq \frac{1}{2}$. If $n > m$ then

$$\| Tx_m - Tx_n \| = \| x_m - ((I - T)x_m + x_n - (I - T)x_n) \| \geq \tfrac{1}{2},$$

because, by (2), we have $(I - T)x_m + x_n - (I - T)x_n \in \mathscr{R}_{m+1}$. Since (Tx_n) must have a convergent subsequence we have derived a contradiction and, therefore, $\mathscr{R}_m = \mathscr{R}_{m+1}$ for some non-negative integer m. Let μ be the least such integer. Since $\mathscr{R}_{n+1} = (I - T)(\mathscr{R}_n)$ it follows easily that $\mathscr{R}_{n+1} = \mathscr{R}_n$ for all $n \geq \mu$ so μ satisfies the conditions of the theorem. □

6.6.7 THEOREM *Let μ be the integer defined in Theorem 6.6.6. Then $E = \mathscr{N}_\mu \oplus \mathscr{R}_\mu$, $T(\mathscr{N}_\mu) \subseteq \mathscr{N}_\mu$ and $T(\mathscr{R}_\mu) \subseteq \mathscr{R}_\mu$. Further $(I - T)|\mathscr{R}_\mu$ is a regular element of $L(\mathscr{R}_\mu)$. In particular, if $\mu = 0$ then $I - T$ is regular (in $L(E)$).*

PROOF Let $x \in E$. By Theorem 6.6.6 we have $\mathscr{R}_\mu = \mathscr{R}_{2\mu}$ so

$$(I - T)^\mu x = (I - T)^{2\mu} y$$

for some $y \in E$. Thus $(I - T)^\mu (x - (I - T)^\mu y) = 0$ and

$$x = (x - (I - T)^\mu y) + (I - T)^\mu y.$$

This shows $E = \mathscr{N}_\mu + \mathscr{R}_\mu$. Suppose next that $m \geq 0$, $z \in \mathscr{N}_m \cap \mathscr{R}_\mu$ and $z \neq 0$. By Theorem 6.6.6 we have $\mathscr{R}_n = \mathscr{R}_\mu$ for $n \geq \mu$. Therefore for each integer $n \geq \mu$ there is a point $z_n \in E$ with $z = (I - T)^n z_n$ and, consequently, since $z \in \mathscr{N}_m$ and $z \neq 0$ we have $z_n \in \mathscr{N}_{m+n} \sim \mathscr{N}_n$. This contradicts Theorem 6.6.5 so we must have $\mathscr{N}_m \cap \mathscr{R}_\mu = \{0\}$ for $m \geq 0$. In particular $\mathscr{N}_\mu \cap \mathscr{R}_\mu = \{0\}$, so $E = \mathscr{N}_\mu + \mathscr{R}_\mu$.

It is clear that, if $\mu \geq 1$, $(I - T)(\mathscr{N}_\mu) \subseteq \mathscr{N}_{\mu-1} \subseteq \mathscr{N}_\mu$ and, if $\mu = 0$, $\mathscr{N}_\mu = \{0\}$ so

$$T(\mathscr{N}_\mu) = (I - (I - T))(\mathscr{N}_\mu) \subseteq \mathscr{N}_\mu + (I - T)(\mathscr{N}_\mu) \subseteq \mathscr{N}_\mu.$$

Also $(I - T)(\mathscr{R}_\mu) = \mathscr{R}_{\mu+1} = \mathscr{R}_\mu$ (Theorem 6.6.6), from which we obtain $T(\mathscr{R}_\mu) \subseteq \mathscr{R}_\mu$. It is obvious that $T \mid \mathscr{R}_\mu$ is a bounded linear operator on $\mathscr{R}_\mu$. We have just seen that $(I - T)(\mathscr{R}_\mu = \mathscr{R}_\mu$ so $(I - T) \mid \mathscr{R}_\mu$ maps $\mathscr{R}_\mu$ onto itself. Also $(I - T) \mid \mathscr{R}_\mu$ is one-to-one because $\mathscr{N}_1 \cap \mathscr{R}_\mu = \{0\}$. It follows now from Lemma 6.2.3 that, to prove that $(I - T) \mid \mathscr{R}_\mu$ is regular, we have only to prove that its inverse mapping is bounded. That $((I - T) \mid \mathscr{R})_\mu^{-1}$ is bounded follows directly from Lemma 6.6.2 because $\mathscr{R}_\mu \subseteq \mathscr{R}_1$ (notice that, if $\mu = 0$, $\mathscr{R}_\mu = E = \mathscr{R}_1$). □

6.6.8 Corollary *The integers* ν *and* μ *of Theorems* 6.6.5 *and* 6.6.6, *respectively, are equal.*

Proof Let $x \in \mathscr{N}_{\mu+1}$. By Theorem 6.6.7 we have $x = y + z$ with $y \in \mathscr{N}_{\mu}$ and $z \in \mathscr{R}_{\mu}$. Then $(I - T)^{\mu+1}z = (I - T)^{\mu+1}(x - y) = 0$. But, by Theorem 6.6.7, $(I - T)|\mathscr{R}_{\mu}$ is regular; thus $z = 0$ and hence $x \in \mathscr{N}_{\mu}$. This proves that $\mathscr{N}_{\mu+1} = \mathscr{N}_{\mu}$ and it follows that $\mu \geq \nu$. Now, by definition of ν, we have $(I - T)^{\nu}(\mathscr{N}_{\mu}) = \{0\}$ and so, by Theorem 6.6.7,

$$\mathscr{R}_{\nu} = (I - T)^{\nu}(E) = (I - T)^{\nu}(\mathscr{R}_{\mu}) + (I - T)^{\nu}(\mathscr{N}_{\mu}) = \mathscr{R}_{\mu+\nu} = \mathscr{R}_{\mu}.$$

It follows that $\nu \geq \mu$ and consequently $\nu = \mu$. □

The reader may like to compare Theorem 6.6.7 with its counterpart for linear operators on a finite-dimensional space. (See for example Halmos [12, p. 113].)

Theorem 6.6.7 is the main tool for the study of compact linear operators. As the first application of this theorem we shall obtain a generalization of a well-known result on linear operators on a finite-dimensional linear space. We shall need this finite-dimensional result later, so we outline its proof below. If F is a finite-dimensional linear space, dim F will denote the dimension of F.

6.6.9 Lemma *Let S be a linear transformation of a finite-dimensional linear space F into a linear space G. Then $\mathscr{R}(S)$ is finite-dimensional and*

$$\dim F = \dim \mathscr{N}(S) + \dim \mathscr{R}(S).$$

Proof Choose a basis $\{x_1, x_2, \ldots, x_n\}$ for $\mathscr{N}(S)$. Then there exist points $x_{n+1}, x_{n+2}, \ldots, x_{n+m}$ in F such that $\{x_1, x_2, \ldots, x_n, x_{n+1}, \ldots, x_{n+m}\}$ is a basis for F (Halmos [12, p. 19]). It is easy to verify that

$$\{Sx_{n+1}, Sx_{n+2}, \ldots, Sx_{n+m}\}$$

is a basis for $\mathscr{R}(S)$, and consequently

$$\dim F = n + m = \dim \mathscr{N}(S) + \dim \mathscr{R}(S). \quad □$$

Using the concept of codimension we can reformulate Lemma 6.6.9 in terms that generalize to infinite-dimensional spaces.

6.6.10 Definition A linear subspace M of a linear space F is said to have *finite codimension in* F if and only if the quotient space F/M has finite dimension. If M has finite codimension the dimension of F/M is called the *codimension of* M *in* F and is denoted by codim M.

6.6.11 Lemma *A linear subspace M of a linear space F has finite co-*

dimension n in F if and only if there exists an n-dimensional linear subspace N of F such that $F = M \oplus N$.

PROOF Suppose that M has finite codimension n and let

$$\{x_1 + M, x_2 + M, \ldots, x_n + M\}$$

be a basis for F/M. It is easy to see that the set $\{x_1, x_2, \ldots, x_n\}$ is linearly independent and that its linear hull N is such that $F = M \oplus N$.

Suppose conversely that there exists an n-dimensional linear subspace N of F such that $F = M \oplus N$. It is easy to verify that the restriction to N of the canonical mapping $x \to x + M$ is a one-to-one linear mapping of N onto F/M and therefore F/M is n-dimensional by Lemma 6.6.9. □

If F is a finite-dimensional linear space and M and N are linear subspaces of F with $F = M \oplus N$ then $\dim F = \dim M + \dim N$, and hence, by Lemma 6.6.11, we have $\dim F = \dim M + \operatorname{codim} M$. It now follows from Lemma 6.6.9 that if S is a linear operator on F then

$$\operatorname{codim} \mathscr{R}(S) = \dim \mathscr{N}(S). \tag{3}$$

Let us now return to the study of the compact linear operator T on E.

6.6.12 THEOREM *The range $\mathscr{R}(I - T)$ has finite codimension in E and* $\operatorname{codim} \mathscr{R}(I - T) = \dim \mathscr{N}(I - T)$.

PROOF Let ν be the integer defined in Theorem 6.6.5. By Theorem 6.6.7 and Corollary 6.6.8 we have $E = \mathscr{N}_\nu \oplus \mathscr{R}_\nu$, $(I - T)(\mathscr{N}_\nu) \subseteq \mathscr{N}_\nu$ and $(I - T)(\mathscr{R}_\nu) = \mathscr{R}_\nu$ so

$$\begin{aligned}\mathscr{R}(I - T) &= (I - T)(E) \\ &= (I - T)(\mathscr{N}_\nu) \oplus (I - T)(\mathscr{R}_\nu) \\ &= ((I - T)(\mathscr{N}_\nu)) \oplus \mathscr{R}_\nu.\end{aligned} \tag{4}$$

But $\mathscr{N}_\nu$ is finite-dimensional (Theorem 6.6.5) and

$$\mathscr{N}((I - T)|\mathscr{N}_\nu) = \mathscr{N}(I - T)$$

because $\mathscr{N}(I - T) \subseteq \mathscr{N}_\nu$. Consequently by (3) applied to the operator $(I - T)|\mathscr{N}_\nu$, we see that the dimension of $\mathscr{N}(I - T)$ is equal to the codimension of $(I - T)(\mathscr{N}_\nu)$ in $\mathscr{N}_\nu$. By Lemma 6.6.11 there is a linear subspace N of $\mathscr{N}_\nu$, with dimension equal to the codimension of $(I - T)(\mathscr{N}_\nu)$ in $\mathscr{N}_\nu$ such that $\mathscr{N}_\nu = (I - T)(\mathscr{N}_\nu) \oplus N$. Thus

$$E = \mathscr{N}_\nu \oplus \mathscr{R}_\nu = N \oplus (I - T)(\mathscr{N}_\nu) \oplus \mathscr{R}_\nu = N \oplus \mathscr{R}(I - T),$$

where the last step follows from (4). Lemma 6.6.11 now shows that $\mathscr{R}(I - T)$ has finite codimension in E and

$$\operatorname{codim} \mathscr{R}(I - T) = \dim N = \dim \mathscr{N}(I - T). \ \square$$

By Schauder's theorem (6.5.10) the adjoint T^* is also a compact linear operator so all the above results apply equally to T^*. The following theorem is the key to the relationship between T and T^*.

6.6.13 THEOREM *The null-spaces $\mathscr{N}(I - T)$ and $\mathscr{N}(I^* - T^*)$ have the same dimension.*

PROOF
$$\begin{aligned}\dim \mathscr{N}(I - T) &= \operatorname{codim} \mathscr{R}(I - T), && \text{by Theorem 6.6.12,}\\ &= \dim E/\mathscr{R}(I - T), && \text{by definition,}\\ &= \dim (E/\mathscr{R}(I - T))^*, && \text{by Theorem 5.1.3,}\\ &= \dim \mathscr{R}(I - T)^{\perp}, && \text{by Theorem 5.4.5,}\\ &= \dim \mathscr{N}(I^* - T^*), && \text{by Lemma 6.4.8. } \square\end{aligned}$$

Our final theorem is the abstract form of the famous Fredholm alternative for linear integral equations (Theorem 6.8.2).

6.6.14 THEOREM
$$\mathscr{N}(I - T)^{\perp} = \mathscr{R}(I^* - T^*) \text{ and } \mathscr{N}(I^* - T^*)_{\perp} = \mathscr{R}(I - T).$$

PROOF Lemma 6.4.8(c) gives $\mathscr{N}(I^* - T^*)_{\perp} = \mathscr{R}(I - T)$ because $\mathscr{R}(I - T)$ is closed (Lemma 6.6.3). Lemma 6.4.8(a) gives
$$\mathscr{N}(I - T) = \mathscr{R}(I^* - T^*)_{\perp}$$
and hence $\mathscr{R}(I^* - T^*) \subseteq (\mathscr{R}(I^* - T^*)_{\perp})^{\perp} = \mathscr{N}(I - T)^{\perp}$. It remains only to prove that $\mathscr{N}(I - T)^{\perp} \subseteq \mathscr{R}(I^* - T^*)$. This fact lies somewhat deeper than the others. We have
$$\begin{aligned}\dim E^*/\mathscr{R}(I^* - T^*) &= \operatorname{codim} \mathscr{R}(I^* - T^*), && \text{by definition,}\\ &= \dim \mathscr{N}(I^* - T^*), && \text{by Theorem 6.6.12,}\\ &= \dim \mathscr{N}(I - T), && \text{by Theorem 6.6.13,}\\ &= \dim (\mathscr{N}(I - T))^* && \text{by Theorem 5.1.3,}\\ &= \dim E^*/\mathscr{N}(I - T)^{\perp}, && \text{by Theorem 5.4.4.}\end{aligned}$$
It is clear that if we set $S(x^* + \mathscr{R}(I^* - T^*)) = x^* + \mathscr{N}(I - T)^{\perp}$ for all $x^* \in E^*$ then S is a linear transformation of $E^*/\mathscr{R}(I^* - T^*)$ onto $E^*/\mathscr{N}(I - T)^{\perp}$. Lemma 6.6.9 gives
$$\dim E^*/\mathscr{R}(I^* - T^*) = \dim \mathscr{N}(S) + \dim \mathscr{R}(S).$$
Since $\mathscr{R}(S) = E^*/\mathscr{N}(I - T)^{\perp}$ we must have $\dim \mathscr{N}(S) = 0$. Therefore S is one-to-one and it follows from the definition of S that if $x^* \in \mathscr{N}(I - T)^{\perp}$ then $x^* \in \mathscr{R}(I^* - T^*)$. Hence
$$\mathscr{N}(I - T)^{\perp} \subseteq \mathscr{R}(I^* - T^*). \square$$

EXERCISES

(1) The results of this section remain true when the hypothesis that

T is compact is replaced by the hypothesis that $T \in L(E)$ and T^n is compact for some positive integer n. The necessary modifications of the proofs of 6.6.1 to 6.6.6 can be based on the fact that there is an operator $S \in L(E)$ with $I - T^n = S(I - T) = (I - T)S$. Observe that Theorem 6.6.7, Corollary 6.6.8 and Theorems 6.6.12, 6.6.13 and 6.6.14 are true for any operator $T \in L(E)$ that satisfies the conditions of Lemma 6.6.2 and Theorems 6.6.5 and 6.6.6.

(2) (a) Let E be a normed linear space, and $T,S \in L(E)$. Suppose that T is compact and that, for some integer $n \geq 2$, S^{n-1} is not compact and $S^n = 0$. Prove that $(T + S)^{n-1}$ is not compact, but that $(T + S)^n$ is compact.

(b) Let n be an integer ≥ 2. Construct a bounded linear operator S on ℓ^2 such that S^{n-1} is not compact and $S^n = 0$.

6.7 The Spectrum of a Compact Linear Operator

Throughout this section T will denote a compact linear operator on a non-zero complex normed linear space E. We shall analyse the spectrum of T. To avoid having to consider exceptional cases we shall again suppose that E is infinite-dimensional.

In Theorem 6.3.9 we saw that the spectrum of any bounded linear operator is non-empty. For compact linear operators we can always identify at least one point of the spectrum.

6.7.1 LEMMA $0 \in \mathrm{sp}(T)$.

PROOF Suppose that $0 \notin \mathrm{sp}(T)$. Then T is regular. Since $I = T^{-1}T$ Lemma 6.5.6 shows that I is compact. It then follows from Lemma 6.6.1 that $E = \mathcal{N}(I - I)$ is finite-dimensional, which is a contradiction. Thus $0 \in \mathrm{sp}(T)$. □

There are compact operators whose spectrum consists of the single point 0. (See Exercise 6.5(1).) We turn now to the study of non-zero points in $\mathrm{sp}(T)$.

6.7.2 THEOREM *Each non-zero point of* $\mathrm{sp}(T)$ *is an eigenvalue of* T.

PROOF Let λ be a non-zero complex number and suppose that λ is not eigenvalue of T. Then $\mathcal{N}(I - \lambda^{-1}T) = \mathcal{N}(\lambda I - T) = \{0\}$, and it follows from Corollary 6.6.8 and Theorem 6.6.7, applied to $\lambda^{-1}T$, that $I - \lambda^{-1}T$ is regular. Thus $\lambda I - T$ is regular and hence $\lambda \notin \mathrm{sp}(T)$. □

Since T^* is also compact (Theorem 6.5.10) the non-zero points of $\mathrm{sp}(T^*)$ are eigenvalues of T^*.

6.7.3 Corollary *The operators T and T^* have the same non-zero eigenvalues and the same number of linearly independent eigenvectors corresponding to each non-zero eigenvalue.*

Proof Apply Theorem 6.6.13 to $\lambda^{-1}T$, where $\lambda \neq 0$. □

In the following two theorems λ will denote a non-zero eigenvalue of T. Corollary 6.6.8 and Theorem 6.6.7, applied to $\lambda^{-1}T$, show that there is a decomposition $E = \mathscr{N}_\nu \oplus \mathscr{R}_\nu$ where

$$\mathscr{N}_\nu = \{x \in E: (\lambda I - T)^\nu x = 0\}$$

and $\mathscr{R}_\nu = \mathscr{R}((\lambda I - T)^\nu)$ (the integer ν is obtained by applying Theorem 6.6.5 to $\lambda^{-1}T$). Since $T(\mathscr{N}_\nu) \subseteq \mathscr{N}_\nu$ and $T(\mathscr{R}_\nu) \subseteq \mathscr{R}_\nu$ the restrictions of T to $\mathscr{N}_\nu$ and $\mathscr{R}_\nu$ are bounded linear operators on $\mathscr{N}_\nu$ and $\mathscr{R}_\nu$, respectively. We shall determine the spectra of $T|\mathscr{N}_\nu$ and $T|\mathscr{R}_\nu$.

6.7.4 Theorem $\operatorname{sp}(T|\mathscr{N}_\nu) = \{\lambda\}$.

Proof If $\nu = 1$ then $T|\mathscr{N}_\nu = \lambda I$ and obviously $\operatorname{sp}(T|\mathscr{N}_\nu) = \{\lambda\}$. Suppose that $\nu \geq 2$ and let $S = (\lambda I - T)|\mathscr{N}_\nu$. By definition of ν we have $S^\nu = 0$. Consequently if $\mu \neq 0$ then (writing I_0 for the identity operator on $\mathscr{N}_\nu$) we have

$$\begin{aligned}\mu^\nu I_0 = \mu^\nu I_0 - S^\nu &= (\mu I_0 - S)\left(\sum_{j=1}^{\nu} \mu^{\nu-j} S^{j-1}\right) \\ &= \left(\sum_{j=1}^{\nu} \mu^{\nu-j} S^{j-1}\right)(\mu I_0 - S).\end{aligned}$$

This shows that $\mu I_0 - S$ is regular. But

$$(\lambda - \mu) I_0 - (T|N_\nu) = -(\mu I_0 - S)$$

so we have $\lambda - \mu \notin \operatorname{sp}(T|\mathscr{N}_\nu)$. Since μ is an arbitrary non-zero complex number it follows that $\operatorname{sp}(T|\mathscr{N}_\nu) = \{\lambda\}$. □

The behaviour of the operator $T|\mathscr{N}_\nu$ can be analysed in more detail. This analysis depends on the fact that the operator S introduced above satisfies the condition $S^\nu = 0$; such operators are said to be *nilpotent*. The reader will find a discussion of nilpotent operators in the book of Halmos [12, pp. 109–12].

6.7.5 Theorem *The restriction $T|\mathscr{R}_\nu$ is a compact linear operator on $\mathscr{R}_\nu$ and* $\operatorname{sp}(T|\mathscr{R}_\nu) = \operatorname{sp}(T) \sim \{\lambda\}$.

Proof Let $S = T|\mathscr{R}_\nu$ and let (x_n) be a bounded sequence in $\mathscr{R}_\nu$. The sequence (Tx_n) has a convergent subsequence, (Tx_{n_k}) say. Since

$Sx_{n_k} = Tx_{n_k} \in \mathscr{R}_\nu$ and $\mathscr{R}_\nu$ is closed (Theorem 6.6.6) we see that $\lim_{k\to\infty} Sx_{n_k} \in \mathscr{R}_\nu$. This proves that S is a compact linear operator on $\mathscr{R}_\nu$.

Since $\mathscr{N}_\nu$ is finite dimensional (Theorem 6.6.5) and $E = \mathscr{N}_\nu \oplus \mathscr{R}_\nu$ the linear subspace $\mathscr{R}_\nu$ must be infinite-dimensional. Consequently Lemma 6.7.1 shows that $0 \in \mathrm{sp}(T)$ and $0 \in \mathrm{sp}(S)$, and Theorem 6.7.2 shows that the non-zero points of $\mathrm{sp}(T)$ and $\mathrm{sp}(S)$ are eigenvalues of T and S, respectively. It is obvious that each eigenvalue of S is an eigenvalue of T. Suppose that μ is a non-zero eigenvalue of T different from λ and let x be an eigenvector of T corresponding to μ. Since $Tx = \mu x$ we have $(\lambda I - T)x = (\lambda - \mu)x$ and hence

$$(\lambda I - T)^\nu x = (\lambda - \mu)^\nu x.$$

Therefore $x = (\lambda I - T)^\nu((\lambda - \mu)^{-\nu}x) \in \mathscr{R}_\nu$. It follows that μ is an eigenvalue of S. This proves that $\mathrm{sp}(T) \sim \{\lambda\} \subseteq \mathrm{sp}(S) \subseteq \mathrm{sp}(T)$. Theorem 6.6.7, applied to $\lambda^{-1}T$, shows that $(\lambda I - T)|\mathscr{R}_\nu$ is regular and therefore $\lambda \notin \mathrm{sp}(S)$. This proves that $\mathrm{sp}(S) = \mathrm{sp}(T) \sim \{\lambda\}$. □

We now have enough information to complete our analysis of $\mathrm{sp}(T)$.

6.7.6 Theorem *The spectrum of T is a compact subset of the complex plane each non-zero point of which is isolated.*

Proof Suppose that $\lambda \notin \mathrm{sp}(T)$. Then $\lambda I - T$ is regular so there exists $M > 0$ such that $\| (\lambda I - T)x \| \geq M \| x \|$ for all $x \in E$. It follows that, for all $x \in E$,

$$\| (\mu I - T)x \| = \| (\lambda I - T)x - (\lambda - \mu)x \| \geq (M - | \lambda - \mu |) \| x \|.$$

Therefore if $| \lambda - \mu | < M$ then $\mathscr{N}(\mu I - T) = \{0\}$ and hence $\mu I - T$ is regular by Theorem 6.6.7. This proves that $\mathrm{sp}(T)$ is closed.

If λ is an eigenvalue of T and x is an eigenvector of T corresponding to λ then $Tx = \lambda x$ and hence $| \lambda | \| x \| = \| Tx \| \leq \| T \| \| x \|$. Thus $| \lambda | \leq \| T \|$ because $x \neq 0$. Since the non-zero points of $\mathrm{sp}(T)$ are eigenvalues of T (Theorem 6.7.2) this shows that $\mathrm{sp}(T)$ is a bounded set. Since $\mathrm{sp}(T)$ is also closed it is compact by the Borel–Lebesgue theorem (4.2.11).

Finally let $\lambda \in \mathrm{sp}(T)$ and $\lambda \neq 0$. By Theorem 6.7.2 the number λ is an eigenvalue of T and, by Theorem 6.7.5, $\lambda \notin \mathrm{sp}(T|\mathscr{R}_\nu)$ and $T|\mathscr{R}_\nu$ is compact (we are using the notation established on p. 256). By the first part of the proof, applied to $T|\mathscr{R}_\nu$, the set $\mathrm{sp}(T|\mathscr{R}_\nu)$ is closed. Therefore there is a neighbourhood of λ that does not meet $\mathrm{sp}(T|\mathscr{R}_\nu)$. But, by Theorem 6.7.5, $\mathrm{sp}(T|\mathscr{R}_\nu) = \mathrm{sp}(T) \sim \{\lambda\}$, so it follows that $\mathrm{sp}(T)$ contains no point of this neighbourhood other than λ. This proves that each non-zero point of $\mathrm{sp}(T)$ is isolated. □

The reader should notice that the compactness of $\mathrm{sp}(T)$ does not follow from Theorem 6.3.10 because we have not supposed that E is a Banach space.

It is not difficult to see that a subset of a separable metric space which has at most one point of accumulation must be countable (see Exercise 2.4(3)); consequently $\mathrm{sp}(T)$ is a countable set. If $\mathrm{sp}(T)$ is an infinite set then the set of eigenvalues of T is countable and has 0 as its only point of accumulation. In general nothing more can be said about the spectrum of a compact linear operator. (See Exercise (1).)

Exercises

(1) Let (λ_n) be a sequence of complex numbers with $\lim_{n\to\infty} \lambda_n = 0$. Construct a compact linear operator T on some Banach space with $\mathrm{sp}(T) = \{\lambda_n\colon\ n = 1, 2,\ldots\} \cup \{0\}$. (Hint: see Exercises 6.1(2) and 6.3(7).)

(2) Let T be a compact linear operator on a complex normed linear space E. Given a non-zero eigenvalue λ of T let $E = \mathcal{N}(\lambda) \oplus \mathcal{R}(\lambda)$ be the decomposition obtained by applying Theorem 6.6.5 to the operator $\lambda^{-1}T$. Let $\lambda_1, \lambda_2,\ldots, \lambda_n$ be non-zero eigenvalues of T. Prove that there is a closed linear subspace M of E with $T(M) \subseteq M$ and

$$E = \mathcal{N}(\lambda_1) \oplus \mathcal{N}(\lambda_2) \oplus \ldots \oplus \mathcal{N}(\lambda_n) \oplus M.$$

Prove also that $\mathrm{sp}(T|M) = \mathrm{sp}(T) \sim \{\lambda_1, \lambda_2,\ldots, \lambda_n\}$.

(3) The results of this section remain true when the hypothesis that T is compact is replaced by the hypothesis that $T \in L(E)$ and T^n is compact for some positive integer n. (Use Exercise 6.6(1) and observe that only the proofs of Lemma 6.7.1 and Theorem 6.7.5 require modifications.)

6.8 Fredholm Integral Equations

The theory of Section 6.6 was developed by F. Riesz as a tool for the study of linear integral equations. We shall use it to obtain a famous theorem of Fredholm (1903). We need a lemma on the change of the order of integration in a repeated integral.

6.8.1 Lemma *Let k be a complex-valued continuous function on the square $S = \{(s,t)\colon a \le s \le b,\ a \le t \le b\}$. Then*

$$\int_a^b \left(\int_a^b k(s,t)\, ds \right) dt = \int_a^b \left(\int_a^b k(s,t)\, dt \right) ds.$$

PROOF We have seen in Theorem 6.1.4 that the functions

$$t \to \int_a^b k(s,t)\, ds \quad \text{and} \quad s \to \int_a^b k(s,t)\, dt$$

are continuous on $[a,b]$; consequently the repeated integrals are defined. Let $I = \int_a^b \left(\int_a^b k(s,t)\, ds \right) dt$ and $J = \int_a^b \left(\int_a^b k(s,t)\, dt \right) ds$. By Theorem 4.3.8 the function k is uniformly continuous on S so for each $\varepsilon > 0$ there exists $\delta > 0$ such that $| k(s,t) - k(s',t') | < \varepsilon$ for all $(s,t),(s',t') \in S$ with $((s - s')^2 + (t - t')^2)^{1/2} < \delta$. Choose a positive integer n with $n^{-1}(b - a) < \delta/\sqrt{2}$ and let $s_j = t_j = a + jn^{-1}(b - a)$ for $j = 0, 1, 2,\ldots, n + 1$. Let $k_n(s,t) = k(s_j,t_k)$ for $s_j \leq s < s_{j+1}$, $t_k \leq t < t_{k+1}$ and $j,k = 0, 1, 2,\ldots, n$. Then it is clear that

$$\sup \{ | k(s,t) - k_n(s,t) | : (s,t) \in S\} \leq \varepsilon.$$

For each $t \in [a,b]$, the function $s \to k_n(s,t)$ is a step-function on $[a,b]$ and

$$\int_a^b k_n(s,t)\, ds = \sum_{j=0}^{n-1} k(s_j,t_k)(s_{j+1} - s_j)$$

for $t_k \leq t < t_{k+1}$ and $k = 0, 1, 2,\ldots, n$. Therefore the function

$$t \to \int_a^b k_n(s,t)\, ds$$

is a step-function on $[a,b]$ and

$$\int_a^b \left(\int_a^b k_n(s,t)\, ds \right) dt = \sum_{k=0}^{n-1} \sum_{j=0}^{n-1} k(s_j,t_k)(s_{j+1} - s_j)(t_{k+1} - t_k).$$

Let I_n denote the right-hand side of this equation. Then

$$\begin{aligned} | I - I_n | &= \left| \int_a^b \left(\int_a^b (k(s,t) - k_n(s,t))\, ds \right) dt \right| \\ &\leq (b - a) \sup \left\{ \left| \int_a^b (k(s,t) - k_n(s,t))\, ds \right| : a \leq t \leq b \right\} \\ &\leq (b - a)^2 \sup \{ | k(s,t) - k_n(s,t) | : (s,t) \in S\} \\ &\leq \varepsilon(b - a)^2. \end{aligned}$$

Similarly we can prove that $| J - I_n | \leq \varepsilon(b - a)^2$. Therefore $| I - J | \leq 2\varepsilon(b - a)^2$. This shows that $I = J$. □

The reader will find an outline of an alternative proof of the above lemma in Exercise 4.3(13).

6.8.2 THEOREM (the Fredholm alternative) *Let k be a complex-valued continuous function on the square $\{(s,t): a \leq s \leq b,\ a \leq t \leq b\}$ and let μ be a non-zero complex number. The following alternative holds: either* (a) *each of the equations*

$$f(s) = g(s) + \mu \int_a^b k(s,t) f(t)\, dt,\ a \leq s \leq b, \tag{1}$$

and

$$f(s) = g(s) + \mu \int_a^b k(t,s) f(t)\, dt,\ a \leq s \leq b, \tag{2}$$

has a unique solution $f \in C_{\mathbf{C}}([a,b])$ *for each* $g \in C_{\mathbf{C}}([a,b])$,

or (b) *the two homogeneous equations*

$$f(s) = \mu \int_a^b k(s,t) f(t)\, dt,\ a \leq s \leq b, \tag{3}$$

and

$$f(s) = \mu \int_a^b k(t,s) f(t)\, dt,\ a \leq s \leq b, \tag{4}$$

have non-zero solutions $f \in C_{\mathbf{C}}([a,b])$.

If (b) *holds then the equations* (3) *and* (4) *have the same number of linearly independent solutions in* $C_{\mathbf{C}}([a,b])$; *equation* (1) *has solutions* $f \in C_{\mathbf{C}}([a,b])$ *if and only if*

$$\int_a^b g(s) h(s)\, ds = 0$$

for all solutions $h \in C_{\mathbf{C}}([a,b])$ *of* (4); *and equation* (2) *has solutions* $f \in C_{\mathbf{C}}([a,b])$ *if and only if*

$$\int_a^b g(s) h(s)\, ds = 0$$

for all solutions $h \in C_{\mathbf{C}}([a,b])$ *of* (3).

Finally, the set of complex numbers μ *for which* (b) *holds is a countable set with no accumulation points.*

PROOF We shall write $C = C_{\mathbf{C}}([a,b])$. Let K and K' be the Fredholm integral operators on C defined by

$$(Kf)(s) = \int_a^b k(s,t) f(t)\, dt$$

and

$$(K'f)(s) = \int_a^b k(t,s) f(t)\, dt,$$

for $a \leq s \leq b$ and $f \in C$. By Theorem 6.5.9 the operators K and K' are compact. Equations (1), (2), (3) and (4) can be written in the form

$$(I - \mu K) f = g, \tag{5}$$

$$(I - \mu K') f = g, \tag{6}$$

$$(I - \mu K) f = 0, \tag{7}$$

and

$$(I - \mu K') f = 0. \tag{8}$$

The theorem does not follow directly from Theorems 6.6.13 and 6.6.14 (applied to μK) because K' is not the adjoint of K. However K' and K^*

are closely related. To see this we first observe that C can be embedded in C^* as follows. For each $g \in C$ let Jg be the functional on C defined by

$$(Jg)(f) = \int_a^b g(s)f(s)\,ds$$

for all $f \in C$. It is easy to verify that $Jg \in C^*$ and that $g \to Jg$ is a one-to-one linear mapping of C into C^*. We shall prove that $K^* \circ J = J \circ K'$. For all $f,g \in C$ Lemma 6.8.1 gives

$$\begin{aligned}(K^*(Jg))(f) &= (Jg)(Kf)\\ &= \int_a^b g(s)(Kf)(s)\,ds\\ &= \int_a^b g(s)\left(\int_a^b k(s,t)f(t)\,dt\right)ds\\ &= \int_a^b f(t)\left(\int_a^b k(s,t)g(s)\,ds\right)dt\\ &= \int_a^b f(t)(K'g)(t)\,dt\\ &= (J(K'g))(f),\end{aligned}$$

from which it follows that $K^* \circ J = J \circ K'$.

We have now $(I^* - \mu K^*) \circ J = J \circ (I - \mu K')$ and hence, since J is one-to-one,

$$J(\mathcal{N}(I - \mu K')) = \mathcal{N}(I^* - \mu K^*) \cap J(C) \tag{9}$$

from which it follows, again since J is one-to-one, that

$$\begin{aligned}\dim \mathcal{N}(I - \mu K') &= \dim J(\mathcal{N}(I - \mu K'))\\ &= \dim (\mathcal{N}(I^* - \mu K^*) \cap J(C))\\ &\leq \dim \mathcal{N}(I^* - \mu K^*)\\ &= \dim \mathcal{N}(I - \mu K),\end{aligned} \tag{10}$$

where the last step follows from Theorem 6.6.13. Since the relationship between K and K' is symmetrical we can interchange K and K' in (10) and therefore

$$\dim \mathcal{N}(I - \mu K') = \dim \mathcal{N}(I - \mu K). \tag{11}$$

Since μK and $\mu K'$ are compact operators the alternative (a) or (b) follows from (5), (6), (7), (8), (11), Theorem 6.6.7 and Corollary 6.6.8.

Next, (10) and (11) give

$$\dim (\mathcal{N}(I^* - \mu K^*) \cap J(C)) = \dim \mathcal{N}(I^* - \mu K^*)$$

and consequently $\mathcal{N}(I^* - \mu K^*) \subseteq J(C)$. But then (9) gives

$$J(\mathcal{N}(I - \mu K')) = \mathcal{N}(I^* - \mu K^*) \tag{12}$$

and, interchanging K and K', we obtain

$$J(\mathcal{N}(I - \mu K)) = \mathcal{N}(I^* - \mu K'^*). \tag{13}$$

The assertions about the existence of solutions of (1) and (2) when (b) holds follow from (5), (6), (7), (8), (12), (13) and Theorem 6.6.14.

Finally, the set of complex numbers μ for which (b) holds is precisely the set of reciprocals of the non-zero eigenvalues of the compact linear operator K, and is therefore, by Theorem 6.7.6, a countable set with no accumulation points. □

EXERCISES

(1) Let k be a complex-valued continuous function on the square $S = \{(s,t): a \leq s \leq b, a \leq t \leq b\}$. Prove that for each $\varepsilon > 0$ there exists a function p_ε of the form $p_\varepsilon(s,t) = \sum_{n,m=0}^{N} \alpha_{mn} s^m t^n$ such that $| p_\varepsilon(s,t) - k(s,t) | < \varepsilon$ for all $(s,t) \in S$.

(2) Let K be a Fredholm integral operator on $C_{\mathrm{C}}([a,b])$. Prove that there is a sequence (K_n) of Fredholm integral operators on $C_{\mathrm{C}}([a,b])$ such that $\lim_{n\to\infty} K_n = K$ and $\mathscr{R}(K_n)$ is finite-dimensional for $n = 1, 2, \ldots$. Hence give an alternative proof of Theorem 6.5.9. It is possible to base a proof of Theorem 6.8.2 on this result. (cf. RIESZ and NAGY [23, §§ 69–73].) It is an open question whether each compact linear operator on a Banach space can be approximated by a sequence of operators each with finite-dimensional range.

(3) Use (1) to give an alternative proof of Lemma 6.8.1.

CHAPTER 7

The Fréchet Differential Calculus

We shall be concerned with a generalization, due to M. Fréchet (1925), of the classical differential calculus of real-valued functions of a real variable. We recall that a real-valued function f on R has derivative m at a point a of R if and only if for each $\varepsilon > 0$ there exists $\delta > 0$ such that

$$\left| \frac{f(x) - f(a)}{x - a} - m \right| \leq \varepsilon \tag{*}$$

whenever $0 < | x - a | < \delta$. The inequality (*) can be replaced by the equivalent inequality

$$| f(x) - f(a) - m(x - a) | \leq \varepsilon \, | x - a | \tag{**}$$

whenever $| x - a | < \delta$.

Fréchet's generalization of the differential calculus applies to (non-linear) mappings of a real normed linear space E into a real normed linear space F. Let f be such a mapping. The derivative of f at a point a of E will be defined to be a *linear transformation* T of E into F which satisfies the inequality

$$\| f(x) - f(a) - T(x - a) \| \leq \varepsilon \, \| x - a \| \tag{***}$$

whenever $\| x - a \| < \delta$. It is obvious that (***) is a generalization of (**); and the analogy between (***) and (**) becomes clearer when we remark that the mapping $y \to my$ is a linear mapping of R into itself.

In the case when $E = \mathsf{R}^n$ and $F = \mathsf{R}$ the Fréchet differential calculus yields all the results of the classical differential calculus of real-valued functions of n real variables in a notation which is considerably simpler than the classical notation. This is the most important case of the general theory but, as no significant simplification of the proofs can be obtained in this special case, we shall develop the theory in the general (possibly infinite-dimensional) case. There are important applications of the Fréchet differential calculus in infinite-dimensional contexts. (See for instance, Exercise 7.5(8).)

Perhaps the most useful theorem of the classical differential calculus is the Mean Value Theorem. In the Fréchet differential calculus there is no complete analogue of this theorem, but the inequality obtained in Section 7.3 is, for most purposes, an adequate substitute for it.

In Section 7.5 we shall apply the Fréchet differential calculus to obtain generalizations of the important Implicit Function and Inverse Function Theorems of classical analysis.

We shall not touch on the question of higher derivatives in the Fréchet differential calculus, because to do so would involve too much preliminary material. The reader will find a discussion of higher derivatives in DIEUDONNÉ [7, pp. 174–86].

7.1 The Fréchet Derivative

Throughout this section E and F will denote real normed linear spaces, A will denote a non-empty open subset of E and f will denote a mapping of A into F.

7.1.1 DEFINITION The mapping f is said to be *differentiable at a point* $a \in A$ if and only if there is a linear mapping T of E into F which satisfies the following condition: for each $\varepsilon > 0$ there exists $\delta > 0$ such that

$$\| f(x) - f(a) - T(x - a) \| \leq \varepsilon \| x - a \| \tag{1}$$

for all $x \in A$ with $\| x - a \| < \delta$. The mapping f is said to be *differentiable on* A if and only if it is differentiable at each point of A.

The reader should notice that in this definition we do not require the linear transformation T to be bounded.

Since A is open we have $B(a,\delta) \subseteq A$ for all sufficiently small $\delta > 0$; we shall usually suppose without explicit mention that the δ in Definition 7.1.1 has been chosen so that this is true.

7.1.2 LEMMA *If f is differentiable at a point $a \in A$ then there is a unique linear transformation of E into F which satisfies condition* (1).

PROOF Suppose that T and S are linear transformations of E into F that both satisfy condition (1) and let $\varepsilon > 0$. There exist $\delta_1 > 0$ and $\delta_2 > 0$ such that

$$\| f(x) - f(a) - T(x - a) \| \leq \varepsilon \| x - a \|$$

for $\| x - a \| < \delta_1$, and

$$\| f(x) - f(a) - S(x - a) \| \leq \varepsilon \| x - a \|$$

for $\| x - a \| < \delta_2$. Let $\delta = \min\{\delta_1, \delta_2\}$ and $\| x - a \| < \delta$. Then

$$\begin{aligned}\| (T - S)(x - a) \| &= \| T(x - a) - S(x - a) \| \\ &\leq \| T(x - a) - f(x) + f(a) \| + \| f(x) - f(a) - S(x - a) \| \\ &\leq 2\varepsilon \| x - a \|.\end{aligned}$$

If $y \in E$ and $y \neq 0$ then $\| (\frac{1}{2}\delta \| y \|^{-1} y + a) - a \| < \delta$ and so

$$\left\| (T - S)\left(\frac{\delta}{2 \| y \|} y\right) \right\| \leq 2\varepsilon \left\| \frac{\delta}{2 \| y \|} y \right\| = \delta\varepsilon.$$

This gives $\| (T - S)y \| \leq 2\varepsilon \| y \|$. The last inequality is clearly true when $y = 0$ and hence $T - S \in L(E,F)$ and $\| T - S \| \leq 2\varepsilon$. Since $\varepsilon > 0$ is arbitrary $\| T - S \| = 0$ and so $T = S$. □

7.1.3 Definition Let f be differentiable at a point $a \in A$. The unique linear transformation of E into F which satisfies condition (1) is called the *derivative of f at a* and is denoted by $f'(a)$.

The linear transformation $f'(a)$ is often called the *Fréchet derivative* or *Fréchet differential* of f at a. To simplify our notation we shall write $f'(a) \cdot x$ instead of $(f'(a))x$.

7.1.4 Theorem *Let f be differentiable at a point $a \in A$. Then $f'(a) \in L(E,F)$ if and only if f is continuous at a.*

Proof Suppose first that f is continuous at a and let $\varepsilon > 0$. There exist $\delta_1 > 0$ and $\delta_2 > 0$ such that

$$\| f(x) - f(a) \| < \tfrac{1}{2}\varepsilon$$

for $\| x - a \| < \delta_1$, and

$$\| f(x) - f(a) - f'(a) \cdot (x - a) \| \leq \tfrac{1}{2}\varepsilon \| x - a \|$$

for $\| x - a \| < \delta_2$. Let $\delta = \min\{\delta_1, \delta_2, 1\}$ and $\| x - a \| < \delta$. Then

$$\begin{aligned}\| f'(a) \cdot (x - a) \| &\leq \| f'(a) \cdot (x - a) - f(x) + f(a) \| + \| f(x) - f(a) \| \\ &< \tfrac{1}{2}\varepsilon \| x - a \| + \tfrac{1}{2}\varepsilon \\ &< \varepsilon.\end{aligned}$$

This shows that $f'(a)$ is continuous at a. It now follows from Lemma 3.5.2 and Theorem 3.5.4 that $f'(a) \in L(E,F)$.

Suppose conversely that $f'(a) \in L(E,F)$. Then there exists $\delta > 0$ such that

$$\| f(x) - f(a) - f'(a) \cdot (x - a) \| \leq \| x - a \|$$

for $\| x - a \| < \delta$. Let $\| x - a \| < \delta$. Then, using inequality (5) of Section 3.5, we obtain

$$\begin{aligned}\| f(x) - f(a) \| &\leq \| f(x) - f(a) - f'(a)\cdot(x - a) \| + \| f'(a)\cdot(x - a) \| \\ &\leq \| x - a \| + \| f'(a) \| \; \| x - a \| \\ &= (1 + \| f'(a) \|) \| x - a \|\end{aligned}$$

and it follows directly from this inequality that f is continuous at a. □

7.1.5 COROLLARY *Suppose that E is finite-dimensional. Then if f is differentiable at a point $a \in A$ it is continuous at a.*

PROOF Theorem 6.1.2 shows that $f'(a)$, which is a linear transformation of E into F, must be bounded. □

The condition that E be finite-dimensional cannot be dropped from the above corollary. (See Exercise (3).)

7.1.6 DEFINITION Let f be continuous and differentiable on A. Then f is said to be *continuously differentiable on A* if and only if $a \to f'(a)$ is a continuous mapping of A into $L(E,F)$. The mapping $a \to f'(a)$ will be denoted by f', and f' will be called the *derivative of f on A.*

The rest of this section is devoted to a discussion of the Fréchet derivative in some of the special cases that are important in applications.

7.1.7 LEMMA *Let $y \in F$ and suppose that $f(x) = y$ for all $x \in E$. Then f is continuously differentiable on E and $f'(a) = 0$ for all $a \in E$.*

PROOF $f(x) - f(a) = 0$ for all $x,a \in E$. □

7.1.8 LEMMA *Let T be a linear transformation of E into F. Then T is differentiable on E and $T'(a) = T$ for all $a \in E$. Consequently, if T is a bounded linear transformation then T is continuously differentiable on E.*

PROOF $Tx - Ta - T(x - a) = 0$ for all $x,a \in E$. □

In the case when $E = \mathsf{R}$ we now have two, apparently different, concepts of differentiability and consequently of derivative: that of Definition 3.9.2 and that of Definition 7.1.1. The symbol $f'(a)$ has been used to denote both of these derivatives. To avoid confusion we shall adopt the following convention in this chapter: we shall denote the Fréchet derivative of f at a by $f'(a)$ and we shall denote the derivative of f at a in the sense of Definition 3.9.2 by $Df(a)$. The following theorem reconciles the two definitions of differentiability.

7.1.9 THEOREM *Suppose that $E = \mathsf{R}$. Then f is differentiable at a point $a \in A$ in the sense of Definition* 7.1.1 *if and only if it is differentiable* *at a in the sense of Definition* 3.9.2. *Further*

$$f'(a)\cdot x = xDf(a)$$

for all $x \in \mathsf{R}$ whenever one of the derivatives exists.

PROOF Suppose first that f is differentiable at a in the sense of Definition 7.1.1. Let $T = f'(a)$ and let $y = T1$. Given $\varepsilon > 0$ there exists $\delta > 0$ such that

$$\| f(x) - f(a) - T(x - a) \| \leq \varepsilon \mid x - a \mid \tag{2}$$

for $\mid x - a \mid < \delta$. Since

$$T(x - a) = T((x - a)1) = (x - a)T1 = (x - a)y$$

inequality (2) gives

$$\left\| \frac{1}{x - a}(f(x) - f(a)) - y \right\| \leq \varepsilon \tag{3}$$

for $0 < \mid x - a \mid < \delta$. This shows that f is differentiable at a in the sense of Definition 3.9.2 and that $Df(a) = y$. It follows that $f'(a)\cdot x = xf'(a)\cdot 1 = xDf(a)$ for all $x \in \mathsf{R}$.

Suppose conversely that f is differentiable at a in the sense of Definition 3.9.2. Let $y = Df(a)$ and let $Tx = xy$ for all $x \in \mathsf{R}$. It is clear that T is a linear transformation of R into F. Given $\varepsilon > 0$ there exists $\delta > 0$ such that (3) holds. Consequently (2) also holds. This shows that f is differentiable at a in the sense of Definition 7.1.1. □

The reader will notice that when $E = F = \mathsf{R}$ (that is when f is a real-valued function of a real variable) the number $f'(a)\cdot 1$ is simply the derivative of f at a in the classical sense.

The rest of this section is devoted to a study of the Fréchet derivative for mappings of R^n into R^m. It is convenient to use the norm on R^n given by

$$\| x \| = \max \{ \mid \xi_1 \mid, \mid \xi_2 \mid, \ldots, \mid \xi_n \mid \}$$

for $x = (\xi_1, \xi_2, \ldots, \xi_n) \in \mathsf{R}^n$. However it is not difficult to see that the results are independent of the choice of the norm on R^n. (See Exercise (1).)

The following theorem allows us to reduce the study of mappings of E into R^m to that of real-valued functions on E.

7.1.10 THEOREM *Suppose that $F = \mathsf{R}^m$ and let $f_1, f_2, \ldots, f_m$ be the real-valued functions on A defined by*

$$f(x) = (f_1(x), f_2(x), \ldots, f_m(x))$$

for all $x \in A$. Then f is differentiable at a point $a \in A$ if and only if $f_1, f_2, \ldots, f_m$ are differentiable at a. If f is differentiable at a then

$$f'(a)\cdot x = (f_1'(a)\cdot x, f_2'(a)\cdot x, \ldots, f_m'(a)\cdot x)$$

for all $x \in E$. Further f is continuously differentiable on A if and only if $f_1, f_2, \ldots, f_m$ are continuously differentiable on A.

PROOF Suppose first that $f_1, f_2, \ldots, f_m$ are differentiable at a and let

$$Tx = (f_1'(a)\cdot x,\ f_2'(a)\cdot x, \ldots, f_m'(a)\cdot x)$$

for all $x \in E$. It is clear that T is a linear mapping of E into R^m. Also, for all $x \in A$, we have

$$\begin{aligned}\| f(x) - f(a) - T(x-a) \| \\ &= \max \{ | f_j(x) - f_j(a) - f_j'(a)\cdot(x-a) | : j = 1, 2, \ldots, m\}\end{aligned}$$

and so f is differentiable at a and

$$f'(a)\cdot x = (f_1'(a)\cdot x,\ f_2'(a)\cdot x, \ldots, f_m'(a)\cdot x)$$

for all $x \in E$.

Suppose conversely that f is differentiable at a and let

$$f'(a)\cdot x = (T_1 x,\ T_2 x, \ldots, T_m x)$$

for all $x \in E$. It is clear that $T_1, T_2, \ldots, T_m$ are linear mappings of E into R. Also, for all $x \in A$ and $j = 1, 2, \ldots, m$, we have

$$| f_j(x) - f_j(a) - T_j(x-a) | \le \| f(x) - f(a) - f'(a)\cdot(x-a) \|$$

and so $f_1, f_2, \ldots, f_m$ are differentiable at a.

Finally we have

$$\| (f'(a) - f'(b))\cdot x \| = \max \{ | (f_j'(a) - f_j'(b))\cdot x | : j = 1, 2, \ldots, m\}$$

for all $a,b \in A$ and $x \in E$ from which it follows that

$$\begin{aligned}\| f_k'(a) - f_k'(b) \| &\le \| f'(a) - f'(b) \| \\ &\le \max \{ \| f_j'(a) - f_j'(b) \| : j = 1, 2, \ldots, m\}\end{aligned}$$

for all $a,b \in A$ and $k = 1, 2, \ldots, m$. This shows that f is continuously differentiable on A if and only if $f_1, f_2, \ldots, f_m$ are continuously differentiable on A. □

In particular, if we set $E = \mathsf{R}$ in Theorem 7.1.10 and use Theorem 7.1.9, we see that a mapping f of R into R^m is differentiable at a point a of R if and only if the real-valued functions $f_1, f_2, \ldots, f_m$ on R are differentiable at a in the classical sense. Further we have

$$\begin{aligned}Df(a) &= f'(a)\cdot 1 \\ &= (f_1'(a)\cdot 1,\ f_2'(a)\cdot 1, \ldots, f_m'(a)\cdot 1) \\ &= (Df_1(a),\ Df_2(a), \ldots, Df_m(a)).\end{aligned}$$

In the case when $E = \mathsf{R}^n$ and $F = \mathsf{R}^m$ the Fréchet derivative $f'(a)$ is a linear transformation of R^n into R^m and hence can be represented by a matrix when bases for R^n and R^m have been chosen (Theorem 6.1.1). In this chapter we shall always use the basis $\{e_1, e_2, \ldots, e_n\}$ for R^n given by

$$\begin{aligned} e_1 &= (1, 0, 0, \ldots, 0) \\ e_2 &= (0, 1, 0, \ldots, 0) \\ &\cdots\cdots \\ e_n &= (0, 0, 0, \ldots, 1). \end{aligned}$$

Let $\{e_1', e_2', \ldots, e_m'\}$ be the analogous basis for R^m. Then $f'(a)$ is represented by the $m \times n$ matrix (τ_{jk}) of real numbers where

$$f'(a)\cdot e_k = \sum_{j=1}^{m} \tau_{jk} e_j' \tag{4}$$

for $k = 1, 2, \ldots, n$ (see Theorem 6.1.1). We begin by calculating the numbers τ_{jk}.

7.1.11 THEOREM *Suppose that $E = \mathsf{R}^n$ and $F = \mathsf{R}^m$ and let $f_1, f_2, \ldots, f_m$ be the real-valued functions on A defined by*

$$f(x) = (f_1(x), f_2(x), \ldots, f_m(x))$$

for all $x \in A$. Suppose also that f is differentiable at a point $a \in A$. Then $f_1, f_2, \ldots, f_m$ have first partial derivatives at a and the matrix which represents $f'(a)$ is†

$$\begin{pmatrix} D_1f_1(a) & D_2f_1(a) \ldots & D_nf_1(a) \\ D_1f_2(a) & D_2f_2(a) \ldots & D_nf_2(a) \\ \cdot\;\cdot\;\cdot & \cdot\;\cdot\;\cdot & \cdot\;\cdot\;\cdot \\ D_1f_m(a) & D_2f_m(a) \ldots & D_nf_m(a) \end{pmatrix}.$$

PROOF Suppose first that $m = 1$. Let $(\tau_1, \tau_2, \ldots, \tau_n)$ be the matrix representing $f'(a)$. Then

$$f'(a)\cdot x = \tau_1\xi_1 + \tau_2\xi_2 + \ldots + \tau_n\xi_n \tag{5}$$

for all $x = (\xi_1, \xi_2, \ldots, \xi_n) \in \mathsf{R}^n$. Given $\varepsilon > 0$ there exists $\delta > 0$ such that

$$| f(x) - f(a) - f'(a)\cdot(x - a) | \leq \varepsilon \| x - a \|$$

for $\| x - a \| < \delta$ and so (5) gives

$$\left| f(x) - f(a) - \sum_{j=1}^{n} \tau_j(\xi_j - \alpha_j) \right| \leq \varepsilon \| x - a \| \tag{6}$$

for $\| x - a \| < \delta$, where $a = (\alpha_1, \alpha_2, \ldots, \alpha_n)$. Let $1 \leq k \leq n$, $t \in \mathsf{R}$ and $x_k = (\xi_{1k}, \xi_{2k}, \ldots, \xi_{nk})$ where $\xi_{jk} = \alpha_j$ for $j \neq k$ and $\xi_{kk} = t$. Then $\| x_k - a \| = | t - \alpha_k |$, and so, if $| t - \alpha_k | < \delta$, we obtain from (6)

$$| f(x_k) - f(a) - \tau_k(t - \alpha_k) | \leq \varepsilon | t - \alpha_k |.$$

† The real number $D_kf_j(a)$ is the partial derivative of f_j with respect to the kth variable at a in the usual sense of elementary calculus. The notations $\dfrac{\partial f_j}{\partial \xi_k}(a)$ and $(f_j)_{\xi^k}(a)$ are often used instead of $D_kf_j(a)$; they have obvious disadvantages, and we shall avoid them here.

Consequently

$$\left| \frac{f(\alpha_1, \ldots, \alpha_{k-1}, t, \alpha_{k+1}, \ldots, \alpha_n) - f(\alpha_1, \ldots, \alpha_{k-1}, \alpha_k, \alpha_{k+1}, \ldots, \alpha_n)}{t - \alpha_k} - \tau_k \right| \leq \varepsilon$$

for $0 < |t - \alpha_k| < \delta$. This shows that f is differentiable with respect to the kth variable at a, and that $D_k f(a) = \tau_k$. We have now proved the theorem in the case when $m = 1$.

Consider now the general case. By Theorem 7.1.10 the real-valued functions $f_1, f_2, \ldots, f_m$ are (Fréchet) differentiable at a and therefore, by what we have proved above, f_j is differentiable with respect to the kth variable at a for $j = 1, 2, \ldots, m$ and $k = 1, 2, \ldots, n$. It remains only to identify the matrix (τ_{jk}) which represents $f'(a)$.

It follows easily from (4) that

$$f'(a) \cdot x = \left(\sum_{k=1}^{n} \tau_{1k}\xi_k, \sum_{k=1}^{n} \tau_{2k}\xi_k, \ldots, \sum_{k=1}^{n} \tau_{mk}\xi_k \right) \tag{7}$$

for all $x = (\xi_1, \xi_2, \ldots, \xi_n) \in \mathsf{R}^n$. Also Theorem 7.1.10 gives

$$f'(a) \cdot x = (f_1'(a) \cdot x, f_2'(a) \cdot x, \ldots, f_m'(a) \cdot x) \tag{8}$$

for all $x \in \mathsf{R}^n$. Finally by (5) and the first part of the proof we have

$$f_j'(a) \cdot x = \sum_{k=1}^{n} \xi_k D_k f_j(a) \tag{9}$$

for all $x = (\xi_1, \xi_2, \ldots, \xi_n) \in \mathsf{R}^n$. From (7), (8) and (9) we obtain

$$\sum_{k=1}^{n} \tau_{jk}\xi_k = f_j'(a) \cdot x = \sum_{k=1}^{n} \xi_k D_k f_j(a)$$

for $j = 1, 2, \ldots, m$ and all $x = (\xi_1, \xi_2, \ldots, \xi_n) \in \mathsf{R}^n$. Consequently $\tau_{jk} = D_k f_j(a)$ for $j = 1, 2, \ldots, m$ and $k = 1, 2, \ldots, n$. □

The matrix which represents $f'(a)$ is called the *Jacobian matrix of the mapping* f. When $n = m$ the determinant of the Jacobian matrix of f is called the *Jacobian of* f; it is usually denoted by

$$\frac{\partial(f_1, f_2, \ldots, f_n)}{\partial(\xi_1, \xi_2, \ldots, \xi_n)}.$$

When f is a real-valued function of n real variables, we have

$$f'(a) \cdot x = \xi_1 D_1 f(a) + \xi_2 D_2 f(a) + \ldots + \xi_n D_n f(a) \tag{10}$$

for all $x = (\xi_1, \xi_2, \ldots, \xi_n) \in \mathsf{R}^n$. In this case the concept of Fréchet differentiability coincides with the classical concept of total differentiability. The function $(a,x) \to f'(a) \cdot x$ is the classical *total derivative* of f and is denoted by df (see GRAVES [10, pp. 75–7]). In order to rewrite the relation (10) between the total and partial derivatives of f in its traditional form, we have to adopt the traditional practice of allowing one symbol to denote sometimes a function, and sometimes one of the values of the function. If we let ξ_j denote both the jth coordinate of x and the 'coordinate function' $x \to \xi_j$ then the function ξ_j is Fréchet differentiable, and equation (10) gives the value, $d\xi_j$, of its total derivative:

$$d\xi_j = \xi_j.$$

We now obtain from equation (10) the classical formula

$$df = d\xi_1\, D_1 f(a) + d\xi_2\, D_2 f(a) + \ldots + d\xi_n\, D_n f(a)$$

which is usually written in the form

$$df = \frac{\partial f}{\partial \xi_1}\, d\xi_1 + \frac{\partial f}{\partial \xi_2}\, d\xi_2 + \ldots + \frac{\partial f}{\partial \xi_n}\, d\xi_n$$

and which expresses the relationship between the (values of) the total derivative df of f and the total derivatives $d\xi_1, d\xi_2, \ldots, d\xi_n$ of the coordinate functions $\xi_1, \xi_2, \ldots, \xi_n$ in terms of the partial derivatives of f.

The reader should notice that the existence of the partial derivatives $D_k f_j(a)$ for $j = 1, 2, \ldots, m$ and $k = 1, 2, \ldots, n$ is not sufficient to ensure the existence of the Fréchet derivative $f'(a)$. (See Exercises (10) and (12).) The following theorem provides a sufficient condition for the existence of the Fréchet derivative in terms of the partial derivatives.

7.1.12 THEOREM *Suppose that* $E = \mathsf{R}^n$ *and* $F = \mathsf{R}^m$ *and let* $f_1, f_2, \ldots, f_m$ *be the real-valued functions on* A *defined by*

$$f(x) = (f_1(x), f_2(x), \ldots, f_m(x))$$

for all $x \in A$. *Then* f *is continuously differentiable on* A *if and only if* $f_1, f_2, \ldots, f_m$ *have continuous first partial derivatives on* A.

PROOF It is clear from Theorem 7.1.10 that we may suppose that $m = 1$ (that is that f is real-valued).

Suppose first that f is continuously differentiable on A. By Theorem 7.1.11 the function f has first partial derivatives on A. Also for $a,b \in A$ and $k = 1, 2, \ldots, n$ we have, by (10),

$$(f'(a) - f'(b))\cdot e_k = (0, \ldots, 0, D_k f(a) - D_k f(b), 0, \ldots, 0),$$

and consequently

$$|D_k f(a) - D_k f(b)| = \|(f'(a) - f'(b))\cdot e_k\| \leq \|f'(a) - f'(b)\|.$$

This proves that $D_k f$ is continuous on A.

Suppose conversely that f has continuous first partial derivatives on A. Let $a = (\alpha_1, \alpha_2, \ldots, \alpha_n) \in A$ and let

$$Tx = \xi_1 D_1 f(a) + \xi_2 D_2 f(a) + \ldots + \xi_n D_n f(a)$$

for all $x = (\xi_1, \xi_2, \ldots, \xi_n) \in \mathsf{R}^n$. It is clear that T is a linear mapping of R^n into R. We shall prove that T is the (Fréchet) derivative of f at a.

Let $\varepsilon > 0$. Since the functions $D_1 f, D_2 f, \ldots, D_n f$ are continuous at a and A is open there exists $\delta > 0$ such that $B(a,\delta) \subseteq A$ and

$$| D_j f(x) - D_j f(a) | < \frac{\varepsilon}{n} \tag{11}$$

for $x \in B(a,\delta)$ and $j = 1, 2, \ldots, n$. Let $x = (\xi_1, \xi_2, \ldots, \xi_n)$ be any point of $B(a,\delta)$. Then we have

$$f(x) - f(a) = \sum_{j=1}^{n} (f(x_j) - f(x_{j-1})), \tag{12}$$

where $x_0 = a$, $x_n = x$ and $x_j = (\xi_1, \ldots, \xi_j, \alpha_{j+1}, \ldots, \alpha_n)$ for $j = 1, 2, \ldots, n - 1$. (Observe that $\| x_j - a \| \leq \| x - a \|$ so that $x_j \in B(a,\delta) \subseteq A$ for $j = 0, 1, 2, \ldots, n$.) By hypothesis, for $j = 1, 2, \ldots, n$, the function $t \to f(\xi_1, \ldots, \xi_{j-1}, t, \alpha_{j+1}, \ldots, \alpha_n)$ is defined and differentiable on the open interval $(\alpha_j - \delta, \alpha_j + \delta)$ so we can apply the classical mean value theorem to this function to obtain

$$f(x_j) - f(x_{j-1}) = (\xi_j - \alpha_j) D_j f(a_j) \tag{13}$$

where $a_j = (\xi_1, \ldots, \xi_{j-1}, t_j, \alpha_{j+1}, \ldots, \alpha_n)$ and t_j lies between α_j and ξ_j. Clearly $\| a_j - a \| \leq \| x - a \| < \delta$, so from (11), (12) and (13) we obtain

$$\begin{aligned}
| f(x) - f(a) - T(x - a) | \\
&= \Big| \sum_{j=1}^{n} (f(x_j) - f(x_{j-1})) - \sum_{j=1}^{n} (\xi_j - \alpha_j) D_j f(a) \Big| \\
&= \Big| \sum_{j=1}^{n} (\xi_j - \alpha_j)(D_j f(a_j) - D_j f(a)) \Big| \\
&\leq \frac{\varepsilon}{n} \sum_{j=1}^{n} | \xi_j - \alpha_j | \\
&\leq \varepsilon \| x - a \|.
\end{aligned}$$

Since x is an arbitrary point of $B(a,\delta)$ this proves that f is differentiable at a.

Finally let $a,b \in A$. By (10)

$$\begin{aligned}
| (f'(a) - f'(b)) \cdot x | &\leq \sum_{j=1}^{n} | \xi_j | \; | D_j f(a) - D_j f(b) | \\
&\leq \| x \| \sum_{j=1}^{n} | D_j f(a) - D_j f(b) |
\end{aligned}$$

for all $x = (\xi_1, \xi_2, \ldots, \xi_n) \in \mathsf{R}^n$. Consequently

$$\| f'(a) - f'(b) \| \leq \sum_{j=1}^{n} | D_j f(a) - D_j f(b) |.$$

This proves that f' is continuous on A. □

We end this section with some remarks about Fréchet derivatives of mappings defined on complex normed linear spaces. It is convenient first to introduce some terminology. (cf. p. 186.) Let T be a mapping of a complex linear space E into a complex linear space F. We shall say that T is a *complex linear mapping* if and only if $T(\alpha x + \beta y) = \alpha Tx + \beta Ty$ for all $x,y \in E$ and all $\alpha,\beta \in \mathsf{C}$; and we shall say that T is a *real linear mapping* if and only if $T(\alpha x + \beta y) = \alpha Tx + \beta Ty$ for all $x,y \in E$ and all $\alpha,\beta \in \mathsf{R}$. It is clear that each complex linear mapping is also a real linear mapping; the converse is, in general, false. (See Exercise (13).)

Now let f be a mapping of a non-empty open subset A of a complex normed linear space E into a complex normed linear space F. We shall say that f has a *complex derivative* at a point $a \in A$ if and only if there is a complex linear mapping T of E into F, which will be called the *complex derivative* of f at a, that satisfies the condition (1). We shall say that f has a *real derivative* at a point $a \in A$ if and only if there is a real linear mapping T of E into F, which will be called the *real derivative* of f at a, that satisfies the condition (1). It is obvious that if f has a complex derivative at a point $a \in A$ then it also has a real derivative at a and the two derivatives coincide. The converse is, in general, false, and it is this fact that distinguishes the theory of functions of a complex variable from the theory of functions of two real variables. (See Exercise (13).)

To avoid any possibility of confusion between real and complex derivatives we shall, in the rest of this chapter, consider only real normed linear spaces. However all the results, and their proofs, carry over to the complex case.

Exercises

(1) Let E and F be real linear spaces and let $\|\cdot\|_1$ and $\|\cdot\|_2$, and $\|\cdot\|_3$ and $\|\cdot\|_4$ be pairs of equivalent norms on E and F, respectively. Let A be a non-empty open subset of $(E, \|\cdot\|_1)$ and let f be a mapping of A into F. Suppose that f is differentiable at a point $a \in A$ with respect to the norms $\|\cdot\|_1$ and $\|\cdot\|_3$. Prove that f is differentiable at a with respect to the norms $\|\cdot\|_2$ and $\|\cdot\|_4$ and that the two derivatives coincide.

Deduce that when E and F are finite-dimensional the differentiability of a mapping of E into F is independent of the choice of norms on E and F.

(2) Let $\mathscr{G}$ be the set of regular elements of $L(E)$, where E is a real Banach space, and let $f(T) = T^{-1}$ for all $T \in \mathscr{G}$. Prove that f is continuous and differentiable on $\mathscr{G}$ and that $f'(T)\cdot S = -T^{-1}ST^{-1}$ for all $T \in \mathscr{G}$ and all $S \in L(E)$.

(3) Let E be an infinite-dimensional real normed linear space. Exhibit a mapping of E into itself that is differentiable at each point of E but is continuous at no point of E. (Hint: use Exercise 6.2(5) and Lemma 7.1.8.)

(4) Let E,F be real normed linear spaces, A an open subset of E and f a mapping of A into F. Let f be differentiable at $a \in A$, let $x \in E$ and let $A_x = \{t \in \mathsf{R} : a + tx \in A\}$. Prove that A_x is open and that the mapping ϕ_x defined by $\phi_x(t) = f(a + tx)$ for all $t \in A_x$ is differentiable at 0 with $D\phi_x(0) = f'(a)\cdot x$.

(5) (a) Let f be a real-valued function on a non-empty open subset A of a real normed linear space E. The function f is said to have a *relative maximum* [or a *relative minimum*] at a point $a \in A$ if and only if there exists $r > 0$ with $f(x) \leq f(a)$ [or $f(x) \geq f(a)$] for $\| x - a \| < r$. Prove that if f is differentiable at $a \in A$ and has a relative maximum or minimum at a then $f'(a) = 0$.

(b) Let B be a bounded closed subset of R^n with a non-empty interior B° and let f be a continuous real-valued function on B that is differentiable on B°. Prove that if f is constant on the frontier of B then $f'(x) = 0$ for some $x \in B^\circ$. (Use Corollary 4.3.2.) This result in the case $n = 1$ is Rolle's theorem of classical analysis.

(6) Let B be the open unit ball in a real normed linear space E and let f be a differentiable mapping of B into R. Prove that

$$f(\alpha x + (1 - \alpha)y) \leq \alpha f(x) + (1 - \alpha)f(y) \qquad (*)$$

for all $x,y \in B$ and all $\alpha \in [0,1]$ if and only if

$$f(x) - f(y) \geq f'(y)\cdot(x - y)$$

for all $x,y \in B$. Deduce that, when $E = \mathsf{R}$, the function f satisfies (*) for all $x,y \in (-1,1)$ if and only if Df is monotonic increasing on $(-1,1)$.

(7) Let g be a real-valued function on a real normed linear space E and let $g(\alpha x) = \alpha g(x)$ for all $x \in E$ and $\alpha \geq 0$. Prove that g is differentiable at 0 if and only if g is linear. Deduce that the norm $x \to \| x \|$ on E is not differentiable at 0 unless $E = \{0\}$.

(8) Prove that the norm $(\xi_1, \xi_2,\ldots, \xi_n) \to \left(\sum_{k=1}^{n} \xi_k^2\right)^{1/2}$ is differentiable at each non-zero point of R^n and find its derivative at such a point.

(9) (a) Let f be a continuous real-valued function on c_0 that is differentiable at a point $a = (\alpha_n) \in c_0$. For each positive integer n let e_n be as in Example 5.1.5 and let f_n be the real-valued function on R

defined by $f_n(t) = f(a + te_n)$ for all $t \in \mathsf{R}$. Prove that, for each positive integer n, the function f_n is differentiable at 0, and that

$$f'(a)\cdot x = \sum_{n=1}^{\infty} \xi_n Df_n(0)$$

for all $x = (\xi_n) \in c_0$. (Use Example 5.1.5.)

(b) Prove that the norm $x \to \| x \|$ on c_0 is differentiable at the point $a = (\alpha_n) \in c_0$ if and only if there is a positive integer N such that $| \alpha_N | > | \alpha_n |$ for all positive integers $n \neq N$.

(10) Let

$$f(x,y) = \begin{cases} \dfrac{x^3 - xy^2}{x^2 + y^2} & \text{when } (x,y) \neq (0,0), \\ 0 & \text{when } (x,y) = (0,0). \end{cases}$$

At which points do the partial derivatives D_1f and D_2f exist, and at which points does the Fréchet derivative f' exist?

(11) Let

$$f(x,y) = \begin{cases} (x^2 + y^2) \sin \dfrac{1}{\sqrt{(x^2 + y^2)}} & \text{when } (x,y) \neq (0,0), \\ 0 & \text{when } (x,y) = (0,0). \end{cases}$$

Prove that f is differentiable at (0,0) but that D_1f and D_2f are not continuous at (0,0).

(12) Let

$$f(x,y) = \begin{cases} \dfrac{xy}{r} \sin\left(\dfrac{xy}{r^2}\right) & \text{when } (x,y) \neq (0,0), \\ 0 & \text{when } (x,y) = (0,0), \end{cases}$$

where $r = \sqrt{(x^2 + y^2)}$. Prove that the first partial derivatives of f exist at each point of R^2, that the functions $t \to D_1f(t,0)$, $t \to D_1f(0,t)$, $t \to D_2f(t,0)$ and $t \to D_2f(0,t)$ are all continuous at 0, but that f is not differentiable at (0,0).

(13) (a) Let T be a real linear mapping of C into itself with matrix

$$\begin{pmatrix} \tau_{11} & \tau_{12} \\ \tau_{21} & \tau_{22} \end{pmatrix}$$

relative to the basis $\{1, i\}$ for (C,R). Prove that T is a complex linear mapping if and only if $\tau_{11} = \tau_{22}$ and $\tau_{12} = -\tau_{21}$.

(b) Let f be a complex-valued function on a non-empty open subset A of C and let $f(z) = f_1(x,y) + if_2(x,y)$ for all $z = x + iy \in A$. Prove that if f has a complex derivative at $w = u + iv \in A$ then f_1 and f_2 have partial derivatives with respect to both variables at (u,v),

$$D_1 f_1(u,v) = D_2 f_2(u,v) \text{ and } D_2 f_1(u,v) = -D_1 f_2(u,v).$$

(These equations are known as the *Cauchy–Riemann equations.*) Prove that if f_1 and f_2 have continuous partial derivatives with respect to both variables on A, and these partial derivatives satisfy the Cauchy–Riemann equations on A, then f has a complex derivative on A.

(c) Prove that the function $z \to \bar{z}$ has a real derivative at each point of C but has a complex derivative at no point of C.

7.2 Elementary Rules of Differentiation

Throughout this section E, F and G will denote real normed linear spaces. Given mappings f and g of a non-empty subset A of E into F and a real number α we define mappings $f + g$ and αf of A into F by

$$(f + g)(x) = f(x) + g(x)$$

and

$$(\alpha f)(x) = \alpha f(x)$$

for all $x \in A$. The proof of our first theorem is quite elementary; we leave the reader to supply the details.

7.2.1 THEOREM *Let f and g be mappings of a non-empty open subset A of E into F that are differentiable at a point $a \in A$ and let $\alpha \in$ R. Then $f + g$ and αf are differentiable at a,*

$$(f + g)'(a) = f'(a) + g'(a)$$

and

$$(\alpha f)'(a) = \alpha f'(a).$$

7.2.2 THEOREM *Let f be a mapping of a non-empty open subset A of E into an open subset B of F and let g be a mapping of B into G. Suppose that f is continuous and differentiable at a point $a \in A$ and that g is continuous and differentiable at the point $b = f(a)$. Then $g \circ f$ is continuous and differentiable at a and*

$$(g \circ f)'(a) = g'(b) \circ f'(a).$$

PROOF Let $T = f'(a)$ and $S = g'(b)$. Theorem 7.1.4 shows that $T \in L(E,F)$ and $S \in L(F,G)$. By hypothesis, given a real number η with $0 < \eta \leq 1$ there exist $\delta_1 > 0$ and $\delta_2 > 0$ such that

$$\| f(x) - f(a) - T(x - a) \| \leq \eta \| x - a \| \tag{1}$$

for $\| x - a \| < \delta_1$, and

$$\| g(y) - g(b) - S(y - b) \| \leq \eta \| y - b \| \tag{2}$$

for $\| y - b \| < \delta_2$. It follows from (1) that if $\| x - a \| < \delta_1$ then

$$\begin{aligned}\| f(x) - b \| &\leq \| f(x) - f(a) - T(x-a) \| + \| T(x-a) \| \\ &\leq \eta \| x - a \| + \| T \| \ \| x - a \| \\ &\leq (1 + \| T \|) \| x - a \|. \end{aligned} \tag{3}$$

Let $\delta = \min \{\delta_1, \delta_2(1 + \| T \|)^{-1}\}$ and $\| x - a \| < \delta$. Then, by (3), we have $\| f(x) - b \| < \delta_2$ and, consequently, (1), (2) and (3) give

$$\begin{aligned}&\| (g \circ f)(x) - (g \circ f)(a) - (S \circ T)(x - a) \| \\ &\quad = \| g(f(x)) - g(b) - S(T(x - a)) \| \\ &\quad \leq \| g(f(x)) - g(b) - S(f(x) - b) \| \\ &\qquad + \| S(f(x) - b) - S(T(x - a)) \| \\ &\quad \leq \eta \| f(x) - b \| + \| S(f(x) - f(a) - T(x - a)) \| \\ &\quad \leq \eta(1 + \| T \|) \| x - a \| + \| S \| \| f(x) - f(a) - T(x - a) \| \\ &\quad \leq \eta(1 + \| T \| + \| S \|) \| x - a \|.\end{aligned}$$

Let $\varepsilon > 0$ and let $\eta = \min \{1, \varepsilon(1 + \| T \| + \| S \|)^{-1}\}$. Then, by what we have just proved, there exists $\delta > 0$ such that

$$\| (g \circ f)(x) - (g \circ f)(a) - (S \circ T)(x - a) \| \leq \varepsilon \| x - a \|$$

whenever $\| x - a \| < \delta$. This proves that $g \circ f$ is differentiable at a and that $(g \circ f)'(a) = S \circ T$. The continuity of $g \circ f$ at a follows now from Theorem 7.1.4 and the fact that $S \circ T$ is bounded. □

The formula $(g \circ f)'(a) = g'(f(a)) \circ f'(a)$ for the derivative of the composition $g \circ f$ is usually called the *chain-rule*. The following corollary is the chain rule of the classical differential calculus.

7.2.3 COROLLARY *Let g be a real-valued function on a non-empty open subset B of R^m that has continuous first partial derivatives on B and let $f_1, f_2, \ldots, f_m$ be real-valued functions on a non-empty open subset A of R^n that have continuous first partial derivatives on A and are such that $(f_1(x), f_2(x), \ldots, f_m(x)) \in B$ for all $x \in A$. Then the real-valued function h on A, defined by*

$$h(x) = g(f_1(x), f_2(x), \ldots, f_m(x))$$

for all $x \in A$ has continuous first partial derivatives on A and, for $j = 1, 2, \ldots, n$ and all $x \in A$, we have

$$D_jh(x) = D_1g(y)D_jf_1(x) + D_2g(y)D_jf_2(x) + \ldots + D_mg(y)D_jf_m(x), \tag{4}$$

where $y = (f_1(x), f_2(x), \ldots, f_m(x))$.

PROOF Let $f(x) = (f_1(x), f_2(x), \ldots, f_m(x))$ for all $x \in A$. By Theorem 7.1.12 the mapping f is differentiable on A and the mapping g is differentiable on B, and hence, by Corollary 7.1.5, f and g are continuous on A and B respectively. Theorem 7.2.2 now shows that h is differentiable, and that, for all $x \in A$, we have $h'(x) = g'(y) \circ f'(x)$, where

$y = f(x)$. It now follows from Theorem 7.1.11 that h has first partial derivatives on A. Equation (4) follows from the matrix representation of the derivatives $h'(x)$, $g'(y)$ and $f'(x)$ (Theorem 7.1.11), and the fact (which is easily proved) that the matrix that represents $g'(y)\circ f'(x)$ is the product of the matrices that represent $g'(y)$ and $f'(x)$. It follows from equation (4) and Theorems 2.7.4 and 2.8.2 that the partial derivatives of h are continuous on A. □

Equation (4) is often written in the form

$$\frac{\partial h}{\partial \xi_j} = \frac{\partial g}{\partial \eta_1}\frac{\partial \eta_1}{\partial \xi_j} + \frac{\partial g}{\partial \eta_2}\frac{\partial \eta_2}{\partial \xi_j} + \cdots + \frac{\partial g}{\partial \eta_m}\frac{\partial \eta_m}{\partial \xi_j},$$

where $\eta_k = f_k(\xi_1, \xi_2, \ldots, \xi_n)$.

7.2.4 THEOREM *Let f be a homeomorphism of a non-empty open subset A of E onto an open subset B of F and suppose that, for some point $a \in A$, f is differentiable at a and $f'(a)$ is a linear homeomorphism of E onto F. Then f^{-1} is differentiable at the point $b = f(a)$ and $(f^{-1})'(b) = f'(a)^{-1}$.*

PROOF Let $f'(a) = T$ and $S = T^{-1}$. By hypothesis $T \in L(E,F)$ and $S \in L(F,E)$. Given $\eta > 0$ with $\eta \| S \| \leq \frac{1}{2}$ there exists $\delta_1 > 0$ such that

$$\| f(x) - f(a) - T(x-a) \| \leq \eta \| x - a \| \tag{5}$$

for $\| x - a \| < \delta_1$, and there exists $\delta > 0$ such that

$$\| f^{-1}(y) - f^{-1}(b) \| < \delta_1 \tag{6}$$

for $\| y - b \| < \delta$. Let $\| y - b \| < \delta$ and $x = f^{-1}(y)$. Then, by (6), we have $\| x - a \| < \delta_1$ and consequently, by (5),

$$\begin{aligned} \| x - a - S(f(x) - f(a)) \| &= \| S(T(x-a) - f(x) + f(a)) \| \\ &\leq \| S \| \; \| T(x-a) - f(x) + f(a) \| \\ &\leq \eta \| S \| \; \| x - a \|. \end{aligned} \tag{7}$$

From (7) we obtain

$$\begin{aligned} \| x - a \| &\leq \| x - a - S(f(x) - f(a)) \| + \| S(f(x) - f(a)) \| \\ &\leq \eta \| S \| \| x - a \| + \| S \| \| f(x) - f(a) \| \\ &\leq \tfrac{1}{2} \| x - a \| + \| S \| \| y - b \| \end{aligned}$$

which gives

$$\| x - a \| \leq 2 \| S \| \; \| y - b \|. \tag{8}$$

It follows from (5), (7) and (8) that

$$\begin{aligned} \| f^{-1}(y) - f^{-1}(b) - S(y-b) \| &= \| x - a - S(f(x) - f(a)) \| \\ &\leq \eta \| S \| \; \| x - a \| \\ &\leq 2\eta \| S \|^2 \| y - b \|. \end{aligned}$$

Now let $\varepsilon > 0$ and let $\eta = \min \{\frac{1}{2} \| S \|^{-1}, \frac{1}{2}\varepsilon \| S \|^{-2}\}$. By what we have just proved there exists $\delta > 0$ such that

$$\| f^{-1}(y) - f^{-1}(b) - S(y - b) \| \leq \varepsilon \| y - b \|$$

for $\| y - b \| < \delta$. This shows that f^{-1} is differentiable at b and $(f^{-1})'(b) = S$. □

We shall see in Theorem 7.5.8 that, if E and F are Banach spaces and f is a continuously differentiable mapping of A into F such that $f'(a)$ is a linear homeomorphism of E onto F, then f is a homeomorphism of some neighbourhood of a onto a neighbourhood of $f(a)$.

EXERCISES

(1) Let f be a homeomorphism of A onto an open subset B of F and suppose that, for some point $a \in A$, the mapping f is differentiable at a and $f'(a)$ is *not* a linear homeomorphism of E onto F. Prove that f^{-1} is not differentiable at the point $f(a)$.

(2) Let A be a non-empty open subset of an n-dimensional real normed linear space E and let f be a mapping of A into an m-dimensional real normed linear space F. Choose bases $\{e_1, e_2, \ldots, e_n\}$ and $\{e'_1, e'_2, \ldots, e'_m\}$ for E and F, respectively. Let

$$A_0 = \left\{(\xi_1, \xi_2, \ldots, \xi_n) \in \mathsf{R}^n : \sum_{j=1}^{n} \xi_j e_j \in A\right\}$$

and let $f_1, f_2, \ldots, f_m$ be the real-valued functions on A_0 which are defined by $f(x) = \sum_{k=1}^{m} f_k(\xi_1, \xi_2, \ldots, \xi_n) e'_k$ for all $x = \sum_{j=1}^{n} \xi_j e_j \in A$.

Prove that A_0 is open. Prove also that f is differentiable at a point $a = \sum_{j=1}^{n} \alpha_j e_j \in A$ if and only if the mapping

$$(\xi_1, \xi_2, \ldots, \xi_n) \to (f_1(\xi_1, \xi_2, \ldots, \xi_n), \ldots, f_m(\xi_1, \xi_2, \ldots, \xi_n))$$

of A_0 into R^m is differentiable at $(\alpha_1, \alpha_2, \ldots, \alpha_n)$.

Prove that, if f is differentiable at a, then the matrix that represents $f'(a)$ relative to the bases $\{e_1, e_2, \ldots, e_n\}$ and $\{e'_1, e'_2, \ldots, e'_m\}$ is

$$\begin{pmatrix} D_1 f_1(\alpha_1, \ldots, \alpha_n) & \ldots & D_n f_1(\alpha_1, \ldots, \alpha_n) \\ \cdot & \cdots & \cdot \\ D_1 f_m(\alpha_1, \ldots, \alpha_n) & \ldots & D_n f_m(\alpha_1, \ldots, \alpha_n) \end{pmatrix}.$$

7.3 The Mean Value Theorem

Throughout this section f will denote a mapping of a non-empty open subset A of a real normed linear space E into a real normed linear space F that is continuous and differentiable on A. The following theorem is the analogue for the Fréchet derivative of the classical mean value theorem of the differential calculus. The proof is similar to that of Lemma 3.9.8.

7.3.1 THEOREM (the mean value theorem) *Let a and b be distinct points of A and suppose that the line segment*

$$\sigma = \{a + \xi(b - a) : 0 \leq \xi \leq 1\}$$

is contained in A. Suppose also that there is a real number M with $\| f'(x) \| \leq M$ for all $x \in \sigma$. Then

$$\| f(b) - f(a) \| \leq M \| b - a \|.$$

PROOF Let $c = b - a$ and $g(\xi) = f(a + \xi c)$ for $0 \leq \xi \leq 1$. For each $\varepsilon > 0$ let

$$J_\varepsilon = \{\xi \in [0,1] : \| g(\eta) - g(0) \| \leq (M + \varepsilon)\, \eta \, \| c \| \text{ for } 0 \leq \eta \leq \xi\}.$$

We shall prove that $1 \in J_\varepsilon$. It is obvious that $0 \in J_\varepsilon$. Let $\alpha = \sup J_\varepsilon$. We begin by proving that $\alpha \in J_\varepsilon$. This is obvious if $\alpha = 0$ so suppose that $\alpha > 0$. For each number η with $0 \leq \eta < \alpha$ there exists a number $\xi_\eta \in J_\varepsilon$ with $\eta < \xi_\eta$ and, consequently, by definition of J_ε,

$$\| g(\eta) - g(0) \| \leq (M + \varepsilon)\eta \, \| c \|. \tag{1}$$

The mapping $\xi \to a + \xi c$ is continuous on $[0,1]$ so g is continuous on $[0,1]$. Using (1) and the continuity of g we obtain

$$\begin{aligned} \| g(\alpha) - g(0) \| &= \lim_{n \to \infty} \left\| g\left(\alpha - \frac{1}{n}\right) - g(0) \right\| \\ &\leq \lim_{n \to \infty} (M + \varepsilon)\left(\alpha - \frac{1}{n}\right) \| c \| \\ &= (M + \varepsilon)\, \alpha \, \| c \|. \end{aligned} \tag{2}$$

Inequalities (1) and (2) show that $\alpha \in J_\varepsilon$.

We shall prove now that $\alpha = 1$. Suppose, on the contrary, that $\alpha < 1$. Let $x = a + \alpha c$. Then f is differentiable at x and so there exists $\delta > 0$ such that

$$\| f(y) - f(x) - f'(x) \cdot (y - x) \| \leq \varepsilon \| y - x \|$$

for $\| y - x \| < \delta$. Since $x \in \sigma$ we have $\| f'(x) \| \leq M$ and therefore, for $\| y - x \| < \delta$,

$$\begin{aligned} \| f(y) - f(x) \| &\leq \| f(y) - f(x) - f'(x) \cdot (y - x) \| + \| f'(x) \cdot (y - x) \| \\ &\leq \varepsilon \| y - x \| + \| f'(x) \| \, \| y - x \| \\ &\leq (M + \varepsilon) \| y - x \|. \end{aligned} \tag{3}$$

Choose a real number ξ_0 so that $\alpha < \xi_0 \leq 1$ and $(\xi_0 - \alpha) \| c \| < \delta$.

If $\alpha < \eta \leq \xi_0$ and $y = a + \eta c$, then $\| y - x \| = | \eta - \alpha | \| c \| < \delta$ and hence (3) gives

$$\| g(\eta) - g(\alpha) \| = \| f(y) - f(x) \| \leq (M + \varepsilon)(\eta - \alpha) \| c \|. \tag{4}$$

It follows from (2) and (4) that if $\alpha < \eta \leq \xi_0$, then

$$\begin{aligned} \| g(\eta) - g(0) \| &\leq \| g(\eta) - g(\alpha) \| + \| g(\alpha) - g(0) \| \\ &\leq (M + \varepsilon)(\eta - \alpha) \| c \| + (M + \varepsilon)\alpha \| c \| \\ &= (M + \varepsilon)\eta \| c \|. \end{aligned} \tag{5}$$

Inequalities (1), (2) and (5) show that $\xi_0 \in J_\varepsilon$, which contradicts the definition of α. Therefore $\alpha = 1$. This proves that $1 \in J_\varepsilon$.

By definition of J_ε we now have

$$\| f(b) - f(a) \| = \| g(1) - g(0) \| \leq (M + \varepsilon) \| c \| = (M + \varepsilon) \| b - a \|.$$

Since $\varepsilon > 0$ is arbitrary, it follows that $\| f(b) - f(a) \| \leq M \| b - a \|$. □

The following slight extension of the mean value theorem is often more useful than the mean value theorem itself.

7.3.2 Corollary *Let a and b be distinct points of A and suppose that the line segment $\sigma = \{a + \xi(b - a) : 0 \leq \xi \leq 1\}$ is contained in A. Let $T \in L(E,F)$ and suppose that there is a real number M with*

$$\| f'(x) - T \| \leq M$$

for all $x \in \sigma$. Then

$$\| f(b) - f(a) - T(b - a) \| \leq M \| b - a \|.$$

Proof Let $h(x) = f(x) - Tx$ for all $x \in A$. It is clear that h is continuous on A and it follows from Theorem 7.2.1 and Lemma 7.1.8 that h is differentiable on A and $h'(x) = f'(x) - T$ for all $x \in A$. Consequently $\| h'(x) \| \leq M$ for all $x \in \sigma$ and Theorem 7.3.1 gives

$$\| f(b) - f(a) - T(b - a) \| = \| h(b) - h(a) \| \leq M \| b - a \|. \; \square$$

Corollary 7.3.2 will nearly always be used in the following, slightly weaker, form.

7.3.3 Corollary *Let B be an open or closed ball contained in A and let $T \in L(E,F)$. Suppose that there is a real number M with*

$$\| f'(x) - T \| \leq M$$

for all $x \in B$. Then

$$\| f(b) - f(a) - T(b - a) \| \leq M \| b - a \|$$

for all $a,b \in B$.

Proof An elementary calculation shows that the line segment $\sigma = \{a + \xi(b - a) : 0 \leq \xi \leq 1\}$ is contained in B for all $a,b \in B$. □

It is clear that the ball B in Corollary 7.3.3 can be replaced by any convex subset of A.

EXERCISES

(1) Show that Theorem 7.3.1 can be deduced from the classical mean value theorem for real-valued functions of a real variable by considering functions h of the form $h(\xi) = x^*(g(\xi))$ for $0 \leq \xi \leq 1$, where $x^* \in F^*$. (The notation is that used in Theorem 7.3.1.) (Hint: use Corollary 5.3.7.)

(2) Let A be a non-empty connected open subset of a real normed linear space E and let f be a continuous and differentiable mapping of A into a real normed linear space F. If $f'(x) = 0$ for all $x \in A$ prove that f is constant on A. Observe that the condition that A be connected cannot be omitted.

(3) Let E be a real normed linear space and let F be a real Banach space. Let $a \in A$, $r > 0$ and $b \in B(a,r)$. Suppose that (f_n) is a sequence of continuous and differentiable mappings of $B(a,r)$ into F with the following properties: (a) the sequence $(f_n(b))$ converges, and (b) for each $\varepsilon > 0$ there is a positive integer N with $\| f'_n(x) - f'_m(x) \| < \varepsilon$ for all $n,m \geq N$ and all $x \in B(a,r)$. Prove that there is a continuous and differentiable mapping f of $B(a,r)$ into F with $\lim_{n \to \infty} f_n(x) = f(x)$ and $\lim_{n \to \infty} f'_n(x) = f'(x)$ for all $x \in B(a,r)$.

7.4 Partial Derivatives

Throughout this section E_1, E_2 and F will denote real normed linear spaces. It is easy to verify that the set $E_1 \times E_2$ of all ordered pairs (x_1,x_2) with $x_1 \in E_1$ and $x_2 \in E_2$ is a real linear space with respect to the linear space operations defined by

$$(x_1,x_2) + (y_1,y_2) = (x_1 + x_2, y_1 + y_2)$$

and

$$\alpha(x_1,x_2) = (\alpha x_1, \alpha x_2)$$

for all $(x_1,x_2),(y_1,y_2) \in E_1 \times E_2$ and all $\alpha \in \mathsf{R}$. It is also easy to verify that the mapping

$$(x_1,x_2) \to \| (x_1,x_2) \| = \max \{ \| x_1 \|, \| x_2 \| \}$$

is a norm on $E_1 \times E_2$. In this section and the next we shall always assume that $E_1 \times E_2$ is endowed with the linear space operations and norm defined here.

Let A be a non-empty open subset of $E_1 \times E_2$ and let f be a mapping of A into F. Let $(a_1,a_2) \in A$, let $A_1 = \{x_1 \in E_1 : (x_1,a_2) \in A\}$ and let

$g(x_1) = f(x_1,a_2)$ for all $x_1 \in A_1$. It is clear that A_1 is an open subset of E_1.

7.4.1 DEFINITION The mapping f is said to be *differentiable with respect to the first variable at the point* (a_1,a_2) if and only if g is differentiable at a_1. If g is differentiable at a_1 we write $g'(a_1) = f_1'(a_1,a_2)$. The derivative $f_1'(a_1,a_2)$ is called the *partial derivative of f with respect to the first variable at* (a_1,a_2); it is a linear transformation of E_1 into F. The mapping f is said to be *differentiable with respect to the first variable on A* if and only if it is differentiable with respect to the first variable at each point of A.

The mapping $x_1 \to (x_1,a_2)$ is continuous on E_1, so if f is continuous on A then g is continuous on A_1 (Theorem 2.7.4) and hence, by Theorem 7.1.4, $f_1'(a_1,a_2) \in L(E_1,F)$.

Differentiability of f with respect to the second variable at (a_1,a_2), and the partial derivative $f_2'(a_1,a_2)$, are defined similarly. If f is continuous on A then $f_2'(a_1,a_2) \in L(E_2,F)$.

In the case when $E_1 = E_2 = F = \mathsf{R}$ (that is when f is a real-valued function of two real variables) Theorem 7.1.9 shows that the partial derivative $f_1'(a_1,a_2)$ exists if and only if the usual partial derivative $D_1f(a_1,a_2)$ exists, and it shows further that $D_1f(a_1,a_2) = f_1'(a_1,a_2)\cdot 1$. A similar connection exists between the partial derivatives $f_2'(a_1,a_2)$ and $D_2f(a_1,a_2)$. With these remarks in mind, the reader should compare the proofs of the following two theorems with those of Theorems 7.1.11 and 7.1.12.

7.4.2 THEOREM *Let f be a mapping of a non-empty open subset A of $E_1 \times E_2$ into F that is differentiable at a point $(a_1,a_2) \in A$. Then f is differentiable with respect to both variables at (a_1,a_2) and*

$$f'(a_1,a_2)\cdot(x_1,x_2) = f_1'(a_1,a_2)\cdot x_1 + f_2'(a_1,a_2)\cdot x_2 \tag{1}$$

for all $(x_1,x_2) \in E_1 \times E_2$. Further, if f is continuously differentiable on A, then $(x_1,x_2) \to f_1'(x_1,x_2)$ and $(x_1,x_2) \to f_2'(x_1,x_2)$ are continuous mappings of A into $L(E_1,F)$ and $L(E_2,F)$, respectively.

PROOF Let

$$T_1x_1 = f'(a_1,a_2)\cdot(x_1,0)$$

and

$$T_2x_2 = f'(a_1,a_2)\cdot(0,x_2)$$

for all $x_1 \in E_1$ and $x_2 \in E_2$. It is easy to verify that T_1 and T_2 are linear mappings of E_1 and E_2, respectively, into F. Let $\varepsilon > 0$. There exists $\delta > 0$ such that

$$\| f(x_1,x_2) - f(a_1,a_2) - f'(a_1,a_2)\cdot((x_1,x_2) - (a_1,a_2)) \| \leq \varepsilon \| (x_1,x_2) - (a_1,a_2) \| \tag{2}$$

for $\| (x_1,x_2) - (a_1,a_2) \| < \delta$. Let $x_1 \in E_1$ and $\| x_1 - a_1 \| < \delta$. Then $\| (x_1,a_2) - (a_1,a_2) \| = \| (x_1 - a_1,0) \| = \| x_1 - a_1 \| < \delta$ and

$$f'(a_1,a_2)\cdot((x_1,a_2) - (a_1,a_2)) = f'(a_1,a_2)\cdot(x_1 - a_1,0) = T_1(x_1 - a_1),$$

and, therefore, (2) gives

$$\| f(x_1,a_2) - f(a_1,a_2) - T_1(x_1 - a_1) \| \leq \varepsilon \| x_1 - a_1 \|.$$

This shows that f is differentiable with respect to the first variable at (a_1,a_2) and $f_1'(a_1,a_2) = T_1$. Similarly we can prove that f is differentiable with respect to the second variable at (a_1,a_2) and $f_2'(a_1,a_2) = T_2$. It follows now from the definition of T_1 and T_2 that

$$\begin{aligned} f'(a_1,a_2)\cdot(x_1,x_2) &= f'(a_1,a_2)\cdot((x_1,0) + (0,x_2)) \\ &= f'(a_1,a_2)\cdot(x_1,0) + f'(a_1,a_2)\cdot(0,x_2) \\ &= f_1'(a_1,a_2)\cdot x_1 + f_2'(a_1,a_2)\cdot x_2 \end{aligned}$$

for all $(x_1,x_2) \in E_1 \times E_2$.

Suppose now that f is continuously differentiable on A and let (a_1,a_2) and (b_1,b_2) be points of A. For all $x_1 \in E_1$ equation (1) gives

$$\begin{aligned} \| (f_1'(a_1,a_2) - f_1'(b_1,b_2))\cdot x_1 \| &= \| (f'(a_1,a_2) - f'(b_1,b_2))\cdot(x_1,0) \| \\ &\leq \| f'(a_1,a_2) - f'(b_1,b_2) \| \, \| x_1 \|. \end{aligned}$$

Consequently

$$\| f_1'(a_1,a_2) - f_1'(b_1,b_2) \| \leq \| f'(a_1,a_2) - f'(b_1,b_2) \|$$

from which it follows that the mapping $(x_1,x_2) \to f_1'(x_1,x_2)$ is continuous on A. Similarly the mapping $(x_1,x_2) \to f_2'(x_1,x_2)$ is continuous on A. □

7.4.3 THEOREM *Let f be a continuous mapping of a non-empty open subset A of $E_1 \times E_2$ into F. Suppose that f is differentiable with respect to both variables on A and that $(x_1,x_2) \to f_1'(x_1,x_2)$ and $(x_1,x_2) \to f_2'(x_1,x_2)$ are continuous mappings of A into $L(E_1,F)$ and $L(E_2,F)$, respectively. Then f is continuously differentiable on A.*

PROOF Let $(a_1,a_2) \in A$ and let

$$T(x_1,x_2) = f_1'(a_1,a_2)\cdot x_1 + f_2'(a_1,a_2)\cdot x_2$$

for all $(x_1,x_2) \in E_1 \times E_2$. It is clear that $T \in L(E_1 \times E_2,F)$. We shall prove that T is the derivative of f at (a_1,a_2).

We observe first that, for all $(x_1,x_2) \in A$, we have

$$\begin{aligned} &\| f(x_1,x_2) - f(a_1,a_2) - T((x_1,x_2) - (a_1,a_2)) \| \\ &\quad = \| (f(x_1,x_2) - f(a_1,x_2) - f_1'(a_1,a_2)\cdot(x_1 - a_1)) \\ &\qquad + (f(a_1,x_2) - f(a_1,a_2) - f_2'(a_1,a_2)\cdot(x_2 - a_2)) \| \\ &\quad \leq \| f(x_1,x_2) - f(a_1,x_2) - f_1'(a_1,a_2)\cdot(x_1 - a_1) \| \\ &\qquad + \| f(a_1,x_2) - f(a_1,a_2) - f_2'(a_1,a_2)\cdot(x_2 - a_2) \|. \end{aligned} \tag{3}$$

Let $\varepsilon > 0$. By the continuity of the mapping $(x_1,x_2) \to f_1'(x_1,x_2)$ at

(a_1,a_2) there exists $\delta_1 > 0$ such that

$$\| f_1'(x_1,x_2) - f_1'(a_1,a_2) \| < \tfrac{1}{2}\varepsilon \tag{4}$$

for $\| (x_1,x_2) - (a_1,a_2) \| < \delta_1$.

For each (fixed) $x_2 \in E_2$ with $\| x_2 - a_2 \| < \delta_1$ we can apply the inequality (4) and Corollary 7.3.3 to the mapping $x_1 \to f(x_1,x_2)$ and the open ball $B_1 = \{y_1 \in E_1 : \| y_1 - a_1 \| < \delta_1\}$. This gives

$$\| f(x_1,x_2) - f(y_1,x_2) - f_1'(a_1,a_2)\cdot(x_1 - y_1) \| \leq \tfrac{1}{2}\varepsilon \| x_1 - y_1 \| \tag{5}$$

for all $x_1,y_1 \in B_1$ and all $x_2 \in E_2$ with $\| x_2 - a_2 \| < \delta_1$. Also since f is differentiable with respect to the second variable at (a_1,a_2) there exists $\delta_2 > 0$ such that

$$\| f(a_1,x_2) - f(a_1,a_2) - f_2'(a_1,a_2)\cdot(x_2 - a_2) \| \leq \tfrac{1}{2}\varepsilon \| x_2 - a_2 \| \tag{6}$$

for $\| x_2 - a_2 \| < \delta_2$. Let $\delta = \min \{\delta_1, \delta_2\}$ and let $(x_1,x_2) \in E_1 \times E_2$ with $\| (x_1,x_2) - (a_1,a_2) \| < \delta$. Then $x_1 \in B_1$ and $\| x_2 - a_2 \| < \delta_2$. Since $a_1 \in B_1$ inequalities (3), (5) and (6) give

$$\begin{aligned} \| f(x_1,x_2) - f(a_1,a_2) - T((x_1,x_2) - (a_1,a_2)) \| & \\ \leq \tfrac{1}{2}\varepsilon \| x_1 - a_1 \| + \tfrac{1}{2}\varepsilon \| x_2 - a_2 \| & \\ \leq \varepsilon \max \{ \| x_1 - a_1 \| , \| x_2 - a_2 \| \} & \\ = \varepsilon \| (x_1,x_2) - (a_1,a_2) \| . & \end{aligned}$$

This shows that f is differentiable at (a_1,a_2) and that $f'(a_1,a_2) = T$.

Finally, if $(b_1,b_2) \in A$ equation (1) gives

$$\begin{aligned} &\| (f'(a_1,a_2) - f'(b_1,b_2))\cdot(x_1,x_2) \| \\ &= \| (f_1'(a_1,a_2) - f_1'(b_1,b_2))\cdot x_1 + (f_2'(a_1,a_2) - f_2'(b_1,b_2))\cdot x_2 \| \\ &\leq \| (f_1'(a_1,a_2) - f_1'(b_1,b_2))\cdot x_1 \| + \| (f_2'(a_1,a_2) - f_2'(b_1,b_2))\cdot x_2 \| \\ &\leq \| f_1'(a_1,a_2) - f_1'(b_1,b_2) \| \; \| x_1 \| + \| f_2'(a_1,a_2) - f_2'(b_1,b_2) \| \; \| x_2 \| \\ &\leq \| (x_1,x_2) \| (\| f_1'(a_1,a_2) - f_1'(b_1,b_2) \| + \| f_2'(a_1,a_2) - f_2'(b_1,b_2) \|) \end{aligned}$$

for all $(x_1,x_2) \in E_1 \times E_2$. Consequently

$$\begin{aligned} \| f'(a_1,a_2) - f'(b_1,b_2) \| \leq \| f_1'(a_1,a_2) - f_1'(b_1\, b_2) \| & \\ + \| f_2'(a_1,a_2) - f_2'(b_1,b_2) \| . & \end{aligned}$$

Since the mappings $(x_1,x_2) \to f_1'(x_1,x_2)$ and $(x_1,x_2) \to f_2'(x_1,x_2)$ are continuous on A this shows that the mapping $(x_1,x_2) \to f'(x_1,x_2)$ is continuous at (a_1,a_2). □

If we examine the proof of Theorem 7.4.3 we shall see that, if the hypothesis that both the mappings

$$(x_1,x_2) \to f_1'(x_1,x_2) \text{ and } (x_1,x_2) \to f_2'(x_1,x_2)$$

are continuous on A is replaced by the hypothesis that *one* of these mappings is continuous on A, then we can still conclude that f is differentiable on A.

EXERCISES

(1) Let E be a real normed linear space and let p be the mapping of $\mathsf{R} \times E$ into E given by $p(\xi,x) = \xi x$ for all $\xi \in \mathsf{R}$ and all $x \in E$. Prove that p is differentiable on $\mathsf{R} \times E$ with $p'(\alpha,a)\cdot(\xi,x) = \xi a + \alpha x$ for all $\alpha,\xi \in \mathsf{R}$ and all $a,x \in E$.

(2) Let A be a non-empty open subset of E and let f and g be continuously differentiable mappings of A into R and into a real normed linear space F, respectively. Prove that the mapping $x \to f(x)g(x)$ is differentiable on A and find its derivative.

7.5 The Implicit Function Theorem

The classical implicit function theorem gives sufficient conditions for the existence, uniqueness and differentiability of local solutions, $y = u(x)$, of an equation $f(x,y) = 0$, where f is a real-valued function of two real variables. (See GRAVES [10, pp. 138–9].) It is not difficult to extend this theorem to a system of equations involving m real valued functions of $n + m$ real variables. Such a system can again be written in the form $f(x,y) = 0$, where f is now a mapping of $\mathsf{R}^n \times \mathsf{R}^m$ (which is linearly isometric to R^{n+m}) into R^m. In 1927 T. H. Hildebrandt and L. M. Graves extended the theorem further by observing that R^n and R^m can be replaced by arbitrary Banach spaces. The proof of this extended version of the implicit function theorem is essentially the same as that of the classical version. We shall need several lemmas.

7.5.1 LEMMA *Let E, F and G be real normed linear spaces. Then $(S,T) \to S \circ T$ is a continuous mapping of $L(F,G) \times L(E,F)$ into $L(E,G)$.*

PROOF By Lemma 3.5.8 we have $S \circ T \in L(E,G)$ and

$$\| S \circ T \| \leq \| S \| \; \| T \|$$

for all $S \in L(F,G)$ and $T \in L(E,F)$. Given operators $S, S_0 \in L(F,G)$ and $T, T_0 \in L(E,F)$, we have

$$\begin{aligned} \| S \circ T - S_0 \circ T_0 \| &\leq \| S \circ T - S \circ T_0 \| + \| S \circ T_0 - S_0 \circ T_0 \| \\ &= \| S \circ (T - T_0) \| + \| (S - S_0) \circ T_0 \|, \\ &\leq \| S \| \; \| T - T_0 \| + \| S - S_0 \| \; \| T_0 \|. \end{aligned}$$

It follows easily from this inequality that the mapping $(S,T) \to S \circ T$ is continuous at (S_0, T_0). □

The following lemma is a slight extension of Theorems 6.2.10 and 6.2.11.

7.5.2 LEMMA *Let E and F be real Banach spaces and let $\mathscr{H}$ be the set of all linear homeomorphisms of E onto F. Then $\mathscr{H}$ is an open subset of $L(E,F)$ and the mapping $T \to T^{-1}$ is continuous on $\mathscr{H}$.*

PROOF If $\mathscr{H}$ is empty there is nothing to prove. Suppose that $\mathscr{H}$ is not empty and let $T_0 \in \mathscr{H}$. Then, for all $T \in L(E,F)$,

$$\| T_0^{-1} \circ T - I \| = \| T_0^{-1} \circ (T - T_0) \| \leq \| T_0^{-1} \| \, \| T - T_0 \|.$$

Consequently, if $\| T - T_0 \| < \| T_0^{-1} \|^{-1}$ Corollary 6.2.8 shows that $T_0^{-1} \circ T$ is regular in $L(E)$ and hence is a linear homeomorphism of E onto itself by Lemma 6.2.3. But $T = T_0 \circ (T_0^{-1} \circ T)$ so $T \in \mathscr{H}$ whenever $\| T - T_0 \| < \| T_0^{-1} \|^{-1}$. This proves that $\mathscr{H}$ is open.

The proof that the mapping $T \to T^{-1}$ is continuous on $\mathscr{H}$ is the same as the proof of Theorem 6.2.11. □

7.5.3 LEMMA *Let A be a metric space, F a real Banach space and $C_F(A)$ the set of all continuous mappings of A into F which have bounded range. Then $C_F(A)$ is a Banach space with linear space operations given by*

$$(f + g)(x) = f(x) + g(x)$$

and

$$(\alpha f)(x) = \alpha f(x),$$

for all $x \in A$, and norm given by

$$\| f \| = \sup \{ \| f(x) \| : x \in A \}.$$

PROOF It is left to the reader to verify that $C_F(A)$ is a linear space and that $f \to \| f \|$ is a norm on $C_F(A)$. We shall prove that $C_F(A)$ is complete. Let (f_n) be a Cauchy sequence in $C_F(A)$ and let $\varepsilon > 0$. There exists an integer N with $\| f_m - f_n \| < \varepsilon$ for $m,n \geq N$ and therefore, for all $x \in A$,

$$\| f_m(x) - f_n(x) \| \leq \| f_m - f_n \| < \varepsilon \tag{1}$$

for $m,n \geq N$. Inequality (1) shows that $(f_n(x))$ is a Cauchy sequence in F for each $x \in A$. Since F is complete $(f_n(x))$ converges. Let $f(x) = \lim_{n\to\infty} f_n(x)$ for all $x \in A$. Inequality (1) gives, for $n \geq N$ and all $x \in A$,

$$\| f(x) - f_n(x) \| = \lim_{m\to\infty} \| f_m(x) - f_n(x) \| \leq \varepsilon. \tag{2}$$

Consequently $\| f(x) \| \leq \varepsilon + \| f_N(x) \|$ for all $x \in A$, which shows that f has bounded range. Let $x_0 \in A$. Since f_N is continuous on A there exists $\delta > 0$ such that

$$\| f_N(x) - f_N(x_0) \| < \varepsilon \tag{3}$$

for $\| x - x_0 \| < \delta$. Inequalities (2) and (3) give, for $\| x - x_0 \| < \delta$,

$$\begin{aligned}\| f(x) - f(x_0) \| \\ &\leq \| f(x) - f_N(x) \| + \| f_N(x) - f_N(x_0) \| + \| f_N(x_0) - f(x_0) \| \\ &< 3\varepsilon.\end{aligned}$$

This shows that f is continuous at x_0. Thus $f \in C_F(A)$ and, by (2), $\| f - f_n \| = \sup \{ \| f(x) - f_n(x) \| : x \in A \} \leq \varepsilon$ for $n \geq N$, so $\lim_{n \to \infty} f_n = f$. This proves that $C_F(A)$ is complete. □

The next lemma, which is an application of the contraction mapping theorem (2.11.2), is the major step in the proof of the implicit function theorem.

7.5.4 LEMMA *Let E be a real normed linear space, F a real Banach space and g a continuous mapping of a non-empty open subset A of $E \times F$ into F which is differentiable with respect to the second variable on A and is such that the mapping $(x,y) \to g_2'(x,y)$ is continuous on A. Suppose that, for some point $(a,b) \in A$, $g(a,b) = 0$ and $g_2'(a,b) = 0$. Then there are positive real numbers α and β and a continuous mapping u of the open ball*

$$B_\alpha = \{x \in E : \| x - a \| < \alpha\}$$

into the closed ball

$$C_\beta = \{y \in F : \| y - b \| \leq \beta\}$$

such that

(a) $(x,u(x)) \in A$ *for all* $x \in B_\alpha$,
(b) $u(x) = b + g(x,u(x))$ *for all* $x \in B_\alpha$, *and*
(c) $u(a) = b$.

Further, if $0 < \gamma \leq \alpha$ and v is a continuous mapping of B_γ into C_β such that $(x,v(x)) \in A$ and $v(x) = b + g(x,v(x))$ for all $x \in B_\gamma$ then $u(x) = v(x)$ for all $x \in B_\gamma$.

PROOF The mapping $(x,y) \to g_2'(x,y)$ is continuous on A and $g_2'(a,b) = 0$ so there exists $\beta > 0$ such that

$$\| g_2'(x,y) \| < \tfrac{1}{2} \tag{4}$$

for all $(x,y) \in A$ with $\| (x,y) - (a,b) \| \leq \beta$. Since A is open we may suppose, by taking β smaller if necessary, that $(x,y) \in A$ whenever $\| (x,y) - (a,b) \| \leq \beta$. For each (fixed) $x \in B_\beta$ we can apply the inequality (4) and Corollary 7.3.3 (with $T = 0$) to the mapping $y \to g(x,y)$ and the closed ball C_β. This gives

$$\| g(x,y) - g(x,z) \| \leq \tfrac{1}{2} \| y - z \| \tag{5}$$

for all $x \in B_\beta$ and all $y,z \in C_\beta$. Since g is continuous on A and $g(a,b) = 0$ there exists $\alpha_0 > 0$ such that

$$\| g(x,y) \| < \tfrac{1}{2}\beta \tag{6}$$

for all $(x,y) \in A$ with $\| (x,y) - (a,b) \| < \alpha_0$. Let $\alpha = \min \{\alpha_0, \beta\}$. Then, for all $x \in B_\alpha$ and all $y \in C_\beta$, inequalities (5) and (6) give

$$\begin{aligned} \| g(x,y) \| &\leq \| g(x,y) - g(x,b) \| + \| g(x,b) \| \\ &\leq \tfrac{1}{2} \| y - b \| + \tfrac{1}{2}\beta \\ &\leq \beta. \end{aligned} \tag{7}$$

Let X be the set of all continuous mappings of B_α into C_β. Then X is a subset of the Banach space $C_F(B_\alpha)$ defined in Lemma 7.5.3. We shall prove that X is a closed subset of $C_F(B_\alpha)$. Let (u_n) be a sequence in X with $\lim_{n \to \infty} u_n = u$. For all $x \in B_\alpha$ we have† $\| u_n(x) - u(x) \| \leq \| u_n - u \|$. Therefore $\lim_{n \to \infty} u_n(x) = u(x)$ and, since $\| u_n(x) - b \| \leq \beta$, we have $\| u(x) - b \| = \lim_{n \to \infty} \| u_n(x) - b \| \leq \beta$. This shows that $u \in X$ and hence X is closed (Corollary 2.5.6). Since $C_F(B_\alpha)$ is complete it follows from Theorem 2.6.8 that X is also complete.

It follows from the definitions of α, β and X that $(x,u(x)) \in A$ for all $x \in B_\alpha$ and all $u \in X$. Given $u \in X$, we define a mapping Tu of B_α into F by

$$(Tu)(x) = b + g(x,u(x))$$

for all $x \in B_\alpha$. The mapping $x \to (x,u(x))$ is continuous on B_α (because u is continuous on B_α) and thus, by Theorem 2.7.4, the mapping Tu is continuous on B_α. Also, for all $x \in B_\alpha$, (7) gives

$$\| (Tu)(x) - b \| = \| g(x,u(x)) \| \leq \beta,$$

which shows that $Tu \in X$. Now let $u,v \in X$. For all $x \in B_\alpha$ inequality (5) gives

$$\begin{aligned} \| (Tu)(x) - (Tv)(x) \| &= \| g(x,u(x)) - g(x,v(x)) \| \\ &\leq \tfrac{1}{2} \| u(x) - v(x) \| \\ &\leq \tfrac{1}{2} \| u - v \|. \end{aligned}$$

Hence $\| Tu - Tv \| \leq \frac{1}{2} \| u - v \|$. This shows that $u \to Tu$ is a contraction mapping on X. Since X is complete there exists $u \in X$ with $Tu = u$ (Theorem 2.11.2). Clearly $u(x) = b + g(x,u(x))$ for all $x \in B_\alpha$. By (5)

$$\begin{aligned} \| u(a) - b \| &= \| g(a,u(a)) \| \\ &= \| g(a,u(a)) - g(a,b) \| \\ &\leq \tfrac{1}{2} \| u(a) - b \| \end{aligned}$$

from which it follows that $u(a) = b$. Thus u satisfies conditions (a), (b) and (c).

Suppose finally that $0 < \gamma \leq \alpha$ and that v is a continuous mapping

† $\| u_n - u \|$ is the norm of $u_n - u$ in $C_F(B_\alpha)$ which was defined in Lemma 7.5.3.

of B_γ into C_β such that $(x,v(x)) \in A$ and $v(x) = b + g(x,v(x))$ for all $x \in B_\gamma$. Then, again by (5), for all $x \in B_\gamma$ we have

$$\| u(x) - v(x) \| = \| g(x,u(x)) - g(x,v(x)) \| \leq \tfrac{1}{2} \| u(x) - v(x) \|$$

and hence $u(x) = v(x)$. □

7.5.5 THEOREM (the implicit function theorem) *Let E be a real normed linear space, F and G real Banach spaces and f a continuous mapping of a non-empty open subset A of $E \times F$ into G which is differentiable with respect to the second variable on A and such that the mapping $(x,y) \to f_2'(x,y)$ is continuous on A. Suppose that, for some point $(a,b) \in A$, $f(a,b) = 0$ and $f_2'(a,b)$ is a linear homeomorphism of F onto G. Then there is a positive real number α and a continuous mapping u of the open ball $B_\alpha = \{x \in E\colon \| x - a \| < \alpha\}$ into F such that*

(a) $(x,u(x)) \in A$ *for all* $x \in B_\alpha$,
(b) $f(x,u(x)) = 0$ *for all* $x \in B_\alpha$, *and*
(c) $u(a) = b$.

Further, if v is a continuous mapping of a neighbourhood N of a into F such that

(a′) $(x,v(x)) \in A$ *for all* $x \in N$,
(b′) $f(x,v(x)) = 0$ *for all* $x \in N$, *and*
(c′) $v(a) = b$

then there is a number γ, with $0 < \gamma \leq \alpha$, such that $u(x) = v(x)$ for all $x \in B_\gamma$.

PROOF Let $T = f_2'(a,b)$ and $S = T^{-1}$ (so that S is a linear homeomorphism of G onto F). Let

$$g(x,y) = y - b - S(f(x,y)) \tag{8}$$

for all $(x,y) \in A$. We shall verify that g satisfies the conditions of Lemma 7.5.4. It is clear that $g(a,b) = 0$ and that g is continuous on A. By Lemma 7.1.8 the linear mapping S is differentiable on G and $S'(z) = S$ for all $z \in G$. Therefore, by Theorem 7.2.2, the mapping $S \circ f$ is differentiable with respect to the second variable on A and

$$(S \circ f)_2'(x,y) = (S'(f(x,y))) \circ (f_2'(x,y)) = S \circ (f_2'(x,y))$$

for all $(x,y) \in A$. It now follows from Lemmas 7.1.7 and 7.1.8 and Theorem 7.2.1 that g is differentiable with respect to the second variable on A and

$$g_2'(x,y) = I - S \circ (f_2'(x,y))$$

for all $(x,y) \in A$, where I is the identity operator on F. In particular we have $g_2'(a,b) = I - S \circ T = 0$. Since the mapping $(x,y) \to f_2'(x,y)$

is continuous on A so is the mapping $(x,y) \to g_2'(x,y)$. We have now verified that g satisfies the conditions of Lemma 7.5.4.

By Lemma 7.5.4 there exist numbers $\alpha > 0$ and $\beta > 0$ and a continuous mapping u of B_α into the closed ball $C_\beta = \{y \in F: \| y - b \| \leq \beta\}$ with $u(a) = b$ and such that $(x,u(x)) \in A$ and $u(x) = b + g(x,u(x))$ for all $x \in B_\alpha$. It follows now from (8) that $S(f(x,u(x))) = 0$ for all $x \in B_\alpha$. Hence $f(x,u(x)) = 0$ for all $x \in B_\alpha$ because S is a linear homeomorphism. Thus u and α satisfy conditions (a), (b) and (c) of the theorem.

Suppose finally that v is a continuous mapping of a neighbourhood N of a into F which satisfies the conditions (a′), (b′) and (c′). Then (8) gives $g(x,v(x)) = v(x) - b$ for all $x \in N$. Also, since v is continuous and N is a neighbourhood of a, there exists $\gamma > 0$ such that $B_\gamma \subseteq N$ and $\| v(x) - b \| \leq \beta$ for all $x \in B_\gamma$. We may suppose that $\gamma \leq \alpha$ and so it follows from Lemma 7.5.4 that $u(x) = v(x)$ for all $x \in B_\gamma$. □

We shall now show that, under suitable extra conditions, the mapping u of Theorem 7.5.5 is differentiable on a neighbourhood of a. It is easy to verify that if u and f are differentiable at a point x then

$$f_1'(x,u(x)) + (f_2'(x,u(x)))\circ(u'(x)) = 0.$$

Consequently, for such points x we have

$$u'(x) = -(f_2'(x,u(x)))^{-1}\circ(f_1'(x,u(x))) \tag{9}$$

whenever the inverse mapping $(f_2'(x,u(x)))^{-1}$ exists.

The proof of the next theorem consists simply in verifying that (9) does indeed define the derivative of u at x; the calculation is rather long. We shall use the notation of Theorem 7.5.5.

7.5.6 THEOREM *Suppose that f satisfies the hypotheses of Theorem 7.5.5, and suppose in addition that f is differentiable with respect to the first variable on A and the mapping $(x,y) \to f_1'(x,y)$ is continuous on A. Then there exists δ, with $0 < \delta \leq \alpha$, such that u is continuously differentiable on B_δ. Further $u'(x)$ is given by (9) for each $x \in B_\delta$.*

PROOF By Lemma 7.5.2 the set $\mathscr{H}$ of all linear homeomorphisms of F onto G is open in $L(F,G)$ and hence since $T = f_2'(a,b) \in \mathscr{H}$ there exists $\theta > 0$ such that $U \in \mathscr{H}$ whenever $\| T - U \| < \theta$. By Theorem 2.7.4 the mapping $x \to f_2'(x,u(x))$ is continuous on B_α because the mappings $x \to (x,u(x))$ and $(x,y) \to f_2'(x,y)$ are continuous on B_α and A respectively. Thus since $u(a) = b$ there exists δ, with $0 < \delta \leq \alpha$, such that $\| f_2'(x,u(x)) - T \| < \theta$ for all $x \in B_\delta$. Consequently the operator $f_2'(x,u(x))$ is in $\mathscr{H}$ for all $x \in B_\delta$.

Let $c \in B_\delta$ and let $V_1 = f_1'(c,u(c))$ and $V_2 = f_2'(c,u(c))$. We shall prove that $-V_2^{-1}\circ V_1$ is the derivative of u at c. We remark that, by

Theorem 7.4.3, the mapping f is differentiable on A. For all $x \in B_\delta$ we have

$$\begin{aligned} \| u(x) - u(c) + (V_2^{-1} \circ V_1)(x - c) \| &= \| V_2^{-1}(V_2(u(x) - u(c)) + V_1(x - c)) \| \\ &\leq \| V_2^{-1} \| \; \| V_2(u(x) - u(c)) + V_1(x - c) \| \\ &= \| V_2^{-1} \| \; \| f'(c,u(c))\cdot((x,u(x)) - (c,u(c))) \|, \end{aligned} \tag{10}$$

where the last step follows from equation (1) of Section 7.4.

Let $0 < \eta \leq \frac{1}{2} \| V_2^{-1} \|^{-1}$. Since f is differentiable at $(c,u(c))$ there exists $\mu_1 > 0$ with

$$\| f(x,y) - f(c,u(c)) - f'(c,u(c))\cdot((x,y) - (c,u(c))) \| \leq \eta \| (x,y) - (c,u(c)) \| \tag{11}$$

for $\| (x,y) - (c,u(c)) \| < \mu_1$. Since B_δ is open we may suppose that $x \in B_\delta$ whenever $\| x - c \| < \mu_1$. Also, since u is continuous on B_δ, there exists $\mu_2 > 0$ such that

$$\| u(x) - u(c) \| < \mu_1$$

for $x \in B_\delta$ with $\| x - c \| < \mu_2$. Let $\mu = \min \{\mu_1, \mu_2\}$ and let $\| x - c \| < \mu$. Then

$$\| (x,u(x)) - (c,u(c)) \| = \max \{ \| x - c \|, \| u(x) - u(c) \| \} < \mu_1$$

and, since $f(x,u(x)) = f(c,u(c)) = 0$, it follows from (10) and (11) that

$$\begin{aligned} \| u(x) - u(c) + (V_2^{-1} \circ V_1)(x - c) \| &\leq \| V_2^{-1} \| \eta \| (x,u(x)) - (c,u(c)) \| \\ &\leq \| V_2^{-1} \| \eta (\| x - c \| + \| u(x) - u(c) \|). \end{aligned} \tag{12}$$

But $\| V_2^{-1} \| \eta \leq \frac{1}{2}$ so (12) gives

$$\begin{aligned} \| u(x) - u(c) \| &\leq \| u(x) - u(c) + (V_2^{-1} \circ V_1)(x - c) \| + \| (V_2^{-1} \circ V_1)(x - c) \| \\ &\leq \tfrac{1}{2}(\| x - c \| + \| u(x) - u(c) \|) + \| V_2^{-1} \circ V_1 \| \; \| x - c \| \end{aligned}$$

and therefore

$$\| u(x) - u(c) \| \leq (1 + 2 \| V_2^{-1} \circ V_1 \|) \| x - c \|. \tag{13}$$

Finally, from (12) and (13) we obtain

$$\| u(x) - u(c) + (V_2^{-1} \circ V_1)(x - c) \| \leq 2 \| V_2^{-1} \| \eta (1 + \| V_2^{-1} \circ V_1 \|) \| x - c \|.$$

Given $\varepsilon > 0$ let

$$\eta = \min \left\{ \frac{1}{2 \| V_2^{-1} \|}, \frac{\varepsilon}{2 \| V_2^{-1} \| (1 + \| V_2^{-1} \circ V_1 \|)} \right\}.$$

Then, by what we have proved above, there exists $\mu > 0$ such that

$$\| u(x) - u(c) + (V_2^{-1} \circ V_1)(x - c) \| \leq \varepsilon \| x - c \|$$

for $\| x - c \| < \mu$. This proves that u is differentiable at c and that $u'(c) = - V_2^{-1} \circ V_1$.

Finally we have to prove that u' is continuous on B_δ. We shall use Theorem 2.7.4 repeatedly without comment. We have already seen that $x \to f_2'(x,u(x))$ is continuous on B_α. Similarly $x \to f_1'(x,u(x))$ is continuous on B_α. By Lemma 7.5.2 the mapping $U \to U^{-1}$ is continuous on $\mathscr{H}$ and hence the mapping $x \to (f_2'(x,u(x)))^{-1}$ is continuous on B_δ. It follows that $x \to (-(f_2'(x,u(x)))^{-1}, f_1'(x,u(x)))$ is a continuous mapping of B_δ into $L(G,F) \times L(E,G)$. Since, by Lemma 7.5.1, the mapping $(U_1,U_2) \to U_1 \circ U_2$ is continuous on $L(G,F) \times L(E,G)$ it follows that the mapping $x \to -(f_2'(x,u(x)))^{-1} \circ (f_1'(x,u(x)))$ is continuous on B_δ. □

A comparison of Theorems 7.5.5 and 7.5.6 with the following corollary, which is the classical version of the implicit function theorem, will reveal the economy of notation that can be achieved by using Fréchet derivatives.

7.5.7 COROLLARY *Let $f_1, f_2, \ldots, f_m$ be real-valued functions on a non-empty open subset A of R^{n+m} and let $(\alpha_1, \ldots, \alpha_n, \beta_1, \ldots, \beta_m) \in A$. Suppose that $f_1, f_2, \ldots, f_m$ have continuous first partial derivatives on A, that $f_j(\alpha_1, \ldots, \alpha_n, \beta_1, \ldots, \beta_m) = 0$ for $j = 1, 2, \ldots, m$ and that*

$$\det\begin{pmatrix} D_{n+1}f_1(\alpha_1,\ldots,\alpha_n,\beta_1,\ldots,\beta_m)\ldots D_{n+m}f_1(\alpha_1,\ldots,\alpha_n,\beta_1,\ldots,\beta_m) \\ \cdots\cdots\cdots\cdots\cdots\cdots \\ D_{n+1}f_m(\alpha_1,\ldots,\alpha_n,\beta_1,\ldots,\beta_m)\ldots D_{n+m}f_m(\alpha_1,\ldots,\alpha_n,\beta_1,\ldots,\beta_m) \end{pmatrix} \neq 0. \quad (14)$$

Then there is a non-empty open ball B in R^n, with centre $(\alpha_1, \ldots, \alpha_n)$, and continuous real-valued functions $u_1, u_2, \ldots, u_m$ on B such that

$$u_j(\alpha_1, \ldots, \alpha_n) = \beta_j$$

and $$f_j(\xi_1, \ldots, \xi_n, u_1(\xi_1, \ldots, \xi_n), \ldots, u_m(\xi_1, \ldots, \xi_n)) = 0$$

for $j = 1, 2, \ldots, m$ and all $(\xi_1, \ldots, \xi_n) \in B$. Further $u_1, u_2, \ldots, u_m$ have continuous first partial derivatives on B and

$$D_jf_i(\xi_1, \ldots, \xi_n, \eta_1, \ldots, \eta_m) + \sum_{k=1}^{m} D_{n+k}f_i(\xi_1, \ldots, \xi_n, \eta_1, \ldots, \eta_m)D_ju_k(\xi_1, \ldots, \xi_n) = 0 \quad (15)$$

for $i = 1, 2, \ldots, m$, $j = 1, 2, \ldots, n$ and all $(\xi_1, \ldots, \xi_n) \in B$ (where $\eta_i = u_i(\xi_1, \ldots, \xi_n)$).

Further, if $v_1, v_2, \ldots, v_m$ are continuous real-valued functions on a neighbourhood N of $(\alpha_1, \ldots, \alpha_n)$ such that $v_j(\alpha_1, \ldots, \alpha_n) = \beta_j$ and

$$f_j(\xi_1, \ldots, \xi_n, v_1(\xi_1, \ldots, \xi_n), \ldots, v_m(\xi_1, \ldots, \xi_n)) = 0$$

for $j = 1, 2, \ldots, m$ and all $(\xi_1, \ldots, \xi_n) \in N$, then there is a non-empty open ball B_0, with centre $(\alpha_1, \ldots, \alpha_n)$, contained in B and such that

$$v_j(\xi_1, \ldots, \xi_n) = u_j(\xi_1, \ldots, \xi_n)$$

for $j = 1, 2, \ldots, m$ and all $(\xi_1, \ldots, \xi_n) \in B_0$.

PROOF Let A' be the set of points (x,y) in $\mathsf{R}^n \times \mathsf{R}^m$ such that

$$x = (\xi_1,\dots, \xi_n), y = (\eta_1,\dots, \eta_m) \text{ and } (\xi_1,\dots, \xi_n, \eta_1\dots, \eta_m) \in A.$$

The mapping $(x,y) \to (\xi_1,\dots, \xi_n, \eta_1,\dots, \eta_m)$ is a linear isometry of $\mathsf{R}^n \times \mathsf{R}^m$ onto R^{n+m} so the set A' is open in $\mathsf{R}^n \times \mathsf{R}^m$. Let f be the mapping of A' into R^m defined by

$$f(x,y) = (f_1(\xi_1,\dots, \xi_n, \eta_1,\dots, \eta_m),\dots, f_m(\xi_1,\dots, \xi_n, \eta_1,\dots, \eta_m))$$

for all $(x,y) \in A'$, where $x = (\xi_1,\dots, \xi_n)$ and $y = (\eta_1,\dots, \eta_m)$. We shall show that f satisfies the conditions of Theorems 7.5.5 and 7.5.6. Since $\mathsf{R}^n \times \mathsf{R}^m$ and R^{n+m} are linearly isometric it follows easily from Theorem 7.1.12 that f is continuously differentiable on A'. Corollary 7.1.5 shows that f is continuous on A', and Theorem 7.4.2 shows that f is differentiable with respect to both variables on A', and that the mappings $(x,y) \to f_1'(x,y)$ and $(x,y) \to f_2'(x,y)$ are continuous on A'. Also, by Theorem 7.1.11, $f_2'(a,b)$ is represented by the matrix which appears in equation (14) and which, by (14), is non-singular. It follows that $f_2'(a,b)$ is a linear homeomorphism of R^m onto itself. (See p. 232.) It is now easy to check that the remaining conditions of Theorems 7.5.5 and 7.5.6 are satisfied.

By Theorems 7.5.5 and 7.5.6 there is a non-empty open ball B in R^n, with centre a, and a continuously differentiable mapping u of B into R^m, with $u(a) = b$, such that $(x,u(x)) \in A'$ and $f(x,u(x)) = 0$ for all $x \in B$. Let $u(x) = (u_1(x), u_2(x),\dots, u_m(x))$ for all $x \in B$. It follows easily from the continuity of u on B that the functions $u_1, u_2,\dots, u_m$ are continuous on B. Also, since u is continuously differentiable on B, Theorem 7.1.12 shows that the functions $u_1, u_2,\dots, u_m$ have continuous first partial derivatives on B. It follows from equation (9) that

$$f_1'(x,u(x)) + (f_2'(x,u(x)))\circ(u'(x)) = 0$$

for all $x \in B$. Equation (15) follows from the above equation and the matrix representation of the derivatives involved (Theorem 7.1.11). The final assertion of the corollary concerning the uniqueness of the functions $u_1, u_2,\dots, u_m$ is an immediate consequence of the corresponding uniqueness assertion in Theorem 7.5.5. □

Equation (15) is often written in the form

$$\frac{\partial f_i}{\partial \xi_j} + \frac{\partial f_i}{\partial \eta_1}\frac{\partial \eta_1}{\partial \xi_j} + \dots + \frac{\partial f_i}{\partial \eta_m}\frac{\partial \eta_m}{\partial \xi_j} = 0,$$

where $\eta_i = u_i(\xi_1,\dots, \xi_n)$.

Theorem 7.5.5 can be used to prove the inverse function theorem which is concerned with the existence of local solutions, $x = u(y)$, of an equation $y = f(x)$.

7.5.8 THEOREM (the inverse function theorem) *Let E and F be real Banach spaces and f a continuously differentiable mapping of a non-empty open subset A of E into F. Suppose that $f'(a)$ is a linear homeomorphism of E onto F for some point $a \in A$. Then there is a positive real number δ such that*

(a) *$f(B_\delta)$ is an open subset of F, and*
(b) *$f|B_\delta$ is a homeomorphism of B_δ onto $f(B_\delta)$,*

where $B_\delta = \{x \in E: \| x - a \| < \delta\}$.

PROOF Let $T = f'(a)$. Since f is continuous at a there exists $\delta_1 > 0$ such that $\| f'(x) - T \| < \frac{1}{2} \| T^{-1} \|^{-1}$ for $\| x - a \| < \delta_1$. From this inequality and Corollary 7.3.3 we obtain the inequality

$$\| f(x_1) - f(x_2) - T(x_1 - x_2) \| \leq \tfrac{1}{2} \| T^{-1} \|^{-1} \| x_1 - x_2 \|$$

for all $x_1,x_2 \in B_{\delta_1}$. Thus for $x_1,x_2 \in B_{\delta_1}$ we have

$$\begin{aligned} &\| x_1 - x_2 \| \\ &\quad = \| T^{-1}(T(x_1 - x_2)) \| \\ &\quad \leq \| T^{-1} \| \; \| T(x_1 - x_2) \| \\ &\quad \leq \| T^{-1} \| (\| T(x_1 - x_2) - f(x_1) + f(x_2) \| + \| f(x_1) - f(x_2) \|) \\ &\quad \leq \tfrac{1}{2} \| x_1 - x_2 \| + \| T^{-1} \| \; \| f(x_1) - f(x_2) \| \end{aligned}$$

and therefore

$$\| x_1 - x_2 \| \leq 2 \| T^{-1} \| \; \| f(x_1) - f(x_2) \|. \tag{16}$$

Next, by Lemma 7.5.2, there exists $\delta_2 > 0$ such that $f'(x)$ is a linear homeomorphism of E onto F whenever $\| x - a \| < \delta_2$. We set $\delta = \min \{\delta_1, \delta_2\}$. The assertion (b) is an immediate consequence of the inequality (16) so it remains only to prove (a).

Let $c \in f(B_\delta)$ and let b be the unique point of B_δ with $f(b) = c$. Let $A_0 = \{(z,w)\colon z \in F, w \in B_\delta\}$ and let $g(z,w) = f(w) - z$ for all $(z,w) \in A_0$. It is clear that A_0 is an open subset of $F \times E$. We shall verify that g satisfies the conditions of Theorem 7.5.5. It is obvious that $g(c,b) = 0$ and g is continuous on A_0. Also g is differentiable with respect to the second variable on A_0 and $g_2'(z,w) = f'(w)$ for all $(z,w) \in A_0$. Consequently the mapping $(z,w) \to g_2'(z,w)$ is continuous on A_0 and $g_2'(c,b) = f'(b)$ which is a linear homeomorphism of E onto F. This proves that g satisfies the conditions of Theorem 7.5.5 and hence there exists $\alpha > 0$ and a mapping u of $\{z \in F: \| z - c \| < \alpha\}$ into E such that $(z,u(z)) \in A_0$ and $g(z,u(z)) = 0$ for $\| z - c \| < \alpha$. It follows that if $z \in F$ and $\| z - c \| < \alpha$ then $u(z) \in B_\delta$ and $z = f(u(z))$. Therefore

$$\{z \in F : \| z - c \| < \alpha\} \subseteq f(B_\delta).$$

Since $c \in f(B_\delta)$ was arbitrary, this proves that $f(B_\delta)$ is open. □

It follows directly from Theorem 7.2.4 and Lemma 7.5.2 that the mapping $(f|B_\delta)^{-1}$ is continuously differentiable on $f(B_\delta)$ and has derivative $f'(x)^{-1}$ at each point $f(x)$ of $f(B_\delta)$.

The classical version of the inverse function theorem follows from Theorem 7.5.8 just as Corollary 7.5.7 follows from Theorems 7.5.5 and 7.5.6. We leave the reader to supply the proof.

7.5.9 COROLLARY *Let $f_1, f_2,\dots, f_n$ be real-valued functions on a non-empty open subset A of R^n and let $(\alpha_1, \alpha_2,\dots, \alpha_n) \in A$. Suppose that $f_1, f_2,\dots, f_n$ have continuous first partial derivatives on A and that*

$$\det\begin{pmatrix} D_1f_1(\alpha_1, \alpha_2,\dots, \alpha_n) & \dots & D_nf_1(\alpha_1, \alpha_2,\dots, \alpha_n) \\ \cdot & \cdot\;\cdot\;\cdot & \cdot \\ D_1f_n(\alpha_1, \alpha_2,\dots, \alpha_n) & \dots & D_nf_n(\alpha_1, \alpha_2,\dots, \alpha_n) \end{pmatrix} \neq 0.$$

Then there is a non-empty open ball B, with centre $(\alpha_1, \alpha_2,\dots, \alpha_n)$, such that the restriction to B of the mapping

$$(\xi_1, \xi_2,\dots, \xi_n) \to (f_1(\xi_1, \xi_2,\dots, \xi_n),\dots, f_n(\xi_1, \xi_2,\dots, \xi_n))$$

is a homeomorphism of B onto an open subset of R^n.

As an illustration of the use of the implicit function theorem we shall give a justification of Lagrange's method of undetermined multipliers for finding local maxima and minima of functions of several variables when the variables are subject to side conditions. The reader may find the following heuristic description of the proof of Theorem 7.5.10 helpful, particularly in the case when $n = 3$ and $m = 2$. The set A_0 is a 'surface' in R^n and a neighbourhood of a on the surface A_0 is homeomorphic to an open ball B in R^{n-m}. We can regard the restriction of the function f to this neighbourhood of a as a function h on the open ball B. The function h has a local maximum (or minimum) at the centre of B, and consequently the classical partial derivatives of h vanish there. The theorem then follows from the relationship between the (classical) partial derivatives of f and h.

7.5.10 THEOREM *Let $f, g_1, g_2,\dots, g_m$ be real-valued functions on a non-empty open subset A of R^n, where $n > m$. Let*

$$A_0 = \{(\xi_1, \xi_2,\dots, \xi_n) \in \mathsf{R}^n : g_j(\xi_1, \xi_2,\dots, \xi_n) = 0 \text{ for } j = 1, 2,\dots, m\}$$

and let $(\alpha_1, \alpha_2,\dots, \alpha_n) \in A_0$. Suppose that $f, g_1, g_2,\dots, g_m$ have continuous first partial derivatives on A, that

$$\det\begin{pmatrix} D_{n-m+1}g_1(\alpha_1, \alpha_2,\dots, \alpha_n) & \dots & D_ng_1(\alpha_1, \alpha_2,\dots, \alpha_n) \\ \cdot & \cdot\;\cdot\;\cdot & \cdot \\ D_{n-m+1}g_m(\alpha_1, \alpha_2,\dots, \alpha_n) & \dots & D_ng_m(\alpha_1, \alpha_2,\dots, \alpha_n) \end{pmatrix} \neq 0, \quad (17)$$

and that there is a neighbourhood N *of* $(\alpha_1, \alpha_2,\dots, \alpha_n)$ *such that* $f(\xi_1, \xi_2,\dots, \xi_n) \leq f(\alpha_1, \alpha_2,\dots, \alpha_n)$ *for all* $(\xi_1, \xi_2,\dots, \xi_n) \in N \cap A_0$ [*or that* $f(\xi_1, \xi_2,\dots, \xi_n) \geq f(\alpha_1, \alpha_2,\dots, \alpha_n)$ *for all* $(\xi_1, \xi_2,\dots, \xi_n) \in N \cap A_0$]. *Then there are real numbers* $\lambda_1, \lambda_2,\dots, \lambda_m$ *such that*

$$D_kf(\alpha_1, \alpha_2,\dots, \alpha_n) + \sum_{j=1}^{m} \lambda_j D_k g_j(\alpha_1, \alpha_2,\dots, \alpha_n) = 0 \tag{18}$$

for $k = 1, 2,\dots, n$.

Proof It follows from (17) that there are real numbers $\lambda_1, \lambda_2,\dots, \lambda_m$ such that

$$\sum_{j=1}^{m} \lambda_j D_{n-m+k} g_j(\alpha_1, \alpha_2,\dots, \alpha_n) = -D_{n-m+k}f(\alpha_1, \alpha_2,\dots, \alpha_n) \tag{19}$$

for $k = 1, 2,\dots, m$. We have to prove that $\lambda_1, \lambda_2,\dots, \lambda_m$ satisfy the remaining $n - m$ equations in (18).

By Corollary 7.5.7 there is a non-empty open ball B in R^{n-m} and continuous real-valued functions $u_1, u_2,\dots, u_m$ on B which have continuous first partial derivatives on B and are such that

$$u_j(\alpha_1, \alpha_2,\dots, \alpha_{n-m}) = \alpha_{n-m+j}$$

and $\quad g_j(\xi_1,\dots, \xi_{n-m}, u_1(\xi_1,\dots, \xi_{n-m}),\dots, u_m(\xi_1,\dots, \xi_{n-m})) = 0$

for $j = 1, 2,\dots, m$ and all $(\xi_1,\dots, \xi_{n-m}) \in B$. Let

$$h(\xi_1,\dots, \xi_{n-m}) = f(\xi_1,\dots, \xi_{n-m}),u_1(\xi_1,\dots, \xi_{n-m}),\dots, u_m(\xi_1,\dots, \xi_{n-m}))$$

for all $(\xi_1,\dots, \xi_{n-m}) \in B$. It follows from the chain rule (Corollary 7.2.3) that h has continuous first partial derivatives on B and

$$D_kh(\alpha_1,\dots, \alpha_{n-m})$$
$$= D_kf(\alpha_1,\dots, \alpha_n) + \sum_{j=1}^{m} D_{n-m+j}f(\alpha_1,\dots, \alpha_n)D_ku_j(\alpha_1,\dots, \alpha_{n-m}) \tag{20}$$

for $k = 1, 2,\dots, n - m$. Further, from equation (15) we obtain

$$D_kg_j(\alpha_1,\dots, \alpha_n) + \sum_{i=1}^{m} D_{n-m+i}g_j(\alpha_1,\dots, \alpha_n)D_ku_i(\alpha_1,\dots, \alpha_{n-m}) = 0 \tag{21}$$

for $k = 1, 2,\dots, n - m$ and $j = 1, 2,\dots, m$.

It follows easily from the continuity of the functions $u_1, u_2,\dots, u_m$ that there is a neighbourhood N_0 of $(\alpha_1,\dots, \alpha_{n-m})$ in R^{n-m} such that the point $(\xi_1, \xi_2,\dots, \xi_{n-m}, u_1(\xi_1,\dots, \xi_{n-m}),\dots, u_m(\xi_1,\dots, \xi_{n-m}))$ is in N for all $(\xi_1,\dots, \xi_{n-m}) \in N_0$. Consequently $h(\xi_1,\dots, \xi_{n-m}) \leq h(\alpha_1,\dots, \alpha_{n-m})$ for all $(\xi_1,\dots, \xi_{n-m}) \in N_0$ [or $h(\xi_1,\dots, \xi_{n-m}) \geq h(\alpha_1,\dots, \alpha_{n-m})$ for all $(\xi_1,\dots, \xi_{n-m}) \in N_0$]. This shows that h has a local maximum [or local minimum] at $(\alpha_1,\dots, \alpha_{n-m})$ and therefore $D_kh(\alpha_1,\dots, \alpha_{m-n}) = 0$ for

$k = 1, 2,\ldots, n - m$. It now follows from equations (19), (20) and (21) that, for $k = 1, 2,\ldots, n - m$,

$$D_k f(\alpha_1,\ldots,\alpha_n) + \sum_{j=1}^{m} \lambda_j D_k g_j(\alpha_1,\ldots,\alpha_n)$$
$$= -\sum_{i=1}^{m} \Big(D_{n-m+i} f(\alpha_1,\ldots,\alpha_n) + \sum_{j=1}^{m} \lambda_j D_{n-m+i} g_j(\alpha_1,\ldots,\alpha_n)\Big) \times D_k u_i(\alpha_1,\ldots,\alpha_{n-m})$$
$$= 0. \square$$

EXERCISES

(1) Let E and F be real Banach spaces and let f be a mapping of a non-empty open subset A of E into F. Prove that if f is continuously differentiable on A and $f'(x)$ is a linear homeomorphism of E onto F for each $x \in A$, then $f(B)$ is open in F whenever B is open and $B \subseteq A$.

(2) If, in Theorem 7.5.8, the linear spaces E and F are finite dimensional then property (a) can be proved more simply than by using the implicit function theorem. Let $c = f(b) \in f(B_\delta)$, let $y \in F$ and let $g(x) = \| y - f(x) \|$ for $x \in B_\delta$. Prove that there exists $r > 0$ such that if $\| y - c \| < r$ then $g(x_0) = \inf \{g(x) : x \in B_\delta\}$ for some $x_0 \in B_\delta$. Prove that if $x \in B_\delta$ and $g(x) > 0$ then $g(x + \lambda z) < g(x)$ for some λ when $z = T^{-1}(y - f(x))$. Deduce that $f(B_\delta)$ is an open subset of F.

(3) By considering the mapping $(x,y) \to (x,f(x,y))$ of A into $E \times G$, show that if E is a Banach space and f satisfies the conditions of both Theorem 7.5.5 and Theorem 7.5.6 then the conclusion of Theorem 7.5.5 can be deduced from Theorem 7.5.8. Together with Exercise (2) this provides an alternative proof of Corollary 7.5.7 which does not depend on Lemma 7.5.4.

(4) Prove that there is an open neighbourhood U of the point $(1,-1)$ in R^2 and real-valued functions ϕ and ψ, with continuous first partial derivatives on U, which satisfy the conditions:

(a) $\phi(1,-1) = \psi(1,-1) = 1$, and

(b) $2\phi(x,y)\psi(x,y) = x^2 + y^2$ and $\phi(x,y)^3 - \psi(x,y)^3 = x^3 + y^3$ for all $(x,y) \in U$.

Find the partial derivatives of ϕ and ψ on U.

(5) Let $f(x,y,z) = (z \cos xy,\ z \sin xy,\ x + z)$ for all $(x,y,z) \in \mathsf{R}^3$. Prove that there is an open neighbourhood U of the point $(1,0,1)$ in R^3 such that $f|U$ is a homeomorphism of U onto an open subset of R^3.

(6) Let the function f on R be defined by

$$f(t) = \begin{cases} \tfrac{1}{2}t + t^2 \sin 1/t & \text{when } t \neq 0, \\ 0 & \text{when } t = 0. \end{cases}$$

Prove that f is differentiable on R and $Df(0) \neq 0$, but that f is not one-to-one on any neighbourhood of 0. Why does this example not contradict Theorem 7.5.8?

(7) (a) Let f be a differentiable real-valued function on R with $Df(x) \neq 0$ for all $x \in \mathsf{R}$. Prove that f is a one-to-one mapping of R onto $f(\mathsf{R})$. Is $f(\mathsf{R})$ open?

(b) Let $f(x) = (e^x \cos y, e^x \sin y)$ for all $(x,y) \in \mathsf{R}^2$. Prove that f is continuously differentiable on R^2 and that $f'(x,y)$ is regular for each point $(x,y) \in \mathsf{R}^2$ but that f is not one-to-one.

(8) Problems of applied mathematics frequently give rise to non-linear equations. The exercise which follows (the first part of which, on differentiation under an integral, is a preliminary) gives an indication of how the implicit function theorem can be applied to non-linear equations in infinite-dimensional contexts.

(a) Let $I = [a,b]$ and let A be a non-empty open subset of a real normed linear space E. Let f be a continuous mapping of the subset $I \times A$ of the normed linear space $\mathsf{R} \times E$ into a real Banach space F and let

$$g(x) = \int_a^b f(t,x)\, dt$$

for all $x \in A$. Prove that g is a continuous mapping of A into F. Prove further that if f is differentiable with respect to the second variable at each point of $I \times A$ and the mapping $(t,x) \to f_2'(t,x)$ is continuous on $I \times A$ then g is continuously differentiable on A with

$$g'(x) = \int_a^b f_2'(t,x)\, dt$$

for all $x \in A$. (See Section 3.7 for the definition of the integrals and use Exercise 3.7(2) to show that

$$\left(\int_a^b f_2'(t,x)\, dt\right)\cdot y = \int_a^b f_2'(t,x)\cdot y\, dt$$

for all $x \in A$ and $y \in E$.)

(b) Let f be a continuous real-valued function on the subset $B = \{(s,t,\xi,\lambda) : a \leq s \leq b,\ a \leq t \leq b,\ |\,\xi\,| < m,\ |\,\lambda\,| < l\}$ of R^4 and suppose that f satisfies the conditions (a) $f(s,t,0,0) = 0$ for all $s,t \in I = [a,b]$, (b) f is differentiable on B with respect to the third variable, $D_3 f$ is continuous on B and $D_3 f(s,t,0,0) = 0$ for all $s,t \in I$. Prove that there exists $\alpha > 0$ and a continuous mapping $\lambda \to x_\lambda$ of $(-\alpha,\alpha)$ into $C_{\mathsf{R}}([a,b])$ such that, for all $\lambda \in (-\alpha,\alpha)$,

$$x_\lambda(s) = \int_a^b f(s,t,x_\lambda(t),\lambda)\, dt$$

for all $s \in I$.

(Hints: Let $A = \{(\lambda,x) \in \mathsf{R} \times C_{\mathsf{R}}([a,b]) : |\lambda| < l, \|x\| < m\}$ and define a mapping F of A into $C_{\mathsf{R}}([a,b])$ by

$$F(\lambda,x)(s) = \int_a^b f(s,t,x(t),\lambda)\,dt.$$

Let $f(\cdot,t,\xi,\lambda)$ denote the function $s \to f(s,t,\xi,\lambda)$. Show that

$$(t,(\lambda,x)) \to f(\cdot,t,x(t),\lambda)$$

is a continuous mapping of $I \times A$ into $C_{\mathsf{R}}([a,b])$ and that

$$F(\lambda,x) = \int_a^b f(\cdot,t,x(t),\lambda)\,dt.$$

Deduce from the result of (a) that F is continuous on A. Now apply (a) to the mapping $(t,x) \to f(\cdot,t,x(t),\lambda)$ to prove that F is differentiable with respect to the second variable and that $F_2'(\lambda,x)$ is the Fredholm integral operator (see p. 218) with kernel $(s,t) \to D_3 f(s,t,x(t),\lambda)$. Finally apply the implicit function theorem to the mapping

$$(\lambda,x) \to x - F(\lambda,x).)$$

CHAPTER 8

Baire's Theorem and its Applications

In this chapter we shall be concerned with applications of a fundamental property of complete metric spaces. Our applications fall into two groups: direct applications to problems of classical analysis, and applications to the theory of bounded linear transformations. Our main interest lies in the second group of applications, which centre on two basic principles of functional analysis: the principle of uniform boundedness and the open mapping theorem.

8.1 Baire's Theorem

We recall that a subset A of a metric space X is said to be dense in X if and only if the closure of A is X.

8.1.1 DEFINITION A subset A of a metric space X is said to be *nowhere dense in* X if and only if the closure of A has empty interior, that is if and only if $(\bar{A})^\circ = \emptyset$. A subset of X is said to be *somewhere dense in* X if and only if it is not nowhere dense in X.

It is obvious that a subset A of X is somewhere dense in X if and only if $\bar{A}$ contains a non-empty open ball. Also, since $\bar{A}$ is closed, A is nowhere dense in X if and only if $\bar{A}$ is nowhere dense in X. A subset of a nowhere dense set is again nowhere dense.

8.1.2 LEMMA *Let A be a subset of a metric space X. Then A is nowhere dense in X if and only if $X \sim \bar{A}$ is dense in X.*

PROOF By Theorem 2.3.11 $(\bar{A})^\circ = X \sim (X \sim \bar{A})^-$. □

There are many situations in analysis in which it is useful to have some description or measure of the size of a given set. In some metric space situations the appropriate description of size is in terms of density. Thus a set that is dense in X is 'thickly spread over X' and a set that is nowhere dense in X is 'thinly spread over X'. Another important description of the 'size' of a subset of a metric space is provided by the concept of meagreness.

8.1.3 DEFINITION A subset A of a metric space X is said to be *meagre in* X if and only if A is the union of a countable family of subsets of X each of which is nowhere dense in X.

It is clear that a subset of a meagre set is again meagre, and it follows from Theorem 1.4.5 that the union of a countable family of meagre sets is meagre.

A set that is meagre in X is often said to be of the *first category in* X and a set that is not meagre in X is said to be of the *second category in* X. The 'category' terminology, which has historical precedence, is, fortunately, falling into disuse. However, arguments that depend on the notion of a meagre set are still often called category arguments.

8.1.4 THEOREM *Let X be a complete metric space and $\{A_n : n = 1, 2,...\}$ a countable family of open subsets of X each of which is dense in X. Then $\bigcap_{n=1}^{\infty} A_n$ is also dense in X.*

PROOF We have to show that, for each $x \in X$ and each $r > 0$, the intersection of the open ball $B(x,r)$ with the set $\bigcap_{n=1}^{\infty} A_n$ is non-empty. Let $x_1 \in X$ and $r_1 > 0$. We shall define inductively a sequence (x_n) in X and a sequence (r_n) of positive real numbers such that, for $n = 1, 2,...,$

(a) $B'(x_{n+1},r_{n+1}) = \{x : d(x_{n+1},x) \leq r_{n+1}\} \subseteq A_n \cap B(x_n,r_n)$, and
(b) $r_n \leq 1/n$.

Suppose that points $x_2, x_3,..., x_n$ in X, and positive real numbers $r_2, r_3,..., r_n$ have been chosen to satisfy (a) and (b). The ball $B(x_n,r_n)$ is a non-empty open set and A_n is an open set that is dense in X; consequently $A_n \cap B(x_n,r_n)$ is a non-empty open set. Thus there is a non-empty open ball, $B(x_{n+1},2r_{n+1})$ say, with $B(x_{n+1},2r_{n+1}) \subseteq A_n \cap B(x_n,r_n)$. Obviously we may take $r_{n+1} \leq (n + 1)^{-1}$. Clearly x_{n+1} and r_{n+1} then satisfy (a) and (b). This completes the definition of the sequences (x_n) and (r_n). It follows from (a) and (b) that $d(x_m,x_n) \leq 1/n$ whenever $m \geq n$ and therefore (x_n) is a Cauchy sequence. Since X is complete (x_n) converges. Let $x_0 = \lim_{n\to\infty} x_n$. For each positive integer m, (a) shows that $x_n \in B'(x_m, r_m)$ for $n \geq m$ and hence, since $B'(x_m, r_m)$ is closed, we have $x_0 \in B'(x_m, r_m)$. It follows now from (a) that $x_0 \in B(x_1, r_1) \cap \bigcap_{n=1}^{\infty} A_n$. □

Theorem 8.1.4 can be reformulated in terms of meagre sets.

8.1.5 Corollary *Let A be a subset of a complete metric space X that is meagre in X. Then $X \sim A$ is dense in X.*

Proof Since A is meagre in X there is a countable family, $\{A_n : n = 1, 2, \ldots\}$ say, of nowhere dense subsets of X with $A = \bigcup_{n=1}^{\infty} A_n$. The open set $X \sim \bar{A}_n$ is dense in X by Lemma 8.1.2 and therefore, by Theorem 8.1.4, $\bigcap_{n=1}^{\infty} (X \sim \bar{A}_n)$ is also dense in X. But

$$X \sim A = X \sim \bigcup_{n=1}^{\infty} A_n = \bigcap_{n=1}^{\infty} (X \sim A_n) \supseteq \bigcap_{n=1}^{\infty} (X \sim \bar{A}_n)$$

and hence $X \sim A$ is dense in X. □

8.1.6 Corollary *A closed subset of a complete metric space X that is meagre in X is nowhere dense in X.*

Proof Let A be a closed subset of X that is meagre in X. By Corollary 8.1.5 the set $X \sim A$ is dense in X and, since $A = \bar{A}$, it then follows from Lemma 8.1.2 that A is nowhere dense in X. □

8.1.7 Corollary *Let X be a complete metric space and let*

$$\{A_n : n = 1, 2, \ldots\}$$

be a countable family of subsets of X with $X = \bigcup_{n=1}^{\infty} A_n$. Then at least one of the sets A_n is somewhere dense in X.

Proof Suppose that each A_n is nowhere dense in X. Then X is meagre in itself and consequently, by Corollary 8.1.5, $\emptyset = X \sim X$ is dense in X, which is absurd. □

Theorem 8.1.4 and Corollaries 8.1.5 and 8.1.7 have all been called Baire's theorem in the literature on metric spaces. They arose from R. Baire's study of continuous functions which was made at the end of the last century. Our first application of Baire's theorem is simple but entertaining.

Every set X is the union of its subsets that contain only one point: $X = \bigcup_{x \in X} \{x\}$. In a metric space X a set consisting of a single point is nowhere dense unless the point is an isolated point of X and consequently, if X has no isolated points, every countable subset of X is meagre in X. It now follows from Corollary 8.1.7 that if X is a complete metric space with no isolated points then X must be an uncountable set. In particular the set of all real numbers is uncountable. (cf. Theorem 1.6.6.)

EXERCISES

(1) Prove that if f is either a continuous or a monotonic real-valued function defined on R then the graph $\{(x,f(x)) : x \in \mathsf{R}\}$ of f is a nowhere dense subset of R^2.

(2) The *diameter* of a bounded subset A of a metric space X is defined to be $d(A) = \sup \{d(x,y) : x,y \in A\}$. Prove that a metric space X is complete if and only if each countable family $\{A_n : n = 1, 2,...\}$ of closed bounded subsets of X with $A_{n+1} \subseteq A_n$ for $n = 1, 2,...$ and $\lim_{n\to\infty} d(A_n) = 0$ has a non-empty intersection. Show that the condition that $\lim_{n\to\infty} d(A_n) = 0$ cannot be omitted. Examine the proof of Theorem 8.1.4 in the light of this result.

(3) Prove that R is not the union of any countable family of pairwise disjoint proper closed subsets of R. (Hint: suppose that $\mathsf{R} = \bigcup_{n=1}^{\infty} E_n$ with $E_m \cap E_n = \emptyset$ for $m \neq n$ and consider the subspace $F = \mathsf{R} \sim \bigcup_{n=1}^{\infty} E_n^{\circ}$ of R.)

(4) Prove that an infinite-dimensional Banach space cannot be expressed as a countable union of compact sets.

(5) Prove that the set of rational numbers cannot be expressed as a countable intersection of open subsets of R.

8.2 Nowhere Differentiable Continuous Functions

In the early development of analysis, mathematical intuition was relatively naïve. The common notion of a continuous function was in terms of a graph which could be drawn, and it was therefore for long assumed implicitly that a continuous function was differentiable except possibly at a set of isolated exceptional points. Early in the nineteenth century unsuccessful attempts were made to prove this assumption, and the question was finally settled by Weierstrass, who showed that the function defined by

$$f(x) = \sum_{n=0}^{\infty} a^n \cos (b^n x)$$

is nowhere differentiable when b is an odd integer, $0 < a < 1$ and $ab > 1 + (3/2)\pi$. This example was first published in 1875 by du Bois Reymond who reproduced Weierstrass's proof. Since then simpler examples have been constructed. The reader will find an account of one of the simplest, given by van der Waarden in 1930, in RIESZ and NAGY [23, §1], and he will find a systematic discussion of nowhere differentiable functions in HOBSON [16, pp. 401–12].

The proof that we shall give of the existence of a continuous function that is nowhere differentiable is non-constructive in the sense that it does not yield a concrete example of such a function but it does show that, in a sense, a continuous function is more likely to be nowhere differentiable than to be differentiable at even one point.

We recall that the set $C_{\mathsf{R}}(\mathsf{R})$ of all bounded continuous real-valued functions on R is a complete metric space with the uniform metric given by

$$d(f,g) = \| f - g \| = \sup \{ | f(t) - g(t) | : t \in \mathsf{R}\}$$

(Theorem 2.8.3). The existence, and relative abundance, of functions in $C_{\mathsf{R}}(\mathsf{R})$ that are nowhere differentiable follows immediately from the following theorem and Corollary 8.1.5.

8.2.1 THEOREM *The set of functions in* $C_{\mathsf{R}}(\mathsf{R})$ *that are differentiable at at least one point of* R *is meagre in* $C_{\mathsf{R}}(\mathsf{R})$.

PROOF Let A be the set of all functions in $C_{\mathsf{R}}(\mathsf{R})$ that are differentiable at at least one point of R. Let $f \in A$ and let $t_0 \in \mathsf{R}$ be such that f is differentiable at t_0. Then $\lim\limits_{h \to 0} (f(t_0 + h) - f(t_0))/h$ exists. Also, since f is bounded on R we have $\lim\limits_{h \to \pm\infty} (f(t_0 + h) - f(t_0))/h = 0$. It now follows easily that there exists δ with $0 < \delta < 1$ such that the function

$$h \to (f(t_0 + h) - f(t_0))/h$$

is bounded on each of the open intervals $(-\infty, -1/\delta)$, $(-\delta,0)$, $(0,\delta)$ and $(1/\delta, +\infty)$. The function $h \to (f(t_0 + h) - f(t_0))/h$ is continuous on the closed intervals $[-1/\delta, -\delta]$ and $[\delta, 1/\delta]$ and hence is bounded there (Corollary 4.3.2). This proves that the set

$$\{(f(t_0 + h) - f(t_0))/h : h \neq 0\}$$

is bounded.

For each pair of positive integers m and n, let A_{mn} be the set of all functions f in $C_{\mathsf{R}}(\mathsf{R})$ with the property that there exists $t \in [-n,n]$ with

$$\left| \frac{f(t + h) - f(t)}{h} \right| \leq m$$

for all $h \neq 0$. It follows from what we have already proved that $A \subseteq \bigcup_{m,n=1}^{\infty} A_{mn}$. Since a subset of a meagre set is meagre, to complete the proof of the theorem we have only to show that each A_{mn} is nowhere dense in $C_{\mathsf{R}}(\mathsf{R})$.

Let m and n be positive integers. We shall prove first that A_{mn} is closed. Let (f_k) be a sequence in A_{mn} with $\lim_{k\to\infty} f_k = f$. Then for $k = 1, 2,\ldots$ there exists $t_k \in [-n,n]$ with

$$\left| \frac{f_k(t_k + h) - f_k(t_k)}{h} \right| \leq m \tag{1}$$

for all $h \neq 0$. By the Bolzano–Weierstrass theorem (1.6.3) the sequence (t_k) has a convergent subsequence so, replacing (f_k) by a suitable subsequence, we may suppose that (t_k) converges to a limit $t \in [-n,n]$. Now consider numbers $\varepsilon > 0$ and $h \neq 0$. There exists an integer K such that

$$\| f - f_k \| < \tfrac{1}{4}\varepsilon \,|\, h \,|\text{ and } |\, t - t_k \,| < \min\left\{ \frac{\varepsilon \,|\, h \,|}{4m}, \tfrac{1}{2} \,|\, h \,| \right\} \tag{2}$$

for all $k \geq K$. If $t_k = t$ for all $k \geq K$ we see at once, by letting $k \to \infty$ in (1), that $f \in A_{mn}$; otherwise choose $k \geq K$ with $t_k \neq t$. By (2) we have $|\, t + h - t_k \,| \geq |\, h \,| - |\, t - t_k \,| > \frac{1}{2} \,|\, h \,| > 0$ so $t + h - t_k \neq 0$. Consequently we obtain from (1) and (2)

$$\begin{aligned}
\left| \frac{f(t+h) - f(t)}{h} \right| &\leq \left| \frac{f(t+h) - f_k(t+h)}{h} \right| + \left| \frac{f_k(t+h) - f_k(t_k)}{h} \right| \\
&\quad + \left| \frac{f_k(t_k) - f_k(t)}{h} \right| + \left| \frac{f_k(t) - f(t)}{h} \right| \\
&\leq \left| \frac{f_k(t_k + t + h - t_k) - f_k(t_k)}{t + h - t_k} \right| \left| 1 + \frac{t - t_k}{h} \right| \\
&\quad + \left| \frac{f_k(t_k) - f_k(t_k + t - t_k)}{t - t_k} \right| \left| \frac{t - t_k}{h} \right| \\
&\quad + \frac{2}{|\, h \,|} \| f - f_k \| \\
&< \tfrac{1}{2}\varepsilon + m\left(1 + \frac{2 \,|\, t - t_k \,|}{|\, h \,|}\right) \\
&< m + \varepsilon.
\end{aligned}$$

This proves that $|\,(f(t+h) - f(t))/h \,| < m + \varepsilon$ for all $h \neq 0$ and all $\varepsilon > 0$. Consequently $|\,(f(t+h) - f(t))/h \,| \leq m$ for all $h \neq 0$. This proves that $f \in A_{mn}$ and hence that A_{mn} is closed.

We shall prove next that $C_{\mathsf{R}}(\mathsf{R}) \sim A_{mn}$ is dense in $C_{\mathsf{R}}(\mathsf{R})$. Let

$f \in C_{\mathsf{R}}(\mathsf{R})$ and $\varepsilon > 0$. The function f is uniformly continuous on $[-n,n]$ (Theorem 4.3.8) so there exists $\delta > 0$ such that

$$| f(s) - f(t) | < \tfrac{1}{2}\varepsilon \tag{3}$$

for all $s,t \in [-n,n]$ with $| s - t | < \delta$. Choose a positive integer K so that $2(n+1) < \delta K$ and let $t_k = -(n+1) + 2(n+1)K^{-1}k$ for $k = 0, 1, 2, \ldots, K$. Let $m_k = \frac{1}{2}K(n+1)^{-1}(f(t_k) - f(t_{k-1}))$ and

$$m' = \max \{ | m_1 |, | m_2 |, \ldots, | m_K | \}.$$

We define a function g_1 on R, which is piecewise linear on the interval $[-(n+1), n+1]$, by setting

$$g_1(t) = \begin{cases} m_k(t - t_{k-1}) + f(t_{k-1}) & \text{for } t_{k-1} \leq t \leq t_k, \text{ and } k = 1, 2, \ldots, K, \\ f(t) & \text{for } t \notin [-(n+1), n+1]. \end{cases}$$

It is obvious that $g_1 \in C_{\mathsf{R}}(\mathsf{R})$ and it follows from (3) that $\| f - g_1 \| \leq \frac{1}{2}\varepsilon$. It is easy to see that for each point $t \in [-n,n]$

$$\left| \frac{g_1(t+h) - g_1(t)}{h} \right| \leq m' \tag{4}$$

for all sufficiently small non-zero numbers h. To obtain a function in $C_{\mathsf{R}}(\mathsf{R}) \sim A_{mn}$ that approximates to f we have only to add to g_1 a small 'saw-tooth' function with sufficiently 'sharp' teeth. Let g_2 be the function whose graph is the polygon joining the points

$$(\tfrac{1}{4}k\varepsilon(m + m')^{-1}, \tfrac{1}{2}(-1)^k\varepsilon),$$

where $k = 0, \pm 1, \pm 2, \ldots$. It is clear that $g_2 \in C_{\mathsf{R}}(\mathsf{R})$ and $\| g_2 \| = \frac{1}{2}\varepsilon$. A simple calculation shows that, for each $t \in \mathsf{R}$,

$$\left| \frac{g_2(t+h) - g_2(t)}{h} \right| = 2(m + m') \tag{5}$$

for all sufficiently small non-zero numbers h.

Let $g = g_1 + g_2$. Then $g \in C_{\mathsf{R}}(\mathsf{R})$ and

$$\| f - g \| \leq \| f - g_1 \| + \| g_2 \| \leq \varepsilon.$$

Also, for each $t \in [-n,n]$, inequalities (4) and (5) give

$$\begin{aligned} \left| \frac{g(t+h) - g(t)}{h} \right| &= \left| \frac{g_1(t+h) - g_1(t)}{h} + \frac{g_2(t+h) - g_2(t)}{h} \right| \\ &\geq \left| \frac{g_2(t+h) - g_2(t)}{h} \right| - \left| \frac{g_1(t+h) - g_1(t)}{h} \right| \\ &\geq 2(m + m') - m' \\ &> m \end{aligned}$$

for all sufficiently small non-zero numbers h. This proves that $g \in C_{\mathsf{R}}(\mathsf{R}) \sim A_{mn}$. Consequently $C_{\mathsf{R}}(\mathsf{R}) \sim A_{mn}$ is dense in $C_{\mathsf{R}}(\mathsf{R})$, and it follows from Lemma 8.1.2 that A_{mn} is nowhere dense in $C_{\mathsf{R}}(\mathsf{R})$. □

8.3 Pointwise Limits of Continuous Functions

Our second application of Baire's theorem is again mainly of historical interest. Baire himself, in 1899, gave a complete characterization of those functions on an interval that are pointwise limits of sequences of continuous functions. One half of Baire's characterization is an immediate consequence of the following theorem.

8.3.1 THEOREM *Let X be a metric space, let (f_n) be a sequence of continuous real-valued functions on X that converges pointwise on X and let $f(x) = \lim_{n\to\infty} f_n(x)$ for all $x \in X$. Then the set of points at which f is not continuous is meagre in X.*

PROOF Let D be the set of points of X at which f is not continuous. Then $x \in D$ if and only if there exists $\varepsilon > 0$ with the property that, for each $\delta > 0$, there exists $x_\delta \in B(x,\delta)$ such that $|f(x_\delta) - f(x)| \geq \varepsilon$. For each positive integer n let D_n be the set of all points $x \in X$ with the property that, for each $\delta > 0$, there exists $x_\delta \in B(x,\delta)$ such that

$$|f(x_\delta) - f(x)| \geq \frac{1}{n}.$$

Clearly $D = \bigcup_{n=1}^{\infty} D_n$. We shall prove that each D_n is meagre in X, and it will then follow that D is also meagre in X. (See the remarks after Definition 8.1.3.) Let n be a positive integer and, for $m = 1, 2, \ldots$, let

$$A_m = \left\{x \in X : |f_k(x) - f(x)| < \frac{1}{8n} \text{ for all } k \geq m\right\}. \tag{1}$$

We have $X = \bigcup_{m=1}^{\infty} A_m$ because $f(x) = \lim_{k\to\infty} f_k(x)$ for all $x \in X$, and hence $D_n = \bigcup_{m=1}^{\infty} (D_n \cap A_m)$. We shall prove that $D_n \cap A_m$ is nowhere dense for each m.

Suppose on the contrary that, for some positive integer m, the set $D_n \cap A_m$ is somewhere dense in X and choose an interior point x_0 of $(D_n \cap A_m)^-$. Then there exists $\delta_0 > 0$ such that

$$B(x_0,\delta_0) \subseteq (D_n \cap A_m)^-. \tag{2}$$

Since f_m is continuous on X we may also choose δ_0 so that

$$|f_m(x) - f_m(x_0)| < \frac{1}{8n} \tag{3}$$

for all $x \in B(x_0,\delta_0)$. For all $x \in A_m$ and all $k \geq m$, (1) gives

$$|f_k(x) - f_m(x)| \leq |f_k(x) - f(x)| + |f(x) - f_m(x)| < \frac{1}{4n}. \tag{4}$$

For all $x,y \in B(x_0,\delta_0)$ inequality (3) gives

$$| f_m(x) - f_m(y) | \leq | f_m(x) - f_m(x_0) | + | f_m(x_0) - f_m(y) | < \frac{1}{4n}. \quad (5)$$

By (4) and (5), for all $x,y \in A_m \cap B(x_0,\delta_0)$ and all $k \geq m$, we have

$$\begin{aligned} | f_k(x) - f_k(y) | &\leq | f_k(x) - f_m(x) | + | f_m(x) - f_m(y) | \\ &\quad + | f_m(y) - f_k(y) | \\ &< \frac{3}{4n}. \end{aligned} \quad (6)$$

Let $x,y \in B(x_0,\delta_0)$. By (2) there are sequences (x_r) and (y_r) in $D_n \cap A_m$ with $\lim_{r\to\infty} x_r = x$ and $\lim_{r\to\infty} y_r = y$. Since $B(x_0,\delta_0)$ is a neighbourhood of x and of y there is an integer R such that $x_r \in B(x_0,\delta_0)$ and $y_r \in B(x_0,\delta_0)$ for all $r \geq R$, from which it follows incidentally that

$$B(x_0,\delta_0) \subseteq (B(x_0,\delta_0) \cap D_n)^-.$$

Since each of the functions f_k is continuous on X inequality (6) and Theorem 2.7.7 give, for $k \geq m$,

$$| f_k(x) - f_k(y) | = \lim_{r\to\infty} | f_k(x_r) - f_k(y_r) | \leq \frac{3}{4n}$$

and hence

$$| f(x) - f(y) | = \lim_{k\to\infty} | f_k(x) - f_k(y) | \leq \frac{3}{4n}. \quad (7)$$

Since $B(x_0,\delta_0)$ is open, for each $x \in B(x_0,\delta_0)$ there exists $\delta > 0$ such that $B(x,\delta) \subseteq B(x_0,\delta_0)$ and it therefore follows from (7) that $D_n \cap B(x_0,\delta_0) = \emptyset$. But we have already remarked that $B(x_0,\delta_0) \subseteq (D_n \cap B(x_0,\delta_0))^-$, so we have derived a contradiction. This proves that $D_n \cap A_m$ is nowhere dense in X. □

Theorem 8.3.1 and Corollary 8.1.5 together show that if X is a complete metric space then the set of points of continuity of a pointwise limit of a sequence of real-valued continuous functions on X is dense in X. It is easy to give examples of discontinuous functions that are pointwise limits of sequences of continuous functions, for example the function f given by

$$f(x) = \lim_{n\to\infty} \frac{1 - x^{2n}}{1 + x^{2n}}$$

is discontinuous at the points ± 1.

In 1899 Baire proved that a real-valued function f on a bounded closed interval $[a,b]$ is the pointwise limit of a sequence of real-valued continuous functions on $[a,b]$ if and only if, for each closed subset A of $[a,b]$ without isolated points, the set of points of A at which $f|A$ is discontinuous is meagre in A. The necessity of Baire's condition is a special case of Theorem 8.3.1; for a proof of the sufficiency of the condition the reader is referred to HOBSON [16, pp. 264–74].

8.4 The Principle of Uniform Boundedness

The techniques used in the preceding applications of Baire's theorem are typical of those used in 'classical' applications. From the point of view of functional analysis, the importance of Baire's theorem derives from the fact that it is the basis of the proofs of two fundamental principles of functional analysis: the principle of uniform boundedness and the open mapping theorem. The rest of this chapter is devoted to proving these principles and giving some of their applications in classical analysis and in the theory of bounded linear transformations.

8.4.1 THEOREM (the principle of uniform boundedness) *Let E be a Banach space, let F be a normed linear space over the same field and let $\mathscr{T}$ be a subset of $L(E,F)$ with the property that $\{Tx : T \in \mathscr{T}\}$ is a bounded subset of F for each $x \in E$. Then $\mathscr{T}$ is a bounded subset of $L(E,F)$.*

PROOF For $n = 1, 2,\ldots$ let

$$A_n = \{x \in E : \| Tx \| \leq n \text{ for all } T \in \mathscr{T}\}$$

By hypothesis $E = \bigcup_{n=1}^{\infty} A_n$. But E is complete so, by Corollary 8.1.7, there is a positive integer n_0 such that A_{n_0} is somewhere dense in E. We shall show that A_{n_0} is also closed. Let (x_m) be a sequence in A_{n_0} with $\lim_{m\to\infty} x_m = x$. Then, for each $T \in \mathscr{T}$, we have $\| Tx_m \| \leq n_0$ for $m = 1, 2,\ldots$ and $\lim_{m\to\infty} \| Tx_m \| = \| Tx \|$, from which it follows that $\| Tx \| \leq n_0$. This shows that $x \in A_{n_0}$ and hence that A_{n_0} is closed. Since A_{n_0} is somewhere dense, there is a non-empty open ball, $B(x_0,r_0)$ say, contained in A_{n_0}. Let $x \in E$ and $\| x \| < r_0$. Then $x + x_0 \in A_{n_0}$ and hence, for all $T \in \mathscr{T}$,

$$\begin{aligned} \| Tx \| &= \| T(x + x_0) - Tx_0 \| \\ &\leq \| T(x + x_0) \| + \| Tx_0 \| \\ &\leq 2n_0 \\ &\leq \frac{2n_0}{r_0} \| x \|. \end{aligned}$$

This shows that $\| T \| \leq 2n_0/r_0$ for all $T \in \mathscr{T}$. □

There is a special case of Theorem 8.4.1 which is sufficiently important to merit a separate statement.

8.4.2 Corollary *Let A be a subset of a normed linear space E with the property that the set $\{x^*(x) : x \in A\}$ is bounded for each $x^* \in E^*$. Then A is a bounded subset of E.*

Proof There is a canonical mapping $x \to \hat{x}$ of E into E^{**} given by $\hat{x}(x^*) = x^*(x)$ for all $x^* \in E^*$ (Theorem 5.3.9). By hypothesis the set $\{\hat{x}(x^*) : x \in A\}$ is bounded for each $x^* \in E^*$ and therefore, since E^* is a Banach space (see Section 5.1), Theorem 8.4.1 shows that the set $\{\hat{x} : x \in A\}$ is bounded. Since $\| \hat{x} \| = \| x \|$ the set A is also bounded. □

We shall consider two applications of the principle of uniform boundedness in classical analysis.

8.4.3 Theorem *Let (α_n) be a sequence of complex numbers, let $p \geq 1$ and suppose that the series $\sum_n \alpha_n \xi_n$ converges for each sequence (ξ_n) in ℓ^p. Then if $p = 1$ the sequence (α_n) is in m and if $p > 1$ the sequence (α_n) is in ℓ^q, where $q = p/(p-1)$.*

Proof Let $x_k^*(x) = \sum_{n=1}^{k} \alpha_n \xi_n$ for all $x = (\xi_n) \in \ell^p$ and $k = 1, 2, \ldots$. It follows from Examples 5.1.4 and 5.1.6 that $x_k^* \in (\ell^p)^*$ and

$$\| x_k^* \| = \begin{cases} \max \{ |\alpha_1|, |\alpha_2|, \ldots, |\alpha_k| \}, & \text{if } p = 1, \\ \left(\sum_{n=1}^{k} |\alpha_n|^q \right)^{1/q}, & \text{if } p > 1. \end{cases} \quad (1)$$

Since the series $\sum_n \alpha_n \xi_n$ converges for each $x = (\xi_n) \in \ell^p$ the set $\{x_k^*(x) : k = 1, 2, \ldots\}$ is bounded for each $x \in \ell^p$. By Theorem 8.4.1 the set $\{x_k^* : k = 1, 2, \ldots\}$ is bounded and therefore, by (1), $(\alpha_n) \in m$ if $p = 1$ and $(\alpha_n) \in \ell^q$ if $p > 1$. □

The second application is perhaps more interesting—and surprising. We recall that a real-valued function f on R is said to be periodic, with period 2π, if and only if $f(t + 2\pi) = f(t)$ for all $t \in \mathsf{R}$. We shall denote by $\tilde{C}(2\pi)$ the set of all continuous real-valued functions on R that are periodic with period 2π. Obviously $\tilde{C}(2\pi) \subseteq C_{\mathsf{R}}(\mathsf{R})$ and if $f \in \tilde{C}(2\pi)$ then

$$\| f \| = \sup \{ |f(t)| : -\pi < t \leq \pi \}.$$

It is not difficult to see that $\tilde{C}(2\pi)$ is a closed linear subspace of $C_{\mathsf{R}}(\mathsf{R})$ and therefore by Theorems 2.8.3 and 2.6.8 $\tilde{C}(2\pi)$ is a Banach space.

Given $f \in \tilde{C}(2\pi)$ let

$$\alpha_n = \frac{1}{\pi} \int_{-\pi}^{\pi} f(t) \cos nt \, dt$$

and
$$\beta_n = \frac{1}{\pi}\int_{-\pi}^{\pi} f(t) \sin nt \, dt$$

for $n = 0, 1, 2, \ldots$. The series

$$\tfrac{1}{2}\alpha_0 + \sum_{n=1}^{\infty} (\alpha_n \cos nt + \beta_n \sin nt)$$

is called the Fourier series of f. It is important to know whether the Fourier series of f converges to f in any sense. The next theorem shows that the Fourier series of f need not converge pointwise to f.

8.4.4 THEOREM *There is a function in $\tilde{C}(2\pi)$ whose Fourier series fails to converge at at least one point of* R.

PROOF Given $f \in \tilde{C}(2\pi)$ let

$$S_n(t;f) = \tfrac{1}{2}\alpha_0 + \sum_{k=1}^{n} (\alpha_k \cos kt + \beta_k \sin kt)$$

for $t \in \mathsf{R}$ and $n = 1, 2, \ldots$. (The numbers α_k and β_k are defined above.) We shall prove that the sequence $(S_n(0;f))$ fails to converge for some $f \in \tilde{C}(2\pi)$.

Suppose, on the contrary, that $(S_n(0;f))$ converges for each $f \in \tilde{C}(2\pi)$. Let

$$x_n^*(f) = S_n(0;f) = \tfrac{1}{2}\alpha_0 + \sum_{k=1}^{n} \alpha_k$$

for all $f \in \tilde{C}(2\pi)$ and $n = 1, 2, \ldots$. An elementary calculation gives

$$x_n^*(f) = \frac{1}{\pi}\int_{-\pi}^{\pi} f(t) D_n(t) \, dt, \tag{2}$$

where $D_n(t) = \frac{1}{2} + \sum_{k=1}^{n} \cos kt$. Another elementary calculation gives

$$D_n(t) = \frac{\sin (n + \frac{1}{2})t}{2 \sin (t/2)} \tag{3}$$

for $0 < |t| < \pi$ and $n = 1, 2, \ldots$. It follows from (2) that x_n^* is a linear functional on $\tilde{C}(2\pi)$ and

$$|x_n^*(f)| \le \frac{1}{\pi} \|f\| \int_{-\pi}^{\pi} |D_n(t)| \, dt$$

for all $f \in \tilde{C}(2\pi)$. Thus x_n^* is a bounded linear functional and

$$\|x_n^*\| \le \frac{1}{\pi}\int_{-\pi}^{\pi} |D_n(t)| \, dt. \tag{4}$$

By hypothesis the sequence $(x_n^*(f))$ converges, and hence is bounded, for each $f \in \tilde{C}(2\pi)$. Since $\tilde{C}(2\pi)$ is a Banach space Theorem 8.4.1 shows that the sequence $(\| x_n^* \|)$ is also bounded. We shall obtain a contradiction by showing that

$$\| x_n^* \| = \frac{1}{\pi} \int_{-\pi}^{\pi} | D_n(t) | \, dt \tag{5}$$

and then estimating the size of the integral $\int_{-\pi}^{\pi} | D_n(t) | \, dt$.

Let us prove (5). By (4) it is sufficient to show that for each $\varepsilon > 0$ there exists a function $f_n \in \tilde{C}(2\pi)$ with $\| f_n \| = 1$ and

$$| x_n^*(f_n) | > \frac{1}{\pi} \int_{-\pi}^{\pi} | D_n(t) | \, dt - \varepsilon. \tag{6}$$

Let

$$\operatorname{sgn}(x) = \begin{cases} 1 & \text{if } x > 0, \\ 0 & \text{if } x = 0, \\ -1 & \text{if } x < 0. \end{cases}$$

Then $\int_{-\pi}^{\pi} \operatorname{sgn}(D_n(t)) D_n(t) \, dt = \int_{-\pi}^{\pi} | D_n(t) | \, dt$. The function

$$t \to \operatorname{sgn}(D_n(t))$$

has discontinuities in the interval $[-\pi,\pi]$ at the points

$$\pm 2\pi/(2n+1), \ \pm 4\pi/(2n+1), \ldots, \ \pm 2n\pi/(2n+1).$$

By modifying this function slightly in suitably small neighbourhoods of each of these points and extending it outside the interval $[-\pi,\pi]$ by periodicity, we can obtain a function $f_n \in \tilde{C}(2\pi)$ with $\| f_n \| = 1$ and

$$\left| \int_{-\pi}^{\pi} f_n(t) \, D_n(t) \, dt - \int_{-\pi}^{\pi} | D_n(t) | \, dt \right| < \varepsilon.$$

Such a function f_n will obviously satisfy (6). We leave the details of the explicit construction of the function f_n to the reader.

To complete the proof of the theorem we have only to show that the sequence of integrals $\left(\int_{-\pi}^{\pi} | D_n(t) | \, dt \right)$ is unbounded. Since

$$2 \sin \tfrac{1}{2}t \leq t \text{ for } 0 \leq t \leq \pi$$

equation (3) gives

$$\begin{aligned} \int_{-\pi}^{\pi} | D_n(t) | \, dt &= 2 \int_0^{\pi} | D_n(t) | \, dt \\ &\geq 2 \int_0^{\pi} \frac{| \sin (n + \frac{1}{2})t |}{t} \, dt \\ &= 2 \int_0^{\pi/2} \frac{| \sin (2n + 1)t |}{t} \, dt \end{aligned}$$

$$\geq 2\sum_{k=0}^{n-1}\int_{\frac{k\pi}{2n+1}}^{\frac{(k+1)\pi}{2n+1}}\frac{|\sin(2n+1)t|}{t}\,dt$$

$$\geq 2\sum_{k=0}^{n-1}\frac{2n+1}{(k+1)\pi}\int_{\frac{k\pi}{2n+1}}^{\frac{(k+1)\pi}{2n+1}}|\sin(2n+1)t|\,dt$$

$$=\frac{4}{\pi}\sum_{k=0}^{n-1}\frac{1}{k+1}. \tag{7}$$

Since the series $\sum_n 1/n$ diverges (5) and (7) contradict the fact that the sequence $(\|x_n^*\|)$ is bounded. □

Theorem 8.4.4 was proved in 1876 by du Bois Reymond. His proof was constructive; he exhibited a function, in $\tilde{C}(2\pi)$, whose Fourier series fails to converge at the point 0. Several other (simpler) examples of such functions have since been found. (See ZYGMUND [27, p. 299].) The proof of Theorem 8.4.4 given above is non-constructive; it is essentially due to H. Lebesgue, although the general principle of uniform boundedness in the form we have stated it was not available to him. The numbers $\frac{1}{\pi}\int_{-\pi}^{\pi}|D_n(t)|\,dt$ are called Lebesgue's constants. It is interesting to notice that they are the bounds of the linear functionals x_n^* on $\tilde{C}(2\pi)$.

A careful examination of the proofs of Theorems 8.4.1 and 8.4.4 will show that the techniques used in these proofs will give more information about the convergence of Fourier series than is contained in Theorem 8.4.4. (Cf. Exercises (3) and (6).)

EXERCISES

(1) Let E be the linear hull in ℓ of the set $\{e_n : n = 1, 2,\ldots\}$ of Example 3.4.9. For $m = 1, 2,\ldots$ and $x = (\xi_n) \in E$ let $x_m^*(x) = m\xi_m$. Prove that the set $\{x_m^*(x) : m = 1, 2,\ldots\}$ is bounded for each $x \in E$ but that $\{x_m^* : m = 1, 2,\ldots\}$ is an unbounded subset of E^*.

(2) Let E and F be Banach spaces and let (T_n) be a sequence in $L(E,F)$ with the property that $(T_n x)$ converges for each $x \in E$. Prove that if $Tx = \lim_{n\to\infty} T_n x$ for all $x \in E$ then $T \in L(E,F)$. This result is known as the Banach-Steinhaus theorem.

(3) Let E be a Banach space, F a normed linear space, and $\mathscr{T}$ a subset of $L(E,F)$. Suppose that the set $\{Tx_0 : T \in \mathscr{T}\}$ is unbounded for some point $x_0 \in E$. Prove that the subset of E consisting of those points $x \in E$ for which the set $\{Tx : T \in \mathscr{T}\}$ is bounded is meagre in E. (Hint: examine the proof of Theorem 8.4.1.)

(4) Let (α_n) be a sequence of complex numbers. Prove that if the series $\sum_n \alpha_n \xi_n$ converges for all sequences $(\xi_n) \in c$ then the series $\sum_n \alpha_n$ is absolutely convergent.

(5) Prove that there is no sequence (ξ_n) of complex numbers with the following property: a series $\sum_n \alpha_n$ of complex numbers is absolutely convergent if and only if the sequence $(\xi_n \alpha_n)$ is bounded.

(6) (a) Prove that for each point $t \in \mathsf{R}$ the set of functions $f \in \tilde{C}(2\pi)$ which are such that the sequence $(S_n(t;f))$ is bounded, is meagre in $\tilde{C}(2\pi)$. (Use Theorem 8.4.4 and Exercise (3).)

(b) Prove that there is a function in $\tilde{C}(2\pi)$ whose Fourier series fails to converge at each point of a dense subset of R.

(7) Let $A = (\alpha_{mn})$ be a double sequence of real numbers. A sequence (ξ_n) of real numbers is said to be *A-summable* to the limit ξ if and only if the series $\sum_n \alpha_{mn}\xi_n$ converges for each positive integer m and $\lim_{m\to\infty} \sum_{n=1}^{\infty} \alpha_{mn}\xi_n = \xi$. An infinite series is said to be *A-summable* to the sum ξ if and only if its sequence of partial sums is A-summable to ξ. A double sequence $A = (\alpha_{mn})$ is said to determine a *regular method of summability* if and only if each convergent sequence (ξ_n) of real numbers is A-summable to $\lim_{n\to\infty} \xi_n$.

Prove the following result, which is due to Toeplitz: a double sequence (α_{mn}) determines a regular method of summability if and only if

(a) the series $\sum_n \alpha_{mn}$ is absolutely convergent for $m = 1, 2,\ldots,$ and the set $\left\{ \sum_{n=1}^{\infty} | \alpha_{mn} | : m = 1, 2,\ldots \right\}$ is bounded,

(b) $\lim_{m\to\infty} \alpha_{mn} = 0$ for $n = 1, 2,\ldots,$

(c) $\lim_{m\to\infty} \sum_{n=1}^{\infty} \alpha_{mn} = 1.$

(8) If $\alpha_{mn} = 1/m$ for $n = 1, 2,\ldots, m$, and $m = 1, 2,\ldots,$ and $\alpha_{mn} = 0$ for $n = m + 1, m + 2,\ldots,$ and $m = 1, 2,\ldots,$ prove that (α_{mn}) determines a regular method of summability. This method is called Cesàro $(C,1)$-summability. It is an important result in the theory of Fourier series that the Fourier series of any function $f \in \tilde{C}(2\pi)$ is $(C,1)$-summable to $f(t)$ at each point t of R. (See ZYGMUND [27, p. 89].)

8.5 The Open Mapping Theorem

It is convenient to break the proof of the open mapping theorem into two parts each of which is presented as a lemma. The second lemma will be used again in the next section. The open mapping theorem was obtained by Banach in 1929.

8.5.1 LEMMA *Let T be a linear mapping of a normed linear space E onto a Banach space F over the same field. Then there exists $\alpha > 0$ such that*

$$\{y \in F : \| y \| \leq 1\} \subseteq T(B_\alpha)^-,$$

where $B_\alpha = \{x \in E : \| x \| \leq \alpha\}$.

PROOF Since $T(E) = F$ we have $F = \bigcup_{n=1}^{\infty} T(B_n)$. The space F is complete so, by Corollary 8.1.7, there is an integer m such that $T(B_m)$ is somewhere dense in F. Therefore there is a point $y_0 \in F$ and a number $\beta > 0$ with

$$\{y \in F : \| y - y_0 \| \leq \beta\} \subseteq T(B_m)^-.$$

Since $-x \in B_m$ whenever $x \in B_m$ we have $-y \in T(B_m)$ whenever $y \in T(B_m)$, and it follows easily that $-y \in T(B_m)^-$ whenever $y \in T(B_m)^-$. A similar argument shows that $\frac{1}{2}(y_1 + y_2) \in T(B_m)^-$ whenever $y_1, y_2 \in T(B_m)^-$.

Now let $y \in F$ with $\| y \| \leq \beta$. Then $\| (y + y_0) - y_0 \| \leq \beta$ and $\| (y_0 - y) - y_0 \| \leq \beta$ and hence $y + y_0 \in T(B_m)^-$ and $y_0 - y \in T(B_m)^-$. Consequently, by what we have proved above,

$$y = \tfrac{1}{2}((y + y_0) - (y_0 - y)) \in T(B_m)^-.$$

This proves that $\{y \in F : \| y \| \leq \beta\} \subseteq T(B_m)^-$ and it follows easily that $\{y \in F : \| y \| \leq 1\} \subseteq T(B_\alpha)^-$, where $\alpha = m/\beta$. □

8.5.2 LEMMA *Let T be a bounded linear mapping of a Banach space E into a normed linear space F over the same field and suppose that there exists $\alpha > 0$ such that*

$$\{y \in F : \| y \| \leq 1\} \subseteq T(B_\alpha)^-. \tag{1}$$

Then there exists $\beta > 0$ such that

$$\{y \in F : \| y \| \leq 1\} \subseteq T(B_\beta),$$

where $B_\alpha = \{x \in E : \| x \| \leq \alpha\}$.

PROOF Let $y \in F$ with $\| y \| \leq 1$. We shall define inductively a sequence (y_n) in $T(B_\alpha)$ with

$$\left\| y - \sum_{k=1}^{n} \frac{1}{2^{k-1}} y_k \right\| \leq \frac{1}{2^n} \tag{2}$$

for $n = 1, 2, \ldots$. By (1) there is a point $y_1 \in T(B_\alpha)$ with $\| y - y_1 \| \leq \frac{1}{2}$. Suppose that $y_1, y_2, \ldots, y_n$ have been chosen to satisfy (2). Then

$$\left\| 2^n\left(y - \sum_{k=1}^{n} \frac{1}{2^{k-1}} y_k\right) \right\| \leq 1$$

and therefore by (1) there is a point $y_{n+1} \in T(B_\alpha)$ with

$$\left\| 2^n\left(y - \sum_{k=1}^{n} \frac{1}{2^{k-1}} y_k\right) - y_{n+1} \right\| \leq \tfrac{1}{2}.$$

It is obvious that $y_1, y_2, \ldots, y_n, y_{n+1}$ satisfy (2), so this completes the definition of the sequence (y_n). It is clear from (2) that $y = \sum_{k=1}^{\infty} 2^{-k+1} y_k$.

For $k = 1, 2, \ldots$, choose $x_k \in B_\alpha$ with $Tx_k = y_k$. Since

$$\| 2^{-k+1} x_k \| \leq \alpha 2^{-k+1}$$

the series $\sum_k 2^{-k+1} x_k$ converges absolutely and consequently, since E is complete, the series $\sum_k 2^{-k+1} x_k$ converges (Lemma 3.3.8). Let

$$x = \sum_{k=1}^{\infty} 2^{-k+1} x_k.$$

By Lemma 3.3.8 again we have

$$\| x \| \leq \sum_{k=1}^{\infty} \frac{1}{2^{k-1}} \| x_k \| \leq \sum_{k=1}^{\infty} \frac{\alpha}{2^{k-1}} = 2\alpha.$$

Since T is continuous and linear we also have

$$Tx = \sum_{k=1}^{\infty} \frac{1}{2^{k-1}} Tx_k = \sum_{k=1}^{\infty} \frac{1}{2^{k-1}} y_k = y.$$

This shows that $y \in T(B_\beta)$ where $\beta = 2\alpha$. Thus

$$\{y \in F : \| y \| \leq 1\} \subseteq T(B_\beta). \ \square$$

We remark that the operator T in Lemma 8.5.2 maps E onto F. In fact it follows from the lemma that if $y \in F$ and $y \neq 0$ then there is a point $x \in B_\beta$ with $\| y \|^{-1} y = Tx$ and hence $y = T(\| y \| x)$.

8.5.3 Theorem (the open mapping theorem) *Let T be a bounded linear mapping of a Banach space E onto a Banach space F over the same field. Then $T(U)$ is an open subset of F whenever U is an open subset of E.*

Proof Let U be an open subset of E. Since $T(\emptyset) = \emptyset$ we may suppose that $U \neq \emptyset$. Let $x_0 \in U$. Then there exists $\gamma > 0$ with

$$\{x \in E : \| x - x_0 \| \leq \gamma\} \subseteq U.$$

Consequently

$$B_\gamma = \{x \in E : \| x \| \le \gamma\} \subseteq U - x_0.$$

It follows from Lemmas 8.5.1 and 8.5.2 that there exists $\beta > 0$ with

$$\{y \in F : \| y \| \le \gamma/\beta\} \subseteq T(B_\gamma)$$

and therefore, since $T(B_\gamma) \subseteq T(U - x_0) = T(U) - Tx_0$, we have

$$\{y \in F : \| y - Tx_0 \| \le \gamma/\beta\} \subseteq T(U).$$

This shows that $T(U)$ is a neighbourhood of Tx_0 and hence that $T(U)$ is open. □

The reader will notice that in the open mapping theorem the completeness of E and the completeness of F are needed for quite different reasons. In Lemma 8.5.1 the completeness of F is needed to ensure that Baire's theorem is applicable to F; in Lemma 8.5.2 the completeness of E is used directly (through Lemma 3.3.8).

Many important applications of the open mapping theorem depend directly on the following special case of the theorem.

8.5.4 THEOREM *A one-to-one continuous linear mapping of a Banach space E onto a Banach space F over the same field is a linear homeomorphism of E onto F.*

PROOF Let T be a one-to-one continuous linear mapping of E onto F. Then, by the open mapping theorem (8.5.3), $T(U)$ is an open subset of F whenever U is an open subset of E. Since T is one-to-one

$$(T^{-1})^{-1}(U) = T(U)$$

and consequently it follows from Theorem 2.7.3 that T^{-1} is continuous. □

Theorem 8.5.4 is sufficiently important to merit our restating it in a slightly different form. It asserts that for a continuous linear mapping T of one Banach space onto another the existence of the inverse mapping T^{-1} automatically entails the continuity of T^{-1}.

From Theorem 8.5.4 we obtain the following simple criterion for the regularity of a bounded linear operator T on a Banach space E (cf. Definition 6.2.1 and Lemma 6.2.3): T is regular if and only if it is a one-to-one mapping of E onto itself. From this criterion we obtain the following characterization of the spectrum of a bounded linear operator T on a non-zero complex Banach space E (cf. Definition 6.3.1): $\lambda \in \mathrm{C} \sim \mathrm{sp}(T)$ if and only if $\lambda I - T$ is a one-to-one mapping of E onto itself. Consequently if $\lambda \in \mathrm{sp}\,(T)$ then either $\lambda I - T$ is not one-to-one, in which case λ is an eigenvalue of T, or $\lambda I - T$ is one-to-one and $\mathscr{R}(\lambda I - T) \neq E$. In the latter case $\lambda I - T$, regarded as a linear mapping of E onto $\mathscr{R}(\lambda I - T)$, has an inverse mapping which will be continuous on $\mathscr{R}(\lambda I - T)$ if and only if $\mathscr{R}(\lambda I - T)$ is closed (Theorem 2.6.8, Lemma 2.9.4 and Theorem 8.5.4). Thus, if λ is not an eigenvalue of T and if $\mathscr{R}(\lambda I - T)$ is closed, the solution of the equation

$$y = (\lambda I - T)x$$

depends continuously on y (for $y \in \mathscr{R}(\lambda I - T)$).

We note one further immediate consequence of Theorem 8.5.4. Let $\|\cdot\|_1$ and $\|\cdot\|_2$ be norms on a linear space E and suppose that $(E, \|\cdot\|_1)$ and $(E, \|\cdot\|_2)$ are both Banach spaces, and that there is a real number M with $\| x \|_1 \leq M \| x \|_2$ for all $x \in E$. Then, by Theorem 8.5.4 applied to the identity mapping I on E regarded as a mapping of $(E, \|\cdot\|_1)$ onto $(E, \|\cdot\|_2)$, the norms $\|\cdot\|_1$ and $\|\cdot\|_2$ are equivalent. (See p. 110.)

The reader will find further, less immediate, applications of Theorem 8.5.4 in the following three sections.

EXERCISES

(1) Let T be defined by $(Tf)(t) = \int_0^t f(s)\, ds$ for all $f \in C_{\mathsf{R}}([0,1])$. Prove that T is a one-to-one bounded linear mapping of $C_{\mathsf{R}}([0,1])$ onto the linear subspace $F = \{g \in C^1_{\mathsf{R}}([0,1]) : g(0) = 0\}$ of $C_{\mathsf{R}}([0,1])$. Prove also that T^{-1} is not continuous on F.

(2) Let E be a Banach space. It is possible to classify those operators $T \in L(E)$ that are not regular in terms of the algebra $L(E)$ in a way which does not involve the space E itself. Prove that, (a) T is not one-to-one if and only if there exists $S \in L(E)$ with $S \neq 0$ and $TS = 0$, (b) the range of T is not dense in E if and only if there exists $S \in L(E)$ with $S \neq 0$ and $ST = 0$, (c) if T is not regular but is one-to-one with its range dense in E then there is a sequence (S_n) in $L(E)$ with $\| S_n \| = 1$ for $n = 1, 2,\ldots$ and $\lim_{n\to\infty} TS_n = 0$.

(3) Let E and F be Banach spaces and let $T \in L(E,F)$ be such that $\mathscr{R}(T)$ has finite codimension in F. Prove that $\mathscr{R}(T)$ is closed. (Hint: Let $N = \{x \in E : Tx = 0\}$ and define $T_1 \in L(E/N,F)$ by

$$T_1(x + N) = Tx.$$

Let $F = \mathscr{R}(T) \oplus G$ (Lemma 6.6.11) and define S in $L((E/N)\times G,F)$ by

$$S(x + N,y) = T(x + N) + y.$$

Apply Theorem 8.5.4 to S.)

8.6 The Adjoint of a Bounded Linear Transformation

In this section we shall use the open mapping theorem to extend the results of Lemma 6.4.8. We begin by taking a closer look at the definition of the adjoint of a bounded linear transformation.

Let E and F be normed linear spaces over the same field and let T be a bounded linear transformation of E into F. We recall that the adjoint T^* of T is the mapping of F^* into E^* defined by

$$T^*y^* = y^* \circ T$$

for all $y^* \in F^*$. Now let G be a linear subspace of F with $\mathscr{R}(T) \subseteq G$. Clearly T is a bounded linear transformation of E into G and consequently has an adjoint, T' say, which is the mapping of G^* into E^* defined by

$$T'z^* = z^* \circ T$$

for all $z^* \in G^*$. Thus T has one adjoint corresponding to each linear subspace of F that contains $\mathscr{R}(T)$. This fact is generally not emphasized in the definition of the adjoint transformation; indeed it is obscured by the usual notation T^* which does not express the fact that the adjoint depends not only on T but also on F.

In the following lemma we shall use the notation established above.

8.6.1 LEMMA $\mathscr{R}(T') = \mathscr{R}(T^*)$.

PROOF Let $x^* \in \mathscr{R}(T^*)$ and choose $y^* \in F^*$ with $T^*y^* = x^*$. Let $z^* = y^*|G$. Then $z^* \in G^*$ and, for all $x \in E$,

$$(T'z^*)(x) = z^*(Tx) = y^*(Tx) = (T^*y^*)(x) = x^*(x),$$

which shows that $x^* = T'z^*$. Therefore $\mathscr{R}(T^*) \subseteq \mathscr{R}(T')$.

On the other hand, let $x^* \in \mathscr{R}(T')$ and choose $z^* \in G^*$ with $T'z^* = x^*$. By Theorem 5.3.2 there is a functional $y^* \in F^*$ with $y^*|G = z^*$ and hence, as above, we obtain $x^* = T^*y^*$. This shows that the inclusion $\mathscr{R}(T') \subseteq \mathscr{R}(T^*)$ also holds. □

Lemma 8.6.1 shows that if we are interested only in the set $\mathscr{R}(T^*)$ we can replace the adjoint T^* determined by F by the adjoint T' determined by any linear subspace of F that contains $\mathscr{R}(T)$. This observation greatly simplifies the proof of the following theorem.

8.6.2 THEOREM *Let E and F be Banach spaces over the same field and let $T \in L(E,F)$. Then $\mathscr{R}(T)$ is closed if and only if $\mathscr{R}(T^*)$ is closed and if $\mathscr{R}(T)$ is closed then*

$$\mathscr{N}(T)^{\perp} = \mathscr{R}(T^*)$$

and

$$\mathscr{N}(T^*)_{\perp} = \mathscr{R}(T).$$

PROOF It follows from Lemma 6.4.8(c) that if $\mathscr{R}(T)$ is closed then $\mathscr{R}(T) = \mathscr{N}(T^*)_\perp$. Consequently we have only to prove that $\mathscr{R}(T)$ is closed if and only if $\mathscr{R}(T^*)$ is closed, and that if $\mathscr{R}(T)$ is closed then $\mathscr{R}(T^*) = \mathscr{N}(T)^\perp$. The proof is in two stages. First we shall prove that if $\mathscr{R}(T)$ is closed then $\mathscr{R}(T^*) = \mathscr{N}(T)^\perp$ from which it follows by Lemma 5.4.2 that $\mathscr{R}(T^*)$ is closed. Secondly we shall prove that if $\mathscr{R}(T^*)$ is closed then so is $\mathscr{R}(T)$.

The following observation will simplify the proof considerably. Lemma 8.6.1 shows that we may replace F by $\mathscr{R}(T)^-$ (which is a Banach space by Theorem 2.6.8), and therefore we may suppose throughout the proof the $\mathscr{R}(T)$ is dense in F.

Suppose first that $\mathscr{R}(T)$ is closed. Then, since $\mathscr{R}(T)$ is dense in F, we have $\mathscr{R}(T) = F$. We have to prove that $\mathscr{R}(T^*) = \mathscr{N}(T)^\perp$. By Lemma 6.4.8(a) we have $\mathscr{N}(T) = \mathscr{R}(T^*)_\perp$ and hence $\mathscr{R}(T^*) \subseteq \mathscr{N}(T)^\perp$. Let $x^* \in \mathscr{N}(T)^\perp$. We shall define a functional $y^* \in F^*$ such that $T^*y^* = x^*$. We observe that if $x_1, x_2 \in E$ and $Tx_1 = Tx_2$ then $x^*(x_1) = x^*(x_2)$. Consequently, since $F = \mathscr{R}(T)$, we can define without ambiguity a mapping y^* of F into K by

$$y^*(Tx) = x^*(x)$$

for all $x \in E$. It is obvious that y^* is linear. Since $\mathscr{R}(T) = F$ we can apply Lemmas 8.5.1 and 8.5.2 to obtain a number $\beta > 0$ such that $\{y \in F : \| y \| \leq 1\} \subseteq T(B_\beta)$ where $B_\beta = \{x \in E : \| x \| \leq \beta\}$. Let $y \in F$ with $\| y \| \leq 1$. Then $y = Tx$ for some $x \in B_\beta$ and hence, by definition of y^*,

$$| y^*(y) | = | x^*(x) | \leq \| x^* \| \| x \| \leq \beta \| x^* \|.$$

This shows that the linear functional y^* is bounded. Also, for all $x \in E$, we have $(T^*y^*)(x) = y^*(Tx) = x^*(x)$ so $T^*y^* = x^*$ as required. This proves that $\mathscr{N}(T)^\perp \subseteq \mathscr{R}(T^*)$ and hence that $\mathscr{R}(T^*) = \mathscr{N}(T)^\perp$.

Suppose now that $\mathscr{R}(T^*)$ is closed. Then, by Theorem 2.6.8, $\mathscr{R}(T^*)$ is a Banach space. Lemma 6.4.8(b) gives $\mathscr{N}(T^*) = \mathscr{R}(T)^\perp$ and hence, since we are also supposing that $\mathscr{R}(T)$ is dense in F, it follows from Corollary 5.3.5 that $\mathscr{N}(T^*) = \{0\}$. Thus T^* is a one-to-one continuous linear mapping of the Banach space F^* onto the Banach space $\mathscr{R}(T^*)$ and consequently, by Theorem 8.5.4, T^* is a linear homeomorphism. Therefore there exists $m > 0$ such that

$$\| T^*y^* \| \geq m \| y^* \| \tag{1}$$

for all $y^* \in F^*$ (Lemma 3.5.9). We shall prove that

$$\{y \in F : \| y \| \leq 1\} \subseteq T(B_\alpha)^-,$$

where $\alpha = 2/m$.

Let $y_0 \in F \sim T(B_\alpha)^-$. We have to prove that $\| y_0 \| > 1$. It is easy to verify that $T(B_\alpha)$ is a convex set and that $\lambda y \in T(B_\alpha)$ whenever $y \in T(B_\alpha)$ and $\lambda \in \mathsf{K}$ with $|\lambda| = 1$. Therefore, by Theorem 5.5.4, there is a point $y_0^* \in F^*$ with

$$|y_0^*(y_0)| = 1 \tag{2}$$

and

$$|y_0^*(y)| < 1 \tag{3}$$

for all $y \in T(B_\alpha)$. If $x \in B_\alpha$ then (3) gives $|(T^*y_0^*)(x)| = |y_0^*(Tx)| < 1$ and hence

$$\begin{aligned} \| T^*y_0^* \| &= \sup \{ |(T^*y_0^*)(x)| : \|x\| \leq 1\} \\ &= \sup \left\{ \left| (T^*y_0^*)\left(\frac{1}{\alpha}x\right) \right| : x \in B_\alpha \right\} \\ &\leq \frac{1}{\alpha} \\ &< m. \end{aligned} \tag{4}$$

The inequalities (1) and (4) show that $\| y_0^* \| < 1$, and the equality (2) then gives

$$1 = |y_0^*(y_0)| \leq \| y_0^* \| \, \| y_0 \| < \| y_0 \|.$$

This proves that $\{y \in F : \| y \| \leq 1\} \subseteq T(B_\alpha)^-$.

It now follows from Lemma 8.5.2 and the remark immediately following that lemma that $\mathscr{R}(T) = F$. This proves that $\mathscr{R}(T)$ is closed. □

Theorem 8.6.2 lies quite deep; two basic principles of functional analysis, the Hahn–Banach theorem and the open mapping theorem, play an essential role in its proof.

EXERCISES

(1) Let E and F be Banach spaces. Suppose that E is reflexive and that there exists $T \in L(E,F)$ with $\mathscr{R}(T) = F$. Prove that F is reflexive.

(2) Let M and N be closed linear subspaces of a Banach space E. Prove that $M + N$ is closed if and only if $M^\perp + N^\perp$ is closed in E^* and that if $M + N$ is closed then $M^\perp + N^\perp = (M \cap N)^\perp$. (Hint: let π be the canonical mapping of E onto E/N and apply Theorem 8.6.2 to the mapping $\pi|M$ of M into E/N.)

8.7 Perturbation Theorems

Throughout this section E and F will denote Banach spaces over the same field and T will denote a bounded linear transformation of E into F. We shall write $\alpha(T) = \dim \mathscr{N}(T)$ when $\mathscr{N}(T)$ has finite dimension, and $\alpha(T) = \infty$ otherwise; and $\beta(T) = \operatorname{codim} \mathscr{R}(T)$ when $\mathscr{R}(T)$ has finite codimension in F, and $\beta(T) = \infty$ otherwise. We shall call $\alpha(T)$ the *nullity of* T and $\beta(T)$ the *deficiency of* T. If T has finite nullity and finite deficiency we shall write $\varkappa(T) = \alpha(T) - \beta(T)$. The number $\varkappa(T)$ is called the *index of* T.

8.7.1 DEFINITION The linear transformation T is said to be a *Φ-operator* if and only if $\mathscr{R}(T)$ is closed and $\alpha(T)$ and $\beta(T)$ are both finite; it is said to be a *Φ^+-operator* if and only if $\mathscr{R}(T)$ is closed, $\alpha(T)$ is finite and $\beta(T) = \infty$; and it is said to be a *Φ^--operator* if and only if $\mathscr{R}(T)$ is closed, $\alpha(T) = \infty$ and $\beta(T)$ is finite.

It follows from Lemmas 6.6.3 and 6.6.1 and Theorem 6.6.12 that, if S is a compact linear operator on E and $\lambda \neq 0$, then $\lambda I - S$ is a Φ-operator with zero index.

Our first theorem is a corollary of Theorem 8.6.2.

8.7.2 THEOREM *If $\mathscr{R}(T)$ is closed then $\alpha(T) = \beta(T^*)$ and $\beta(T) = \alpha(T^*)$. In particular T is a Φ-, Φ^+-, or Φ^--operator if and only if T^* is a Φ-, Φ^--, or Φ^+-operator, respectively. Also if T is a Φ-operator then $\varkappa(T^*) = -\varkappa(T)$.*

PROOF By Theorem 8.6.2 $\mathscr{R}(T)$ is closed if and only if $\mathscr{R}(T^*)$ is closed and, if $\mathscr{R}(T)$ is closed, then $\mathscr{R}(T^*) = \mathscr{N}(T)^{\perp}$. Also, by Theorem 5.4.4, the normed linear spaces $\mathscr{N}(T)^*$ and $E^*/\mathscr{N}(T)^{\perp}$ are linearly isometric. It follows from Theorems 5.1.3 and 5.3.9 that $\mathscr{N}(T)$ and $\mathscr{N}(T)^*$ are either both infinite-dimensional or both finite-dimensional, and, in the latter case, $\dim \mathscr{N}(T) = \dim \mathscr{N}(T)^*$. This proves that $\alpha(T) = \beta(T^*)$. Similarly we can prove that $\beta(T) = \alpha(T^*)$. □

Perturbation theorems are concerned with the description of how the addition of a small bounded linear transformation to a given bounded linear transformation affects the properties of the given transformation. Our aim in this section is to show that small perturbations of Φ-, Φ^+- and Φ^--operators are again Φ-, Φ^+- and Φ^--operators, respectively.

We shall need several lemmas on finite-dimensional linear subspaces of a normed linear space. In these lemmas G will denote a normed linear space. Our first lemma asserts that each finite-dimensional linear subspace of G has a complementary closed linear subspace. (cf. Definition 3.4.11 and the remarks that follow it.)

8.7.3 LEMMA *Let N be a finite-dimensional linear subspace of G. Then there is a closed linear subspace M of G with $G = N \oplus M$.*

PROOF Let $\{x_1, x_2, \ldots, x_n\}$ be a basis for N and let $\{x_1^*, x_2^*, \ldots, x_n^*\}$ be the corresponding dual basis for N^*. (See Theorem 5.1.3.) It follows from

Theorems 5.1.3 and 5.3.2 that there are functionals $y_1^*, y_2^*, \ldots, y_n^*$ in G^* such that $y_k^*|N = x_k^*$ for $k = 1, 2, \ldots, n$. Let

$$M = \{x \in G : y_k^*(x) = 0 \text{ for } k = 1, 2, \ldots, n\}.$$

It is obvious that M is a linear subspace of G. Since $y_1^*, y_2^*, \ldots, y_n^*$ are continuous on G and $M = \bigcap_{k=1}^{n} \{x \in G : y_k^*(x) = 0\}$ the linear subspace M is closed. Let $x \in G$. Then, for $j = 1, 2, \ldots, n$,

$$\begin{aligned} y_j^*\left(x - \sum_{k=1}^{n} y_k^*(x)x_k\right) &= y_j^*(x) - \sum_{k=1}^{n} y_k^*(x)y_j^*(x_k) \\ &= y_j^*(x) - \sum_{k=1}^{n} y_k^*(x)x_j^*(x_k) \\ &= y_j^*(x) - y_j^*(x) \\ &= 0, \end{aligned}$$

because $x_j^*(x_k) = \delta_{jk}$. Thus $x - \sum_{k=1}^{n} y_k^*(x)x_k \in M$. Since $\sum_{k=1}^{n} y_k^*(x)x_k \in N$ and

$$x = \sum_{k=1}^{n} y_k^*(x)x_k + \left(x - \sum_{k=1}^{n} y_k^*(x)x_k\right),$$

it follows that $G = N + M$.

Finally let $x \in N \cap M$. Then $x = \sum_{k=1}^{n} \alpha_k x_k$, where $\alpha_1, \alpha_2, \ldots, \alpha_n$ are in K, and, for $j = 1, 2, \ldots, n$, $0 = y_j^*(x) = \sum_{k=1}^{n} \alpha_k y_j^*(x_k) = \alpha_j$. This proves that $N \cap M = \{0\}$ and hence that $N \oplus M = G$. □

8.7.4 Lemma *Let N be a finite-dimensional linear subspace of G and let M be a closed linear subspace of G. Then $N + M$ is a closed linear subspace of G.*

Proof It is obvious that $N + M$ is a linear subspace of G. Since N is finite-dimensional and $N \cap M$ is a linear subspace of N there is a linear subspace N_1 of N with $N = (N \cap M) \oplus N_1$. (See Halmos [12, p. 31, Theorem 2].) It is clear that

$$N + M = N_1 + M \text{ and } N_1 \cap M = N_1 \cap (N \cap M) = \{0\}.$$

Thus we may suppose that $N \cap M = \{0\}$.

Let (x_n) be a sequence in $N + M$ with $\lim_{n\to\infty} x_n = x$ and let $x_n = y_n + z_n$ where $y_n \in N$ and $z_n \in M$. We shall show first that the sequence (y_n) is bounded. Suppose on the contrary that (y_n) is unbounded. Then we may suppose, replacing (y_n) by a suitable subsequence if necessary, that $\| y_n \| \geq n$ for $n = 1, 2, \ldots$. The sequence $(\| y_n \|^{-1} y_n)$ is bounded, and so has a subsequence which converges in N to the limit $y_0 \in N$, say. (See the remark following Corollary 4.4.4.) It is obvious that $\| y_0 \| = 1$. On the other hand, since the sequence (x_n) converges and $\| y_n \| \geq n$ we have

$$0 = \lim_{n \to \infty} \| y_n \|^{-1} x_n = \lim_{n \to \infty} \| y_n \|^{-1}(y_n + z_n).$$

Therefore a subsequence of the sequence $(- \| y_n \|^{-1} z_n)$ converges to y_0. Thus $y_0 \in M$ because $z_n \in M$ and M is closed. This is a contradiction because $N \cap M = \{0\}$ and $\| y_0 \| = 1$. Consequently the sequence (y_n) is bounded.

Another application of Corollary 4.4.4 now shows that the sequence (y_n) has a subsequence, (y_{n_k}) say, that converges in N. Let $y = \lim_{k \to \infty} y_{n_k}$. Then $x - y = \lim_{k \to \infty} (x_{n_k} - y_{n_k}) = \lim_{k \to \infty} z_{n_k}$ and hence $x - y \in M$ because $z_{n_k} \in M$ and M is closed. Since $x = y + (x - y)$ this proves that $x \in N + M$. Hence $N + M$ is closed. □

The proof of Lemma 8.7.4 is a typical application of Corollary 4.4.4. Lemmas 8.7.5 to 8.7.8 form a group, of which only the last is used in the sequel.

8.7.5 Lemma *Let N be a finite-dimensional linear subspace of G and let $x \in G$. Then there is a point $y \in N$ with $d(x,N) = \| x - y \|$.*

Proof There is a sequence (y_n) in N with $\| x - y_n \| < d(x,N) + n^{-1}$ for $n = 1, 2, \ldots$. Obviously (y_n) is bounded, so it follows from Corollary 4.4.4 that (y_n) has a subsequence, (y_{n_k}) say, that converges in N. Let $y = \lim_{k \to \infty} y_{n_k}$. Then $y \in N$ and $\| x - y \| = \lim_{k \to \infty} \| x - y_{n_k} \| \leq d(x,N)$. Thus $\| x - y \| = d(x,N)$. □

8.7.6 Lemma *Let N and M be linear subspaces of G. Suppose that N is finite-dimensional and that there is a number k with $0 < k < \frac{1}{2}$ such that $d(x,N) \leq k \| x \|$ for all $x \in M$. Then M is also finite-dimensional.*

Proof Suppose that M is infinite-dimensional and let $2k < \delta < 1$. We shall define inductively a sequence (y_n) in M such that, for each positive integer n, $\| y_n \| = 1$ and $\| y_n - y_m \| \geq \delta$ for $m = 1, 2, \ldots, n - 1$. Choose $y_1 \in M$ with $\| y_1 \| = 1$, and suppose that points $y_2, y_3, \ldots, y_n$ of M have been chosen so that, for each integer j with $1 \leq j \leq n$, $\| y_j \| = 1$ and $\| y_j - y_m \| \geq \delta$ for $m = 1, 2, \ldots, j - 1$. By Corollary 4.4.3 the linear hull, M_n say, of the set $\{y_1, y_2, \ldots, y_n\}$ is closed and, by hypothesis, $M_n \neq M$. Therefore, by Lemma 4.4.5, there is a point

$y_{n+1} \in M$ with $\| y_{n+1} \| = 1$ and $d(y_{n+1},M) \geq \delta$. It is clear that $\| y_{n+1} - y_m \| \geq \delta$ for $m = 1, 2,..., n$, so this completes the inductive definition of the sequence (y_n). By Lemma 8.7.5 for each positive integer n there is a point $x_n \in N$ with $\| y_n - x_n \| = d(y_n,N)$. Thus we have $\| y_n - x_n \| \leq k \| y_n \| = k$ and hence $\| x_n \| \leq \| y_n \| + \| y_n - x_n \| \leq 1 + k$ for $n = 1, 2,....$ This shows that the sequence (x_n) is bounded. It follows from Corollary 4.4.4 that the sequence (x_n) has a subsequence that converges in N. On the other hand, if $m \neq n$, then

$$\begin{aligned} \| x_m - x_n \| &= \| (x_m - y_m) + (y_m - y_n) + (y_n - x_n) \| \\ &\geq \| y_m - y_n \| - \| x_m - y_m \| - \| y_n - x_n \| \\ &\geq \delta - 2k \\ &> 0. \end{aligned}$$

This shows that no subsequence of (x_n) can converge. This contradiction proves that M is finite-dimensional. □

We remark that a much stronger result than Lemma 8.7.6 is true: the condition $d(x,N) \leq k \| x \|$ for all $x \in M$, where $0 < k < 1$, is sufficient to ensure that $\dim M \leq \dim N$ whenever N is finite-dimensional. Unfortunately there does not seem to be a proof of this result that lies within the scope of this book. However Lemma 8.7.6 is sufficient for our needs.

8.7.7 LEMMA *Let N and M be linear subspaces of G and suppose that there is a number $k > 0$ such that $d(x,N) \leq k \| x \|$ for all $x \in M$. Then $d(x^*,M^\perp) \leq k \| x^* \|$ for all $x^* \in N^\perp$.*

PROOF Let $x \in M$ and $x^* \in N^\perp$. Then, for all $y \in N$,

$$| x^*(x) | = | x^*(x - y) | \leq \| x^* \| \; \| x - y \|$$

and hence

$$\begin{aligned} | x^*(x) | &\leq \| x^* \| \inf \{ \| x - y \| : y \in N\} \\ &= \| x^* \| \, d(x,N) \\ &\leq k \| x^* \| \; \| x \|. \end{aligned}$$

It follows from the last inequality that $\| x^*|M \| \leq k \| x^* \|$. By Theorem 5.4.4 $\| x^*|M \| = \| x^* + M^\perp \|$ and, by definition

$$\| x^* + M^\perp \| = d(x^*,M^\perp),$$

so we have $d(x^*,M^\perp) \leq k \| x^* \|$. □

8.7.8 LEMMA *Let N and M be closed linear subspaces of G. Suppose that G/N is infinite-dimensional and that there is a number k with $0 < k < \frac{1}{2}$ such that $d(x,N) \leq k \| x \|$ for all $x \in M$. Then G/M is infinite-dimensional.*

PROOF By Theorem 5.4.5 $(G/N)^*$ and $(G/M)^*$ are linearly isometric to $N^\perp$ and $M^\perp$, respectively, and it follows from Theorems 5.1.3 and 5.3.9 that a normed linear space is finite-dimensional if and only if its dual space is finite-dimensional. Therefore $N^\perp$ is infinite-dimensional and we have to prove that $M^\perp$ is infinite-dimensional. But this follows directly from Lemma 8.7.6 (with $N^\perp$ and $M^\perp$ in place of M and N, respectively) and Lemma 8.7.7 which show that, if $M^\perp$ is finite dimensional, then so also is $N^\perp$. □

The next two lemmas are purely algebraic results.

8.7.9 LEMMA *Let N and M be linear subspaces of G and suppose that N is finite-dimensional and that $N + M$ has finite codimension in G. Then M has finite codimension in G and*

$$\operatorname{codim} M = \dim N - \dim (N \cap M) + \operatorname{codim} (N + M).$$

PROOF By Lemma 6.6.11 there is a finite-dimensional linear subspace L of G such that $G = (N + M) \oplus L$ and $\dim L = \operatorname{codim} (N + M)$. Since $N \cap M$ is a linear subspace of N there is a linear subspace N_1 of N with $N = N_1 \oplus (N \cap M)$. (See HALMOS [12, p. 31, Theorem 2].) We have also

$$\dim N = \dim N_1 + \dim (N \cap M).$$

(See HALMOS [12, p. 30, Theorem 1].) Further, since

$$N + M = N_1 \oplus M$$

we have $G = M \oplus (N_1 \oplus L)$. It follows now from Lemma 6.6.11 that M has finite codimension in G and

$$\begin{aligned} \operatorname{codim} M &= \dim (N_1 \oplus L) \\ &= \dim N_1 + \dim L \\ &= \dim N - \dim (N \cap M) + \operatorname{codim} (N + M). \ \square \end{aligned}$$

8.7.10 LEMMA *Let M and N be linear subspaces of G and suppose that M has finite codimension in G and that $N \cap M = \{0\}$. Then N is finite-dimensional and* $\dim N \leq \operatorname{codim} M$.

PROOF Let π be the canonical mapping of G onto G/M given by $\pi x = x + M$ for all $x \in G$. Since $N \cap M = \{0\}$ it is clear that $\pi|N$ is a one-to-one linear mapping of N onto $\pi(N)$. By hypothesis G/M is finite-dimensional so $\pi(N)$ is finite-dimensional and hence Lemma 6.6.9 (applied to the mapping $(\pi|N)^{-1}$) shows that N is finite-dimensional and $\dim N = \dim \pi(N) \leq \dim G/M = \operatorname{codim} M$. □

We come now to our first perturbation theorem. We recall that E and F are Banach spaces and $T \in L(E,F)$.

8.7.11 THEOREM *Suppose that $\mathscr{R}(T)$ is closed and that T has finite nullity. Then there exists $\delta > 0$ such that, if $S \in L(E,F)$ and $\| S \| < \delta$, then $\mathscr{R}(T+S)$ is closed and $T+S$ has finite nullity with*

$$\alpha(T+S) \leq \alpha(T).$$

Suppose further that T is a Φ^+-operator. Then there exists $\delta > 0$ such that, if $S \in L(E,F)$ and $\| S \| < \delta$, then $T+S$ is a Φ^+-operator and

$$\alpha(T+S) \leq \alpha(T).$$

PROOF Suppose first that $\mathscr{R}(T)$ is closed and that T has finite nullity. Let $N = \mathscr{N}(T)$. By hypothesis N is finite-dimensional so, by Lemma 8.7.3, there is a closed linear subspace M of E with $N \oplus M = E$. Then $\mathscr{R}(T) = T(N) + T(M) = T(M)$, because $T(N) = \{0\}$, and hence $T|M$ is a one-to-one continuous linear mapping of M onto $\mathscr{R}(T)$. By hypothesis $\mathscr{R}(T)$ is closed so, by Theorem 2.6.8, both M and $\mathscr{R}(T)$ are Banach spaces. Consequently, by Theorem 8.5.4, $T|M$ is a linear homeomorphism and thus there exists $m > 0$ such that

$$\| Tx \| \geq m \| x \| \tag{1}$$

for all $x \in M$ (Lemma 3.5.9).

Let $S \in L(E,F)$ with $\| S \| < m$. We shall prove that $\mathscr{R}(T+S)$ is closed. For all $x \in M$, inequality (1) gives

$$\| (T+S)x \| \geq \| Tx \| - \| Sx \| \geq (m - \| S \|) \| x \|. \tag{2}$$

Therefore, by Lemma 3.5.9, $(T+S)|M$ is a linear homeomorphism of M onto $(T+S)(M)$. Thus $(T+S)(M)$ is complete because M is complete. (See the remarks on p. 110.) Theorem 2.6.8 now shows that $(T+S)(M)$ is closed. Since $E = N \oplus M$ and $T(N) = \{0\}$ we have

$$\mathscr{R}(T+S) = (T+S)(N) + (T+S)(M) = S(N) + (T+S)(M). \tag{3}$$

By Lemma 6.6.9 $S(N)$ is finite-dimensional because N is finite-dimensional, and therefore by (3) and Lemma 8.7.4 $\mathscr{R}(T+S)$ is closed.

We shall prove next that $T+S$ has finite nullity and

$$\alpha(T+S) \leq \alpha(T).$$

Inequality (2) shows that $M \cap \mathscr{N}(T+S) = \{0\}$ and Lemma 6.6.11 shows that M has finite codimension in E with codim $M = \dim N$. Thus Lemma 8.7.10 shows that $\mathscr{N}(T+S)$ is finite-dimensional and

$$\begin{aligned} \alpha(T+S) &= \dim \mathscr{N}(T+S) \\ &\leq \operatorname{codim} M \\ &= \dim N. \\ &= \alpha(T). \end{aligned}$$

We have now proved that the first assertion of the theorem is true with $\delta = m$.

Suppose next that T is a Φ^+-operator. Then, in particular, $\mathscr{R}(T)$ is closed and T has finite nullity so the first part of the proof applies to T. Let $S \in L(E,F)$ with $\| S \| < m/3$. We shall prove that $T + S$ is a Φ^+-operator and $\alpha(T + S) \leq \alpha(T)$. It follows from the first part of the proof that $\mathscr{R}(T + S)$ is closed and $T + S$ has finite nullity with $\alpha(T + S) \leq \alpha(T)$, so it remains only to prove that $\beta(T + S) = \infty$.

Suppose on the contrary that $\beta(T + S)$ is finite. Let $R_1 = T(M)$ and $R_2 = (T + S)(M)$. Since $T(N) = \{0\}$ we have

$$\mathscr{R}(T) = T(N) + T(M) = R_1.$$

Also, inequality (2) gives, for all $x \in M$,

$$\| (T + S)x - Tx \| = \| Sx \| \leq \| S \| \, \| x \| \leq \frac{\| S \|}{m - \| S \|} \| (T + S)x \|,$$

from which it follows that

$$d(y,R_1) \leq \frac{\| S \|}{m - \| S \|} \| y \| \tag{4}$$

for all $y \in R_2$. Now $\| S \| (m - \| S \|)^{-1} < \frac{1}{2}$ and F/R_1 is infinite-dimensional because T is a Φ^+-operator. Therefore inequality (4) and Lemma 8.7.8 show that F/R_2 is also infinite-dimensional. On the other hand (3) gives $\mathscr{R}(T + S) = S(N) + R_2$ and consequently, since we have seen above that $S(N)$ is finite-dimensional, and since we are supposing that $\mathscr{R}(T + S)$ has finite codimension in F, Lemma 8.7.9 shows that R_2 has finite codimension in F. We have now obtained a contradiction so we must have $\beta(T + S) = \infty$. Thus $T + S$ is a Φ^+-operator, and this shows that the second assertion of the theorem is true with $\delta = m/3$. □

Our next perturbation theorem is obtained from Theorem 8.7.11 by using the duality between an operator and its adjoint.

8.7.12 THEOREM *Suppose that $\mathscr{R}(T)$ is closed and that T has finite deficiency. Then there exists $\delta > 0$ such that, if $S \in L(E,F)$ and $\| S \| < \delta$, then $\mathscr{R}(T + S)$ is closed and $T + S$ has finite deficiency with*

$$\beta(T + S) \leq \beta(T).$$

Suppose further that T is a Φ^--operator. Then there exists $\delta > 0$ such that, if $S \in L(E,F)$ and $\| S \| < \delta$, then $T + S$ is a Φ^--operator and

$$\beta(T + S) \leq \beta(T).$$

PROOF Suppose that $\mathscr{R}(T)$ is closed and has finite deficiency. Then by Theorem 8.6.2 $\mathscr{R}(T^*)$ is closed and, by Theorem 8.7.2, T^* has finite nullity. Consequently by Theorem 8.7.11 there exists $\delta > 0$ such that, if $U \in L(F^*,E^*)$ and $\| U \| < \delta$, then $\mathscr{R}(T^* + U)$ is closed and $T^* + U$ has finite nullity with $\alpha(T^* + U) \leq \alpha(T^*)$. If $S \in L(E,F)$

then, by Theorem 6.4.2, $S^* \in L(F^*,E^*)$ and $\| S^* \| = \| S \|$. Using Theorems 8.6.2 and 8.7.2 again, we see that if $S \in L(E,F)$ and $\| S \| < \delta$ then $\mathscr{R}(T + S)$ is closed and $T + S$ has finite deficiency with $\beta(T + S) = \alpha(T^* + S^*) \leq \alpha(T^*) = \beta(T)$.

The proof of the second assertion of the theorem also follows from Theorems 8.7.2 and 8.7.11. We leave the details to the reader. □

Theorems 8.7.11 and 8.7.12 can be described as 'stability theorems' for Φ^+- and Φ^--operators. Our last perturbation theorem, (8.7.15), is a stability theorem for Φ-operators. It is possible to give a proof of this theorem which does not depend on results about the adjoint operator. However, for brevity, we shall deduce it from the preceding perturbation theorems by using Theorem 8.7.2 (thereby reversing the historical order of discovery). We need two more lemmas, the first of which is purely algebraic.

8.7.13 LEMMA *Let M and N be linear subspaces of E and suppose that M has finite codimension in E. Then $M + N$ has finite codimension in E.*

PROOF We can define without ambiguity a mapping S of E/M onto $E/(M + N)$ by $S(x + M) = x + (M + N)$ for all $x \in E$. It is clear that S is linear. Since E/M is finite-dimensional, Lemma 6.6.9 shows that $\mathscr{R}(S) = E/(M + N)$ is also finite-dimensional. Thus $M + N$ has finite codimension in E. □

8.7.14 LEMMA *Let M be a closed linear subspace of E with finite codimension in E and suppose that $T|M$ is a Φ-operator on M with $\alpha(T|M) = 0$. Then T is a Φ-operator on E and*

$$\varkappa(T) = \operatorname{codim} M + \varkappa(T|M).$$

PROOF Since $\alpha(T|M) = 0$ we have $M \cap \mathscr{N}(T) = \{0\}$ and consequently Lemma 8.7.10 shows that $\mathscr{N}(T)$ is finite-dimensional. Lemma 8.7.13 shows that $M + \mathscr{N}(T)$ has finite codimension in E and Lemma 8.7.9 then shows that

$$\operatorname{codim}(M + \mathscr{N}(T)) = \operatorname{codim} M - \dim \mathscr{N}(T) \tag{5}$$

because $M \cap \mathscr{N}(T) = \{0\}$. By Lemma 6.6.11 there is a finite dimensional linear subspace L of E with $(M + \mathscr{N}(T)) \oplus L = E$ and

$$\dim L = \operatorname{codim}(M + \mathscr{N}(T)). \tag{6}$$

It is clear that

$$\mathscr{R}(T) = T(M) + T(L). \tag{7}$$

Since $T|M$ is a Φ-operator $T(M) = \mathscr{R}(T|M)$ is closed and, since

$T|L$ is a one-to-one linear mapping of L onto $T(L)$, Lemma 6.6.9 shows that $T(L)$ is finite-dimensional and

$$\dim T(L) = \dim L. \tag{8}$$

It now follows from (7) and Lemma 8.7.4 that $\mathscr{R}(T)$ is closed and, from (7) and Lemma 8.7.13, that $\mathscr{R}(T)$ has finite codimension in F. We observe that $T(M) \cap T(L) = \{0\}$ because, if $Tx = Ty$ with $x \in M$ and $y \in L$ then $z = x - y \in \mathscr{N}(T)$ and hence

$$y = x - z \in L \cap (M + \mathscr{N}(T)) = \{0\}.$$

Thus from (7) and Lemma 8.7.9 we obtain

$$\operatorname{codim} \mathscr{R}(T) = \operatorname{codim} T(M) - \dim T(L). \tag{9}$$

This proves that T is a Φ-operator, and from (5), (6), (8) and (9) we obtain

$$\begin{aligned} \varkappa(T) &= \alpha(T) - \beta(T) \\ &= \dim \mathscr{N}(T) - \operatorname{codim} \mathscr{R}(T) \\ &= \operatorname{codim} M - \operatorname{codim} T(M) \\ &= \operatorname{codim} M - \beta(T|M) \\ &= \operatorname{codim} M + \varkappa(T|M). \ \square \end{aligned}$$

8.7.15 THEOREM *Suppose that T is a Φ-operator. Then there exists $\delta > 0$ such that, if $S \in L(E,F)$ and $\| S \| < \delta$, then $T + S$ is a Φ-operator, $\alpha(T + S) \leq \alpha(T)$, $\beta(T + S) \leq \beta(T)$ and $\varkappa(T + S) = \varkappa(T)$.*

PROOF We begin by proving that there exists $\delta_1 > 0$ such that, if $S \in L(E,F)$ and $\| S \| < \delta_1$, then $T + S$ is a Φ-operator,

$$\beta(T + S) \leq \beta(T) \text{ and } \varkappa(T + S) \geq \varkappa(T).$$

By Lemma 8.7.3 there is a closed linear subspace M of E with $M \oplus \mathscr{N}(T) = E$ and, by Lemma 6.6.11, M has finite codimension in E with $\operatorname{codim} M = \dim \mathscr{N}(T) = \alpha(T)$. Obviously $\mathscr{R}(T|M) = T(M) = \mathscr{R}(T)$ so $\mathscr{R}(T|M)$ is closed and has finite codimension in F. Also $\mathscr{N}(T|M) = \mathscr{N}(T) \cap M = \{0\}$. This shows that $T|M$ is a Φ-operator on M with $\alpha(T|M) = 0$ and $\beta(T|M) = \beta(T)$. Since M is closed and E is a Banach space M is also a Banach space (Theorem 2.6.8). Consequently we may apply the first parts of Theorems 8.7.11 and 8.7.12 to the operator $T|M$ to obtain a number $\delta_1 > 0$ with the property that, if $S_0 \in L(M,F)$ and $\| S_0 \| < \delta_1$, then $(T|M) + S_0$ is a Φ-operator with

$$\alpha((T|M) + S_0) \leq \alpha(T|M) = 0$$

and

$$\beta((T|M) + S_0) \leq \beta(T|M) = \beta(T).$$

Let $S \in L(E,F)$ with $\| S \| < \delta_1$. Then $\| S|M \| \leq \| S \| < \delta_1$ and $(T|M) + (S|M) = (T + S)|M$ so $(T + S)|M$ is a Φ-operator with $\alpha((T + S)|M) = 0$ and $\beta((T + S)|M) \leq \beta(T)$. It follows now from Lemma 8.7.14 that $T + S$ is a Φ-operator and

$$\begin{aligned}\varkappa(T+S) &= \text{codim } M + \varkappa((T+S)|M) \\ &= \alpha(T) - \beta((T+S)|M) \\ &\geq \alpha(T) - \beta(T) \\ &= \varkappa(T).\end{aligned}$$

Further, we have

$$\begin{aligned}\mathscr{R}(T+S) &= (T+S)(M) + (T+S)(\mathscr{N}(T)) \\ &= \mathscr{R}((T+S)|M) + S(\mathscr{N}(T))\end{aligned}$$

and hence, since $S(\mathscr{N}(T))$ is finite-dimensional (Lemma 6.6.9), it follows from Lemma 8.7.9 that

$$\begin{aligned}\beta(T+S) &= \text{codim } \mathscr{R}(T+S) \\ &\leq \text{codim } \mathscr{R}((T+S)|M) \\ &= \beta((T+S)|M) \\ &\leq \beta(T).\end{aligned}$$

Applying what we have just proved to T^*, which is a Φ-operator by Theorem 8.7.2, we obtain a number $\delta_2 > 0$ with the property that, if $S \in L(E,F)$ and $\| S^* \| = \| S \| < \delta_2$, then $T^* + S^*$ is a Φ-operator, $\beta(T^* + S^*) \leq \beta(T^*)$ and $\varkappa(T^* + S^*) \geq \varkappa(T^*)$.

Let $\delta = \min \{\delta_1, \delta_2\}$ and let $S \in L(E,F)$ with $\| S \| < \delta$. Then $T + S$ is a Φ-operator, $\beta(T + S) \leq \beta(T)$ and, by Theorem 8.7.2,

$$\alpha(T+S) = \beta(T^* + S^*) \leq \beta(T^*) = \alpha(T).$$

Also $\varkappa(T + S) \geq \varkappa(T)$ and, by Theorem 8.7.2 again,

$$\varkappa(T+S) = -\varkappa(T^* + S^*) \leq -\varkappa(T^*) = \varkappa(T)$$

so $\varkappa(T + S) = \varkappa(T)$. □

We can obtain some information about the spectrum of a bounded linear operator directly from the last theorem. We shall suppose that E is a non-zero complex Banach space and that $F = E$.

8.7.16 DEFINITION A complex number λ is said to be a Φ-, Φ^+- or *Φ^--point of* T if and only if $\lambda I - T$ is a Φ-, Φ^+- or Φ^--operator, respectively.

It is obvious that the Φ^+- and Φ^--points of T are in the spectrum of T. Also a Φ-point of T is in the spectrum of T if and only if at least one of the numbers $\alpha(\lambda I - T)$ and $\beta(\lambda I - T)$ is non-zero.

The following theorem is a simple consequence of Theorems 8.7.11, 8.7.12 and 8.7.15; we leave the details to the reader.

8.7.17 THEOREM *The sets of Φ-, Φ^+- and Φ^--points of T are open subsets of the complex plane. Further the index $\varkappa(\lambda I - T)$ is constant on each connected component of the set of Φ-points of T.*

Theorems 8.7.11, 8.7.12 and 8.7.15 are the simplest of a family of perturbation theorems, and can be considerably generalized and extended. They are included in this book to illustrate how the tools developed in this chapter and in Chapter 6 can be used in the study of bounded linear operators.

In particular the proofs of Theorems 8.7.12 and 8.7.15 provide examples of a way in which the adjoint operator can be used to prove a result which, in its statement, does not involve the adjoint operator at all.

The perturbation theorems of the type considered in this section have developed over a considerable period of time. They arose first in the specific context of certain classes of linear integral equations; their abstract formulation is relatively recent, and is the work of a number of mathematicians. The first comprehensive account of these perturbation theorems was given in 1957 by I. Ts. Gokhberg and M. G. Krein. They introduced the Φ-notation used above, and their paper includes an historical introduction and several applications. The reader who is interested in further developments in the theory of perturbation theorems should consult the treatise of T. KATO [17].

The reader will find some exercises on Φ-operators at the end of the next section.

8.8 The Closed Graph Theorem

Throughout this section E and F will denote normed linear spaces over the same field.

8.8.1 DEFINITION Let T be a linear mapping of a linear subspace M of E into F. Then the set $G = \{(x,Tx) : x \in M\}$ is called the *graph of T*.

Logically a mapping and its graph are identical. (See p. 4.) However, many redundant terminologies are inevitable in mathematics because its historical development is not always (even not often) strictly logical.

The graph of T is a subset of the product $E\times F$. We recall that $E\times F$ is a normed linear space with linear space operations and norm given by

$$(x_1,x_2) + (y_1,y_2) = (x_1 + y_1,x_2 + y_2),$$
$$\alpha(x_1,x_2) = (\alpha x_1,\alpha x_2),$$
and
$$\| (x_1,x_2) \| = \max\{ \| x_1 \|, \| x_2 \|\}.$$

(See p. 282.) Since T is linear its graph is a linear subspace of $E\times F$.

8.8.2 DEFINITION A linear mapping of a linear subspace of E into F is said to be *closed* if and only if its graph is a closed subset of $E \times F$.

It is convenient to have a sequential characterization of closed linear mappings.

8.8.3 LEMMA *Let T be a linear mapping of a linear subspace M of E into F. Then T is closed if and only if the following condition is satisfied: if (x_n) is a sequence in M with $\lim_{n\to\infty} x_n = x$ and $\lim_{n\to\infty} Tx_n = y$ then $x \in M$ and $Tx = y$.*

PROOF Let G be the graph of T. Suppose first that T is closed and let (x_n) be a sequence in M with $\lim_{n\to\infty} x_n = x$ and $\lim_{n\to\infty} Tx_n = y$. Then, since

$$\| (x_n, Tx_n) - (x,y) \| = \max \{ \| x_n - x \|, \| Tx_n - y \| \}, \tag{1}$$

we have $\lim_{n\to\infty} (x_n, Tx_n) = (x,y)$. Since G is closed it follows that $(x,y) \in G$. Consequently, by definition of G, we have $x \in M$ and $Tx = y$. This proves that the condition of the lemma is satisfied.

Suppose, on the other hand, that the condition of the lemma is satisfied. Let $((x_n, Tx_n))$ be a sequence in G with $\lim_{n\to\infty} (x_n, Tx_n) = (x,y)$. It follows from equation (1) that $\lim_{n\to\infty} x_n = x$ and $\lim_{n\to\infty} Tx_n = y$ and therefore, by hypothesis, $x \in M$ and $Tx = y$. Thus $(x,y) = (x, Tx) \in G$ and it follows that G, and hence T, is closed. □

It follows immediately from Lemma 8.8.3 and Theorem 2.7.7 that each continuous linear mapping of E into F is closed. The following example shows that a closed linear mapping may fail to be continuous.

8.8.4 EXAMPLE Let M be the linear subspace of $C_{\mathsf{R}}([0,2\pi])$ consisting of those functions on $[0,2\pi]$ that have continuous first derivatives on $[0,2\pi]$ and let T be the linear mapping of M into $C_{\mathsf{R}}([0,2\pi])$ given by

$$(Tf)(t) = f'(t) + f(t)$$

for all $f \in M$ and all $t \in [0, 2\pi]$. We have already studied the mapping T in Example 6.1.3 where we saw that T is not a continuous mapping of M into $C_{\mathsf{R}}([0,2\pi])$ (the norm on M being the uniform norm). We shall prove that T is closed.

Let (f_n) be a sequence in M with $\lim_{n\to\infty} f_n = f$ and $\lim_{n\to\infty} Tf_n = g$. By Theorem 3.9.9

$$\int_0^t f_n'(s)\, ds = f_n(t) - f_n(0)$$

for $0 \le t \le 2\pi$ so, using Lemma 3.7.9, we obtain, for $0 \le t \le 2\pi$,

$$\begin{aligned}
\int_0^t g(s)\,ds &= \lim_{n\to\infty} \int_0^t (Tf_n)(s)\,ds \\
&= \lim_{n\to\infty} \int_0^t f_n'(s)\,ds + \lim_{n\to\infty} \int_0^t f_n(s)\,ds \\
&= \lim_{n\to\infty} (f_n(t) - f_n(0)) + \int_0^t f(s)\,ds \\
&= f(t) - f(0) + \int_0^t f(s)\,ds,
\end{aligned}$$

where the last step follows from the fact that uniform convergence on $[0,2\pi]$ entails pointwise convergence on $[0,2\pi]$. (See p. 49.) It follows now from Theorem 3.9.7 that $f \in M$ and $Tf = g$. By Lemma 8.8.3 this proves that T is closed.

For linear transformations, being closed is next best to being continuous, and there is an extensive theory of closed linear transformations. We shall not pursue the study of closed linear transformations in this book; our aim here is simply to prove a famous theorem of Banach which is yet another important corollary of the open mapping theorem. We need a preliminary lemma.

8.8.5 LEMMA *Suppose that E and F are Banach spaces. Then $E \times F$ is also a Banach space.*

PROOF Let $((x_n, y_n))$ be a Cauchy sequence in $E \times F$. The equation

$$\max \{\| x_m - x_n \|, \| y_m - y_n \|\} = \| (x_m, y_m) - (x_n, y_n) \|$$

shows that (x_n) and (y_n) are Cauchy sequences in E and F, respectively, and hence (x_n) and (y_n) converge. Let $x = \lim_{n\to\infty} x_n$ and $y = \lim_{n\to\infty} y_n$. The equation

$$\| (x_n, y_n) - (x, y) \| = \max \{\| x_n - x \|, \| y_n - y \|\}$$

then shows that $\lim_{n\to\infty} (x_n, y_n) = (x, y)$. This proves that $E \times F$ is complete. □

8.8.6 THEOREM (the closed graph theorem) *A closed linear mapping of a Banach space E into a Banach space F is continuous on E.*

PROOF Let T be a closed linear mapping of E into F and let G be the graph of T. Then G is a closed linear subspace of the Banach space $E \times F$ and hence is itself a Banach space (Theorem 2.6.8). Let S be the mapping of G into E defined by $S(x, Tx) = x$ for all $x \in E$. It is clear that S is a one-to-one linear mapping of G onto E. Also

$$\| S(x, Tx) \| = \| x \| \le \max \{\| x \|, \| Tx \|\} = \| (x, Tx) \|$$

for all $x \in E$ so S is continuous. Therefore, by Theorem 8.5.4, S is a linear homeomorphism and it follows that, for each $x \in E$,

$$\| Tx \| \leq \max \{\| x \|, \| Tx \|\} = \| (x,Tx) \| = \| S^{-1} x \| \leq \| S^{-1} \| \| x \|.$$

This proves that T is continuous. □

Theorem 8.8.6 provides a powerful tool for establishing continuity of a linear transformation. (See Theorem 8.8.10.)

To illustrate the use of the closed graph theorem, we shall prove a theorem on the existence of direct sum decompositions of the form $E = M \oplus N$, where E is a Banach space and M and N are closed linear subspaces of E. We have already remarked on p. 99 that, given a closed linear subspace M of E, it may happen that there is no closed linear subspace N of E such that $E = M \oplus N$.

We begin with some purely algebraic considerations.

8.8.7 DEFINITION A linear operator P on E is said to be a *projection* if and only if $P^2 = P$.

Notice that we do not require that a projection be a bounded linear operator.

8.8.8 LEMMA *Let P be a projection on E. Then*

(a) *$I - P$ is a projection on E,*
(b) $\mathscr{R}(P) = \{x \in E : Px = x\}$,
(c) $\mathscr{R}(P) = \mathscr{N}(I - P)$,
(d) $E = \mathscr{R}(P) \oplus \mathscr{R}(I - P)$,
(e) *if P is bounded then $\mathscr{R}(P)$ and $\mathscr{R}(I - P)$ are closed.*

PROOF (a) $(I - P)^2 = I - 2P + P^2 = I - P$.

(b) It is obvious that $\{x \in E: Px = x\} \subseteq \mathscr{R}(P)$. On the other hand let $x \in \mathscr{R}(P)$. Then $x = Py$ for some $y \in E$ and hence $Px = P^2y = Py = x$. This proves that $\{x \in E: Px = x\} = \mathscr{R}(P)$.

(c) This follows from (b) and the observation that $(I - P)x = 0$ if and only if $x = Px$.

(d) For each $x \in E$ we have $x = Px + (I - P)x$. Thus

$$E = \mathscr{R}(P) + \mathscr{R}(I - P).$$

If $x \in \mathscr{R}(P) \cap \mathscr{R}(I - P)$ then (b), applied to P and $I - P$, gives

$$x = Px = (I - P)x$$

and hence

$$x = Px = P((I - P)x) = (P - P^2)x = 0.$$

This shows that $\mathscr{R}(P) \cap \mathscr{R}(I - P) = \{0\}$ and hence that

$$E = \mathscr{R}(P) \oplus \mathscr{R}(I - P).$$

(e) By (c), applied to P and $I - P$, we have $\mathscr{R}(P) = \mathscr{N}(I - P)$ and $\mathscr{R}(I - P) = \mathscr{N}(P)$ and hence $\mathscr{R}(P)$ and $\mathscr{R}(I - P)$ are closed because the null-space of any bounded linear operator is closed. □

Part (d) of the above lemma shows that there is a direct sum decomposition of E associated with each projection on E. Our next lemma shows that every direct sum decomposition of E arises in this way from a projection on E.

8.8.9 Lemma *Let M and N be linear subspaces of E with $E = M \oplus N$. Then there is a unique projection P on E with*

$$\mathscr{R}(P) = M \text{ and } \mathscr{R}(I - P) = N.$$

Proof Let $x \in E$. Then there are unique points $y \in M$ and $z \in N$ with $x = y + z$. Let $Px = y$. This defines a mapping P of E into itself and it is easy to verify that P is linear and that $\mathscr{R}(P) = M$ and $\mathscr{N}(P) = N$. For each $x \in E$ we have $Px \in M$ and therefore $P(Px) = Px$. This shows that $P^2 = P$ and hence P is a projection on E. We have seen that $\mathscr{R}(P) = M$ and $\mathscr{N}(P) = N$. By Lemma 8.8.8(c), applied to $I - P$, we have $\mathscr{N}(P) = \mathscr{R}(I - P)$ so $\mathscr{R}(I - P) = N$.

Finally let Q be a projection on E with $\mathscr{R}(Q) = M$ and $\mathscr{R}(I - Q) = N$. For each $x \in E$ we have $x = Qx + (I - Q)x$, $Qx \in M$ and $(I - Q)x \in N$ so, by definition of P, we must have $Px = Qx$. This proves that $P = Q$. □

Parts (d) and (e) of Lemma 8.8.8 show that if P is a bounded projection on E, then E has a direct sum decomposition

$$E = \mathscr{R}(P) \oplus \mathscr{R}(I - P)$$

where $\mathscr{R}(P)$ and $\mathscr{R}(I - P)$ are closed linear subspaces of E. Our final theorem, which uses the closed graph theorem, shows that every direct sum decomposition of E formed from closed linear subspaces of E arises in this way from a bounded projection on E.

8.8.10 Theorem *Let E be a Banach space and let M and N be closed linear subspaces of E with $E = M \oplus N$. Then there is a unique bounded projection P on E such that $\mathscr{R}(P) = M$ and $\mathscr{R}(I - P) = N$.*

Proof By Lemma 8.8.9 there is a unique projection P on E such that $\mathscr{R}(P) = M$ and $\mathscr{R}(I - P) = N$. We shall prove that P is closed. Let (x_n) be a sequence in E such that $\lim_{n\to\infty} x_n = x$ and $\lim_{n\to\infty} Px_n = y$. Since $Px_n \in M$ and M is closed we have $y \in M$. Also $x_n - Px_n = (I - P)x_n \in N$ and N is closed so $x - y = \lim_{n\to\infty} (x_n - Px_n) \in N$. By Lemma 8.8.8(b) $Py = y$ and by Lemma 8.8.8(c), applied to $I - P$, we have

$P(x - y) = 0$. Consequently $y = Py = Px$. This proves that P is closed, and therefore, by Theorem 8.8.6, P is continuous. □

The reader will find further properties of projections in the Exercises below.

Exercises

(1) A linear space E is a Banach space with respect to each of two norms $\|\cdot\|_1$ and $\|\cdot\|_2$ and if a sequence (x_n) in E is convergent to the limits y and z in $(E, \|\cdot\|_1)$ and in $(E, \|\cdot\|_2)$ then $y = z$. Prove that the two norms are equivalent.

(2) In the text Theorem 8.5.4 was obtained from the open mapping theorem (8.5.3), and the closed graph theorem (8.8.6) from Theorem 8.5.4. These three results are essentially equivalent. Deduce Theorem 8.5.4 from Theorem 8.8.6 and Theorem 8.5.3 from Theorem 8.5.4. (Hint: to obtain Theorem 8.5.3 let π be the canonical mapping of E onto the quotient space E/N, express $T \in L(E,F)$ as $T = T_1 \circ \pi$ with $T_1 \in L(E/N,F)$, where $N = \mathcal{N}(T)$, and use Exercise 3.4(2).)

(3) Let E be a closed linear subspace of $C_{\mathsf{R}}([0,1])$ and suppose that each function in E is continuously differentiable on $[0,1]$. Prove that E is finite-dimensional. (Hint: the operation of differentiation is continuous on E.)

(4) Let T be a linear mapping of a Banach space E into a Banach space F. If $y^* \circ T \in E^*$ for all $y^* \in F^*$ prove that $T \in L(E,F)$.

(5) A closed linear subspace M of a normed linear space E is said to be *invariant* under an operator $T \in L(E)$ if and only if $T(M) \subseteq M$. Closed linear subspaces M and N are said to *reduce* $T \in L(E)$ if and only if $E = M \oplus N$ and M and N are invariant under T.

It is an open question whether every bounded linear operator on a Banach space E has closed invariant subspaces other than the trivial ones $\{0\}$ and E. The determination of the family of all closed invariant subspaces of a given bounded linear operator is an important and seldom easy problem.

(a) Prove that if P is a bounded projection on a normed linear space E, $M = \mathcal{R}(P)$ and $N = \mathcal{R}(I - P)$ then M and N reduce T if and only if $PT = TP$.

(b) Prove that if E is a complex Banach space, $T \in L(E)$ and T is reduced by closed linear subspaces M and N then

$$\operatorname{sp}(T) = \operatorname{sp}(T|M) \cup \operatorname{sp}(T|N).$$

(6) Let P be a bounded projection on a complex normed linear space E. Determine the spectrum of P and, for $\lambda \in \mathsf{C} \sim \operatorname{sp}(P)$, express $(\lambda I - P)^{-1}$ in terms of P.

(7) Let E be a Banach space and let M and N be closed linear subspaces of E with $M \cap N = \{0\}$. Prove that $M \oplus N$ is closed in E if and only if there is a $d > 0$ such that $\| x - y \| \geq d$ for all $x \in M$ and $y \in N$ with $\| x \| = \| y \| = 1$.

(8) Let N be a closed linear subspace of a normed linear space E with $N \neq \{0\}$ and $N \neq E$. Prove that if N is either of finite dimension or of finite codimension, then N is the range of infinitely many distinct bounded projections on E. (See Lemmas 6.6.11 and 8.7.3.)

(9) Prove that a bounded projection on a normed linear space is compact if and only if its range is finite-dimensional.

(10) Let P be a bounded projection on a normed linear space E. Prove that P^* is a bounded projection on E^* with $\mathscr{R}(P^*) = \mathscr{N}(P)^\perp$ and $\mathscr{N}(P^*) = \mathscr{R}(P)^\perp$. Prove further that if $\mathscr{R}(P)$ is finite-dimensional then so is $\mathscr{R}(P^*)$ and $\dim \mathscr{R}(P) = \dim \mathscr{R}(P^*)$.

(11) Let P and Q be projections on a linear space E. Prove the following statements:

(a) PQ is a projection if and only if

$$P(\mathscr{R}(Q)) \subseteq \mathscr{R}(Q) \oplus [\mathscr{N}(Q) \cap \mathscr{N}(P)]$$

(b) $PQ = QP$ if and only if

$$\mathscr{R}(Q) = [\mathscr{R}(Q) \cap \mathscr{R}(P)] \oplus [\mathscr{R}(Q) \cap \mathscr{N}(P)]$$

and

$$\mathscr{N}(Q) = [\mathscr{N}(Q) \cap \mathscr{R}(P)] \oplus [\mathscr{N}(Q) \cap \mathscr{N}(P)].$$

(c) If $PQ = QP$ then PQ is a projection, $\mathscr{R}(PQ) = \mathscr{R}(P) \cap \mathscr{R}(Q)$, and $\mathscr{N}(PQ)$ is the linear hull of $\mathscr{N}(P) \cup \mathscr{N}(Q)$.

(d) If $PQ = QP$ then $P + Q - PQ$ is a projection, $\mathscr{R}(P + Q - PQ)$ is the linear hull of $\mathscr{R}(P) \cup \mathscr{R}(Q)$ and $\mathscr{N}(PQ) = \mathscr{N}(P) \cap \mathscr{N}(Q)$.

(e) $PQ = Q$ if and only if $\mathscr{R}(Q) \subseteq \mathscr{R}(P)$.

(f) $QP = Q$ if and only if $\mathscr{N}(P) \subseteq \mathscr{N}(Q)$.

(g) $P - Q$ is a projection if and only if $PQ = QP = Q$.

(h) If $P - Q$ is a projection then $\mathscr{R}(P - Q) = \mathscr{R}(P) \cap \mathscr{N}(Q)$ and $\mathscr{N}(P - Q) = \mathscr{N}(P) \oplus \mathscr{R}(Q)$.

(12) Let $P_1, P_2, \ldots, P_n$ be projections on a linear space E with $P_j P_k = 0$ for $j \neq k$. Prove that $P = P_1 + P_2 + \ldots + P_n$ is a projection with

$$\mathscr{R}(P) = \mathscr{R}(P_1) \oplus \mathscr{R}(P_2) \oplus \ldots \oplus \mathscr{R}(P_n)$$

and

$$\mathscr{N}(P) = \mathscr{N}(P_1) \cap \mathscr{N}(P_2) \cap \ldots \cap \mathscr{N}(P_n).$$

(13) The exercises which follow give further results concerning Φ-operators and lead to a more direct proof of Theorem 8.7.15.

(a) If E,F are normed linear spaces and $T \in L(E,F)$ then T is said to be *left invertible* if there exists $V \in L(F,E)$ such that $V \circ T = I_E$, the identity operator on E. The operator V is then said to be a *left inverse* of T. *Right invertibility* and *right inverses* are defined similarly. Prove that if E and F are Banach spaces and $T \in L(E,F)$ is a Φ-operator with $\alpha(T) = 0$ then T is left invertible. (Use Exercise (8).)

(b) Prove that the condition $\alpha(T|M) = 0$ in the hypotheses of Lemma 8.7.14 can be omitted.

(c) Prove that if E, F and G are Banach spaces and $T \in L(E,F)$ and $S \in L(F,G)$ are Φ-operators then $S \circ T$ is a Φ-operator and

$$\varkappa(S \circ T) = \varkappa(S) + \varkappa(T).$$

(d) A linear transformation T of a normed linear space E into a normed linear space F is said to be *compact* if $T(A)$ is a relatively sequentially compact subset of F for every bounded subset A of E. This extends Definition 6.5.1.

Prove that if E and F are Banach spaces and $T \in L(E,F)$ is a Φ-operator then there is $\delta > 0$ such that if $S \in L(E,F)$ and *either* $\| S \| < \delta$ *or* S is compact then $T + S$ is a Φ-operator and $\varkappa(T + S) = \varkappa(T)$. (Hint: Reduce to the case $\alpha(T) = 0$ and write $T + S = (I_F + S \circ V) \circ T$ where V is a left inverse of T. Then use (c) and the results of Section 6.6.)

CHAPTER 9

Spectral Theory in Hilbert Spaces

A Banach space whose norm arises in a certain way, to be defined in Section 9.1, is called a Hilbert space. Hilbert spaces are natural generalizations of the finite-dimensional spaces R^n and C^n and, from many points of view, they are the most important examples of Banach spaces.

The spectral theory of certain classes of linear operators on Hilbert spaces has been extensively, and successfully, studied since it was initiated between 1904 and 1910 by D. Hilbert. This spectral theory has important applications to problems of classical analysis – particularly to the study of differential equations. It is also an indispensable tool in the study of algebras of operators on Hilbert spaces. Such algebras have been used in attempts to discover a mathematically satisfactory basis for quantum mechanics.

In this chapter our aim is to illustrate the development of spectral theory in Hilbert spaces by proving *the spectral theorem for self-adjoint bounded linear operators* (9.9.4). Self-adjoint bounded linear operators are generalizations of Hermitian symmetric matrices of complex numbers, and the spectral theorem is a generalization of the well-known diagonalization theorem for such matrices.

We have left out almost everything concerning Hilbert spaces that is not essential for the proof of the spectral theorem. The reader who needs a more leisurely introduction to Hilbert spaces should consult the books of HALMOS [11, §§1–34], AHIEZER and GLAZMAN [1] and BERBERIAN [4].

In Sections 9.1 to 9.4 we develop those parts of the general theory of Hilbert spaces that we shall need in our discussion of the spectral theorem. In Section 9.5 we show how the spectral theory of compact operators, which was developed in Section 6.7, can be carried further in the case of compact self-adjoint operators on Hilbert spaces. In

Sections 9.6, 9.7 and 9.8 we set up the technical machinery that is used in the proof of the spectral theorem in Section 9.9.

9.1 Hermitian Symmetric Forms

Throughout this section E will denote a linear space over K.

9.1.1 DEFINITION A mapping B of $E \times E$ into K is said to be a *Hermitian symmetric form on E* if and only if

(a) $B(\alpha x + \beta y, z) = \alpha B(x,z) + \beta B(y,z)$ for all $x,y,z \in E$ and all $\alpha,\beta \in \mathsf{K}$, and

(b) $B(x,y) = \overline{B(y,x)}$ for all $x,y \in E$.

It follows from (a) and (b) that

$$B(x, \alpha y + \beta z) = \bar{\alpha} B(x,y) + \bar{\beta} B(x,z)$$

for all $x,y,z \in E$ and all $\alpha,\beta \in \mathsf{K}$, and it follows from (b) that $B(x,x)$ is real for all $x \in E$. A Hermitian symmetric form is sometimes called a *symmetric bilinear form*; this terminology conflicts, when $\mathsf{K} = \mathsf{C}$, with established terminology for bilinear forms, and we shall not use it.

A Hermitian symmetric form B is said to be *non-negative* if and only if

(c) $B(x,x) \geq 0$ for all $x \in E$;

and it is said to be *positive* if and only if it is non-negative and

(d) $B(x,x) = 0$ if and only if $x = 0$.

A positive Hermitian symmetric form on E is called an *inner product on E* or a *scalar product on E*.

The simplest example of an inner product is obtained by taking $E = \mathsf{K}$ and $B(x,y) = x\bar{y}$. The following examples are less trivial; the verification of conditions (a) to (d) is left to the reader.

9.1.2 EXAMPLE The equation

$$B(x,y) = \sum_{k=1}^{n} \xi_k \bar{\eta}_k,$$

where $x = (\xi_1, \xi_2, \ldots, \xi_n)$ and $y = (\eta_1, \eta_2, \ldots, \eta_n)$ are in K^n, defines an inner product on K^n.

9.1.3 EXAMPLE The equation

$$B(f,g) = \int_0^1 f(t)\,\overline{g(t)}\,dt,$$

where $f,g \in C_{\mathsf{K}}([0,1])$, defines an inner product on $C_{\mathsf{K}}([0,1])$.

9.1.4 EXAMPLE The equation

$$B(x,y) = \sum_{n=1}^{\infty} \xi_n \bar{\eta}_n,$$

where $x = (\xi_n)$ and $y = (\eta_n)$ are in ℓ^2, defines an inner product on ℓ^2. (Hölder's inequality (3.2.7) shows that the series $\sum_n \xi_n \bar{\eta}_n$ converges.)

9.1.5 THEOREM (Schwarz's inequality) *Let B be a non-negative Hermitian symmetric form on E. Then*

$$| B(x,y) |^2 \leq B(x,x)B(y,y)$$

for all $x,y \in E$.

PROOF Let $x,y \in E$. For all real numbers t and all $\alpha \in \mathsf{K}$ with $| \alpha | = 1$ we have

$$B(t\alpha x + y,\ t\alpha x + y) \geq 0$$

from which it follows that

$$t^2 B(x,x) + 2t \operatorname{Re}(\alpha B(x,y)) + B(y,y) \geq 0. \tag{1}$$

Since inequality (1) holds for all real numbers t we must have

$$(\operatorname{Re}(\alpha B(x,y)))^2 \leq B(x,x)B(y,y) \tag{2}$$

for all $\alpha \in \mathsf{K}$ with $| \alpha | = 1$. By choosing α so that $\alpha B(x,y) = | B(x,y) |$ we obtain from (2) the inequality $| B(x,y) |^2 \leq B(x,x)B(y,y)$. □

9.1.6 THEOREM *Let B be a non-negative Hermitian symmetric form on E. Then*

$$B(x+y,x+y)^{1/2} \leq B(x,x)^{1/2} + B(y,y)^{1/2}$$

for all $x,y \in E$.

PROOF Let $x,y \in E$. Using Schwarz's inequality we obtain

$$\begin{aligned} B(x+y,x+y) &= B(x,x) + 2\operatorname{Re} B(x,y) + B(y,y) \\ &\leq B(x,x) + 2\,| B(x,y) | + B(y,y) \\ &\leq B(x,x) + 2B(x,x)^{1/2}B(y,y)^{1/2} + B(y,y) \\ &= (B(x,x)^{1/2} + B(y,y)^{1/2})^2. \ \square \end{aligned}$$

9.1.7 COROLLARY *Let B be an inner product on E. Then the mapping $x \to B(x,x)^{1/2}$ is a norm on E.*

PROOF This follows directly from Theorem 9.1.6 and the definition of an inner product. □

Let B be an inner product on E and let $\| x \| = B(x,x)^{1/2}$. Then a simple calculation shows that

$$\| x + y \|^2 + \| x - y \|^2 = 2 \| x \|^2 + 2 \| y \|^2 \tag{3}$$

for all $x,y \in E$. The identity (3) is often called the *parallelogram law* because, when $E = \mathsf{R}^2$ (or $E = \mathsf{C}$), it expresses the well-known geometrical theorem that the sum of the squares of the sides of a parallelogram is equal to the sum of the squares of the diagonals of the parallelogram. It can be proved that if a norm on a linear space E satisfies the parallelogram law (3) then the norm is derived from an inner product on E as above. (See Exercise (1).)

The following two lemmas will be used repeatedly in the rest of this chapter.

9.1.8 LEMMA *Let B be an inner product on E and let $x,y \in E$ be such that $B(x,z) = B(y,z)$ for all $z \in E$. Then $x = y$.*

PROOF $B(x - y,x - y) = B(x,x - y) - B(y,x - y) = 0.$ □

9.1.9 LEMMA *Let B be an inner product on E and let $\| \cdot \|$ be the norm on E defined by $\| x \| = B(x,x)^{1/2}$. Then $(x,y) \to B(x,y)$ is a continuous mapping of $E \times E$ into K.*

PROOF We recall that $E \times E$ is a normed linear space with norm defined by $\| (x,y) \| = \max \{\| x \| , \| y \|\}$. (See p. 282.)

Using Schwarz's inequality we obtain, for all $x,y,x_0,y_0 \in E$,

$$\begin{aligned} | B(x,y) - B(x_0,y_0) | &= | B(x - x_0,y) + B(x_0,y - y_0) | \\ &\leq | B(x - x_0,y) | + | B(x_0,y - y_0) | \\ &\leq \| x - x_0 \| \; \| y \| + \| x_0 \| \; \| y - y_0 \| . \end{aligned}$$

Thus if $\| y - y_0 \| \leq 1$ we have

$$| B(x,y) - B(x_0,y_0) | \leq (1 + \| x_0 \| + \| y_0 \|) \max \{\| x - x_0 \| , \| y - y_0 \|\},$$

from which it follows that the mapping $(x,y) \to B(x,y)$ is continuous at (x_0,y_0). □

We shall often use Lemma 9.1.9 in the following form: if (x_n) and (y_n) are convergent sequences in E with $\lim\limits_{n \to \infty} x_n = x$ and $\lim\limits_{n \to \infty} y_n = y$ then $\lim\limits_{n \to \infty} B(x_n,y_n) = B(x,y)$.

9.1.10 DEFINITION A *Hilbert space* is a linear space H over K together with an inner product B such that the associated normed linear space $(H, \| \cdot \|)$, where $\| x \| = B(x,x)^{1/2}$ for all $x \in H$, is complete.

Briefly, we can say that a Hilbert space is a Banach space in which the norm is defined by an inner product. The spaces of Examples 9.1.2

and 9.1.4 are Hilbert spaces (see Corollary 4.4.2 and Example 3.3.3); but $C_K([0,1])$ is not complete in the norm defined by the inner product of Example 9.1.3. (cf. Exercise 2.8(2).) Thus the only important example of an infinite dimensional Hilbert space readily available to us in this book is the space ℓ^2. This was the space used by Hilbert in his pioneering work on quadratic forms in infinitely many variables; the abstract concept of Hilbert space was formulated in 1929 by J. von Neumann. The examples of Hilbert spaces that are most important for applications cannot be discussed without some knowledge of the Lebesgue integral. (See RIESZ and NAGY [23, Chapters 2 and 4].)

Let F be a closed linear subspace of a Hilbert space H. It is clear that the restriction to F of the inner product on H is an inner product on F and that the norm on F derived from this inner product on F is simply the restriction to F of the norm on H. Consequently, by Theorem 2.6.8, F is a Hilbert space. We shall use this observation in Section 9.5.

In order to simplify our notation we shall denote the inner product $B(x,y)$ on a Hilbert space by $\langle x,y \rangle$.

EXERCISES

(1)(a) Let B be an inner product on a real linear space E and let $\| x \| = B(x,x)^{1/2}$. Prove that

$$B(x,y) = \tfrac{1}{4}(\| x + y \|^2 - \| x - y \|^2) \qquad (*)$$

for all $x,y \in E$.

(b) Let B be an inner product on a complex linear space E and let $\| x \| = B(x,x)^{1/2}$. Prove that

$$B(x,y) = \tfrac{1}{4}(\| x + y \|^2 - \| x - y \|^2) - \tfrac{1}{4}i(\| ix + y \|^2 - \| ix - y \|^2) (**)$$

for all $x,y \in E$.

(c) Let E be a normed linear space over K and suppose that the norm on E satisfies the parallelogram law (3). Prove that the function B on $E \times E$ defined by (*) when $K = R$ and by (**) when $K = C$ is an inner product on E and that $\| x \| = B(x,x)^{1/2}$.

(2) Let B be an inner product on a linear space E. Prove that there is a Hilbert space H, with inner product $\langle \cdot , \cdot \rangle$, and a linear isometry U of E onto a dense linear subspace of H such that $B(x,y) = \langle Ux, Uy \rangle$ for all $x,y \in E$. (Hint: cf. Exercise 3.6(1).)

9.2 Orthogonality

Throughout this section H will denote a Hilbert space with inner product $\langle \cdot , \cdot \rangle$ and norm $\| \cdot \|$ (related of course by $\| x \| = \langle x,x \rangle^{1/2}$). Our first theorem has far-reaching consequences.

9.2.1 THEOREM *Let K be a non-empty closed convex subset of H and let $x_0 \in H$. Then there is a unique point $k_0 \in K$ with*

$$d(x_0,K) = \| x_0 - k_0 \|.$$

PROOF Let $\delta = d(x_0,K)$ and choose a sequence (k_n) in K with $\lim_{n\to\infty} \| k_n - x_0 \| = \delta$. We shall prove that (k_n) is a Cauchy sequence. Using the parallelogram law (equation (3) of Section 9.1) we obtain

$$2 \| k_m - x_0 \|^2 + 2 \| k_n - x_0 \|^2 = \| (k_m + k_n) - 2x_0 \|^2 + \| k_m - k_n \|^2 \quad (1)$$

for $m,n = 1, 2,\ldots$. Since K is convex $\frac{1}{2}(k_m + k_n) \in K$ and so

$$\| (k_m + k_n) - 2x_0 \| = 2 \| \tfrac{1}{2}(k_m + k_n) - x_0 \| \geq 2\delta \quad (2)$$

for $m,n = 1, 2,\ldots$. Further, since $\lim_{n\to\infty} \| k_n - x_0 \| = \delta$, given $\varepsilon > 0$ there exists an integer N with

$$\| k_n - x_0 \| < \left(\delta^2 + \frac{\varepsilon^2}{4}\right)^{1/2} \quad (3)$$

for all $n \geq N$. It follows from (1), (2) and (3) that $\| k_m - k_n \| < \varepsilon$ for all $m,n \geq N$. This shows that (k_n) is a Cauchy sequence. Since H is complete the sequence (k_n) converges. Let $k_0 = \lim_{n\to\infty} k_n$. We have $k_0 \in K$ because $k_n \in K$ and K is closed. Also

$$\| k_0 - x_0 \| = \lim_{n\to\infty} \| k_n - x_0 \| = \delta.$$

It remains to prove that k_0 is unique. Suppose that $k_0' \in K$ and $\| k_0' - x_0 \| = \delta$. Let (h_n) be the sequence defined by $h_{2n-1} = k_0$ and $h_{2n} = k_0'$ for $n = 1, 2,\ldots$. Then $h_n \in K$ and $\lim_{n\to\infty} \| h_n - x_0 \| = \delta$, so, by what we have already proved, the sequence (h_n) converges. This is possible only when $k_0 = k_0'$. □

Theorem 9.2.1 is, in general, false in a Banach space. (See Exercise 4.4(2).)

9.2.2 DEFINITION A point $x \in H$ is said to be *orthogonal* to a point $y \in H$ if and only if $\langle x,y\rangle = 0$.

Since $\langle x,y\rangle = \overline{\langle y,x\rangle}$ we see that x is orthogonal to y if and only if y is orthogonal to x. Thus we may say without ambiguity that x and y are orthogonal. Given a non-empty subset A of H we shall write

$$A^\perp = \{x \in H : \langle x,y\rangle = 0 \text{ for all } y \in A\}.$$

The set $A^\perp$ is called the *orthogonal complement of A in H*. It is clear that $0 \in A^\perp$ for any subset A of H.

Until now we have used the symbol $X^\perp$ to denote the annihilator of a subset X of a normed linear space E; the annihilator $X^\perp$ is a subset of the dual space E^*. That the symbol $^\perp$ is used with two meanings will cause no confusion. (The relation between an annihilator and an orthogonal complement is described on p. 349.)

9.2.3 LEMMA *Let A be a non-empty subset of H. Then $A^\perp$ is a closed linear subspace of H and $A \subseteq (A^\perp)^\perp$.*

PROOF Let $x,y \in A^\perp$ and $\alpha,\beta \in \mathsf{K}$. Then, for all $z \in A$, we have

$$\langle \alpha x + \beta y, z\rangle = \alpha\langle x,z\rangle + \beta\langle y,z\rangle = 0.$$

This shows that $\alpha x + \beta y \in A^\perp$ and hence that $A^\perp$ is a linear subspace of H.

Let (x_n) be a convergent sequence in $A^\perp$ with $\lim_{n\to\infty} x_n = x$. Then, for all $y \in A$, it follows from Lemma 9.1.9 that $\langle x,y\rangle = \lim_{n\to\infty} \langle x_n,y\rangle = 0$. This shows that $x \in A^\perp$ and hence that $A^\perp$ is closed. It is obvious that $A \subseteq (A^\perp)^\perp$. □

The following theorem justifies the use of the term 'orthogonal complement' for $A^\perp$.

9.2.4 THEOREM *Let A be a closed linear subspace of H. Then*

$$H = A \oplus A^\perp.$$

PROOF We have to prove that $H = A + A^\perp$ and $A \cap A^\perp = \{0\}$. It is obvious that $A \cap A^\perp = \{0\}$. Let $x \in H$. Since A is closed and convex Theorem 9.2.1 shows that there is a unique point $x_1 \in A$ with $\| x - x_1 \| = d(x,A)$. We shall prove that $x - x_1 \in A^\perp$. Suppose, on the contrary, that $x - x_1 \notin A^\perp$ and choose a point $y \in A$ with $\langle x - x_1, y\rangle \neq 0$. Since A is a linear subspace of H we may suppose, by replacing y by αy for some suitable $\alpha \in \mathsf{K}$, that $\langle x - x_1, y\rangle$ is real. For each real number t we have

$$\begin{aligned} \| x - x_1 - ty \|^2 &= \langle x - x_1 - ty, x - x_1 - ty\rangle \\ &= \| x - x_1 \|^2 - 2t\langle x - x_1, y\rangle + t^2 \| y \|^2. \end{aligned} \tag{4}$$

Let $t_0 = \| y \|^{-2}\langle x - x_1, y\rangle$. (It is obvious that $y \neq 0$.) Equation (4) gives

$$\| x - x_1 - t_0 y \|^2 = \| x - x_1 \|^2 - \| y \|^{-2}\langle x - x_1, y\rangle^2 < \| x - x_1 \|^2.$$

Since $x_1 + t_0 y \in A$ this contradicts the fact that $\| x - x_1 \| = d(x,A)$. Consequently we must have $x - x_1 \in A^\perp$. Since $x = x_1 + (x - x_1)$ this proves that $H = A + A^\perp$. □

9.2.5 COROLLARY *Let A be a closed linear subspace of H. Then*

$$A = (A^\perp)^\perp.$$

PROOF It is obvious that $A \subseteq (A^{\perp})^{\perp}$. Let $x \in (A^{\perp})^{\perp}$. By Theorem 9.2.4 we have $x = y + z$ with $y \in A$ and $z \in A^{\perp}$. Since $\langle y,z\rangle = 0$ we have

$$\langle z,z\rangle = \langle y,z\rangle + \langle z,z\rangle = \langle x,z\rangle = 0$$

and hence $z = 0$. This proves that $(A^{\perp})^{\perp} \subseteq A$. □

Theorem 9.2.4 shows that every closed linear subspace of a Hilbert space has at least one complementary closed linear subspace, and it gives an explicit description of one such subspace. The reader should note that in some Banach spaces a closed linear subspace may fail to have a complementary closed linear subspace. (See p. 99.)

We shall now obtain a representation for the bounded linear functionals on a Hilbert space. It follows from Schwarz's inequality (9.1.5) that, for each fixed $y \in H$, the mapping $x \to \langle x,y\rangle$ is a bounded linear functional on H. We shall show that every bounded linear functional on H is of this form. For certain special Hilbert spaces this was proved independently in 1907 by M. Fréchet and F. Riesz.

9.2.6 THEOREM *Let x^* be a bounded linear functional on H. Then there exists a unique point $y \in H$ such that*

$$x^*(x) = \langle x,y\rangle$$

for all $x \in H$. Further $\| x^* \| = \| y \|$.

PROOF If $x^* = 0$ we can obviously take $y = 0$. Suppose that $x^* \neq 0$ and let $N = \{x \in H: x^*(x) = 0\}$. Since x^* is a non-zero bounded linear functional N is a closed linear subspace of H and $N \neq H$. By Theorem 9.2.4 we have $H = N \oplus N^{\perp}$ so we must have $N^{\perp} \neq \{0\}$. We observe that if $x \in N^{\perp}$ and $x^*(x) = 0$ then $x \in N \cap N^{\perp} = \{0\}$ and so $x = 0$. Choose $z \in N^{\perp}$ with $z \neq 0$ and let $\lambda = x^*(z)$. By our observation above $\lambda \neq 0$. Let $x \in H$. By Theorem 9.2.4 we have $x = u + v$ with $u \in N$ and $v \in N^{\perp}$. It is clear that $v - \lambda^{-1}x^*(v)z \in N^{\perp}$ and that

$$x^*(v - \lambda^{-1}x^*(v)z) = 0$$

so we have $v = \lambda^{-1}x^*(v)z$. It follows that

$$\langle x,z\rangle = \langle u,z\rangle + \langle v,z\rangle = \frac{1}{\lambda}x^*(v)\langle z,z\rangle$$

and hence

$$x^*(x) = x^*(u) + x^*(v) = \frac{\lambda}{\| z \|^2}\langle x,z\rangle.$$

This shows that $x^*(x) = \langle x,y\rangle$ where $y = \bar{\lambda}\| z \|^{-2}z$. The uniqueness of y follows from Lemma 9.1.8.

Finally, from Schwarz's inequality we obtain

$$| x^*(x) | = | \langle x,y \rangle | \leq \| x \| \, \| y \|$$

for all $x \in H$ and therefore $\| x^* \| \leq \| y \|$. On the other hand

$$\| y \|^2 = \langle y,y \rangle = | x^*(y) | \leq \| x^* \| \, \| y \|$$

so $\| y \| \leq \| x^* \|$ (because $y \neq 0$). This shows that $\| x^* \| = \| y \|$. □

Let H^* be the set of all bounded linear functionals on H, that is H^* is the dual of the Banach space H. It follows from Theorem 9.2.6 and the remark immediately preceding it that there is a one-to-one mapping J of H^* onto H that satisfies

$$x^*(x) = \langle x, Jx^* \rangle$$

for all $x \in H$ and all $x^* \in H^*$. It is easy to verify that

$$J(x^* + y^*) = Jx^* + Jy^* \text{ and } J(\alpha x^*) = \bar{\alpha}\, Jx^*$$

for all $x^*,y^* \in H^*$ and all $\alpha \in \mathsf{K}$. Also, for all $x^*,y^* \in H^*$,

$$\| Jx^* - Jy^* \| = \| J(x^* - y^*) \| = \| x^* - y^* \|$$

and so J is an isometry of H^* onto H. The reader should notice that, when $\mathsf{K} = \mathsf{C}$, J is not a linear mapping. It is easy to verify that $J^{-1}(A^{\perp})$ is the annihilator in H^* of the subset A of H. (See p. 347.)

9.2.7 Definition A non-empty subset A of H is said to be *orthonormal* if and only if

(a) $\langle x,y \rangle = 0$ for all $x,y \in A$ with $x \neq y$, and
(b) $\langle x,x \rangle = 1$ for all $x \in A$.

A countable orthonormal set is often called an *orthonormal sequence.*

9.2.8 Lemma *Each orthonormal subset of H is linearly independent.*

Proof Let A be an orthonormal subset of H and let $\{x_1, x_2,..., x_n\}$ be a finite subset of A. If $\lambda_1, \lambda_2,..., \lambda_n$ are elements of K with

$$\lambda_1 x_1 + \lambda_2 x_2 + \ldots + \lambda_n x_n = 0$$

then, for $m = 1, 2,..., n$, we have

$$0 = \lambda_1 \langle x_1,x_m \rangle + \lambda_2 \langle x_2,x_m \rangle + \ldots + \lambda_n \langle x_n,x_m \rangle = \lambda_m.$$

This shows that A is linearly independent. □

9.2.9 Theorem *Let A be a countable linearly independent subset of H. Then there is a countable orthonormal subset B of H that has the same linear hull as A.*

Proof We shall prove the theorem in the case when A is infinite; the modifications needed when A is finite are obvious. Let $n \to x_n$ be an enumeration of A. We shall construct inductively a countable orthonormal set $\{u_n: n = 1, 2,...\}$ in H such that, for $n = 1, 2,...,$

$$u_n = \alpha_{1n}x_1 + \alpha_{2n}x_2 + \ldots + \alpha_{nn}x_n \tag{5}$$

and

$$x_n = \beta_{1n}u_1 + \beta_{2n}u_2 + \ldots + \beta_{nn}u_n, \tag{6}$$

where $\alpha_{1n}, \alpha_{2n},..., \alpha_{nn}, \beta_{1n}, \beta_{2n},..., \beta_{nn}$ are in K. Let $u_1 = \| x_1 \|^{-1} x_1$. Suppose that points $u_1, u_2,..., u_m$ of H have been chosen so that the set $\{u_1, u_2,..., u_m\}$ is orthonormal and the conditions (5) and (6) are satisfied for $n = 1, 2,..., m$. Let

$$y = x_{m+1} - \langle x_{m+1},u_1\rangle u_1 - \langle x_{m+1},u_2\rangle u_2 - \ldots - \langle x_{m+1},u_m\rangle u_m.$$

Then $\langle y,u_n\rangle = 0$ for $n = 1, 2,..., m$. Also, since the set $\{x_1, x_2,..., x_{m+1}\}$ is linearly independent it follows from (5) that $y \neq 0$. Let

$$u_{m+1} = \| y \|^{-1}y.$$

It is clear that the set $\{u_1, u_2,..., u_{m+1}\}$ is orthonormal and that the conditions (5) and (6) are satisfied for $n = 1, 2,..., m + 1$. This completes the inductive definition of the sequence (u_n). The relations (5) and (6) show that the orthonormal set $B = \{u_n: n = 1, 2,...\}$ has the same linear hull as A. □

The orthogonalization process described in the proof of the last theorem is usually called the *Gram-Schmidt orthogonalization process.* The proof of the following corollary of Theorem 9.2.9 is obvious.

9.2.10 COROLLARY *Suppose that H is n-dimensional. Then H has at least one orthonormal subset with n elements and each orthonormal subset of H with n elements is a basis for H.*

9.2.11 COROLLARY *Suppose that H is infinite dimensional. Then H has an infinite orthonormal subset.*

PROOF It follows from Theorem 9.2.9 that it is sufficient to prove that H contains an infinite linearly independent subset. Let x_1 be any non-zero point of H. Suppose that points $x_1, x_2,..., x_n$ of H have been chosen so that the set $\{x_1, x_2,..., x_n\}$ is linearly independent. If, for each point $y \in H$, the set $\{x_1, x_2,..., x_n, y\}$ is not linearly independent, then H is n-dimensional. Consequently there is at least one point $x_{n+1} \in H$ such that the set $\{x_1, x_2,..., x_n, x_{n+1}\}$ is linearly independent. We have now defined inductively a sequence (x_n) in H such that the set

$$\{x_n: n = 1, 2,...\}$$

is linearly independent. □

9.2.12 Definition An orthonormal subset of H is said to be *maximal* if and only if it is not a proper subset of any other orthonormal subset of H.

To establish the existence of maximal orthonormal subsets of an arbitrary Hilbert space we have to appeal to Zorn's lemma. We shall have no occasion to use the existence of maximal orthonormal subsets in this chapter, so the reader may, if he wishes, omit the next theorem.

9.2.13 Theorem *Each orthonormal subset of H is contained in a maximal orthonormal subset of H.*

Proof Let A be an orthonormal subset of H and let $\mathscr{A}$ be the class of all orthonormal subsets of H that contain A. The relation $\subseteq$ of set inclusion is a partial order relation on $\mathscr{A}$. Let $\mathscr{A}_0$ be a totally ordered subset of $\mathscr{A}$ and let $A_0 = \bigcup_{A' \in \mathscr{A}_0} A'$. We shall prove that $A_0 \in \mathscr{A}$. It is obvious that $A \subseteq A_0$. Let $x,y \in A_0$. Then $x \in A'$ and $y \in A''$ for some $A',A'' \in \mathscr{A}_0$. Since $\mathscr{A}_0$ is totally ordered either $A' \subseteq A''$ or $A'' \subseteq A'$ and in either case it follows that $\langle x,y \rangle = 1$ if $x = y$ and $\langle x,y \rangle = 0$ if $x \neq y$, because A' and A'' are orthonormal. This shows that A_0 is orthonormal and hence that $A_0 \in \mathscr{A}_0$. It now follows from Zorn's lemma (1.5.2) that $\mathscr{A}_0$ has a maximal element. It is obvious that any maximal element of $\mathscr{A}_0$ is a maximal orthonormal subset of H that contains A. □

If $H \neq \{0\}$ then it contains at least one point x with $\| x \| = 1$ and, consequently, since the set $\{x\}$ is orthonormal it follows from the last theorem that H has at least one maximal orthonormal subset. We remark that if H is separable we can prove Theorem 9.2.13 without using Zorn's lemma. (See Exercise (5).)

9.2.14 Lemma *Let $\{u_1, u_2, \ldots, u_n\}$ be a finite orthonormal subset of H, let $x \in H$, and let*

$$s = \sum_{k=1}^{n} \langle x,u_k \rangle u_k.$$

Then

$$\| s \|^2 = \sum_{k=1}^{n} | \langle x,u_k \rangle |^2$$

and

$$\| x \|^2 - \| s \|^2 = \| x - s \|^2.$$

PROOF We have

$$\begin{aligned}\| s \|^2 &= \langle s,s\rangle \\ &= \sum_{k,j=1}^{n} \langle x,u_k\rangle \overline{\langle x,u_j\rangle} \langle u_k,u_j\rangle \\ &= \sum_{k=1}^{n} | \langle x,u_k\rangle |^2.\end{aligned}$$

Also

$$\begin{aligned}\langle s,x\rangle &= \sum_{k=1}^{n} \langle x,u_k\rangle \langle u_k,x\rangle \\ &= \sum_{k=1}^{n} | \langle x,u_k\rangle |^2 \\ &= \| s \|^2.\end{aligned}$$

Consequently

$$\begin{aligned}\| x - s \|^2 &= \langle x - s, x - s\rangle \\ &= \langle x,x\rangle - \langle s,x\rangle - \langle x,s\rangle + \langle s,s\rangle \\ &= \| x \|^2 - \| s \|^2. \ \square\end{aligned}$$

9.2.15 LEMMA *Let f be a non-negative real-valued function on a non-empty set X and suppose that there is a real number K such that*

$$\sum_{k=1}^{n} f(x_k) \leq K$$

for all finite subsets $\{x_1, x_2,..., x_n\}$ of X. Then the set $\{x \in X : f(x) \neq 0\}$ is countable.

PROOF For $n = 1, 2,...$ let $X_n = \{x \in X: f(x) \geq K/n\}$. It is obvious that the set X_n can have at most n elements and so by Theorem 1.4.5 the set $\bigcup_{n=1}^{\infty} X_n$ is countable. Clearly $\{x \in X: f(x) \neq 0\} = \bigcup_{n=1}^{\infty} X_n$. $\square$

9.2.16 THEOREM *Let M be a maximal orthonormal subset of H, let $x \in H$ and let $M_x = \{u \in M : \langle x,u\rangle \neq 0\}$. Then the set M_x is countable. Further, if $n \to u_n$ is any enumeration† of the set M_x then*

$$x = \sum_{n=1}^{\infty} \langle x,u_n\rangle u_n$$

and

$$\| x \|^2 = \sum_{n=1}^{\infty} | \langle x,u_n\rangle |^2.$$

† For simplicity we have supposed that the set M_x is infinite. The interpretation of the statement of the theorem and the necessary modifications to its proof when the set M_x is finite are obvious.

PROOF For each finite subset $\{v_1, v_2, \ldots, v_n\}$ of M it follows from Lemma 9.2.14 that $\sum_{k=1}^{n} | \langle x,v_k \rangle |^2 \leq \| x \|^2$. Consequently by Lemma 9.2.15 the set M_x is countable.

Let $n \to u_n$ be an enumeration of M_x and let

$$s_n = \sum_{k=1}^{n} \langle x,u_k \rangle u_k$$

for $n = 1, 2, \ldots$. Let m and n be positive integers with $m > n$. Then $s_m - s_n = \sum_{k=n+1}^{m} \langle x,u_k \rangle u_k$ and from Lemma 9.2.14 we obtain

$$\| s_m - s_n \|^2 = \sum_{k=n+1}^{m} | \langle x,u_k \rangle |^2. \tag{7}$$

Lemma 9.2.14 also gives, for $n = 1, 2, \ldots$,

$$\begin{aligned} \sum_{k=1}^{n} | \langle x,u_k \rangle |^2 &= \| s_n \|^2 \\ &= \| x \|^2 - \| x - s_n \|^2 \\ &\leq \| x \|^2. \end{aligned} \tag{8}$$

It follows from (8) that the series $\sum_{k} | \langle x,u_k \rangle |^2$ converges, and it then follows from (7) that the sequence (s_n) is a Cauchy sequence in H. Since H is complete, the sequence (s_n) converges. Let $y = \lim_{n\to\infty} s_n$. For each $u \in M$, it follows from Lemma 9.1.9 that

$$\begin{aligned} \langle y,u \rangle &= \lim_{n\to\infty} \langle s_n,u \rangle \\ &= \lim_{n\to\infty} \sum_{k=1}^{n} \langle x,u_k \rangle \langle u_k,u \rangle. \end{aligned} \tag{9}$$

If $u \in M_x$ then $u = u_j$ for some integer j and (9) shows that $\langle y,u \rangle = \langle x,u \rangle$; if $u \in M \sim M_x$ then $u \neq u_k$ for $k = 1, 2, \ldots$ and (9) shows that $\langle y,u \rangle = 0 = \langle x,u \rangle$. Thus $\langle y - x,u \rangle = 0$ for all $u \in M$ and hence $y - x \in M^\perp$. However $M^\perp = \{0\}$, because M is a maximal orthonormal subset of H, so $y = x$. Thus $x = \sum_{k=1}^{\infty} \langle x,u_k \rangle u_k$ and it follows from (8) that

$$\begin{aligned} \| x \|^2 &= \lim_{n\to\infty} \| s_n \|^2 \\ &= \lim_{n\to\infty} \sum_{k=1}^{n} | \langle x,u_k \rangle |^2 \\ &= \sum_{k=1}^{\infty} | \langle x,u_k \rangle |^2. \ \square \end{aligned}$$

It follows from the above theorem that if M is a maximal orthonormal subset of H then the closed linear hull of M is H. Conversely it is easy to see that each orthonormal subset of H whose closed linear hull is H is maximal (Exercise (2)).

EXERCISES

(1) (a) Prove that if H is a Hilbert space and M and N are closed linear subspaces of H such that $\langle x,y\rangle = 0$ for all $x \in M$ and $y \in N$ then the linear subspace $M + N$ of H is closed. (It can be shown that if H is an inner product space that is not complete then the result does not hold.)

(b) Give an example of closed linear subspaces M and N of a Hilbert space H such that the linear subspace $M + N$ of H is not closed. (Use Exercise 8.8(7).)

(2) Let M be an orthonormal subset of a Hilbert space H. Prove that the set $M_x = \{u \in M\colon \langle u,x\rangle \neq 0\}$ is countable for each $x \in H$. Prove further that the following five conditions are all equivalent:

(a) M is fundamental in H,

(b) M is a maximal orthonormal subset of H,

(c) for each $x \in H$, $x = \sum_{n=1}^{\infty} \langle x,u_n\rangle u_n$, where $n \to u_n$ is any enumeration of M_x,

(d) for each pair $x,y \in H$, $\langle x,y\rangle = \sum_{n=1}^{\infty} \langle x,v_n\rangle \overline{\langle y,v_n\rangle}$, where $n \to v_n$ is any enumeration of $M_x \cup M_y$,

(e) for each $x \in H$, $\| x \|^2 = \sum_{n=1}^{\infty} | \langle x,u_n\rangle |^2$, where $n \to u_n$ is any enumeration of M_x.

(If M_x and M_y are finite, the equations in (c), (d) and (e) must be suitably modified. The equality in (d) is called *Parseval's identity*.)

(3) Prove that the set $\{e_n\colon n = 1, 2,\ldots\}$ is a maximal orthonormal subset of ℓ^2. (The points e_n are defined in Example 3.4.9.)

(4) Prove that an orthonormal subset of a separable Hilbert space is countable.

(5) Prove Theorem 9.2.13 for separable Hilbert spaces, without using Zorn's lemma.

(6) Prove that a Hilbert space is separable if and only if it has a countable maximal orthonormal subset.

(7) (a) Let $\{u_n\colon n = 1, 2,\ldots\}$ be a countable orthonormal set in a Hilbert space H and let (ξ_n) be a sequence in K. Prove that the series

$\sum_n \xi_n u_n$ converges in H if and only if $(\xi_n) \in \ell^2$. Prove also that if $(\xi_n) \in \ell^2$ then $\left\| \sum_{n=1}^{\infty} \xi_n u_n \right\| = \left(\sum_{n=1}^{\infty} | \xi_n |^2 \right)^{1/2}$.

(b) Prove that each n-dimensional Hilbert space is linearly isometric to the Hilbert space of Example 9.1.2.

(c) Prove that each separable, infinite dimensional Hilbert space is linearly isometric to ℓ^2.

9.3 The Hilbert Space Adjoint

The term 'adjoint operator' has two well-established meanings: one for operators on Banach spaces (see Definition 6.4.1) and the other for operators on Hilbert spaces. The definition of the Hilbert space adjoint depends on the following theorem which, in turn, depends essentially on Theorem 9.2.6.

Throughout this section H will denote a Hilbert space.

9.3.1 THEOREM *Associated with each operator $T \in L(H)$ there is a unique operator $T^* \in L(H)$ such that*

$$\langle Tx,y \rangle = \langle x,T^*y \rangle \tag{1}$$

for all $x,y \in H$.

PROOF Let $T \in L(H)$. For each $y \in H$ let x_y^* be the functional on H defined by

$$x_y^*(x) = \langle Tx,y \rangle$$

for all $x \in H$. It is clear that the functional x_y^* is linear. Also, by Schwarz's inequality,

$$| x_y^*(x) | = | \langle Tx,y \rangle | \leq \| Tx \| \; \| y \| \leq \| T \| \; \| x \| \; \| y \|$$

for all $x,y \in H$. So x_y^* is bounded and $\| x_y^* \| \leq \| T \| \; \| y \|$. Consequently, by Theorem 9.2.6, corresponding to each point $y \in H$ there is a unique point $T^*y \in H$ with

$$x_y^*(x) = \langle x,T^*y \rangle$$

for all $x \in H$. This defines a mapping T^* of H into itself which satisfies (1). It remains to prove that $T^* \in L(H)$ and that T^* is unique.

Let $y,z \in H$ and $\alpha,\beta \in \mathsf{K}$. Then, for all $x \in H$, we have

$$\begin{aligned} \langle x,T^*(\alpha y + \beta z) \rangle &= \langle Tx,\alpha y + \beta z \rangle \\ &= \bar{\alpha} \langle Tx,y \rangle + \bar{\beta} \langle Tx,z \rangle \\ &= \bar{\alpha} \langle x,T^*y \rangle + \bar{\beta} \langle x,T^*z \rangle \\ &= \langle x,\alpha T^*y + \beta T^*z \rangle \end{aligned}$$

and hence, by Lemma 9.1.8, $T^*(\alpha y + \beta z) = \alpha T^*y + \beta T^*z$.

It was shown above that $\| x_y^* \| \leq \| T \| \; \| y \|$ for all $y \in H$. Also by Theorem 9.2.6 we have $\| x_y^* \| = \| T^*y \|$ so $\| T^*y \| \leq \| T \| \; \| y \|$ for all $y \in H$. This proves that $T^* \in L(H)$ and $\| T^* \| \leq \| T \|$. The uniqueness of T^* follows directly from Lemma 9.1.8. □

9.3.2 DEFINITION Let $T \in L(H)$. The unique operator $T^* \in L(H)$ that satisfies

$$\langle Tx,y \rangle = \langle x,T^*y \rangle$$

for all $x,y \in H$ is called the *Hilbert space adjoint of T*.

We remark that

$$\langle T^*x,y \rangle = \overline{\langle y,T^*x \rangle} = \overline{\langle Ty,x \rangle} = \langle x,Ty \rangle$$

for all $x,y \in H$.

Let $T \in L(H)$. Then T has a Banach space adjoint defined in Definition 6.4.1. This adjoint, which we have denoted by T^* in Chapters 6 and 8, is a bounded linear operator on the dual space H^* of H. Throughout this chapter the symbol T^* will be used only to denote the Hilbert space adjoint operator of Definition 9.3.2. The Banach space and Hilbert space adjoints of T are closely related. For the moment let T' denote the Banach space adjoint of T. Then we have $T'x^* = x^* \circ T$ for all $x^* \in H^*$. (See Definition 6.4.1.) We saw in Section 9.2 that there is a one-to-one mapping J of H^* onto H that satisfies $x^*(x) = \langle x,Jx^* \rangle$ for all $x \in H$ and all $x^* \in H^*$. Consequently, for all $x \in H$ and all $x^* \in H^*$, we have

$$\begin{aligned}
\langle x,(J \circ T')x^* \rangle &= \langle x,J(T'x^*) \rangle \\
&= (T'x^*)(x) \\
&= x^*(Tx) \\
&= \langle Tx,Jx^* \rangle \\
&= \langle x,T^*(Jx^*) \rangle \\
&= \langle x,(T^* \circ J)x^* \rangle .
\end{aligned}$$

Therefore $J \circ T' = T^* \circ J$ and hence $T^* = J \circ T' \circ J^{-1}$. This relation can be used to obtain properties of the Hilbert space adjoint T^* from the corresponding properties of the Banach space adjoint T'; however it is usually simpler to work directly in terms of the Hilbert space adjoint T^*.

9.3.3 EXAMPLE Let T be a linear operator on the Hilbert space K^n of Example 9.1.2, and let $\{e_1, e_2, \ldots, e_n\}$ be an orthonormal basis for K^n (such a basis exists by Corollary 9.2.10). Let (τ_{jk}) and (τ_{jk}^*) be the matrices that represent T and T^*, respectively, relative to this basis.

Then we have

$$Te_k = \sum_{j=1}^{n} \tau_{jk} e_j$$

for $k = 1, 2, \ldots, n$, and so

$$\langle Te_k, e_j \rangle = \tau_{jk}$$

for $j,k = 1, 2, \ldots, n$. Similarly

$$\langle T^* e_k, e_j \rangle = \tau^*_{jk}.$$

Therefore, for $j,k = 1, 2, \ldots, n$, we have

$$\tau^*_{jk} = \langle T^* e_k, e_j \rangle = \langle e_k, Te_j \rangle = \bar{\tau}_{kj}$$

This shows that the matrix (τ^*_{jk}) is the Hermitian transpose of the matrix (τ_{jk}). The reader should compare this example with Example 6.4.5.

The rest of this section is devoted to the derivation of some elementary properties of the Hilbert space adjoint that we shall need later.

9.3.4 LEMMA *Let $T,S \in L(H)$ and $\alpha \in$ K. Then*

(a) $(T + S)^* = T^* + S^*$,
(b) $(\alpha T)^* = \bar{\alpha} T^*$,
(c) $(TS)^* = S^* T^*$,
(d) $(T^*)^* = T$,
(e) $I^* = I$,
(f) *T is regular if and only if T^* is regular, and, if T is regular, then* $(T^*)^{-1} = (T^{-1})^*$.

PROOF Assertions (a) to (e) follow directly from Definition 9.3.2; the details are left to the reader. Let us prove (f). Suppose that T is regular. Then $I = TT^{-1} = T^{-1}T$ and hence using (c) and (e) we obtain

$$I = (TT^{-1})^* = (T^{-1})^* T^*$$

and

$$I = (T^{-1}T)^* = T^*(T^{-1})^*.$$

This shows that T^* is regular and that $(T^*)^{-1} = (T^{-1})^*$. Finally, if T^* is regular then, by (d) and what we have just proved, $T = (T^*)^*$ is also regular. □

9.3.5 COROLLARY *For each $T \in L(H)$ we have $\| T^* \| = \| T \|$.*

PROOF Let $T \in L(H)$. We saw in the proof of Theorem 9.3.1 that $\| T^* \| \leq \| T \|$. From this inequality and Lemma 9.3.4(d) we obtain $\| T \| = \| (T^*)^* \| \leq \| T^* \|$. □

In the next Corollary we shall suppose that H is a *complex* Hilbert space.

9.3.6 COROLLARY *For each $T \in L(H)$, we have*

$$\mathrm{sp}(T^*) = \{\bar{\lambda} : \lambda \in \mathrm{sp}(T)\}.$$

PROOF This follows at once from Lemma 9.3.4(f) and the relation $(\lambda I - T)^* = \bar{\lambda} I - T^*$ which holds for all complex numbers λ. □

9.3.7 THEOREM *For each $T \in L(H)$ we have*

$$\| TT^* \| = \| T^*T \| = \| T \|^2.$$

PROOF Let $T \in L(H)$. Using Corollary 9.3.5 we obtain

$$\| TT^* \| \leq \| T \| \; \| T^* \| = \| T \|^2.$$

On the other hand, using Schwarz's inequality, we obtain

$$\begin{aligned} \| T \|^2 &= \sup \{\| Tx \|^2 : \| x \| \leq 1\} \\ &= \sup \{| \langle Tx, Tx \rangle | : \| x \| \leq 1\} \\ &= \sup \{| \langle (T^*T)x, x \rangle | : \| x \| \leq 1\} \\ &\leq \sup \{\| (T^*T)x \| \; \| x \| : \| x \| \leq 1\} \\ &\leq \| T^*T \|. \end{aligned}$$

This proves that $\| T^*T \| = \| T \|^2$. Applying this equality to the operator T^* we obtain

$$\| TT^* \| = \| (T^*)^*T^* \| = \| T^* \|^2 = \| T \|^2. \; \square$$

9.3.8 COROLLARY *Let $T \in L(H)$ be such that $TT^* = T^*T$. Then $\| T^2 \| = \| T \|^2$ and, consequently, $\lim_{n \to \infty} \| T^n \|^{1/n} = \| T \|$.*

PROOF The existence of the limit $\lim_{n \to \infty} \| T^n \|^{1/n}$ was established in Lemma 6.2.6. It is clear that we may suppose that $T \neq 0$. Three applications of Theorem 9.3.7 give

$$\begin{aligned} \| T \|^4 &= \| T^*T \|^2 \\ &= \| (T^*T)(T^*T)^* \| \\ &= \| (T^*T)(T^*T) \| \\ &= \| (T^2)^*T^2 \| \\ &= \| T^2 \|^2. \end{aligned}$$

Thus $\| T \|^2 = \| T^2 \|$. It follows by induction that $\| T^{2^k} \| = \| T \|^{2^k}$ for $k = 1, 2, \ldots$, and hence we have

$$\lim_{n \to \infty} \| T^n \|^{1/n} = \lim_{k \to \infty} \| T^{2^k} \|^{2^{-k}} = \| T \|. \; \square$$

An operator $T \in L(H)$ with $TT^* = T^*T$ is said to be *normal*. It follows from Theorem 6.3.10 and Corollary 9.3.8 that if T is a normal operator on a complex Hilbert space then $\mathrm{sp}(T)$ is a closed subset of the disc $\{\lambda : |\lambda| \leq \|T\|\}$ and there is a point $\lambda_0 \in \mathrm{sp}(T)$ with $|\lambda_0| = \|T\|$.

EXERCISES

(1) A Hermitian symmetric form B on a Hilbert space H is said to be bounded if and only if $\{B(x,y) : \|x\| \leq 1, \|y\| \leq 1\}$ is a bounded subset of K.

(a) Let $T \in L(H)$. Prove that the mapping $(x,y) \to \langle Tx,y\rangle$ is a bounded Hermitian symmetric form on H and that

$$\|T\| = \sup\{\,|\langle Tx,y\rangle| \;:\|x\| \leq 1, \|y\| \leq 1\}.$$

(b) Let B be a bounded Hermitian symmetric form on H. Prove that there is a unique operator $T \in L(H)$ with $B(x,y) = \langle Tx,y\rangle$ for all $x,y \in H$.

(2) (a) Let $T \in L(\ell^2)$ and let $\alpha_{mn} = \langle Te_n,e_m\rangle$ for $m,n = 1, 2,\ldots$. Prove that, for each $x = (\xi_n) \in \ell^2$, the series $\sum_n \alpha_{mn}\xi_n$ converges for each integer $m = 1, 2,\ldots$ and that $Tx = (\sum_{n=1}^{\infty} \alpha_{mn}\xi_n)_{m\geq 1}$.

(b) Let (α_{mn}) be a double sequence in K. Prove that there is an operator $T \in L(\ell^2)$ with $\alpha_{mn} = \langle Te_n,e_m\rangle$ for $m,n = 1, 2,\ldots$ if and only if there is a real number $K > 0$ such that

$$\left|\sum_{m=1}^{M}\sum_{n=1}^{N} \alpha_{mn}\xi_n\bar{\eta}_m\right| \leq K\Big(\sum_{n=1}^{N} |\xi_n|^2\Big)^{1/2}\Big(\sum_{m=1}^{M} |\eta_m|^2\Big)^{1/2} \qquad (*)$$

for all choices of integers M and N and points $\xi_1, \xi_2,\ldots, \xi_N, \eta_1, \eta_2,\ldots \eta_M$ in K.

(c) Prove that the correspondence $T \sim (\alpha_{mn})$ between operators $T \in L(\ell^2)$ and double sequences (α_{mn}) which satisfy (*) is one-to-one and has the following properties: if $T \sim (\alpha_{mn})$ and $S \sim (\beta_{mn})$ then $T + S \sim (\alpha_{mn} + \beta_{mn})$, $\lambda T \sim (\lambda\alpha_{mn})$, $T^* \sim (\alpha^*_{mn})$ where $\alpha^*_{mn} = \bar{\alpha}_{nm}$, and $TS \sim \left(\sum_{j=1}^{\infty} \alpha_{mj}\beta_{jn}\right)$.

9.4 Self-adjoint Bounded Linear Operators

In this section H will again denote a Hilbert space.

9.4.1 DEFINITION An operator $T \in L(H)$ is said to be *self-adjoint* if and only if $T = T^*$. We shall denote by $\mathscr{S}$ the set of all self-adjoint bounded linear operators on H. The elements of $\mathscr{S}$ are also known as *symmetric* or *Hermitian* bounded linear operators.

It follows from the definition of the adjoint T^* of T (9.3.2) that $T \in L(H)$ is self-adjoint if and only if

$$\langle Tx,y \rangle = \langle x,Ty \rangle$$

for all $x,y \in H$. It also follows from the definitions that a self-adjoint bounded linear operator is normal (p. 358).

The spectral theory of self-adjoint bounded linear operators on a complex Hilbert space, which is developed in the remainder of this chapter, generalizes the elementary theory of the diagonalization of Hermitian symmetric matrices. We begin by studying briefly the relation between the two.

Consider a self-adjoint linear operator T on the Hilbert space C^n of Example 9.1.2. Let (τ_{jk}) be the matrix which represents T relative to some orthonormal basis for C^n. Example 9.3.3 shows that $\bar{\tau}_{jk} = \tau_{kj}$ for $j,k = 1, 2,\ldots,n$. Thus the matrix (τ_{jk}) is Hermitian symmetric.

It now follows from the elementary theory of diagonalization of Hermitian symmetric matrices that there is an orthonormal basis $\{u_1, u_2,\ldots, u_n\}$ for C^n relative to which T is represented by a diagonal matrix. Thus there are complex numbers $\lambda_1, \lambda_2,\ldots, \lambda_n$ such that $Tu_j = \lambda_j u_j$ for $j = 1, 2,\ldots, n$. For $x \in \mathsf{C}^n$ we have, by Theorem 9.2.16, $x = \sum_{j=1}^{n} \langle x,u_j \rangle u_j$ and therefore

$$Tx = \sum_{j=1}^{n} \lambda_j \langle x,u_j \rangle u_j. \tag{*}$$

A bounded linear operator on a complex Hilbert space which is both self-adjoint and compact can be represented in a form (see Theorem 9.5.2) which is an immediate generalization of (*).

It is possible to rewrite (*) in a form that lends itself to further generalization. Let $E_0, E_1,\ldots, E_n$ be the self-adjoint projections on C^n which are defined by the equations

$$E_0 x = 0$$

and

$$E_k x = \sum_{j=1}^{k} \langle x,u_j \rangle u_j,$$

for $x \in \mathsf{C}^n$ and $k = 1, 2,\ldots, n$. Then (*) gives

$$T = \sum_{j=1}^{n} \lambda_j (E_j - E_{j-1}). \tag{**}$$

The spectral theorem (9.9.4) for general self-adjoint bounded linear operators on complex Hilbert spaces gives a representation for such operators which is a direct generalization of (**).

The rest of this section is devoted to the elementary properties of self-adjoint bounded linear operators on H.

9.4.2 THEOREM *Let $T,S \in \mathscr{S}$ and $\alpha,\beta \in \mathsf{R}$. Then $\alpha T + \beta S \in \mathscr{S}$. Also $TS \in \mathscr{S}$ if and only if $TS = ST$.*

PROOF It follows directly from Lemma 9.3.4(a) and (b) that $\alpha T + \beta S \in \mathscr{S}$. Further, by Lemma 9.3.4(c), we have

$$(TS)^* = S^*T^* = ST$$

so $(TS)^* = TS$ if and only if $TS = ST$. □

9.4.3 THEOREM *Suppose that H is a complex Hilbert space and let $T \in L(H)$. Then T is self-adjoint if and only if $\langle Tx,x\rangle$ is real for each $x \in H$.*

PROOF If T is self-adjoint then, for all $x \in H$, we have

$$\langle Tx,x\rangle = \langle x,Tx\rangle = \overline{\langle Tx,x\rangle}$$

and so $\langle Tx,x\rangle$ is real.

Suppose, on the other hand, that $\langle Tx,x\rangle$ is real for each $x \in H$. Then

$$\langle Tx,x\rangle = \overline{\langle Tx,x\rangle} = \langle x,Tx\rangle \tag{1}$$

for all $x \in H$. Two elementary calculations show that, for all $x,y \in H$,

$$\begin{aligned} 4\langle Tx,y\rangle &= \langle T(x+y),x+y\rangle - \langle T(x-y),x-y\rangle \\ &\quad + i\langle T(x+iy),x+iy\rangle - i\langle T(x-iy),x-iy\rangle \end{aligned} \tag{2}$$

$$\text{and } \begin{aligned} 4\langle x,Ty\rangle &= \langle x+y,T(x+y)\rangle - \langle x-y,T(x-y)\rangle \\ &\quad + i\langle x+iy,T(x+iy)\rangle - i\langle x-iy,T(x-iy)\rangle. \end{aligned} \tag{3}$$

Equations (1), (2) and (3) show that $\langle Tx,y\rangle = \langle x,Ty\rangle$ for all $x,y \in H$. Thus T is self-adjoint. □

9.4.4 THEOREM *For each $T \in \mathscr{S}$ we have*

$$\| T \| = \sup \{ | \langle Tx,x\rangle | : \| x \| \leq 1\}.$$

PROOF Let $T \in \mathscr{S}$ and let $\alpha = \sup \{ | \langle Tx,x\rangle | : \| x \| \leq 1\}$. Using Schwarz's inequality we obtain

$$| \langle Tx,x\rangle | \leq \| Tx \| \; \| x \| \leq \| T \| \; \| x \|^2$$

for all $x \in H$ and hence we have $\alpha \leq \| T \|$.

It follows directly from the definition of α that

$$| \langle Tx,x\rangle | \leq \alpha \| x \|^2 \tag{4}$$

for all $x \in H$. A simple calculation shows that

$$\langle T(x+y),x+y\rangle - \langle T(x-y),x-y\rangle = 4\,\mathrm{Re}\,\langle Tx,y\rangle \tag{5}$$

for all $x,y \in H$. Using (4), (5) and the parallelogram law we obtain

$$\begin{aligned} 4\,|\,\mathrm{Re}\,\langle Tx,y\rangle\,| &\leq |\,\langle T(x+y),x+y\rangle\,| + |\,\langle T(x-y),x-y\rangle\,| \\ &\leq \alpha(\,\|x+y\|^2 + \|x-y\|^2) \\ &= 2\alpha(\,\|x\|^2 + \|y\|^2) \end{aligned} \tag{6}$$

for all $x,y \in H$. Let $x \in H$ with $\|x\| \leq 1$ and $Tx \neq 0$. Then, setting $y = \|Tx\|^{-1}Tx$ in (6), we obtain

$$\|Tx\| = \mathrm{Re}\,\langle Tx, \|Tx\|^{-1}Tx\rangle \leq \tfrac{1}{4}\,2\alpha(\,\|x\|^2 + 1) \leq \alpha.$$

The last inequality is clearly true when $Tx = 0$, so we have

$$\|T\| = \sup\{\,\|Tx\| : \|x\| \leq 1\} \leq \alpha. \;\square$$

The proofs of Theorems 9.4.3 and 9.4.4 depend essentially on the fact that the values of $\langle Tx,y\rangle$ as x and y range over H are determined by the values of $\langle Tx,x\rangle$ as x ranges over H (cf. equations (2) and (5) above).

9.4.5 LEMMA *Suppose that H is a complex Hilbert space and let $T \in \mathscr{S}$. Then the eigenvalues, if any, of T are real and the eigenvectors of T corresponding to distinct eigenvalues are mutually orthogonal.*

PROOF Let λ be an eigenvalue of T and let x be an eigenvector corresponding to the eigenvalue λ. Then $\langle Tx,x\rangle = \lambda\langle x,x\rangle$ and hence λ is real because $\langle Tx,x\rangle$ is real and $\langle x,x\rangle$ is positive.

Now let μ be an eigenvalue of T with $\mu \neq \lambda$ and let y be an eigenvector corresponding to the eigenvalue μ. Since λ and μ are real we have

$$\lambda\langle x,y\rangle = \langle Tx,y\rangle = \langle x,Ty\rangle = \langle x,\mu y\rangle = \mu\langle x,y\rangle.$$

Thus $\langle x,y\rangle = 0$, because $\lambda \neq \mu$. $\square$

9.4.6 LEMMA *Let $T \in \mathscr{S}$ and let F be a linear subspace of H with $T(F) \subseteq F$. Then $T(F^\perp) \subseteq F^\perp$.*

PROOF We have $\langle x,Ty\rangle = \langle Tx,y\rangle = 0$ for all $x \in F$ and all $y \in F^\perp$. $\square$

EXERCISES

(1) Let T and S be linear transformations of H into itself such that $\langle Tx,y\rangle = \langle x,Sy\rangle$ for all $x,y \in H$. Prove that $T \in L(H)$ and $S = T^*$. (Use the closed graph theorem.) Consequently if T is a linear trans-

formation of H into itself then $T \in \mathscr{S}$ if and only if $\langle Tx,y\rangle = \langle x,Ty\rangle$ for all $x,y \in H$.

(2) Use the result of Exercise 6.3(12) to show that if H is a complex Hilbert space and $T \in \mathscr{S}$ then the spectrum of T is real.

9.5 Self-adjoint Compact Linear Operators

In this section T will denote a compact and self-adjoint linear operator on a non-zero complex Hilbert space H. For such operators the spectral theory developed in Section 6.7 can be carried a little further. To avoid having to consider exceptional cases we shall suppose throughout this section that the Hilbert space H is infinite-dimensional.

It follows from Theorems 6.7.6 and 6.7.2 and Lemma 9.4.5 that the set $\mathrm{sp}(T)$ consists of the point 0 together with a countable set of non-zero real eigenvalues of T whose only possible accumulation point is 0. To simplify our notation we shall consider only the case when T has infinitely many eigenvalues; we leave the reader to provide the obvious modifications needed in the case when T has only a finite number of eigenvalues.

Let $n \to \lambda_n$ be an enumeration of the non-zero eigenvalues of T with $|\lambda_n| \geq |\lambda_{n+1}|$ for $n = 1, 2,...$, and let

$$N_n = \mathscr{N}(\lambda_n I - T) = \{x \in H: Tx = \lambda_n x\}$$

for $n = 1, 2,....$ The set N_n is a linear subspace of H consisting of the point 0 together with the eigenvectors of T corresponding to the eigenvalue λ_n. It follows from Lemma 6.6.1 that N_n is finite-dimensional and so, by Corollary 9.2.10, we can choose an orthonormal basis for N_n, say $\{u_{m_{n-1}+1}, u_{m_{n-1}+2},..., u_{m_n}\}$ (where $m_0 = 1$). Lemma 9.4.5 shows that if $n \neq m$ then each point of N_n is orthogonal to each point of N_m. It follows that the set $\{u_n: n = 1, 2,...\}$ is orthonormal. We shall denote by F the closed linear hull of the set $\{u_n: n = 1, 2,...\}$.

9.5.1 LEMMA *$T(F) \subseteq F$ and $T|F^{\perp} = 0$.*

PROOF Since each point of the set $\{u_n: n = 1, 2,...\}$ is an eigenvector of T it is easy to verify that $T(F) \subseteq F$, and it follows then from Lemma 9.4.6 that $T(F^{\perp}) \subseteq F^{\perp}$. It is now clear that $T|F^{\perp}$ is a self-adjoint bounded linear operator on the Hilbert space $F^{\perp}$. Consequently, by Corollary 9.3.8 and Theorem 6.3.10, there is a point $\lambda \in \mathrm{sp}(T|F^{\perp})$ with $|\lambda| = \|T|F^{\perp}\|$. Suppose that $T|F^{\perp} \neq 0$. It is easy to verify that the operator $T|F^{\perp}$ is compact. (cf. the proof of Theorem 6.7.5.) Therefore, by Theorem 6.7.2, λ is an eigenvalue of $T|F^{\perp}$. Let x be an eigenvector of $T|F^{\perp}$ corresponding to λ. Then $Tx = (T|F^{\perp})x = \lambda x$ and so λ is an eigenvalue of T. Thus we must have $\lambda = \lambda_n$ for some positive integer n and hence $x \in N_n$. Since $N_n \subseteq F$ and $x \in F^{\perp}$ we have $x = 0$, which is a contradiction. This proves that $T|F^{\perp} = 0$. □

It is easy to see that $F^{\perp} = \mathcal{N}(T)$.

Now let $\mu_n = \lambda_k$ for $n = m_{k-1} + 1, m_{k-1} + 2, \ldots, m_k$ and $k = 1, 2, \ldots$. Then obviously we have $Tu_n = \mu_n u_n$ for $n = 1, 2, \ldots$. The following theorem generalizes the elementary theorem on the diagonalization of Hermitian symmetric matrices described on p. 360.

9.5.2 THEOREM *For each $x \in H$ we have $Tx = \sum_{n=1}^{\infty} \mu_n \langle x,u_n\rangle u_n$.*

PROOF Let $x \in H$. By Theorem 9.2.4 we can write $x = y + z$, where $y \in F$ and $z \in F^{\perp}$. By Lemma 9.5.1 we have $Tx = Ty + Tz = Ty \in F$. The set $\{u_n : n = 1, 2, \ldots\}$ is a maximal orthonormal subset of F because its closed linear hull is F. (See Exercise 9.2(2).) Consequently it follows from Theorem 9.2.16 that

$$\begin{aligned} Tx &= \sum_{n=1}^{\infty} \langle Tx,u_n\rangle u_n \\ &= \sum_{n=1}^{\infty} \langle x,Tu_n\rangle u_n \\ &= \sum_{n=1}^{\infty} \mu_n \langle x,u_n\rangle u_n. \ \square \end{aligned}$$

9.5.3 COROLLARY *For each positive integer n we have*
$$|\mu_n| = \sup \{ |\langle Tx, x\rangle| : \|x\| \leq 1 \text{ and } \langle x,u_m\rangle = 0 \text{ for } 1 \leq m \leq n-1\}.$$

PROOF It is obvious that $\langle Tu_n,u_n\rangle = \mu_n$. Also, by Theorem 9.5.2, if $x \in H$ and $\langle x,u_m\rangle = 0$ for $m = 1, 2, \ldots, n-1$ then

$$Tx = \sum_{m=n}^{\infty} \mu_m \langle x,u_m\rangle u_m$$

and hence, by Lemma 9.1.9,

$$\langle Tx,x\rangle = \sum_{m=n}^{\infty} \mu_m \, |\langle x,u_m\rangle|^2.$$

It follows from Lemma 9.2.14 that the series $\sum_n |\langle x,u_n\rangle|^2$ converges and that

$$\sum_{n=1}^{\infty} | \langle x,u_n \rangle |^2 \leq \| x \|^2.$$

Therefore, since $| \mu_m | \leq | \mu_n |$ for $m \geq n$, we obtain

$$\begin{aligned} | \langle Tx,x \rangle | &\leq \sum_{m=n}^{\infty} | \mu_m | \, | \langle x,u_m \rangle |^2 \\ &\leq | \mu_n | \sum_{m=n}^{\infty} | \langle x,u_m \rangle |^2 \\ &\leq | \mu_n | \, \| x \|^2. \ \square \end{aligned}$$

Corollary 9.5.3 can be used to calculate numerically the eigenvalues of certain self-adjoint compact operators.

There is a more direct approach to the spectral theory of self-adjoint compact operators, due to F. Riesz, which does not use the Riesz-Schauder theory developed in Sections 6.6 and 6.7. The first step in this approach is to prove directly, without appealing to Theorem 6.7.2, that there is an eigenvalue λ of T with $| \lambda | = \| T \|$. Next an inductive argument using this fact and Lemmas 9.4.5 and 9.4.6 shows that there is a sequence (μ_n) of real numbers and a countable orthonormal set $\{u_n : n = 1, 2, \ldots\}$ in H that satisfy the condition $Tu_n = \mu_n u_n$ and the condition of Corollary 9.5.3. Finally it is shown that all the eigenvalues of T occur at least once in the sequence (μ_n) and that Theorem 9.5.2 holds. The reader will find a more detailed outline of this approach to the spectral theory of self-adjoint compact operators in the Exercises below.

Our final theorem describes the resolvent $(\lambda I - T)^{-1}$. Its proof illustrates how easily information about T can be obtained from Theorem 9.5.2.

9.5.4 THEOREM *Let $\lambda \in \mathsf{C} \sim \mathrm{sp}(T)$. Then for all $x \in H$*

$$(\lambda I - T)^{-1} x = \frac{1}{\lambda} x + \frac{1}{\lambda} \sum_{n=1}^{\infty} \frac{\mu_n}{\lambda - \mu_n} \langle x,u_n \rangle u_n.$$

PROOF Let $x \in H$ and $y = (\lambda I - T)^{-1} x$. Then, by Theorem 9.5.2,

$$x = (\lambda I - T) y = \lambda y - \sum_{n=1}^{\infty} \mu_n \langle y,u_n \rangle u_n$$

and hence

$$y - \frac{1}{\lambda} x = \sum_{n=1}^{\infty} \frac{\mu_n}{\lambda} \langle y,u_n \rangle u_n. \tag{1}$$

For $m = 1, 2, \ldots$, equation (1) gives

$$\begin{aligned}\langle y,u_m\rangle - \frac{1}{\lambda}\langle x,u_m\rangle &= \langle y - \frac{1}{\lambda}x,u_m\rangle \\ &= \sum_{n=1}^{\infty}\frac{\mu_n}{\lambda}\langle y,u_n\rangle\langle u_n,u_m\rangle \\ &= \frac{\mu_m}{\lambda}\langle y,u_m\rangle,\end{aligned}$$

and consequently

$$\langle y,u_m\rangle = \frac{1}{\lambda - \mu_m}\langle x,u_m\rangle. \tag{2}$$

The result follows from (1) and (2). □

EXERCISES

In this sequence of exercises T will denote a compact self-adjoint linear operator on an infinite-dimensional complex Hilbert space H. The exercises provide an alternative development of the theory of such operators.

(1) Prove that T has an eigenvalue λ with $|\lambda| = \|T\|$. (Use Theorem 9.4.4 to find a real number λ and a sequence (x_n) in H with $\|x_n\| = 1$, $\lim_{n\to\infty}\langle Tx_n,x_n\rangle = \lambda$ and $|\lambda| = \|T\|$. Prove that $\lim_{n\to\infty}(Tx_n - \lambda x_n) = 0$.)

(2) Prove that there is a sequence (μ_n) of real numbers and a countable orthonormal set $\{u_n : n = 1, 2,\ldots\}$ in H such that $Tu_n = \mu_n u_n$ and $|\mu_n| = \sup\{\|Tx\| : \|x\| = 1 \text{ and } \langle x,u_m\rangle = 0 \text{ for } 1 \leq m \leq n-1\}$ for $n = 1, 2, \ldots$. Deduce from these properties of (μ_n) that $(|\mu_n|)$ is a non-increasing sequence and that $\lim_{n\to\infty}\mu_n = 0$.

(3) Prove that $Tx = \sum_{n=1}^{\infty}\mu_n\langle x,u_n\rangle u_n$ for all $x \in H$. Deduce that if λ is a non-zero eigenvalue of T then $\lambda = \mu_m$ for some integer m, that the set of integers m with $\lambda = \mu_m$ is finite and that $\mathcal{N}(\lambda I - T)$ is the linear hull of $\{u_m : \mu_m = \lambda\}$.

(4) Prove that $\mathrm{sp}(T) = \{\mu_n : n = 1, 2,\ldots\} \cup \{0\}$ by showing that if λ is non-zero and is not an eigenvalue of T then $\lambda \notin \mathrm{sp}(T)$ and

$$(\lambda I - T)^{-1}x = \frac{1}{\lambda}x + \frac{1}{\lambda}\sum_{n=1}^{\infty}\frac{\mu_n}{\lambda - \mu_n}\langle x,u_n\rangle u_n$$

for all $x \in H$.

The reader should note that it is not assumed in these exercises that T has an infinite number of non-zero eigenvalues. It may happen that $\mu_n = 0$ for all but a finite set of integers n.

(5) Let (μ_n) be a sequence of real numbers with $\lim_{n\to\infty}\mu_n = 0$ and let

$\{u_n : n = 1, 2,...\}$ be a countable, orthonormal set in H. Prove that the series $\sum_n \mu_n \langle x,u_n \rangle u_n$ converges in H for each $x \in H$ and that if

$$Tx = \sum_{n=1}^{\infty} \mu_n \langle x,u_n \rangle u_n$$

for all $x \in H$ then T is a compact self-adjoint linear operator on H. (Hint: Use Theorems 6.5.7 and 6.5.8.)

9.6 Positive Linear Operators

In this section H will again denote a Hilbert space. Our aim now is to state and prove the spectral theorem for general self-adjoint bounded linear operators. As we have remarked in Section 9.4, the spectral theorem is a generalization of the classical theory of diagonalization of Hermitian symmetric matrices. We have seen, in Theorem 9.5.2, the form this generalization takes in the case of self-adjoint operators which are also compact. In the absence of compactness the situation is more complicated. In particular, the spectrum of a general self-adjoint bounded linear operator, although it is a subset of the real line (Lemma 9.6.9), may not be a countable set. This fact is reflected in the representation of self-adjoint operators. Whereas, in Theorem 9.5.2, a compact self-adjoint operator is represented by an infinite series, in Theorem 9.9.4 a general self-adjoint operator is represented by an integral. This integral representation involves self-adjoint operators which are called *orthogonal projections* (Definition 9.7.1). In this section we are concerned with properties of self-adjoint operators that are fundamental to the construction of the orthogonal projections which occur in the integral representation of Theorem 9.9.4.

The approach to the spectral theorem presented in the rest of this chapter is due to F. Riesz (1934); it exploits the fact that there is a natural relation of partial order defined on the set $\mathscr{S}$ of all self-adjoint bounded linear operators on H. This partial order relation has some of the properties of the order relation on the real number system. (See in particular Theorem 9.6.14.)

9.6.1 Definition An operator $T \in \mathscr{S}$ is said to be *positive* if and only if $\langle Tx,x \rangle \geq 0$ for all $x \in H$. We shall denote by $\mathscr{S}^+$ the set of all positive operators in $\mathscr{S}$.

We remark that, when H is a complex Hilbert space, Theorem 9.4.3 shows that an operator $T \in L(H)$ is in $\mathscr{S}^+$ if and only if $\langle Tx,x \rangle \geq 0$ for all $x \in H$.

9.6.2 LEMMA *Let $T,S \in \mathscr{S}^+$ and let α be a non-negative real number. Then*

(a) $T + S \in \mathscr{S}^+$,
(b) $\alpha T \in \mathscr{S}^+$, *and*
(c) $T^n \in \mathscr{S}^+$ *for* $n = 1, 2, \ldots$.

PROOF Assertions (a) and (b) are obvious. It follows by induction from Theorem 9.4.2 that $T^n \in \mathscr{S}$. Also we have

$$\langle T^{2n}x,x\rangle = \langle T^n x, T^n x\rangle \geq 0$$

and

$$\langle T^{2n+1}x,x\rangle = \langle T(T^n x), T^n x\rangle \geq 0$$

for all $x \in H$. This proves (c). □

9.6.3 LEMMA *For each $T \in \mathscr{S}$ we have $T^2 \in \mathscr{S}^+$.*

PROOF $\langle T^2x,x\rangle = \langle Tx,Tx\rangle \geq 0$ for all $x \in H$. □

It is obvious that the operators 0 and I are positive. There are, in general, many other positive operators. In fact, if $T \in L(H)$ then $T^*T \in \mathscr{S}^+$ because $\langle T^*Tx,x\rangle = \langle Tx,Tx\rangle \geq 0$.

9.6.4 DEFINITION Given $T,S \in \mathscr{S}$ we shall write $T \leq S$ if and only if $S - T \in \mathscr{S}^+$.

It is obvious that $\mathscr{S}^+ = \{T \in \mathscr{S} : T \geq 0\}$. Also $T \leq S$ if and only if $\langle Tx,x\rangle \leq \langle Sx,x\rangle$ for all $x \in H$.

9.6.5 LEMMA *The relation $\leqslant$ is a partial order on $\mathscr{S}^+$. Further*

(a) *if $T_1 \leq S_1$ and $T_2 \leq S_2$ then $T_1 + T_2 \leq S_1 + S_2$,*
(b) *if $0 \leq T \leq S$ and $0 \leq \alpha \leq \beta$ then $0 \leq \alpha T \leq \beta S$, and*
(c) *$T \leq S$ if and only if $-S \leq -T$.*

PROOF The only difficulty arises in showing that, if $T,S \in \mathscr{S}$ with $T \leq S$ and $S \leq T$, then $T = S$. If $T \leq S$ and $S \leq T$ then we have $\langle Tx,x\rangle = \langle Sx,x\rangle$ and hence $\langle (T - S)x,x\rangle = 0$ for all $x \in H$. Theorem 9.4.4 now shows that $T - S = 0$. The rest of the proof of the lemma is left to the reader. □

9.6.6 LEMMA *Let $T \in \mathscr{S}$, let $m = \inf\{\langle Tx,x\rangle : \|x\| = 1\}$ and let $M = \sup\{\langle Tx,x\rangle : \|x\| = 1\}$. Then $mI \leq T \leq MI$ and*

$$\|T\| = \max\{M, -m\}.$$

PROOF For all $x \in H$ we have

$$m\langle x,x\rangle = m\|x\|^2 \leq \langle Tx,x\rangle \leq M\|x\|^2 = M\langle x,x\rangle$$

which shows that $mI \le T \le MI$. The equality $\| T \| = \max\{M, -m\}$ follows directly from Theorem 9.4.4. □

It follows from Lemma 9.6.6 that if a and b are real numbers with $a \le -\| T \|$ and $b \ge \| T \|$ then $aI \le T \le bI$. The proof of the next lemma is elementary, and we omit it.

9.6.7 LEMMA *Let $T \in \mathscr{S}^+$ and let $B(x,y) = \langle Tx,y\rangle$ for all $x,y \in H$. Then B is a non-negative Hermitian symmetric form on H.*

9.6.8 LEMMA *For each $T \in \mathscr{S}^+$ we have*

$$| \langle Tx,y\rangle |^2 \le \langle Tx,x\rangle \langle Ty,y\rangle$$

for all $x,y \in H$.

PROOF Apply Schwarz's inequality to the non-negative Hermitian symmetric form B of Lemma 9.6.7. □

We shall now show that, when H is a complex Hilbert space, the positive operators in $\mathscr{S}$ are precisely those operators in $\mathscr{S}$ whose spectrum is contained in $[0, \infty)$. This is a simple consequence of the following lemma which gives some information about the location of the spectrum of a bounded self-adjoint operator.

9.6.9 LEMMA *Let H be a complex Hilbert space, let $T \in \mathscr{S}$, let $m = \inf\{\langle Tx,x\rangle : \| x \| = 1\}$ and let $M = \sup\{\langle Tx,x\rangle : \| x \| = 1\}$. Then $\operatorname{sp}(T) \subseteq [m,M]$ and the end points m and M are both in $\operatorname{sp}(T)$.*

PROOF Let $\lambda \in \mathsf{C} \sim [m,M]$. Then $\delta = d(\lambda,[m,M]) > 0$ and for each $x \in H$ with $\| x \| = 1$ we have

$$\| (\lambda I - T)x \| \ge | \langle(\lambda I - T)x,x\rangle | = | \lambda - \langle Tx,x\rangle | \ge \delta. \tag{1}$$

Inequality (1) shows that $\lambda I - T$ is a linear homeomorphism of H onto $\mathscr{R}(\lambda I - T)$ (Lemma 3.5.9). Consequently $\mathscr{R}(\lambda I - T)$ is complete (see p. 110) and hence is closed. Suppose that $\mathscr{R}(\lambda I - T)^{\perp} \ne \{0\}$ and choose $z \in \mathscr{R}(\lambda I - T)^{\perp}$ with $\| z \| = 1$. Then $\langle(\lambda I - T)z,z\rangle = 0$, which contradicts (1). Thus $\mathscr{R}(\lambda I - T)^{\perp} = \{0\}$ and, by Theorem 9.2.4, $\mathscr{R}(\lambda I - T) = H$. This shows that $\lambda \notin \operatorname{sp}(T)$ and therefore we have $\operatorname{sp}(T) \subseteq [m,M]$.

We prove finally that $m \in \operatorname{sp}(T)$. By Lemma 9.6.6 $T - mI \in \mathscr{S}^+$ and so by Lemma 9.6.8 we have, for all $x \in H$ with $\| x \| = 1$,

$$\begin{aligned} \| (T - mI)x \|^4 &= | \langle(T - mI)x,(T - mI)x\rangle |^2 \\ &\le \langle(T - mI)x,x\rangle \langle(T - mI)^2x,(T - mI)x\rangle \\ &\le \| T - mI \|^3(\langle Tx,x\rangle - m). \end{aligned}$$

Therefore, by definition of m, $\inf\{ \| (T - mI)x \| : \| x \| = 1\} = 0$ and hence, by Lemma 3.5.9, we must have $m \in \mathrm{sp}(T)$. Similarly we can prove that $M \in \mathrm{sp}(T)$. □

9.6.10 Theorem *Let H be a complex Hilbert space and let $T \in \mathscr{S}$. Then $T \in \mathscr{S}^+$ if and only if* $\mathrm{sp}(T) \subseteq [0, \infty)$.

Proof Let $m = \inf\{\langle Tx,x\rangle : \| x \| = 1\}$. If $T \in \mathscr{S}^+$ then $m \geq 0$ and so, by Lemma 9.6.9, $\mathrm{sp}(T) \subseteq [0,\infty)$. If, conversely, $\mathrm{sp}(T) \subseteq [0,\infty)$ then $m \geq 0$, because $m \in \mathrm{sp}(T)$ by Lemma 9.6.9, and hence $T \in \mathscr{S}^+$. □

One of the most important properties of the real number system is the fact that every bounded monotonic sequence of real numbers converges. This property extends to the partially ordered set $\mathscr{S}$ provided that the term 'convergence' is interpreted in the correct sense.

9.6.11 Definition A sequence (T_n) in $L(H)$ is said to *converge strongly* to an operator $T \in L(H)$ if and only if $\lim_{n\to\infty} T_n x = Tx$ for all $x \in H$.

It would perhaps be more natural to say 'converge pointwise on H' instead of 'converge strongly' in the above definition (cf. the discussion on p. 49); however the term 'strong convergence' is firmly established in general usage.

It is clear that a sequence in $L(H)$ can converge strongly to at most one operator in $L(H)$. If (T_n) converges strongly to T, then T is called the *strong limit* of the sequence (T_n).

The inequality $\| T_n x - Tx \| \leq \| T_n - T \| \, \| x \|$ shows that convergence in the normed linear space $L(H)$ entails strong convergence to the same limit. On the other hand it is not difficult to construct a sequence (T_n) in $L(H)$ which converges strongly to an operator $T \in L(H)$, but does not converge to T in the space $L(H)$. (See the Exercise below.)

9.6.12 Lemma *Let (T_n) be a sequence in $\mathscr{S}$ that converges strongly to an operator $T \in L(H)$. Then $T \in \mathscr{S}$. Further, if $T_n \in \mathscr{S}^+$ for $n =$ 1, 2,... then also $T \in \mathscr{S}^+$.*

Proof Let $x,y \in H$. Then using Lemma 9.1.9 and the fact that $\lim_{n\to\infty} T_n x = Tx$ and $\lim_{n\to\infty} T_n y = Ty$ we obtain

$$\langle Tx,y\rangle = \lim_{n\to\infty} \langle T_n x,y\rangle = \lim_{n\to\infty} \langle x,T_n y\rangle = \langle x,Ty\rangle.$$

This shows that $T \in \mathscr{S}$. If, further, $T_n \in \mathscr{S}^+$ for $n = 1, 2,\ldots$ then

$$\langle Tx,x\rangle = \lim_{n\to\infty} \langle T_n x,x\rangle \geq 0$$

and so $T \in \mathscr{S}^+$. □

9.6.13 LEMMA *The sets $\mathscr{S}$ and $\mathscr{S}^+$ are closed in $L(H)$.*

PROOF This follows from Lemma 9.6.12 and the observation that convergence in the space $L(H)$ entails strong convergence to the same limit. □

Our next theorem provides the crucial tool for Riesz's proof of the spectral theorem. We recall that a sequence (T_n) in $L(H)$ is bounded if and only if there exists $M > 0$ with $\| T_n \| \leq M$ for $n = 1, 2, \ldots$. Using Lemma 9.6.6 we see that the sequence (T_n) is bounded if and only if there exists $M > 0$ with $-MI \leq T_n \leq MI$ for $n = 1, 2, \ldots$.

9.6.14 THEOREM *Let (T_n) be a bounded sequence in $\mathscr{S}$ with $T_n \leq T_{n+1}$ for $n = 1, 2, \ldots$. Then the sequence (T_n) converges strongly to an operator T in $\mathscr{S}$.*

PROOF By the remark preceding the statement of the theorem, there exists $M > 0$ with $-MI \leq T_n \leq MI$ for $n = 1, 2, \ldots$. It is clear that if the sequence $(\frac{1}{2}M^{-1}(T_n + MI))$ converges strongly to an operator $S \in \mathscr{S}$ then the sequence (T_n) converges strongly to the operator $2MS - MI$ which is also in $\mathscr{S}$. Thus, replacing T_n by $\frac{1}{2}M^{-1}(T_n + MI)$ if necessary, we may suppose that $0 \leq T_n \leq I$ for $n = 1, 2, \ldots$.

Let m and n be positive integers with $m > n$ and let $T_{mn} = T_m - T_n$. Then $0 \leq T_{mn} \leq I$ and hence by Lemma 9.6.6 we have $\| T_{mn} \| \leq 1$. Let $x \in H$. Using Lemma 9.6.8 (with $y = T_{mn}x$) and Schwarz's inequality we obtain

$$\begin{aligned} \| T_m x - T_n x \|^4 &= \| T_{mn} x \|^4 \\ &= | \langle T_{mn}x, T_{mn}x \rangle |^2 \\ &\leq \langle T_{mn}x, x \rangle \langle T_{mn}^2 x, T_{mn}x \rangle \\ &\leq \langle T_{mn}x, x \rangle \| T_{mn}^2 x \| \, \| T_{mn}x \| \\ &\leq \langle T_{mn}x, x \rangle \| T_{mn} \|^3 \| x \|^2 \\ &\leq \| x \|^2 (\langle T_m x, x \rangle - \langle T_n x, x \rangle). \end{aligned} \tag{2}$$

Since $0 \leq T_k \leq T_{k+1} \leq I$ for $k = 1, 2, \ldots$ the sequence $(\langle T_k x, x \rangle)$ is a bounded increasing sequence of real numbers, and hence converges. Inequality (2) now shows that $(T_n x)$ is a Cauchy sequence in H. Let $Tx = \lim_{n \to \infty} T_n x$. This defines a mapping T of H into itself. It is easy to verify that T is linear. Also, for each $x \in H$, we have

$$\| Tx \| = \lim_{n \to \infty} \| T_n x \| \leq \| x \|$$

because $\| T_n \| \leq 1$. This shows that $T \in L(H)$. It is clear that (T_n) converges strongly to T and hence, by Lemma 9.6.12, $T \in \mathscr{S}$. □

EXERCISE

Let $\{u_n : n = 1, 2, \ldots\}$ be a countable orthonormal set in H. Define operators P_m and P on H by $P_m x = \sum_{n=1}^{m} \langle x,u_n\rangle u_n$ and $Px = \sum_{n=1}^{\infty} \langle x,u_n\rangle u_n$ for all $x \in H$. Prove that (P_m) converges strongly to P but that $\| P_m - P \| = 1$ for $m = 1, 2, \ldots$.

9.7 Orthogonal Projections

The purpose of this section is to introduce a certain class of projections on Hilbert space that plays an essential role in the spectral theorem for self-adjoint bounded linear operators. Throughout the section H will denote a Hilbert space. We recall that a projection on H is a linear operator P on H with $P^2 = P$. (See Section 8.8.)

9.7.1 DEFINITION An *orthogonal projection on* H is a projection on H that is also a self-adjoint bounded linear operator on H.

We shall see that the orthogonal projections on H are precisely those projections associated with direct sum decompositions of the form $H = F \oplus F^\perp$, where F is a closed linear subspace of H. (cf. Lemma 8.8.9.)

9.7.2 LEMMA *Let P be an orthogonal projection on H. Then $\mathscr{R}(P)^\perp = \mathscr{R}(I - P)$.*

PROOF Let $x \in \mathscr{R}(P)^\perp$. Then, for all $y \in H$, we have

$$\langle Px,y\rangle = \langle x,Py\rangle = 0$$

and hence, by Lemma 9.1.8, $Px = 0$. Thus $x = (I - P)x \in \mathscr{R}(I - P)$.

On the other hand let $x \in \mathscr{R}(I - P)$. Then $Px = 0$ (by Lemma 8.8.8(b) with P and $I - P$ interchanged) and so, for all $y \in H$, we have $\langle x,Py\rangle = \langle Px,y\rangle = 0$. Thus $x \in \mathscr{R}(P)^\perp$. □

9.7.3 THEOREM *Let F be a closed linear subspace of H. Then there is a unique orthogonal projection P on H with $\mathscr{R}(P) = F$.*

PROOF By Theorem 9.2.4 we have $H = F \oplus F^\perp$ and consequently by Lemma 8.8.9 there is a unique projection P on H with $\mathscr{R}(P) = F$ and $\mathscr{R}(I - P) = F^\perp$. For each $x \in H$, we have $x = Px + (I - P)x$ and $\langle Px,(I - P)x\rangle = 0$ so

$$\begin{aligned} \| x \|^2 &= \langle x,x\rangle \\ &= \langle Px,Px\rangle + \langle (I - P)x,(I - P)x\rangle \\ &= \| Px \|^2 + \| (I - P)x \|^2, \end{aligned}$$

and therefore $\| Px \| \leq \| x \|$. This proves that P is bounded and that $\| P \| \leq 1$.

Let $x,y \in H$. Then since $\langle Px,(I - P)y\rangle = \langle (I - P)x,Py\rangle = 0$ we have

$$\begin{aligned}\langle Px,y\rangle &= \langle Px,Py + (I - P)y\rangle \\ &= \langle Px,Py\rangle \\ &= \langle Px + (I - P)x,Py\rangle \\ &= \langle x,Py\rangle.\end{aligned}$$

This shows that P is self-adjoint and hence that P is an orthogonal projection.

Suppose that Q is an orthogonal projection on H with $\mathscr{R}(Q) = F$. Then, by Lemma 9.7.2, $\mathscr{R}(I - Q) = \mathscr{R}(Q)^{\perp} = F^{\perp}$ and consequently, by the uniqueness of P, we have $P = Q$. □

Some of the more important properties of orthogonal projections are noted in the following lemmas. The reader will find further properties in the Exercises below.

9.7.4 LEMMA *Let P be an orthogonal projection on H. Then*

(a) $P \in \mathscr{S}^{+}$,
(b) $\langle Px,x\rangle = \| Px \|^2$ *for all* $x \in H$, *and*
(c) $\| P \| = 1$ *unless* $P = 0$.

PROOF For all $x \in H$, we have

$$\langle Px,x\rangle = \langle P^2x,x\rangle = \langle Px,Px\rangle = \| Px \|^2 \geq 0.$$

This proves (a) and (b). We saw in the proof of Theorem 9.7.3 that $\| P \| \leq 1$. Suppose that $P \neq 0$ and choose $x \in H$ with $Px \neq 0$. Then $\| Px \| = \| P(Px) \| \leq \| P \| \; \| Px \|$ which gives $\| P \| \geq 1$. This proves (c). □

9.7.5 THEOREM *Let P and Q be orthogonal projections on H. Then the following four conditions are equivalent:*

(a) $\mathscr{R}(P) \subseteq \mathscr{R}(Q)$,
(b) $P \leq Q$,
(c) $P = PQ$, *and*
(d) $P = QP$.

PROOF Suppose that $\mathscr{R}(P) \subseteq \mathscr{R}(Q)$. Then, for all $x \in H$, we have $Px \in \mathscr{R}(Q)$ and hence $Px = Q(Px)$. This shows that $P = QP$. Hence (a) implies (d).

Suppose that $P = QP$. Then $P = P^* = (QP)^* = P^*Q^* = PQ$. Thus (d) implies (c).

Suppose that $P = PQ$. Then, by Lemma 9.7.4(b) and (c), we have, for all $x \in H$,

$$\langle Px,x\rangle = \| Px \|^2 = \| PQx \|^2 \leq \| P\|^2 \| Qx \|^2 \leq \| Qx \|^2 = \langle Qx,x\rangle.$$

This shows that $P \leq Q$. Hence (c) implies (b).

Suppose finally that $P \leq Q$ and let $x \in \mathscr{R}(P)$. Then

$$\langle x,x\rangle = \langle Px,x\rangle \leq \langle Qx,x\rangle$$

and hence $\langle (I - Q)x,x\rangle \leq 0$. By Lemma 9.7.4(a), we have

$$\langle (I - Q)x,x\rangle \geq 0.$$

Consequently, by Lemma 9.7.4(b), we have

$$\| (I - Q)x \|^2 = \langle (I - Q)x,x\rangle = 0.$$

Thus $x = Qx + (I - Q)x = Qx \in \mathscr{R}(Q)$. This proves that $\mathscr{R}(P) \subseteq \mathscr{R}(Q)$. Hence (b) implies (a). □

We remark that if P and Q are orthogonal projections on H that satisfy any one of the conditions in Theorem 9.7.5 then $P = PQ = QP$.

EXERCISES

In Exercises (1)–(4) *P and Q will denote orthogonal projections on a Hilbert space H.*

(1) Prove that PQ is an orthogonal projection if and only if $PQ = QP$, and that, if $PQ = QP$, then $\mathscr{R}(PQ) = \mathscr{R}(P) \cap \mathscr{R}(Q)$.

(2) Prove that if $PQ = QP$ then $P + Q - PQ$ is an orthogonal projection and $\mathscr{R}(P + Q - PQ) = \mathscr{R}(P) + \mathscr{R}(Q)$.

(3) Prove that $P - Q$ is an orthogonal projection if and only if $Q \leq P$. Prove that if $Q \leq P$ then $\mathscr{R}(P - Q) = \mathscr{R}(P) \cap \mathscr{R}(I - Q)$.

(4) Prove that the following three conditions are equivalent: (a) each point of $\mathscr{R}(P)$ is orthogonal to each point of $\mathscr{R}(Q)$, (b) $PQ = 0$, (c) $QP = 0$.

(5) Let $P_1, P_2, \ldots, P_n$ be orthogonal projections and let

$$P = P_1 + P_2 + \ldots + P_n.$$

Prove that P is an orthogonal projection if and only if $P_jP_k = 0$ for $j \neq k$. If P is an orthogonal projection prove that

$$\mathscr{R}(P) = \mathscr{R}(P_1) \oplus \mathscr{R}(P_2) \oplus \ldots \oplus \mathscr{R}(P_n).$$

(6) Prove that if Q_n is an orthogonal projection for $n = 1, 2, \ldots$ and the sequence (Q_n) converges strongly to $Q \in L(H)$ then Q is an orthogonal projection.

(7) Let (P_n) be a sequence of orthogonal projections on H such that $P_mP_n = 0$ whenever $m \neq n$. Prove that the equation $Px = \sum_{n=1}^{\infty} P_nx$

defines an orthogonal projection P and that the range of P is the closed linear hull of $\bigcup_{n=1}^{\infty} \mathscr{R}(P_n)$.

9.8 Functions of a Self-adjoint Bounded Linear Operator

Throughout this section T will denote a self-adjoint bounded linear operator on a complex Hilbert space H and m and M will denote real numbers with $mI \leq T \leq MI$ (such numbers exist by Lemma 9.6.6). To avoid trivialities we shall suppose that $m < M$. The somewhat elaborate machinery set up in this section will be used in the following section to define the orthogonal projections needed for the spectral theorem.

We have already used in Section 6.3 the mapping $p \to p(T)$ of the set of all polynomial functions into the algebra $L(H)$. (See p. 233.) This mapping is our starting point in this section; however it is convenient slightly to modify the definition given in Section 6.3.

We shall denote by $\mathscr{P}$ the linear space of all real-valued functions p on the interval $[m,M]$ that are of the form

$$p(t) = \alpha_0 + \alpha_1 t + \ldots + \alpha_n t^n \tag{1}$$

for $m \leq t \leq M$, where n is a non-negative integer (depending on p) and $\alpha_0, \alpha_1, \ldots, \alpha_n$ are real numbers. Since $m < M$ each function $p \in \mathscr{P}$ has a unique representation in the form (1). If $p \in \mathscr{P}$ is given by (1) we shall write

$$p(T) = \alpha_0 I + \alpha_1 T + \ldots + \alpha_n T^n.$$

It is clear that $p \to p(T)$ is a linear mapping of $\mathscr{P}$ into $\mathscr{S}$ that satisfies

$$(pq)(T) = p(T)q(T)$$

for all $p, q \in \mathscr{P}$.

The mapping $p \to p(T)$ has one further important property which is not so obvious. We shall denote by $\mathscr{P}^+$ the set of all $p \in \mathscr{P}$ with $p \geq 0$.

9.8.1 Lemma *If $p \in \mathscr{P}^+$ then $p(T) \in \mathscr{S}^+$.*

Proof We may suppose that p is non-constant. By Theorem 9.4.2 we have $p(T) \in \mathscr{S}$. Thus, by Theorem 9.6.10, it is sufficient to prove that $\operatorname{sp}(p(T)) \subseteq [0, \infty)$. Let $\lambda \in \operatorname{sp}(p(T))$ and let

$$\lambda - p(t) = \mu_0(\mu_1 - t)(\mu_2 - t) \ldots (\mu_n - t)$$

for $m \leq t \leq M$, where $\mu_0, \mu_1, \ldots, \mu_n$ are complex numbers and $\mu_0 \neq 0$. Then

$$\lambda I - p(T) = \mu_0(\mu_1 I - T)(\mu_2 I - T) \ldots (\mu_n I - T)$$

and, consequently, since $\lambda I - p(T)$ is not regular, $\mu_j I - T$ must fail to be regular for at least one integer j with $1 \leq j \leq n$. Thus $\mu_j \in \text{sp}(T)$, and by Lemma 9.6.9 we have $m \leq \mu_j \leq M$. Therefore $\lambda = p(\mu_j) \geq 0$. □

9.8.2 Definition We shall denote by $\mathscr{L}^+$ the class of all real-valued functions f on $[m,M]$ with the property that there is a sequence (p_n) in $\mathscr{P}^+$ with

(a) $0 \leq p_{n+1} \leq p_n$ for $n = 1, 2,\ldots$, and
(b) $\lim_{n\to\infty} p_n(t) = f(t)$ for each $t \in [m,M]$.

It is clear that if $f,g \in \mathscr{L}^+$ and α is a non-negative real number then $f + g \in \mathscr{L}^+$ and $\alpha f \in \mathscr{L}^+$. It is also clear that $\mathscr{L}^+ \subseteq B_{\mathsf{R}}([m,M])$.

We shall denote by $\mathscr{L}$ the class of all functions in $B_{\mathsf{R}}([m,M])$ of the form $f - g$ with $f,g \in \mathscr{L}^+$. It is obvous that $\mathscr{L}$ is a linear subspace of $B_{\mathsf{R}}([m,M])$. Also $\mathscr{L}$ contains $\mathscr{P}$ because if $p \in \mathscr{P}$ then $p + \alpha 1 \in \mathscr{P}^+$ for some positive number α and $p = (p + \alpha 1) - \alpha 1$.

Our aim is to extend the mapping $p \to p(T)$ of $\mathscr{P}$ into $\mathscr{S}$ to a mapping of $\mathscr{L}$ into $\mathscr{S}$ in such a way that the algebraic properties of the mapping $p \to p(T)$ and the property of Lemma 9.8.1 are preserved. The following two lemmas provide the crucial steps in the extension process.

9.8.3 Lemma *Let (p_n) be a sequence in $\mathscr{P}^+$ with $p_{n+1} \leq p_n$ for $n = 1, 2,\ldots$. Then the sequence $(p_n(T))$ converges strongly to an operator in $\mathscr{S}^+$.*

Proof By Lemma 9.8.1 we have $0 \leq p_{n+1}(T) \leq p_n(T)$ for $n = 1, 2,\ldots$ so Theorem 9.6.14, applied to the sequence $(-p_n(T))$, shows that the sequence $(p_n(T))$ converges strongly to an operator S in $\mathscr{S}$. Lemma 9.6.12 shows that $S \in \mathscr{S}^+$. □

9.8.4 Lemma *Let (p_n) and (q_n) be sequences in $\mathscr{P}^+$ with $p_{n+1} \leq p_n$ and $q_{n+1} \leq q_n$ for $n = 1, 2,\ldots$ and let S_1 and S_2 be the strong limits of the sequences $(p_n(T))$ and $(q_n(T))$, respectively. Then, if*

$$\lim_{n\to\infty} p_n(t) \leq \lim_{n\to\infty} q_n(t)$$

for $m \leq t \leq M$, we have $S_1 \leq S_2$.

Proof Notice that for each $t \in [m,M]$ the sequences $(p_n(t))$ and $(q_n(t))$ are decreasing sequences of non-negative real numbers, so they converge.

Let k be a fixed positive integer. For each $t \in [m,M]$ we have

$$\lim_{n\to\infty} (p_n(t) - q_k(t)) \leq \lim_{n\to\infty} p_n(t) - \lim_{n\to\infty} q_n(t) \leq 0. \tag{2}$$

Let $r_n = (p_n - q_k) \vee 0$ for $n = 1, 2, \ldots$. It follows easily from inequality (2) that $\lim_{n\to\infty} r_n(t) = 0$ for each $t \in [m,M]$. It is also easy to see that $r_{n+1} \leq r_n$ for $n = 1, 2, \ldots$. Hence by Dini's theorem (4.3.9) the sequence (r_n) converges uniformly to 0 on the interval $[m,M]$. Thus given $\varepsilon > 0$ there is an integer N with $0 \leq r_n(t) < \varepsilon$ for all $t \in [m,M]$ and all integers $n \geq N$. It follows that $p_n(t) - q_k(t) < \varepsilon$ for all $t \in [m,M]$ and all $n \geq N$, and therefore, by Lemma 9.8.1, we have

$$p_n(T) \leq q_k(T) + \varepsilon I$$

for all $n \geq N$. It now follows from Lemma 9.6.12 that $S_1 \leq q_k(T) + \varepsilon I$. Since k is an arbitrary positive integer, another application of Lemma 9.6.12 gives $S_1 \leq S_2 + \varepsilon I$. Finally, since ε is an arbitrary positive number, yet another application of Lemma 9.6.12 gives $S_1 \leq S_2$. □

We are now in a position to extend the mapping $p \to p(T)$ to a mapping of $\mathscr{L}^+$ into $\mathscr{S}^+$.

9.8.5 Definition Let $f \in \mathscr{L}^+$ and choose a sequence (p_n) in $\mathscr{P}^+$ with $p_{n+1} \leq p_n$ for $n = 1, 2, \ldots$ and $\lim_{n\to\infty} p_n(t) = f(t)$ for each $t \in [m,M]$. By Lemma 9.8.3 the sequence $(p_n(T))$ converges strongly to an operator in $\mathscr{S}^+$ which we shall denote by $f(T)$.

To show that the definition of $f(T)$ is free from ambiguity we must verify that it is independent of the choice of the sequence (p_n). Let (q_n) be a sequence in $\mathscr{P}^+$ with $q_{n+1} \leq q_n$ for $n = 1, 2, \ldots$ and $\lim_{n\to\infty} q_n(t) = f(t)$ for each $t \in [m,M]$, and let S be the strong limit of the sequence $(q_n(T))$. It follows from Lemma 9.8.4 that $S \leq f(T)$ and $f(T) \leq S$ so, by Lemma 9.6.5, we have $S = f(T)$.

Let $p \in \mathscr{P}^+$. By considering the sequence (p_n), where $p_n = p$ for $n = 1, 2, \ldots$, we see that $p \in \mathscr{L}^+$ and that the operator $p(T)$ defined in Definition 9.8.5 coincides with the operator $p(T)$ defined at the beginning of this section.

9.8.6 Lemma *The mapping $f \to f(T)$ of $\mathscr{L}^+$ into $\mathscr{S}^+$ has the following properties:*

(a) $(f + g)(T) = f(T) + g(T)$ *for all* $f,g \in \mathscr{L}^+$,
(b) $(\alpha f)(T) = \alpha f(T)$ *for all* $f \in \mathscr{L}^+$ *and all* $\alpha \geq 0$,
(c) $(fg)(T) = f(T)g(T)$ *for all* $f,g \in \mathscr{L}^+$, *and*
(d) *if* $f,g \in \mathscr{L}^+$ *and* $f \leq g$ *then* $f(T) \leq g(T)$.

Proof Properties (a) and (b) follow directly from the corresponding properties of the mapping $p \to p(T)$ of $\mathscr{P}^+$ into $\mathscr{S}^+$. Property (d) follows from Lemma 9.8.4. Let us prove property (c). Let $f,g \in \mathscr{L}^+$ and choose sequences (p_n) and (q_n) in $\mathscr{P}^+$ with $p_{n+1} \leq p_n$ and $q_{n+1} \leq q_n$ for $n = 1, 2, \ldots,$ and with $\lim_{n\to\infty} p_n(t) = f(t)$ and $\lim_{n\to\infty} q_n(t) = g(t)$ for all $t \in [m,M]$. Obviously $p_{n+1}q_{n+1} \leq p_n q_n$ and

$$\lim_{n\to\infty} (p_n q_n)(t) = \lim_{n\to\infty} p_n(t)q_n(t) = f(t)g(t) = (fg)(t)$$

for all $t \in [m,M]$. Thus $(fg)(T)$ is the strong limit of the sequence $((p_n q_n)(T))$. Now $f(T) \in \mathscr{S}$ and $p_n(T) \in \mathscr{S}$ so, by Lemma 9.1.9, for all $x,y \in H$ we have

$$\begin{aligned}
\langle (f(T)g(T))x,y\rangle &= \langle f(T)(g(T)x),y\rangle \\
&= \langle g(T)x,f(T)y\rangle \\
&= \lim_{n\to\infty} \langle q_n(T)x,p_n(T)y\rangle \\
&= \lim_{n\to\infty} \langle p_n(T)(q_n(T)x),y\rangle \\
&= \lim_{n\to\infty} \langle (p_n(T)q_n(T))x,y\rangle \\
&= \lim_{n\to\infty} \langle (p_n q_n)(T)x,y\rangle \\
&= \langle (fg)(T)x,y\rangle.
\end{aligned}$$

Thus, by Lemma 9.1.8, $f(T)g(T)x = (fg)(T)x$ for all $x \in H$ and so $f(T)g(T) = (fg)(T)$. □

It is now easy to extend the mapping $f \to f(T)$ to the whole of $\mathscr{L}$.

9.8.7 DEFINITION Let $f \in \mathscr{L}$ and choose $g,h \in \mathscr{L}^+$ with $f = g - h$. We define the operator $f(T)$ by $f(T) = g(T) - h(T)$. It is clear that $f(T) \in \mathscr{S}$. The definition of $f(T)$ is free from ambiguity because if also $f = g' - h'$ with $g',h' \in \mathscr{L}^+$, then $g + h' = h + g'$, and hence by Lemma 9.8.6

$$g(T) + h'(T) = (g + h')(T) = (h + g')(T) = h(T) + g'(T)$$

which gives $g(T) - h(T) = g'(T) - h'(T)$. It is clear that if $f \in \mathscr{L}^+$ then the two definitions of $f(T)$ (9.8.7 and 9.8.5) coincide.

9.8.8 THEOREM *The mapping $f \to f(T)$ defined above is a linear mapping of $\mathscr{L}$ into $\mathscr{S}$ that also satisfies the conditions:*

(a) *$(fg)(T) = f(T)g(T)$ for all $f,g \in \mathscr{L}$, and*
(b) *if $f,g \in \mathscr{L}$ and $f \leq g$ then $f(T) \leq g(T)$.*

PROOF The proof is straightforward, and uses Lemma 9.8.6 and parts (a) and (c) of Lemma 9.6.5. The reader is left to supply the details. □

Our final theorem shows that the class $\mathscr{L}$ is considerably larger than the class $\mathscr{P}$ from which we started.

9.8.9 THEOREM $C_{\mathsf{R}}([m,M]) \subseteq \mathscr{L}$.

PROOF Let $f \in C_{\mathsf{R}}([m,M])$. Then $f = f^+ - f^-$, where $f^+ = f \vee 0$ and $f^- = (-f) \vee 0$. Since $f^+ \geq 0$ and $f^- \geq 0$, and, by Theorem 2.8.2, f^+ and f^- are in $C_{\mathsf{R}}([m,M])$, we may clearly suppose that $f \geq 0$. Then, using the Weierstrass approximation theorem (4.5.1), we can find a sequence (p_n) in $\mathscr{P}$ with

$$f(t) + \frac{1}{n+1} < p_n(t) < f(t) + \frac{1}{n}$$

for all $t \in [m,M]$ and $n = 1, 2,\ldots$. It is clear that $p_n \in \mathscr{P}^+$, $p_{n+1} \leq p_n$ for $n = 1, 2,\ldots$ and $\lim_{n\to\infty} p_n(t) = f(t)$ for all $t \in [m,M]$. Thus $f \in \mathscr{L}^+$. □

The theory of the mapping $f \to f(t)$ constructed above is an example of what is often called an *operational calculus for* T (or a *functional calculus for* T). This operational calculus was developed by F. Riesz in 1934; it provides a powerful tool for the study of self-adjoint bounded linear operators.

The technique of studying the spectral theory of a linear operator through an operational calculus can be used in a more general setting; for details the reader may refer to the treatise of DUNFORD and SCHWARTZ [8, pp. 566–77].

EXERCISES

(1) Let $f \in C_{\mathsf{R}}([m,M])$ and $f = \lim_{n\to\infty} p_n$ with $p_n \in \mathscr{P}$ for $n = 1, 2,\ldots$. Prove that $f(T) = \lim_{n\to\infty} p_n(T)$ and that $Sf(T) = f(T)S$ whenever $S \in L(H)$ and $ST = TS$.

(2) Let $T \in \mathscr{S}^+$. Prove that there is a $U \in \mathscr{S}^+$ such that $U^2 = T$ and $US = SU$ whenever $S \in L(H)$ and $ST = TS$. (It can be shown that the positive square root U of T is unique. See Exercise 9.9(4).)

(3) Use the result of Exercise (2) to prove that if $S,T \in \mathscr{S}^+$ and $ST = TS$ then $ST \in \mathscr{S}^+$.

(4)(a) Prove that if T is a normal operator (see p. 358) and $\lambda I - T$ is regular then $\| (\lambda I - T)^{-1} \| = d(\lambda,\text{sp}\,(T))^{-1}$.

(b) Let T be a normal operator. Prove that if $S \in L(H)$ and $d(\lambda,\text{sp}(T)) > \| S - T \|$ then $\lambda I - S$ is regular.

(c) Prove that if $T \in \mathscr{S}$ and $f \in C_{\mathsf{R}}([m,M])$ then

$$\text{sp}(f(T)) = \{f(\lambda) : \lambda \in \text{sp}(T)\}.$$

(Use Exercise (1) and Exercise 6.3(11).)

9.9 The Spectral Theorem

As in the last section T will denote a self-adjoint bounded linear operator on a complex Hilbert space H; and m and M will denote real numbers with $mI \leq T \leq MI$ and $m < M$. Our aim is to show that T can be approximated, in a manner to be defined below, by linear combinations of certain orthogonal projections on H. These orthogonal projections are obtained from the operational calculus developed in the last section.

With each real number s we associate a real-valued function e_s on $[m,M]$ as follows: if $m \leq s < M$ we set

$$e_s(t) = \begin{cases} 1 \text{ for } m \leq t \leq s, \\ 0 \text{ for } s < t \leq M; \end{cases}$$

if $s < m$ we set $e_s = 0$; and if $s \geq M$ we set $e_s = 1$.

9.9.1 LEMMA *For each real number s we have $e_s \in \mathscr{L}^+$.*

PROOF Let s be a real number. It is obvious that $e_s \in \mathscr{L}^+$ if $s < m$ or $s \geq M$. Suppose that $m \leq s < M$ and let N be the least positive integer with $s + N^{-1} \leq M$. For $n \geq N$ let

$$f_n(t) = \begin{cases} 1 & \text{for } m \leq t \leq s, \\ -nt + ns + 1 & \text{for } s < t < s + \dfrac{1}{n}, \\ 0 & \text{for } s + \dfrac{1}{n} \leq t \leq M. \end{cases}$$

It is easy to see that $f_n \in C_{\mathrm{R}}([m,M])$ and that $\lim_{n\to\infty} f_n(t) = e_s(t)$ for each $t \in [m,M]$. We have also $0 \leq f_{n+1} \leq f_n$ for $n = N, N+1, \ldots$. Using the Weierstrass approximation theorem (4.5.1) we can find a sequence (p_n) in $\mathscr{P}$ with $f_n(t) + 2^{-n-1} < p_n(t) < f_n(t) + 2^{-n}$ for each $t \in [m,M]$ and $n = N, N+1, \ldots$. It is clear that $p_n \in \mathscr{P}^+$, $p_{n+1} \leq p_n$ for $n = N, N+1, \ldots$ and that $\lim_{n\to\infty} p_n(t) = e_s(t)$ for each $t \in [m,M]$. Thus $e_s \in \mathscr{L}^+$. □

It is possible to construct explicitly a sequence (p_n) in $\mathscr{P}^+$ with $p_{n+1} \leq p_n$ and $\lim_{n\to\infty} p_n(t) = e_s(t)$ for each $t \in [m,M]$. (See Exercise (1).) Using Definition 9.8.5 we can now define, for each real number t, an operator $E(t) \in \mathscr{S}^+$ by $E(t) = e_t(T)$.

9.9.2 LEMMA *For each real number t, the operator $E(t)$ is an orthogonal projection on H and $E(t)T = TE(t)$. Further*

(a) $E(s) \leq E(t)$ *for $s \leq t$,*
(b) $E(t) = 0$ *for $t < m$, and*
(c) $E(t) = I$ *for $t \geq M$.*

PROOF This follows from Lemma 9.8.6 and the following facts: $e_t^2 = e_t$, $e_s \leq e_t$ for $s \leq t$, $e_t = 0$ for $t < m$ and $e_t = 1$ for $t \geq M$. $\square$

We remark that, by the construction used in Definition 9.8.5, each orthogonal projection $E(t)$ is the strong limit of a sequence of polynomials in T; this fact is sometimes useful. The reader will find some further properties of the projections $E(t)$ in the Exercises below.

We shall now specify the manner in which T is to be approximated by linear combinations of the projections $E(t)$. The reader should compare the following definition with Definition 5.6.1.

9.9.3 DEFINITION Let a and b be real numbers with $a < b$. A finite subset P of $[a,b]$ which contains a and b will be called a partition of $[a,b]$. When P is a partition of $[a,b]$ and we write $P = \{s_0, s_1,..., s_n\}$ it will be understood that $a = s_0 < s_1 < \ldots < s_n = b$. Let f be a real-valued function on $[a,b]$. We shall say that f is *E-integrable* if and only if there is an operator $S \in \mathscr{S}$ with the following property: for each $\varepsilon > 0$ there exists $\delta > 0$ such that

$$\left\| S - \sum_{k=1}^{n} f(t_k)(E(s_k) - E(s_{k-1})) \right\| < \varepsilon \tag{1}$$

for all partitions $\{s_0, s_1,..., s_n\}$ of $[m,M]$ with $s_k - s_{k-1} < \delta$ for $k = 1, 2,..., n$ and all real numbers $t_1, t_2,..., t_n$ with $s_{k-1} \leq t_k \leq s_k$ for $k = 1, 2,..., n$.

It is clear that, if f is E-integrable, then there is exactly one operator $S \in \mathscr{S}$ that satisfies (1); this operator will be denoted by $\int_a^b f(t)\, dE(t)$.

9.9.4 THEOREM (the spectral theorem) *Let a and b be real numbers with $a < m$ and $b \geq M$. Then the mapping $t \to t$ of $[a,b]$ into* R *is E-integrable and*

$$T = \int_a^b t\, dE(t).$$

PROOF It is easy to verify that if s and u are real numbers with $s < u$ then

$$e_u(t) - e_s(t) = \begin{cases} 1 \text{ for } t \in (s,u] \cap [m,M], \\ 0 \text{ for } t \in [m,M] \sim (s,u]. \end{cases}$$

It follows that

$$s(e_u(t) - e_s(t)) \leq t(e_u(t) - e_s(t)) \leq u(e_u(t) - e_s(t))$$

for all $t \in [m,M]$ and consequently, by Theorem 9.8.8 and the definition of $E(s)$ and $E(u)$, we have

$$s(E(u) - E(s)) \leq T(E(u) - E(s)) \leq u(E(u) - E(s)). \tag{2}$$

Let $\varepsilon > 0$ and let $\{s_0, s_1, \ldots, s_n\}$ be a partition of $[a,b]$ with $s_k - s_{k-1} < \varepsilon$ for $k = 1, 2, \ldots, n$. Then using the inequalities (2) we obtain

$$\sum_{k=1}^{n} s_{k-1}(E(s_k) - E(s_{k-1})) \leq T \sum_{k=1}^{n} (E(s_k) - E(s_{k-1}))$$
$$\leq \sum_{k=1}^{n} s_k(E(s_k) - E(s_{k-1})).$$

But

$$\sum_{k=1}^{n} (E(s_k) - E(s_{k-1})) = E(b) - E(a) = I, \tag{3}$$

because $a < m$ and $b \geq M$, so

$$\sum_{k=1}^{n} s_{k-1}(E(s_k) - E(s_{k-1})) \leq T \leq \sum_{k=1}^{n} s_k(E(s_k) - E(s_{k-1})). \tag{4}$$

Now let $t_1, t_2, \ldots, t_n$ be real numbers with $s_{k-1} \leq t_k \leq s_k$. Then from (4) and Lemma 9.6.5 we obtain

$$\sum_{=1}^{n} (s_{k-1} - t_k)(E(s_k) - E(s_{k-1})) \leq T - \sum_{k=1}^{n} t_k(E(s_k) - E(s_{k-1}))$$
$$\leq \sum_{k=1}^{n} (s_k - t_k)(E(s_k) - E(s_{k-1})). \tag{5}$$

But $s_{k-1} - t_k > -\varepsilon$ and $s_k - t_k < \varepsilon$ and, by Lemma 9.9.2, we have $E(s_k) - E(s_{k-1}) \geq 0$, so Lemma 9.6.5 gives

$$(s_{k-1} - t_k)(E(s_k) - E(s_{k-1})) \geq -\varepsilon(E(s_k) - E(s_{k-1})) \tag{6}$$

and

$$(s_k - t_k)(E(s_k) - E(s_{k-1})) \leq \varepsilon(E(s_k) - E(s_{k-1})). \tag{7}$$

From (3), (5), (6) and (7) we obtain

$$-\varepsilon I \leq T - \sum_{k=1}^{n} t_k(E(s_k) - E(s_{k-1})) \leq \varepsilon I,$$

from which, using Lemma 9.6.6, we obtain

$$\left\| T - \sum_{k=1}^{n} t_k(E(s_k) - E(s_{k-1})) \right\| \leq \varepsilon.$$

This proves that

$$T = \int_a^b t\, dE(t). \ \square$$

The reader should consider the relation between Theorem 9.9.4 and Theorem 9.5.2. If T is a compact self-adjoint linear operator the orthogonal projections $E(t)$ corresponding to T can be determined in terms of the numbers μ_n and the orthonormal set $\{u_n : n = 1, 2, \ldots\}$ defined in Section 9.5. (See Exercise (6).) When this has been done it is not difficult to see that Theorem 9.5.2 is equivalent to Theorem 9.9.4 in this case.

The spectral theorem is an important tool in the study of bounded linear operators on Hilbert spaces. Its importance derives from the fact that it allows questions about self-adjoint bounded linear operators to be reduced to similar questions about orthogonal projections, which are usually more easily answered.

Lack of space precludes us from pursuing further the study of bounded linear operators on Hilbert space. The reader who wishes to continue this study should consult the books of RIESZ and NAGY [23], AHIEZER and GLAZMAN [1], and DUNFORD and SCHWARTZ [8, Part II].

EXERCISES

(1) Let $m \leq s < M$ and let $p(t) = 1 + (M - m)^{-1}(s - t)$ for $m \leq t \leq M$. Let (p_n) be the sequence of polynomials defined inductively by $p_1 = p$ and $p_{n+1} = p_n(1 - \frac{1}{4}(p_n - 1)^2)$ for $n = 1, 2,\ldots$. Prove that $0 \leq p_{n+1} \leq p_n$ for $n = 1, 2,\ldots$ and that $\lim_{n\to\infty} p_n(t) = e_s(t)$ for $m \leq t \leq M$.

(2)(a) Let $t \to E(t)$ be a mapping of R into the set of orthogonal projections on H that satisfies conditions (a), (b) and (c) of Lemma 9.9.2. Let $g_x(t) = E(t)x$ for all $x \in H$ and $t \in \mathsf{R}$. Prove that the mapping $t \to g_x(t)$ is regulated on R for each $x \in H$. For each $t \in \mathsf{R}$ let $E(t,+)$ and $E(t,-)$ be defined by $E(t,+)x = g_x(t,+)$ and $E(t,-)x = g_x(t,-)$ for all $x \in H$. Prove that $E(t,+)$ and $E(t,-)$ are orthogonal projections on H and that

$$E(u) \leq E(t,-) \leq E(t) \leq E(t,+) \leq E(v)$$

for $u < t < v$.

(b) Prove that if $t \to E(t)$ is the mapping constructed in the text then $E(t,+) = E(t)$ for all $t \in \mathsf{R}$. (Hint: modify the construction of Lemma 9.9.1 to obtain a sequence (p_n) in $\mathscr{P}$ with $p_{n+1} \leq p_n$, $e_{s+n^{-1}} \leq p_n$ and $\lim_{n\to\infty} p_n(t) = e_s(t)$ for $t \in [m,M]$.)

A mapping $t \to E(t)$ of R into the set of orthogonal projections on H, which satisfies conditions (a), (b) and (c) of Lemma 9.9.2 and the further condition that $E(t,+) = E(t)$ for all $t \in \mathsf{R}$, is called a *resolution of the identity*. The next exercise shows that there is just one resolution of the identity which is related to the operator $T \in \mathscr{S}$ as in Theorem 9.9.4.

(3) Let $t \to F(t)$ be any mapping of R into the set of orthogonal projections on H that satisfies conditions (a), (b) and (c) of Lemma 9.9.2 and has $T = \int_a^b t\, dF(t)$. Let $f \in C_{\mathsf{R}}([a,b])$ and $f_0 = f|[m,M]$.

(a) Prove that f is F-integrable and that $f_0(T) = \int_a^b f(t)\, dF(t)$. (Prove the result first for f a polynomial and then use Exercise 9.8(1).)

(b) Let $m \leq s < M$ and let f_n, $n = N, N+1, \ldots$, be the functions defined in the proof of Lemma 9.9.1. Prove that the sequence

$$\left(\int_a^b f_{+m}(t)\, dF(t) \right)_{m \geq 1}$$

converges strongly to $F(s,+)$.

(c) Prove that if the mapping $t \to F(t)$ has the further property that it is a resolution of the identity then $F(t) = E(t)$ for all $t \in \mathsf{R}$.

(4) Prove that if $T \in \mathscr{S}^+$ then there is a unique $U \in \mathscr{S}^+$ such that $U^2 = T$. (See Exercise 9.8(2). Let $U \in \mathscr{S}^+$, with $U^2 = T$, consider the resolutions of the identity for U and T and use Exercise (3).)

(5) This exercise shows how the spectrum of T can be described in terms of the resolution of the identity $t \to E(t)$. The first part of the exercise is a preliminary.

(a) Prove that for each $x \in H$ the function $t \to \langle E(t)x,x \rangle$ is increasing and $\langle f_0(T)x,x \rangle = \int_a^b f(t)\, d\langle E(t)x,x \rangle$, in the sense of Definition 5.6.1, where f and f_0 are as in Exercise (3).

(b) Let λ be a real number and suppose that there exists $\delta > 0$ such that $E(s) = E(t)$ for $\lambda - \delta \leq s \leq t \leq \lambda + \delta$ (which is certainly true if $\lambda < m$ or $\lambda > M$). Prove that $\lambda \notin \mathrm{sp}(T)$ and that if f is any function such that $f(t) = (\lambda - t)^{-1}$ for $t \notin [a,b] \cap [\lambda - \delta, \lambda + \delta]$ then

$$(\lambda I - T)^{-1} = \int_a^b f(t)\, dE(t).$$

(c) Prove that if λ is a real number then $\lambda \notin \mathrm{sp}(T)$ if and only if there exists $\delta > 0$ such that $E(s) = E(t)$ for $\lambda - \delta \leq s \leq t \leq \lambda + \delta$.

(d) Prove that a real number λ is an eigenvalue of T if and only if $E(\lambda,-) \neq E(\lambda)$ and that $\mathscr{N}(\lambda I - T) = \mathscr{R}(E(\lambda) - E(\lambda,-))$. (Hints: To obtain the result of (b) use Exercise (3)(a) and Theorem 9.8.8. For (c) and (d) use (a) to prove that if λ is a real number, $E(s)y = 0$ and $E(t)y = y$ then $\| (\lambda I - T)y \|^2 \leq \max\{(\lambda - s)^2, (\lambda - t)^2\} \| y \|^2$.)

(6) Let T now be a compact self-adjoint operator. The notation will be that of Section 9.5. For $t < 0$ let $E(t)$ be the orthogonal projection on H whose range is the closed linear hull of the set $\{0\} \cup \{u_k : \mu_k \leq t\}$; for $t \geq 0$ let $E(t)$ be the orthogonal projection on H whose range is the closed linear hull of the set $\{0\} \cup \{u_k : \mu_k \leq t\} \cup \mathscr{N}(T)$. Prove that the mapping $t \to E(t)$ is a resolution of the identity and that

$$T = \int_a^b t\, dE(t).$$

(The reader should observe that if T has only a finite number of non-zero eigenvalues the integral can be replaced by a finite sum.)

References

This list of references is not a bibliography; all the books which are listed have been cited in the text. We single out four of them for their special significance and value in relation to functional analysis. The book of S. BANACH [3] was the first systematic account of the early work on normed linear spaces and linear operators; it is one of the milestones in the development of the subject. F. RIESZ must be regarded as one of the principal creators of functional analysis, and the book which he wrote with B. SZ.-NAGY [23] is an elegant introduction which beautifully conveys the spirit of the subject. The principal work in English on general topology, without which much of the subject matter of this book cannot reasonably be carried further, is that of J. L. KELLEY [18]. The treatise of N. DUNFORD and J. T. SCHWARTZ [8] is very much more than its title *Linear Operators* suggests; in addition to its encyclopaedic discussion of the theory of linear operators it contains a concise development of the tools of functional analysis and a very extensive general bibliography—it is an indispensable work of reference.

[1] AHIEZER, N. I. and GLAZMAN, I. M. *Theory of Linear Operators in Hilbert Space* (two volumes), Ungar, New York (1961, 1963). An English translation of the 1950 Russian edn.

[2] AHLFORS, L. V. *Complex Analysis* (2nd edn), McGraw-Hill, New York (1966).

[3] BANACH, S. *Théorie des opérations linéaires*, Monografje Matematyczne, Warsaw (1932). Reprinted by Chelsea, New York (1955).

[4] BERBERIAN, S. K. *Introduction to Hilbert Space*, Oxford University Press, New York (1961).

[5] BERGE, C. *Topological Spaces*, Oliver and Boyd, Edinburgh (1963). An English translation of *Espaces topologiques, fonctions multivoques*, Dunod, Paris, (1959).

[6] BOURBAKI, N. *Éléments de Mathématique*, Livre VI, *Intégration*, Chapitres 1, 2, 3 et 4 (deuxième edition), Hermann et Cie, Actualités Scientifiques et Industrielles, No. 1175, Paris (1965).

[7] DIEUDONNÉ, J. *Foundations of Modern Analysis*, Academic Press, New York and London (1960).

[8] DUNFORD, N. and SCHWARTZ, J. T. *Linear Operators*, Part I: General Theory, Part II: Spectral Theory, Interscience, New York (1958 and 1964).

[9] GLEASON, A. M. *Fundamentals of Abstract Analysis*, Addison-Wesley, Reading, Mass. (1966).

[10] GRAVES, L. M. *The Theory of Functions of Real Variables* (2nd edn), McGraw-Hill, New York (1956).

[11] HALMOS, P. R. *Introduction to Hilbert Space and the Theory of Spectral Multiplicity* (2nd edn), Chelsea, New York (1957).

[12] —— *Finite Dimensional Vector Spaces* (2nd edn), Van Nostrand, Princeton, New Jersey (1958).

[13] —— *Naïve Set Theory*, Van Nostrand, Princeton, New Jersey (1960).

[14] HILLE, E. *Analytic Function Theory*, Volume 1, Ginn and Co., Boston (1959).

[15] HILLE, E. and PHILLIPS, R. S. *Functional Analysis and Semi-Groups*, American Mathematical Society Colloquium Publications, Volume 31, Providence, R.I. (1957).

[16] HOBSON, E. W. *The Theory of Functions of a Real Variable*, Volume 2 (2nd edn), Cambridge University Press (1926).

[17] KATO, T. *Perturbation Theory for Linear Operators*, Springer-Verlag, Berlin (1966).

[18] KELLEY, J. L. *General Topology*, Van Nostrand, Princeton, New Jersey (1955).

[19] KELLOGG, O. D. *Foundations of Potential Theory*, Springer-Verlag, Berlin (1929). Reprinted by Dover, New York (1953).

[20] NACHBIN, L. *The Haar Integral*, Van Nostrand, Princeton, New Jersey (1965).

[21] NAIMARK, M. A. *Normed Rings* (2nd revised edn), P. Noordhoff, N.V., Groningen (1964). Translation of 1st Russian edn, Moscow (1956).

[22] RICKART, C. E. *General Theory of Banach Algebras*, Van Nostrand, Princeton, New Jersey (1960).

[23] RIESZ, F. and SZ.-NAGY, B. *Functional Analysis*, Ungar, New York (1955). An English translation of *Leçons d'analyse fonctionelle* (deuxième édition), Akadémiai Kiadó, Budapest (1953).

[24] RUDIN, W. *Principles of Mathematical Analysis* (2nd edn), McGraw-Hill, New York (1964).

[25] SIMMONS, G. F. *Introduction to Topology and Modern Analysis*, McGraw-Hill, New York (1963).

[26] TAYLOR, A. E. *Introduction to Functional Analysis*, Wiley, New York (1958).

[27] ZYGMUND, A. *Trigonometric Series* (2nd edn), Volume 1, Cambridge University Press (1968).

Index of Symbols

Index